Euler and Modern Science

Leonhard Euler (1707–1783)
Portrait by E. Handmann, 1753

Euler and Modern Science

Edited by

N. N. Bogolyubov, G. K. Mikhaïlov, and A. P. Yushkevich

Translated from Russian by

Robert Burns

Published and Distributed by
The Mathematical Association of America

Originally published in Russian as
"Razvitie ideĭ Leonharda Eulera i sovremennaya nauka"
© 1988 Izdatel′stvo Nauka

English edition
© *2007 by*
The Mathematical Association of America (Incorporated)

Library of Congress Catalog Card Number 2007929594

ISBN: 978-0-88385-564-5

Printed in the United States of America

Current Printing (last digit):
10 9 8 7 6 5 4 3 2 1

SPECTRUM SERIES

The Spectrum Series of the Mathematical Association of America was so named to reflect its purpose: to publish a broad range of books including biographies, accessible expositions of old or new mathematical ideas, reprints and revisions of excellent out-of-print books, popular works, and other monographs of high interest that will appeal to a broad range of readers, including students and teachers of mathematics, mathematical amateurs, and researchers.

777 Mathematical Conversation Starters, by John de Pillis

99 Points of Intersection: Examples—Pictures—Proofs, by Hans Walser. Translated from the original German by Peter Hilton and Jean Pedersen.

All the Math That's Fit to Print, by Keith Devlin

Calculus Gems: Brief Lives and Memorable Mathematics, by George F. Simmons

Carl Friedrich Gauss: Titan of Science, by G. Waldo Dunnington, with additional material by Jeremy Gray and Fritz-Egbert Dohse

The Changing Space of Geometry, edited by Chris Pritchard

Circles: A Mathematical View, by Dan Pedoe

Complex Numbers and Geometry, by Liang-shin Hahn

Cryptology, by Albrecht Beutelspacher

The Early Mathematics of Leonhard Euler, by C. Edward Sandifer

The Edge of the Universe: Celebrating 10 Years of Math Horizons, edited by Deanna Haunsperger and Stephen Kennedy

Euler and Modern Science, edited by N. N. Bogolyubov, G. K. Mikhaĭlov, and A. P. Yushkevich. Translated from Russian by Robert Burns.

Euler at 300: An Appreciation, edited by Robert E. Bradley, Lawrence A. D'Antonio, and C. Edward Sandifer

Five Hundred Mathematical Challenges, Edward J. Barbeau, Murray S. Klamkin, and William O. J. Moser

The Genius of Euler: Reflections on his Life and Work, edited by William Dunham

The Golden Section, by Hans Walser. Translated from the original German by Peter Hilton, with the assistance of Jean Pedersen.

The Harmony of the World: 75 Years of Mathematics Magazine, edited by Gerald L. Alexanderson with the assistance of Peter Ross

How Euler Did It, by C. Edward Sandifer

I Want to Be a Mathematician, by Paul R. Halmos

Journey into Geometries, by Marta Sved

JULIA: a life in mathematics, by Constance Reid

The Lighter Side of Mathematics: Proceedings of the Eugène Strens Memorial Conference on Recreational Mathematics & Its History, edited by Richard K. Guy and Robert E. Woodrow

Lure of the Integers, by Joe Roberts

Magic Numbers of the Professor, by Owen O'Shea and Underwood Dudley

Magic Tricks, Card Shuffling, and Dynamic Computer Memories: The Mathematics of the Perfect Shuffle, by S. Brent Morris

Martin Gardner's Mathematical Games: The entire collection of his Scientific American columns

The Math Chat Book, by Frank Morgan

Mathematical Adventures for Students and Amateurs, edited by David Hayes and Tatiana Shubin. With the assistance of Gerald L. Alexanderson and Peter Ross.

Mathematical Apocrypha, by Steven G. Krantz

Mathematical Apocrypha Redux, by Steven G. Krantz

Mathematical Carnival, by Martin Gardner

Mathematical Circles Vol I: In Mathematical Circles Quadrants I, II, III, IV, by Howard W. Eves

Mathematical Circles Vol II: Mathematical Circles Revisited and Mathematical Circles Squared, by Howard W. Eves

Mathematical Circles Vol III: Mathematical Circles Adieu and Return to Mathematical Circles, by Howard W. Eves

Mathematical Circus, by Martin Gardner

Mathematical Cranks, by Underwood Dudley

Mathematical Evolutions, edited by Abe Shenitzer and John Stillwell

Mathematical Fallacies, Flaws, and Flimflam, by Edward J. Barbeau

Mathematical Magic Show, by Martin Gardner

Mathematical Reminiscences, by Howard Eves

Mathematical Treks: From Surreal Numbers to Magic Circles, by Ivars Peterson

Mathematics: Queen and Servant of Science, by E.T. Bell

Memorabilia Mathematica, by Robert Edouard Moritz

Musings of the Masters: An Anthology of Mathematical Reflections, edited by Raymond G. Ayoub

New Mathematical Diversions, by Martin Gardner

Non-Euclidean Geometry, by H. S. M. Coxeter

Numerical Methods That Work, by Forman Acton

Numerology or What Pythagoras Wrought, by Underwood Dudley

Out of the Mouths of Mathematicians, by Rosemary Schmalz

Penrose Tiles to Trapdoor Ciphers . . . and the Return of Dr. Matrix, by Martin Gardner

Polyominoes, by George Martin

Power Play, by Edward J. Barbeau

The Random Walks of George Pólya, by Gerald L. Alexanderson

Remarkable Mathematicians, from Euler to von Neumann, Ioan James

The Search for E.T. Bell, also known as John Taine, by Constance Reid

Shaping Space, edited by Marjorie Senechal and George Fleck

Sherlock Holmes in Babylon and Other Tales of Mathematical History, edited by Marlow Anderson, Victor Katz, and Robin Wilson

Student Research Projects in Calculus, by Marcus Cohen, Arthur Knoebel, Edward D. Gaughan, Douglas S. Kurtz, and David Pengelley

Symmetry, by Hans Walser. Translated from the original German by Peter Hilton, with the assistance of Jean Pedersen.

The Trisectors, by Underwood Dudley

Twenty Years Before the Blackboard, by Michael Stueben with Diane Sandford

The Words of Mathematics, by Steven Schwartzman

MAA Service Center
P.O. Box 91112
Washington, DC 20090-1112
800-331-1622   FAX 301-206-9789

Introduction

This book is an excellent English translation of a collection of papers in Russian, many of which were presented at a conference that took place in Moscow and St. Petersburg in 1983, 200 years after Euler's death. Two of the Russian papers in the collection are themselves translations from the German. This English version of the Russian volume appears in 2007, 300 years after Euler's birth.

We speak of the age of Euler. An easy justification of this term is provided by a list of scientific terms connected with Euler's name. Another such justification is provided by a glance at the titles of the 25 essays in the present collection. There are essays that deal with Euler's contributions to

> algebra,
> Diophantine equations,
> mechanics and hydromechanics,
> the variational principles of mechanics,
> the mechanics of elastic systems,
> ballistics,
> astronomy and the history of science,
> celestial mechanics,
> physics,
> instruments and technology, and
> a mathematical theory of music.

Euler was also actively involved in the preparation and sale of scientific books, and—while in Berlin—of scientific almanacs—just to mention one of his many administrative activities.

Bourbaki could do many things because he was a committee. But Euler was not a committee, there was a flesh-and-blood Euler.

The essays tell us a great deal about Euler as a person. It is safe to say that he was someone you would enjoy knowing and dealing with. He readily acknowledged the achievements of others. He was deeply religious and could not easily warm up to those who weren't. That is why, while in Berlin, he got along very well with Maupertuis but not with his boss, King Frederick the Great, in whose kingdom—as he put it—everyone was "free to reach Heaven in his own way" (*nach seiner Fasson selig werden*).

There is much we learn about Euler from the brilliant essay (by Yushkevich and Taton) on his correspondence with Clairaut, d'Alembert, and Lagrange. We learn from it that his treatment of d'Alembert and Lagrange was sometimes below par, but such moments were a rare exception rather than the rule.

Let me conclude with what the suspicious will take to be a sales pitch: if you want to acquire a significant and factual knowledge of Euler's contributions to science past and present, then, to my mind, this book is easily the best available choice.

A. Shenitzer
York University

Translator's Note

Of the three addresses delivered on the opening day (October 24, 1983) of the meeting and symposium from which the original collection of essays derives, only the first, by A. P. Aleksandrov, is included in the present English translation. The opening address delivered in St. Petersburg when the conference moved there on October 27 is also omitted. Of the three genealogical contributions appended to the original collection two have been omitted entirely from the English translation (the "Genealogy of Leonhard Euler's descendants", compiled by E. N. Amburger, I. R. Gekker, and G. K. Mikhaĭlov, comprising about seven pages of text followed by a 77-page inventory of the approximately 1000 known descendants of Euler, and "Leonhard Euler's descendants through his son Carl", by M. V. Shestakova, 8 pages). However somewhat over half of the third essay "Leonhard Euler's family and descendants", by I. R. Gekker and A. A. Euler, is included in the translation and may perhaps be taken as representative of all three.

I take the opportunity of thanking those who helped with the translation. First and foremost I thank my friend and mentor Abe Shenitzer for initiating the project, translating one of the essays, and carefully checking the whole translation—among many other things. For help of various kinds involving linguistic, scientific, or musical expertise, or simply an eagle eye, I also thank Jerry Alexanderson, Lydia Burns, Paul Delaney, Vincent Hart, Nina Iukhoveli, and Ryu Sasaki, as well as Don Albers and Jerry Alexanderson (again) of the MAA for making the job so pleasant while it was being done at York University and the University of Queensland.

Robert Burns
York University

Contents

Introduction ix

Translator's Note xi

From the Editors 1

*Opening Speech of the Symposium "Modern Developments of Euler's Ideas"
October 24, 1983*, A. P. Aleksandrov 3

Leonhard Euler: his life and work, A. P. Yushkevich 7

*Leonhard Euler, active and honored member of the Petersburg Academy of
Sciences*, Yu. Kh. Kopelevich 39

*The part played by the Petersburg Academy of Sciences (the Academy of Sciences
of the USSR) in the publication of Euler's collected works*, E. P. Ozhigova 53

Leonhard Euler and the Berlin Academy of Sciences, K. Grau 75

Was Leonhard Euler driven from Berlin by J. H. Lambert?, K.-R. Biermann 87

Euler's Mathematical Notebooks, E. Knobloch 97

On Euler's Surviving Manuscripts and Notebooks, G. P. Matvievskaya 119

The Manuscript Materials of Euler on Number Theory,
G. P. Matvievskaya and E. P. Ozhigova 127

Euler's Contribution to Algebra, I. G. Bashmakova 137

Diophantine Equations in Euler's Works, T. A. Lavrinenko 153

The Foundations of Mechanics and Hydrodynamics in Euler's Works,
G. K. Mikhaïlov and L. I. Sedov 167

Leonhard Euler and the Variational Principles of Mechanics,
V. V. Rumyantsev 183

Leonhard Euler and the Mechanics of Elastic Systems,
N. V. Banichuk and A. Yu. Ishlinskiĭ 213

Euler's Research in Mechanics during the First Petersburg Period,
N. N. Polyakhov 237

The Significance of Euler's Research in Ballistics, A. P. Mandryka 241

Euler and the Development of Astronomy in Russia,
    V. K. Abalakin and E. A. Grebenikov 245

Euler and the Evolution of Celestial Mechanics,
    N. I. Nevskaya and K. V. Kholshevnikov 263

*New Evidence Concerning Euler's Development as an
    Astronomer and Historian of Science*, N. I. Nevskaya 269

*Leonhard Euler in Correspondence with Clairaut, d'Alembert,
    and Lagrange*, A. P. Yushkevich and R. Taton 289

Letters to a German Princess *and Euler's Physics*,
    A. T. Grigor'ian and V. S. Kirsanov 307

Euler and I. P. Kulibin, N. M. Raskin 317

Euler and the History of a Certain Musical-Mathematical Idea, E. V. Gertsman 335

*Euler's Music-Theoretical Manuscripts and the Formation of his
    Conception of the Theory of Music*, S. S. Tserlyuk-Askadskaya 349

An Unknown Portrait of Euler by J. F. A. Darbès,
    G. B. Andreeva and M. P. Vikturina 361

Eulogy in Memory of Leonhard Euler, Nikolaĭ Fuss 369

Leonhard Euler's Family and Descendants, I. R. Gekker and A. A. Euler 397

Index 417

From the Editors

April 15, 1982 marked the 275th anniversary of the birth, and September 18, 1983 the 200th anniversary of the death, of one of the greatest mathematicians and mechanicians of all time, Leonhard Euler (1797–1783), whose scientific activity was for over a half-century inseparably linked to the Petersburg Academy of Sciences. These important dates were widely commemorated by the scientific community in our country, in his motherland Switzerland, and in Berlin, where he worked for 25 years. In this connection the Academy of Sciences of the USSR organized a celebratory meeting and symposium for the period October 24–27 with the title "The development of Euler's ideas in the modern era". In addition to the great number of Soviet scientists attending, there were guest participants from Switzerland, the German Democratic Republic, West Berlin, the Polish People's Republic, as well as other countries.

The celebratory meeting, which took place on October 24 in the Great Hall of the Moscow House of Scientists, was opened by the president of the Academy of Sciences of the USSR, A. P. Aleksandrov. In his opening address he gave a vivid characterization of Euler's contribution to the development of the physico-mathematical sciences, and of the activities of the Petersburg Academy of Sciences in the 18th century. This was followed with welcoming speeches by the deputy of the President of the State Commission for Science and Technology of the USSR, academician G. I. Marchuk, the Minister of Secondary and Tertiary Special Education of the Russian Federation of Soviet Republics, academician I. F. Obraztsov, and the president of the Euler Committee of the Academy of Sciences of the GDR, Professor V. Engel. On that first day the following talks were delivered: "The life and work of Leonhard Euler" (A. P. Yushkevich), "Leonhard Euler and mathematical analysis" (Yu. V. Prokhorov), and "Leonhard Euler and the Berlin Academy" (K. Grau). Next day the symposium continued in Moscow with the talks "The foundations of mechanics and hydrodynamics in Euler's works" (G. K. Mikhaĭlov and L. I. Sedov), "Leonhard Euler and the mechanics of elastic systems" (A. Yu. Ishlinskiĭ), "Leonhard Euler and the variational principles of mechanics" (V. V. Rumyantsev), "Euler's contribution to the development of algebra" (I. G. Bashmakova), "Leonhard Euler and the development of astronomy in Russia" (V. K. Abalakin and E. A. Grebenikov), and "*Letters to a German princess* and Euler's physics" (A. T. Grigor'an and V. S. Kirsanov).

On October 27 the the symposium moved to the old conference hall of the Academy of Sciences in Leningrad. The opening speech by academician L. D. Faddeev was followed by the lectures "Leonhard Euler and the Petersburg Academy of Sciences" (Yu. Kh. Kopelevich), "Mechanics in Euler's works from his first Petersburg period" (N. N. Polyakhov), "Unpublished materials of Euler on number theory" (G. P. Matvievskaya and E. P. Ozhigova), "New evidence concerning Euler's investigations in astronomy" (N. I. Nevskaya), and "The family and descendants of Leonhard Euler" (I. R. Gekker and A. A. Euler).

Concurrently with the meetings in Moscow and Leningrad, there were held exhibitions of old and new editions of Euler's works, writings about him, archival documents, and other illustrative materials. On October 28 the participants in the symposium went on a tour of the places in Leningrad connected in some way or other with Euler's life and work there, and visited his tomb in the Leningrad Mausoleum of the Aleksander Nevskiĭ Monastery.

The present collection contains articles prepared from the notes of the lectures delivered at the symposium, as well as several written expressly for inclusion. We have also included a translation of a talk given by K.-R. Biermann in September 1983 at a conference of the Academy of Sciences of the GDR in Berlin held in that year, and a revised version of an article on Euler's correspondence with A.-C. Clairaut, J. d'Alembert, and J.-L. Lagrange, originally published in the fifth volume of the fourth series of Euler's complete collected works (*L. Euleri Opera omnia*). Taken together these articles very substantially supplement the material contained in the collection published in 1958 by the Academy of Sciences of the USSR, dedicated to Euler on the occasion of the 250th anniversary of the great scientist's birth[1].

By way of complementing the modern articles of the collection, we include an old Russian translation of the eulogy delivered in French at the Academy in 1783, a month and a half after Euler's death, by his closest collaborator and student, the academician N. I. Fuss. This translation, which appeared in the *Academic papers* in 1801 and is now a bibliographic rarity, is reproduced in the present book in its original form—apart from modernization of the spelling. In view of its high quality, Fuss' "Eulogy" fully merits the attention of the modern reader, in addition to which the slightly archaic language of the translation will transport him or her closer to Euler's time.

At the end of the collection we have appended the work entitled "Genealogy of Leonhard Euler's descendants", representing unique sociological material for the history of the family during the ensuing two and a half centuries, as well as two other articles on Euler's descendants[2].

N. N. Bogolyubov, G. K. Mikhaĭlov, A. P. Yushkevich

---

[1] The task of unifying the style of the references to Euler's works turned out to be too difficult. Hence these refer either to first editions or to the *Opera omnia*. In the latter case only the series (in Roman numerals) and the volume number are given. Thus, for example, the reference *Opera* II-8 indicates that the work in question appears in Volume 8 of the second series of the *Opera omnia*. In addition, references to Euler's works are sometimes accompanied by the numbers assigned to them in the well known catalogue compiled by G. Eneström in the years 1910–1913, and reproduced in abridged form in Issue 17 of the *Works of the Archive of the Academy of Sciences of the USSR* for 1962 (or see "The Euler archive" on the internet–*Trans.*); the references to Eneström's catalogue are indicated, in the standard manner, with the letter E before the appropriate numeral. For example, E342 indicates the work of Euler numbered 342 in Eneström's catalogue.–*Eds.*

[2] Of these three papers two have been entirely omitted from the present translation, and only a portion of the third ("Euler's family and descendants" by I. R. Gekker and A. A. Euler) is included (see the "Translator's note").

Opening Speech of the Symposium
"Modern Developments of Euler's Ideas"
October 24, 1983

Academician

A. P. Aleksandrov
President of the Academy of Sciences of the USSR

Dear comrades! Respected guests!

Today we begin the official meeting and symposium marking the 275th anniversary of the birth of Leonhard Euler, and the 200th year since his death.

In 1724 an edict was promulgated ordering the establishment of a Petersburg Academy of Sciences. In 1727 there turned up in that Academy the twenty-year-old Leonhard Euler, native of Basel. He had been invited to join the Academy by its president, L. Blumentrost, at the recommendation of two friends already working there, Nicholas and Daniel Bernoulli. These were sons of the famous professor of Basel University Johann Bernoulli, Euler's mentor in earlier years.

Leonhard Euler was born in 1707, and entered Basel University at the age of thirteen and a half. At that time this was not unusual; Daniel Bernoulli, for instance, enrolled in the university at the same age. Euler at once demonstrated a strong mathematical bent. In the absence of any special program in mathematics, Euler was permitted to turn to Professor Bernoulli for instruction in the more complex questions of the higher branches of the mathematical sciences. Bernoulli agreed to meet with Euler one day a week, on Saturdays, to explain any difficulties that he might encounter in his reading of the specialist literature. These tutorials played a very important part in Euler's scientific apprenticeship. At the age of 18 he was awarded a master's degree and began to look for a sphere in which to apply his skills. At that time in western Europe many young scientists were unable to find work in their specialties in their home country. Such was Euler's situation, and he accepted with alacrity the offer from the Petersburg Academy of Sciences to fill a vacant position in physiology. Although this was not his specialty, he made a determined effort to learn that subject, and arrived in Russia in 1727.

3

Official opening of the symposium
"The development of Euler's ideas in the modern era", October 24, 1983

In the Petersburg Academy Euler was under no compulsion to work in physiology. A gymnasium affiliated with the Academy of Sciences had been opened with the aim of preparing national cadres of specialists, and it was proposed to Euler that he take part in this school, in particular that he write a mathematics textbook for use by the students. He wrote an excellent textbook on arithmetic, which was published in both German and Russian; this was only the second mathematics textbook to appear in Russia after the well known *Arithmetic* of L. F. Magnitskiĭ. At the same time Euler continued the work he had begun in Basel on mathematical analysis, mechanics, and their applications. You may recall that when still in Basel, he had participated in the competition of the Paris Academy of Sciences on the theme of the most efficient arrangement of masts on a sailing ship. In Basel, there were neither seas nor shipbuilding. However Euler was able to propose an original solution of the problem, which, though not awarded a prize, was honored with publication. Later, in Russia, he renewed his investigations into questions to do with shipbuilding and navigation, and, twenty years later, in Berlin, completed his major work on "naval science", published in 1749. And since our present symposium has the influence of Euler's work as one of its themes, it is appropriate to mention that the modern theory of stability of ships, developed so brilliantly by the late academician Alekseĭ Nikolaevich Krylov, owes much to Euler's *Naval science*.

Euler devoted much energy to a successful elaboration of the methods of infinitesimal analysis. He advanced brilliantly the work of Newton and Leibniz, and Leibniz' students Jakob and Johann Bernoulli, in the differential and integral calculus, differential equations, series expansions, the dynamics of a point-mass, and so on. All of this eventually allowed him to begin investigating such complex topics as the motion of the moon and planets. Here he produced a great many works, revisiting celestial mechanics several times during his lifetime. He published two major monographs on the theory of lunar motion and two lengthy works of the motion of Jupiter and Saturn.

Lunar and planetary motion, accurate measurement of the earth's orbit, the causes of the ebb and flow of the tides in the earth's oceans and seas—these all involved real questions which Euler answered, in the process perfecting the theoretical and numerical methods of mathematical analysis.

Euler was a person of unusually wide interests. He conceived the world around him via mathematical expression based on what he observed and the information to hand. For instance, in his theory of music he tried to find the mathematical requirements that music perceived as pleasant must satisfy as against music that sounds unpleasant.

A great number of other questions beset the youthful Euler. Having worked at the Petersburg Academy for 14 years, various external circumstances persuaded him to move to the Berlin Academy, whence, 25 years later, he returned to the Petersburg Academy of Sciences, this time remaining there till his death. He would return to the same questions over and over, applying new methods, exploring new approaches. Supplementary solutions obtained at a later stage often completely changed what had appeared in the first works on a topic.

Euler was an unusually prolific author. He produced over 800 scientific works, including around 20 voluminous monographs, some in two or three volumes. It should be mentioned that in those times—the 18th century—the system of scientific periodicals was very inadequate and correspondence between scientists was of correspondingly great importance for the spread of information. Euler's epistolary legacy comprises a huge number of letters; although many of the original four thousand letters are lost, a large number have survived, and from these we see just how interesting the epistolary exchange of ideas with his contemporaries was, and the role it played in the progress of physico-mathematical science.

In addition to all that, Euler had to participate in investigations of many purely practical problems. For example, he headed the expert committee on the construction of an unsupported, single-span bridge over the Neva. The architect of the project, I. P. Kulibin, had presented results of tests of his model of the bridge, and assumed that these would remain valid for the actual bridge.

In the 1730s Euler was put in charge of the work of drawing a map of the whole of the Russian empire, and in collaboration with other academicians prepared a splendid—for the time—atlas of the empire, published in his absence in 1745. Later he became the first to apply the methods of analytic function theory to cartography.

Although Euler was in Berlin from 1741 to 1766, his connection with the Russian Academy of Sciences remained unbroken; to St. Petersburg he regularly sent his publications and reports on various Russian projects submitted to his scrutiny. He participated in the training of specialists recruited from Berlin to Russia. This was overall a period of active and productive collaboration with the Petersburg Academy of Sciences. Later, when he had returned to Russia, he maintained his connection with the Berlin Academy. Such a link has today been successfully renewed between the Academies of Sciences of the USSR and the German Democratic Republic.

Euler's field of vision encompassed not only the main mathematical interests of the 18th century, but also the theory of numbers, and probability theory together with its applications to demographical questions, pension security, and insurance.

I have already mentioned Euler's work on lunar and planetary motion. Astronomy generally attracted Euler's close attention. He published about 100 works on various aspects of this subject, such as the rings of Saturn and the determination of the orbits of comets, among others. These works have to this day retained their importance.

Euler's work on various problems of mechanics—both theoretical and applied—is of cardinal importance. This work is connected with his investigations into the motion of a body about a fixed point, machine construction—in connection with which he showed that substantial progress can be achieved by introducing measures to reduce friction—, and with his works on nodes of friction and bearings, as well as those on hydrodynamics—where, it is true, he did not take friction into account, this addition to the theory being made later by Stokes. All of this work is of the very greatest interest, and provided the basis for further progress in the field.

It is essential to note also his impressive work on the motion of rigid bodies.

Euler's work in physics is likewise remarkable. He always strove to start with a model that seemed to him best suited to far-reaching mathematical elaboration. In this connection, he generally believed that his models were infallible and fully realistic. Although such convictions were not always justified, they were often the source of later successful research in physics. He himself believed that science progresses via conjectures, by successively rejecting less accurate conjectures in favor of more complete ones.

I think, comrades, that it is time to end my opening remarks, and yield the floor to the specialists, who will tell us about those of Euler's works of significance still today. I will merely remind you of the fact that in modern textbooks on higher mathematics and mechanics, one finds a multitude of methods, formulae, and theorems bearing the name of Euler. In that regard, perhaps, he has practically no equal.

Leonhard Euler: his life and work

A. P. Yushkevich

Of all famous scientists of all times and peoples, there is scarcely one who is recalled more often in today's mathematics courses than Leonhard Euler: In the differential calculus the theorem on homogeneous functions bears his name; in the integral calculus the substitution used to rationalize quadratic irrationalities and the "Euler integrals of the first and second kind" are due to him; in the theory of ordinary differential equations, two classes of linear differential equations with variable coefficients were introduced by him, as was the method of approximate integration serving as the basis for Cauchy's well-known theorem on the existence of solutions; in the calculus of variations there is the differential equation used to find the functions at which a given functional takes on extreme values, and one of the so-called "direct" methods; in the calculus of finite differences the Maclaurin-Euler summation formula; in analytic function theory we have the Cotes-Euler formula relating the exponential function to trigonometric functions, and also the d'Alembert-Euler equations, better known as the Cauchy-Riemann equations; in the theory of infinite series Euler invented one of the methods for summing divergent series and for increasing the rate of convergence of convergent series; in differential geometry we have his formula for the curvature of a normal section of a surface; and in topology the fundamental Euler characteristic of a topological complex. This list is far from complete: for instance one might add to it Euler's function from number theory, Euler's constant, Eulerian numbers, the Euler angles, and so on. We have included here only those instances evoking the great scholar's name in present-day syllabi at the university level or that of higher technical colleges; in fact many of Euler's methods and formulae are usually expounded without acknowledgement, or, worse still, with names of later mathematicians attached. For example, Euler had derived analytically the so-called Fourier formulae of the theory of trigonometric series 30 years before Fourier. To Euler is due the present standard exposition of logarithms and trigonometry in high school. Many of the mathematical symbols suggested by Euler have become firmly established, for instance the imaginary unit i, the natural logarithmic base e, the symbol Δ indicating a finite difference, and Σ denoting summation.

All of these formulae, theorems, methods, and symbols partially reflect Euler's enormous contribution just to mathematics; in the present essay I will for the most part not consider mechanics, astronomy, physics, geography, and technology, in all of which his

View of the town of Riehen
Watercolor by V. I. Perediriǐ (based on an early sketch)

achievements are remarkable. The 18th century is often called the century of enlightenment. As far as the exact sciences are concerned it would be more appropriate to call it the century of Euler, even though there were then also active such outstanding scientists as Clairaut, d'Alembert, Lagrange, Cramer, Lambert, de Moivre, Stirling, Maclaurin, and, in Russia, Lomonosov.

In my talk I shall attempt to describe Euler's life briefly, and to give an overall characterization of his contribution to science, with the emphasis mainly on mathematics.

Euler's life can be divided into four periods: his first 20 years, spent in Basel, the next 14 years working in St. Petersburg, then 25 years in Berlin, and, finally, a second 17-year-long Petersburg period.

Leonhard Euler was born on April 5, 1707, in the city of Basel, into the family of a pastor of modest means, and passed his childhood in the town of Riehen, where his father had obtained a living. His first mathematics lessons were taught him by his father, who had in his student years attended the lectures of Professor Jakob Bernoulli, the nearest mathematical successor to Leibniz. Later a special mathematics tutor was hired, albeit a theologian by profession.

At that time Euler studied the difficult, although by then outdated, *Algebra* of Ch. Rudolff[1] (1522), as revised by M. Stifel[2], another prominent 16th century algebraist. In the autumn of 1720, at the age of thirteen and a half, Euler became a student in the faculty of liberal arts of the University of Basel. The university had at that time no faculty of

---

[1] Christoff Rudolff (end of 15th to early 16th century).–*Trans.*
[2] Michael Stifel (1487–1567).–*Trans.*

The University of Basel in the 17th century
Based on the town plan of Basel prepared by M. Merian in 1615.
(The numeral 1 indicates the Lower College.)

physics and mathematics, and offered no special preparation for a career in mathematics. Students graduating from the liberal arts faculty might continue their education in any of the faculties of divinity, law, or medicine.

The mathematics lectures attended by Euler covered only the elements of mathematics and astronomy. However they were delivered by Johann I Bernoulli, the world's most powerful mathematician of the period—if we exclude the aging Newton—, pupil of his elder brother Jakob, and the latter's successor to the Chair in Basel. Euler was completely absorbed by mathematics, and appealed to his professor to tutor him privately. Bernoulli refused to give separate lessons, but, having noticed the boy's exceptional gift, helped him in a different way. He systematically recommended specific mathematical literature for him to work through, and allowed him to come to his home on Saturdays to discuss difficult questions. For Euler this was the best possible scientific education; in his case no further help was necessary. He soon became close to the sons of his mentor: Nicolaus II, Daniel, and Johann II, and also to Nicolaus I Bernoulli, the nephew of Johann I. These all entertained a lively interest in mathematics, taking it as their calling. Euler was an excellent student, and also actively participated in university life. A year after graduating he gave a lecture in which he compared the views of Newton and Descartes on natural philosophy, following which he was awarded the Master of Arts degree.

His father thought that it would be best for his son to take up, like himself, the career of pastor, and, bowing to his father's wish, the young Euler enrolled in the Faculty of Divinity. However there he was bored, and his studies were not very successful. His father did not try to oppose his son's inclinations further, and Euler totally immersed himself in

Johann Bernoulli, Euler's scientific advisor

mathematics. In 1726-1727 his first two papers were published in the international journal *Acta eruditorum*. These both dealt with problems in analysis, which was then the main focus of research of the Bernoullis and other researchers. It was then also that he took part in the competition of the Paris Academy to find the optimal placement of a ship's masts. Though not awarded the prize, Euler's entry was well received, and was published in Paris in 1728. (I note here parenthetically that between 1738 and 1772 Euler was awarded the prize of the Paris Academy 12 times for a variety of entries on problems of applied mathematics and technology.)

There now arose the problem of finding suitable employment. In Switzerland, and for that matter in European countries generally, departments of mathematics were few and far between, and usually vacancies arose only on the demise of the professorial occupier of a Chair. Taking this situation into account, the Bernoullis often obtained qualifications in addition to their mathematical ones. Thus Nicolaus I Bernoulli, as Doctor of Laws, was able to obtain the Chair in logic in Basel, and then in law. Nicolaus II Bernoulli likewise had as his first appointment the Chair of law in Berne. Daniel Bernoulli, having become a Doctor of Medicine, twice competed unsuccessfully for the Chairs of anatomy and botany, and then of logic. For his part, Euler made an unsuccessful attempt to obtain the position of professor of physics at the University of Basel, when that Chair became vacant in 1727; in fact he was not even permitted to apply, possibly because of his youth. The establishment in St. Petersburg of the Russian Academy of Sciences planned by the emperor Peter the Great, offered the two eldest sons of Johann I, and later Euler—as indeed many other West European scholars—a way out of these difficulties. At that time Russia did not have its own native cadre of scholars, and to the newly founded Academy active, and if possible young, foreign scholars were invited to work under contract. The official opening of the Petersburg Academy took place in August 1725, six months after the death of Peter I. It then boasted 17 professors and adjuncts—i.e., senior and junior academicians—in mathematics, physics, astronomy, chemistry, and other natural sciences, and also in several of the humanities.

Nicolaus II Bernoulli   Daniel Bernoulli

Nicolaus II (in the department of mathematics) and Daniel Bernoulli (in the department of physiology) figured in this first cohort of academicians. On their recommendations, towards the end of 1726 Euler was invited to take up the vacant position of adjunct in the department of physiology, having studied up on this subject specifically to this end. It was understood that he, like Daniel Bernoulli, would occupy himself with applications of mathematics to problems of physiology, in particular to those having to do with the flow of blood in the body's vessels. Following the above-mentioned unsuccessful attempt to obtain a position in his own country, on April 5, 1727 Euler bade farewell to Basel forever, and after 50 days of travel, first on the Rhine, then through Germany, and finally by sea, arrived in St. Petersburg on May 24 of that same year.

At the time when the 20-year-old Euler became adjunct of the Petersburg Academy—where he was immediately given the opportunity of working not in physiology but in the mathematical sciences—, intensive scientific work was already being carried out there. Nowhere else in the world were conditions more favorable for Euler's further scientific development than in St. Petersburg. In the first place, he found himself a member of a scholarly collective with mutual interests and aims, providing stimulus to all concerned. Twice a week scientific colleagues gathered at a collegial conference where they announced and discussed their latest work and letters from foreign scholars, as well as current questions of academic life, such as the content of the *Commentarii* of the Academy, a yearly publication of which the first issue after 1726 appeared in 1728. (The title page of an issue of the *Commentarii* is reproduced below.) Of especial importance for Euler were his regular discussions with Daniel Bernoulli, with whom he shared an apartment until Bernoulli's departure in the spring of 1733. Apart from Bernoulli, there were among the Academicians the noted mathematician J. Hermann, a former student of Jakob Bernoulli, the geometer F.-Ch. Meyer, the physicist and mathematician G. W. Kraft, Ch. Goldbach, of wide education

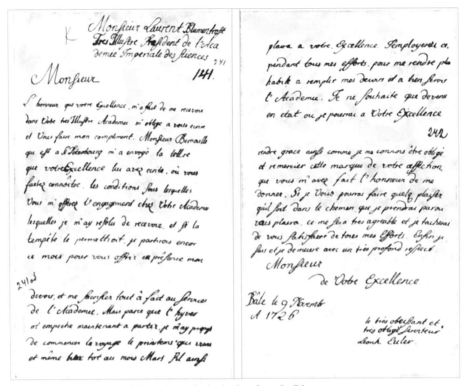

Euler's reply to the invitation from L. Blumentrost,
president of the Petersburg Academy, to work at the Academy
Archive of the Academy of Sciences of the USSR

and extraordinary perspicacity, especially in number theory, the well-known astronomer J. N. Delisle, and others.

Euler's scientific correspondence began immediately to be of importance—especially that with his former teacher Johann Bernoulli, which continued for over 20 years from the autumn of 1727. In view of the almost complete lack of scientific journals and international conferences, which so enhance the scientific life of today's scholars, for those living in different towns or countries in the 17th and 18th centuries scientific correspondence provided the most important means of exchanging information quickly. Many such letters were in effect detailed critiques of current research. When in 1728 Goldbach had to transfer to Moscow for a few years, he and Euler began a correspondence of great substance, comprising 196 letters, which continued until Goldbach's death in 1764. The departure of Daniel Bernoulli likewise led to an epistolary exchange of ideas and problems between him and Euler lasting many years; 90 letters from this correspondence have been preserved. In fact Euler corresponded with nearly all contemporary mathematicians of note: with, in addition to those already mentioned, Stirling, Clairaut, d'Alembert, Lagrange, Cramer, Lambert, and many others. Characteristically, he took great pains over the preservation of his corrrespondence. As he put it: "If anyone should take the trouble to read it [his correspondence], he will find in it many important things whose publication would be more to the public taste than works of the deepest conception". It is worth noting that the Petersburg Academy defrayed for its members the cost of their postage, at that time quite substantial.

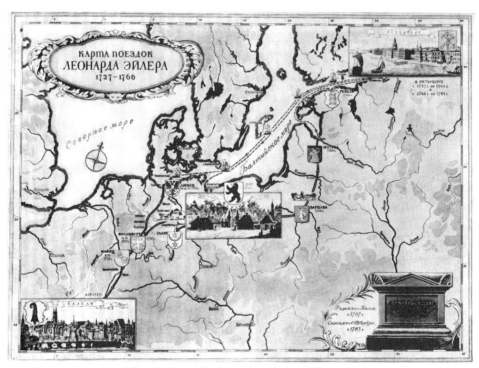

Map of the travels of Leonhard Euler 1727–1766

An exceptionally agreeable condition of Euler's employment in St. Petersburg consisted in the possibility of regular publication of his papers in the Academy's journal *Commentarii* (in Latin, the *lingua franca* of the 18th century scholarly world) and of certain of his books. Euler remains the most productive mathematician of all time, and it is remarkable that his literary output did not decrease with time. I once had occasion to count by decades the number of works he prepared for publication, not distinguishing voluminous books from short articles, and leaving aside a relatively few undated items. Of the overall number of about 850 individual works (including more than 20 large monographs), the distribution of percentages of items prepared for publication in successive decades is as follows:

1726–1734 approx. 5%	1755–1764 approx. 14%
1735–1744 approx. 10%	1765–1774 approx. 18%
1745–1754 approx. 19%	1775–1783 approx. 34%

Incidentally here one should take into account the circumstance that during Euler's second period in St. Petersburg, when he was almost blind, he was greatly helped by first-rate scientific secretaries. Nowhere else in the world would he have been able to publish his work on such a scale as in Russia, where the publications of the Academy were so generously financed by the state.

Of course the above statistics are insufficient for determining the course of Euler's mental development. As is often the case with mathematicians, many of Euler's interests and ideas were formed in his youth, although even in his declining years he invented new approaches and methods and remained fully receptive to the discoveries of his younger contemporaries. Over the span of decades he returned again and again to problems which had at

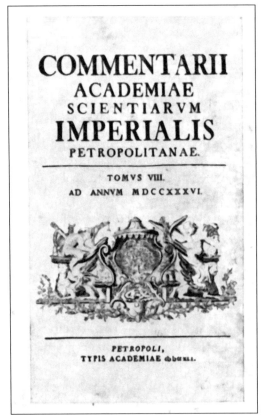

The title page of volume VIII of the *Commentarii* of 1736,
published by the Petersburg Academy of Sciences.

one time piqued his interest, but had for one reason or another been set aside or else solved in a way unsatisfactory to him, and his scientific notebooks, which he kept from 1725 to 1783 (12 notebooks totalling 4000 pages are preserved in the Archive of the Academy of Sciences of the USSR), bear witness to an untiring accumulation of material for further treatment. All his life he was unable to keep pace in written form with the scientific ideas teeming in his mind. Many of the research themes he brought to perfection in the mid-18th century or even later, can be traced back to the early part of his first Petersburg period or even to his years in Basel. For example, it was in Basel that he conceived the project of reformulating in terms of the infinitesimals of the Leibnizian school Newton's mechanics of point-particles, expounded in the latter's *Philosophiae naturalis principia mathematica* (1687) and in Hermann's *Phoronomia* (1716), where, as Newton himself asserted, the proofs are couched in synthetic-geometric terms not useful for obtaining uniform solutions of subsequent problems. Euler realized these intentions ten years later in his two-volume work *Mechanics, that is, the science of motion, expounded analytically* (SPb., 1736), which became the starting point for the whole of the future development of mechanics. The corresponding claim may be made also for his investigations in the theory of music, set out in his *An attempt at a new theory of music, clearly expounded on the basis of the truest elements of harmony* (SPb., 1739).

The building of the Petersburg Academy of Sciences in the 18th century
From a sketch by Lespinas

As a state institution, the Petersburg Academy of Sciences was called upon to provide answers to important practical questions. The university and gymnasium attached to the Academy were engaged in preparing national cadres of scientists; many Russian scholars of the 18th century, including the great M. V. Lomonosov, were students at these academic institutions, which were closed only at the beginning of the 19th century, having become redundant in connection with the reorganization of the whole system of national education. At that time only the graduate program, as we would now call it, remained linked to the Academy, and that only briefly. (It was revived under Soviet rule.) The Academy was also charged with providing technical expertise of many kinds, but especially important among its responsibilities was a complete exploration of the little known regions of the enormous Russian empire, in particular all of Siberia including Kamchatka. The famous academic expeditions of the 18th century were of first importance in this regard.

Euler took an active part in many such enterprises, which incidentally facilitated his learning Russian well, so that, unlike certain other foreign academicians, he was able to communicate with those Russians not familiar with any other language. For the students of the Academy's gymnasium he wrote a *Guide to arithmetic*, published in German in 1738, and subsequently translated into Russian, which greatly influenced the teaching of that subject as well as authors of later texbooks. For several years Euler, in collaboration with J. N. Delisle and the academician G. Heinsius, astronomer and geographer, devoted a considerable amount of time to cartography, even drawing certain maps himself. The importance of having accurate geographical maps, hitherto ignored, had become clear to the government, in particular in connection with the need to define the borders with neighboring foreign countries accurately. Euler's cartographical work was later reflected in his theoretical investigations: 40 years on, in the 1770s, he was the first to apply the theory of functions of a complex variable to cartography.

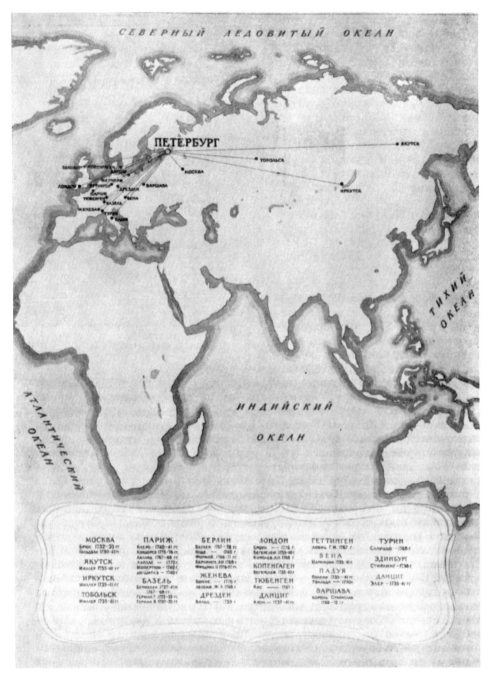

Schema of Euler's correspondence from St. Petersburg

As mentioned earlier, already in Basel Euler had worked on problems related to ship-building. In St. Petersburg in the mid-1730s he returned to this broad theme, and in 1740 undertook to compose a special treatise on the topic. It is hardly necessary to explain the timeliness of this work for Russia, which had emerged under Peter I as a major sea-power. Euler completed the fundamental two-volume work *Naval science, or a treatise on ship-*

building and navigation in Berlin, whence he sent the manuscript back to St. Petersburg where it was published in 1749. This work was of fundamental importance for mechanics as a whole, and not just for fluid mechanics and the kinematics and dynamics of a rigid body. However by the character of its exposition it was not suitable as a textbook, and almost a quarter of a century later Euler wrote for the students of Russian naval colleges the more accessible *The complete theory of shipbuilding and navigation*, which first appeared in French in 1773, and soon afterwards, in 1776, was republished in Paris, and then in Russian, English and Italian translations. The Russian version of 1778 was prepared with useful supplementary explanations by M. E. Golovin, a former student of Euler's and a nephew of Lomonosov.

Euler was able to successfully combine the fulfilment of these and other such commitments with work on purely theoretical problems. He presented his first scientific communication at the Academy's regular conference of August 5th, 1727 and thereafter became its most frequent speaker. It was then that his articles began to appear, in Latin, in the annual of the Academy, which went under a succession of names: *Notes* (*Commentarii*), *New Notes*, *Proceedings*, and *New Proceedings*. From the second volume of the *Commentarii* of 1727, which appeared in 1729, Euler continued throughout his entire life to contribute articles to the Academy's journals, sometimes as many as ten to a single issue. Over his first Petersburg period he prepared over 80 papers and published around 50 on various questions of pure and applied mathematics. Some of these continued or brought to completion investigations begun in Basel; however in the majority of cases these works struck out in new directions in which Euler's predecessors had taken but the first steps; for instance his solution of several problems of the calculus of variations, results on the integration of ordinary differential equations, the introduction of the gamma-, beta-, and zeta-functions, investigations into infinite series, including asymptotic expansions, results on continued fractions, number theory, topology, and so on. This was the period also of his first original researches in astronomy. And of course at the same time he was working on the new problems that would occupy him over the next decade.

Except for one misfortune—the unexpected loss of sight in his right eye in 1738—Euler's personal life was also going well at this time. At the beginning of January 1734 he married Catharina Gsell, the daughter of a painter of the Petersburg Academy of Sciences, also from Switzerland. In that same year their son Johann-Albrecht was born, and in 1740 a second son Karl. Euler acquired a house on Vasil'evskiĭ Island, not far from the Academy, in which the family of his younger brother Heinrich, another artist, also took up residence. It might seem that Euler had put down strong roots in the Russian capital. However after the death in 1740 of the empress Anna Ioannovna and the proclamation of the three-month-old Ioann VI as emperor, the political situation in St. Petersburg became unstable on account of the struggle for power by rival court factions. At first Biron, the favorite of the late empress, became regent, but was very soon sent into exile, being replaced by the mother of Ioann VI, Anna Leopol'dovna. However her hold on the reins of power also turned out to be insecure. The discontent of the Petersburg nobility and guard over the undue influence of foreigners at court presaged further complications. This political uncertainty affected the activities of the Academy of Sciences. Somewhat earlier, in the summer of 1740, Euler had received an invitation from the Prussian king Friedrich II to transfer to Berlin, where he planned to organize an academy of sciences of no less repute than the academies of Paris

First and last pages of a letter written by Euler, in
Russian, to adjunct G. N. Teplov, dated April 9, 1748.

and St. Petersburg. The invitation was renewed in the Winter of 1741, and an anxious Euler answered in the affirmative. He left St. Petersburg, accompanied by his whole family, on June 19th, and by July 25 they were already in Berlin. In the meantime in Russia, at the very beginning of 1742, following on a palace revolution, Elizaveta Petrovna, daughter of Peter I, ascended the throne.

Translation of excerpts of the letter from Euler to Teplov reproduced on these two pages:

"My Dear Sir

I am most humbly grateful to you that the business of Mr. Bernoulli is now almost finished to his satisfaction. I am all the more happy about this in that Mr. Bernoulli had begun to be very angry with me, and had ascribed to me the whole cause of these unpleasant events, because he thought that I had raised your hopes too high concerning his intention, and for this reason he made strong complaints about me to Count von Keigerling(?) and Mr. de Maupertuis. He was so far advanced in his anger towards me that he did not inform me of his father's death, and even now I can hardly assuage it, although I have proved to him that I never wrote to you in firmest terms about his intention—of which there is no need for me to justify myself to you, since he himself showed clearly enough by his decision that I was not in any way in the wrong, in that he gave his father's prohibition as the reason for his refusal, and did not complain that I falsely communicated his intention to you. I was in no doubt when he first wrote to me about this that his intention was not firm. An inclination(?) to imperiousness(?). But, to take the matter more seriously(?), I knew that he had at that time been very sharply rebuked by the mayor of Basel, since in a publicly promulgated decree concerning a certain strike, it had been noted that he had kept company with one of the strikers, and he immediately became so unhappy about this, that he felt he must quit his fatherland. However then the mayor showed that he was more favorably inclined towards him, so that he changed his mind completely, and now ...

....

He is fully competent to be in charge of the printing of my book on navigation and shipbuilding, and for this reason I humbly entreat you to allow me to keep my book here with me while Mr. Oechlitz(?) is on his way to you, since I wanted to elucidate to him personally many circumstances surrounding that book, which would be very difficult to do in writing. I am expecting a response from him this week, which I shall at once communicate to you so that he can leave shortly after receipt of the payment for his trip. And I humbly request that following on his arrival the publication of my book not be delayed, for I fear that that my conceptions will with time cease to be new as French mathematicians are trying very hard in their research on this material, and have already published some important discoveries, which I had made many years earlier than they. I am very unhappy that in the negotiations with Mr. Kies I was unlucky, since if the declared final conditions had arrived within two or three months, he would have accepted them immediately; for Mr. Braun will tell you that at that time he was out of favor with our President, but is now back in favor with him. I asked him to convey his opinion to me in writing, and I am attaching his letter to this. In the meantime I have discouraged his hopes so that he not remain in suspense, since it seems to me that he would certainly not refuse. All of my family humbly bows to you, and I beg that you accept my deepest respect. Convey to his highness the Count my most complete veneration.

My dear Sir

Your most obedient servant

L. Euler.

Berlin

9 April, 1748."

Leonhard Euler at 30
From an engraving by V. Sokolov from a portrait by J. Brucker of 1737
(The Saltykov-Shchedrin State Public Library, Leningrad)

Thus ended the first Petersburg period of Euler's life. In 14 years he had achieved a great deal in furthering science in Russia and the world at large. At the same time he had a clear understanding of the invitation to St. Petersburg as being of decisive significance in his life, and expressed this in one of the letters he sent to the Petersburg Academy in 1749; after all, it was only in St. Petersburg that research in mathematics, which from his youth he saw as his main vocation, could have become his chief occupation and develop so successfully.

Euler arrived in Berlin as a world-renowned 34-year-old scholar, whom his old teacher Johann Bernoulli called with complete justification, in a letter to Euler written in 1745, "princeps mathematicorum"—chief mathematician. In Berlin Euler actively assisted Friedrich in organizing the new Academy of Sciences. Although it is true that that they did not have to start from scratch, since as early as 1700 there had been established in Berlin, on Leibniz' initiative, a Science Society, the members of this Society were not active scientists, and furthermore Leibniz himself lived not in Berlin, but in Hannover. Under the father of Friedrich II—the uneducated and despotic soldier-king Friedrich Wilhelm I— the Berlin Science Society had eked out a pitiful existence. Friedrich II, acceding to the throne in 1740, was as militaristic a ruler as his father—it is enough to recall that 15 of the first 23 years of his reign were spent in warfare: the 8 years of the war of the Austrian succession (1740–1748), and then the seven years' war (1756–1763). However at the same time he was a typical representative of the enlightened absolutism of the 18th century, providing support—partly from considerations of prestige—for science and the arts, and permitting, within bounds consistent with absolute monarchy, moderate philosophical freedom of thought. In his personal tastes Friedrich II was an adherent of French culture, corresponded amiably with Voltaire (but subsequently quarreled with him), wrote chiefly in French, and made French the official language of the new Academy. And while the first president of the Petersburg Academy had been the German doctor to the imperial family L. Blumentrost (born, it is true, in Moscow), to take up the post of first president of the Berlin Academy of Sciences and Literature, which replaced the old Science Society in 1744, Friedrich II invited the French scholar P. L. Moreau de Maupertuis, who especially impressed the king with his courtly manners. Maupertuis took up his duties in 1746, establishing from the start the best of relations with Euler, who in 1744 had been appointed director of the mathematics department. With Friedrich, however, Euler's relations were of another sort. The whole of his upbringing had been such that the king of mathematicians was in his essence a typical *Bürger* of Basel, utterly incapable of playing the role of salon philosopher, moreover a pious *Bürger* and therefore opposed to freethinking, especially the French variant patronized by the king of Prussia. Friedrich II was disdainful of mathematics, unless it proved useful in solving problems of exclusively practical importance. In essence their relations were such that while neither felt any empathy for the other, each tolerated the other out of necessity.

In Berlin Euler had to do a great deal of work of an organizational character, and in addition fulfil a variety of the king's personal commissions. Only his extraordinary capacity for work and ability to rationalize his time allowed him to simultaneously cope with the many duties imposed on him and carry out his scientific research at an ever-increasing rate.

As a member of the directorate of the Berlin Academy, and assuming in effect the function of president during Maupertuis' frequent absences, Euler was occupied with the building and equipping of the observatory, supervising the preparation and publication of maps, the ordering of seeds and plants for the botanical gardens, hiring, firing, and pension arrangements of the Academy's employees, and in addition to all this, the publication of a series of annual calendars, the income from which accounted for most of the Academy's budget. Taking the king's interests into account, Euler translated from English the best text on ballistics of that time, written by B. Robins and published in 1742. Euler provided his German version of *The New Elements of Artillery*, which appeared in 1745, with explanations and fresh applications that significantly increased the size—and price—of the book; these additions were subsequently incorporated into a new English edition and a French translation. I note incidentally that Euler had already become acquainted with problems of ballistics in St. Petersburg, where in 1727 he and D. Bernoulli were present at a testing of artillery weapons; his brief essay on these trials, first printed in 1862, was for a long time preserved in the Archive of the Academy of Sciences of the USSR.

By commission from the king, Euler was obliged to occupy himself also with hydraulic engineering, in particular to act as consultant in connection with the work of leveling the Finow canal between the rivers Havel and Oder, and with the water supply of the royal residence at Potsdam with its many fountains on various levels. Through an exchange of letters with J. A. Segner, professor of mathematics first at Göttingen and then Halle, Euler became familiar with the details of an hydraulic machine invented by Segner, the simplest form of which is the "Segner wheel", familiar to schoolchildren. Over the period 1750–1753 Euler made several significant improvements to this machine, and concurrently laid the foundations of the theory of hydraulic turbines. In conjunction with all of these activities he produced a long series of theoretical articles on the theory of mechanisms and machines, and most importantly on fluid mechanics.

Questions about the mechanics of fluids (or hydrodynamics) had been investigated in ancient times—it is enough to recall Archimedes' law concerning bodies immersed in a liquid. Fresh progress in fluid mechanics was made in the 16th and 17th centuries by Stevin, Galileo, Torricelli, Pascal, Newton, and others, first in hydrostatics, and then in hydrodynamics, ultimately resulting from the need to solve practical problems concerning navigation, and the construction of canals, dams, pumps, water mills, etc. Euler's notebooks show that as early as 1727 he had conceived the idea of writing a substantial work on hydraulics, but then refrained in order not to compete with Daniel Bernoulli, who was already well launched in that direction and while still in St. Petersburg had completed the first version of a large treatise, the second version of which was printed in Strasbourg in 1738 under the title *Hydrodynamics*. In the 1730s Euler was also very much taken up with questions from fluid mechanics, on both the theoretical plane and with applications in view, one outcome of which was the treatise *Naval Science* noted above. However a more profound treatment of fluid mechanics became possible only later, during Euler's time in Berlin, when towards the end of the 1740s, following on the work of d'Alembert, he began to create the apparatus of the theory of partial differential equations. I shall return below to the decisive first step, taken in the context of the theory of elasticity, by d'Alembert and Euler; for now I merely note the series of classic articles on fluid mechanics written by Euler in the period 1757-1761 published in Berlin and St. Petersburg. In these papers Euler provided the foundations of the modern theory of equilibrium and motion of an ideal fluid; they contain the equation of continuity and the general differential equations of both hydrostatics and hydrodynamics, which with complete justification bear the name of Euler.

While I shall not go into detail concerning the other area of mathematical physics close to fluid mechanics, namely aerodynamics, where Euler also made important contributions, I *shall* describe his researches in celestial mechanics. These investigations, including in particular his work on the orbits of Jupiter and Saturn awarded a prize by the Paris Academy, include results of an essentially mathematical nature, for the most part having to do with integration of differential equations, expansions in series (in particular trigonometric series), and with numerical methods in analysis—which incidentally do not figure in Euler's strictly mathematical publications. It is appropriate to dwell on Euler's role in the development of the theory of the moon's motion, which attracted the particular attention of the scientific world because of the discrepancy between the moon's observed orbit and the orbit calculated by Newton using his Universal Law of Gravitation. The disagreement between observation and computation based on theory was in fact so large, that at one time d'Alembert, Clairaut, and Euler all considered that it might be necessary to make a correction to Newton's universal law. However when towards the end of 1748 Clairaut concluded that the discrepancy arose as a result of insufficient accuracy in the basic approximations, Euler advised the Petersburg Academy of Sciences to announce a competition for the best essay on the motion of the moon. At the same time he verified Clairaut's conclusion using a method of his own devising. Clairaut's essay, awarded the prize of the Petersburg Academy on Euler's recommendation, was published in 1752, and a paper by Euler on the same topic appeared in Berlin in 1753, for which he received special compensation from St. Petersburg. In addition to this Euler received a portion of the prize money set aside by the British parliament for the highly accurate lunar tables compiled by the Göttingen astronomer J. T. Mayer, who had used Euler's results with his consent. At that time, and for that matter for a long time thereafter, lunar tables were used to determine longitude on the open sea, and Mayer's tables were included in marine almanacs. Incidentally, the very substantial correspondence between these two scholars has been preserved, and is published.

I note also two areas of Euler's activity in Berlin closely related to his involvement with practical problems. First, there were the calculations he made in connection with the organization of state lotteries—which served as a supplementary source of revenue for the Prussian treasury—, and with problems relating to insurance and demography, involving questions in probability theory and mathematical statistics. Much later a textbook on the insurance business and the organization of lotteries, furnished with tables, was prepared in St. Petersburg under Euler's supervision by his student N. I. Fuss, who will be mentioned several more times below; this book was published in 1776. Secondly, there was his interest in optics, extending over many years. Newton's corpuscular theory of light and colors has the consequence that an increase in the strength of an optical apparatus necessarily results in chromatic aberration of the images of objects and imposes, therefore, an impassable limit to the perfectibility of refracting telescopes, as opposed to the reflecting kind. On the basis of his own theory of light, different from Newton's and also from Huygens' wave theory (but not, it must be said, surviving into modern physics), Euler concluded that achromatic lenses of arbitrary refractive power are after all possible in principle, if made from transparent material having appropriate optical characteristics. Because of technical limitations, Euler's own attempts at manufacturing such instruments were largely unsuccessful, but soon afterwards, in 1758, the Englishman George Dollond was successful in making achromatic lenses of high power from an alloy of crown glass and flint glass. This represented a decisive step forward in the technology of the manufacture of telescopes and microscopes. Euler carried out detailed calculations relating to various dioptrical systems, and expounded the results of this research in a three-volume treatise entitled *Dioptrics*, composed largely in Berlin, but completed in St. Petersburg, where it was published in 1769–1771. Also in St. Petersburg, the talented mechanic and designer I. P. Kulibin took on the task of building a powerful microscope under Euler's supervision, and N. I. Fuss, on the basis of the *Dioptrics*, composed a detailed manual for master opticians, published in 1784.

Before leaving St. Petersburg Euler had reached a formal agreement with the Russian Academy allowing him to retain the title of foreign member, which then carried with it the payment of a large yearly pension of 200 rubles. Of course the value of the ruble in terms of actual goods has gone through a great deal of fluctuation over the past two and a half centuries. Although it would appear to be impossible to provide a single index of prices over such a period, some indication of the worth of 200 rubles at the time in question may be inferred from the fact that, in terms of the price of bread, during the reign of Catherine II one ruble would buy the same amount as 8 rubles at the beginning of the 20th century. Euler undertook to complete certain scientific projects begun in St. Petersburg (in particular his *Naval Science* mentioned earlier), and to submit articles to the St. Petersburg *Commentarii*; he also regularly agreed to fulfil commissions of various kinds, even as ordinary as the acquisition of books and scientific instruments. Notwithstanding the fact of his official position in Berlin, Euler was in essence not so much a foreign member of the Petersburg Academy as an actual member who happened to reside in Berlin—so that he also functioned as an active intermediary between the Petersburg and Berlin Academies. The three volumes of his correspondence with employees of the Petersburg Academy—comprising around 800 letters sent in both directions during his 25-year sojourn in Berlin, i.e., just under three letters a month on average—bear witness to the enduring nature of his connection with Russia. During the seven years' war the volume of this correspondence was substantially reduced without ceasing entirely; the letters were sent via persons residing in German principalities which, unlike Prussia, were non-participants in the war. Euler's correspondence with M. V. Lomonosov is preserved, though not in its entirety; Euler highly esteemed the latter's work in physics, and continued supporting him in the face of transparent hints to the effect that he not do so from influential enemies of Lomonosov in the administration of the Petersburg Academy.

From Berlin Euler edited the mathematical section of the St. Petersburg *Commentarii*, wrote reports on the work—sent to him in Berlin—of adjuncts and Russian students of the Academy, compiled lists of topics for international competitions and wrote reports on submitted essays (the competition on the topic of the moon's motion was mentioned above), communicated news of scientific life in Germany and Western Europe generally, and so on. He was asked several times by the administration of the Petersburg Academy to suggest possible candidates for vacant positions in view of the continuing dearth of suitable specialists in Russia. Thus it was at Euler's suggestion that the eminent physicist F. U. T. Aepinus was invited in 1754, the renowned physiologist C. F. Wolff in 1767, and the astronomer and geographer G. M. Lowitz in 1768. The lengthy mathematical apprenticeship which the three Russian adjuncts of the mathematics department of the Petersburg Academy S. K. Kotel'nikov, S. Ya. Rumovskiĭ, and M. Sofronov served with Euler was of great significance for Russia: under Euler's tutelage these three, together with Euler's eldest son Johann-Albrecht, substantially improved their qualifications in higher mathematics and mechanics, and subsequently became prominent representatives of science and the enlightenment in Russia.

Euler's scientific work at this time is extraordinary for its quantity and for the wide variety of problems tackled in both pure and applied mathematics; on average he published

10 papers a year during this period, approximately half appearing in Latin in the *Commentarii* of the Petersburg Academy and half in French in the corresponding journal of the Berlin Academy. I have already discussed his research in mechanics but to complete the picture of his activity in this area mention should also be made of his fundamental treatise *The theory of motion of rigid bodies*, published in 1765, complementing the *Mechanics* of 1736 devoted to the mechanics of point-particles. The period from the end of the 1740s onwards witnessed a growing interest on Euler's part in mathematical physics—which of course was to undergo a magnificent flowering in the 19th and 20th centuries. This interest was stimulated by rivalry first with d'Alembert and Daniel Bernoulli, and somewhat later with Lagrange. The solution of the new and difficult problems of natural science that arose required herculean efforts in connection with the perfection of analysis in various ways—not only among its several branches, but also in its relations with other areas such as number theory, algebra, and differential geometry. New chapters in the differential calculus appear as part of Euler's output at this time: the differential calculus of functions of several variables, elliptic integrals and an addition theorem for them, special improper integrals, important classes of ordinary differential equations, the theory of integrating factors and many types of partial differential equations of the second and higher order, the calculus of variations, numerical methods, and so on.

Historical interdependencies such as those I have noted glancingly above, rarely occur in mathematics as it is pursued in modern universities. I limit myself to a single example. The Fundamental Theorem of Algebra was discovered in the 17th century in the course of the natural evolution of that subject, but its earliest formulations were both incomplete and unclear in view of the lack of a fully worked-out theory of complex numbers. By the 18th century both this theory and the Fundamental Theorem of Algebra had become essential for solving the problem of integration of rational functions, to which the integration in closed form of many other kinds of functions reduces. At that time no mathematician of note doubted that every polynomial with real coefficients could be decomposed as a product of real first degree and quadratic factors: such was the case for instance with Leibniz, Nicolaus I Bernoulli, and Goldbach. Only d'Alembert and Euler, independently of one another, saw the need for a more thorough investigation of complex numbers in the form $a + b\sqrt{-1}$, and around 1750 produced the first proofs (their proofs were different) of the Fundamental Theorem of Algebra—from our point of view not, it is true, complete, but correct in essence.

The foregoing inventory of Euler's mathematical activities, far from complete, gives some idea of his work only in those areas of mathematics serving the natural sciences. Although in Euler's work the relationship between the so-called pure and applied domains of mathematics was immediate, nevertheless it was in mathematics itself that his deepest interest lay, and from which he derived the greatest satisfaction. More often than not the theorems and methods which he discovered in the course of solving an applied problem later became the starting point of a pertinaceous, systematic, purely theoretical development, and thus turned out to constitute the first links of a new theory, or even a whole discipline. In this respect Euler differed greatly from such contemporaries as for instance Daniel Bernoulli, who, once having found a mathematical solution of some problem or other from the natural sciences, refrained from realizing all the possibilities latent in his solution, even though his mathematical gifts were outstanding. Often Bernoulli confined himself to purely physical arguments, regarding them as sufficient or even preferring them to a precise mathematical treatment of the problem at hand. When Euler was occupied with applied problems, he remained in essence a mathematician; Bernoulli, on the other hand, used mathematics only insofar as it was needed and so remained in essence a physicist. It is characteristic of Euler that at the same time as he was occupied with the mathematical sciences noted above, he was with ever-increasing absorption carrying out investigations in number theory, the subject of three decades of correspondence with Goldbach, and more than a decade with Lagrange. In Berlin, he elaborated, in particular, the theory of Diophantine equations, and of residues of powers. The initial impulse in this direction came to him from Fermat's discoveries in number theory.

Unfortunately, lack of time does not permit me to linger over all of the scientific debates of that period in which Euler participated; I mention just two of them: first, the continuing disagreement with d'Alembert concerning the properties of logarithms of negative numbers, which already at the beginning of the 18th century had been the subject of fruitless polemics exchanged between Leibniz and Johann Bernoulli, at root resulting from the vagueness surrounding the concept of the logarithm itself at the time. Leibniz had suggested that logarithms of negative numbers should be imaginary, in some indefinite sense of the word, while on the other hand Bernoulli tried to prove that the logarithm of a negative quantity must be the same as the logarithm of the absolute value of that quantity. D'Alembert invented more and more reasons supporting Bernoulli's opinion. Euler, however, was the first to give the modern definition of the logarithmic function, and on the basis of his study of complex numbers created the complete theory of this function in the complex domain.

The second debate I wish to mention is the celebrated one concerning the nature of the arbitrary functions that feature in solutions of the equations of mathematical physics, in which Euler, d'Alembert, Daniel Bernoulli, Lagrange—for that matter just about every prominent mathematician of the second half of the 18th century—participated, and which was to have such an enormous influence on the progress not just of mathematical physics, but of analysis as a whole. This debate is often called the "argument over the vibrating string", since it began in 1749 with the analysis proposed by d'Alembert of the problem concerning the transverse vibrations of an ideal string under prescribed initial and boundary conditions, yielding the expression of the general solution of the partial differential equation of the problem as a sum of two arbitrary functions. Without going into great detail, I note merely that d'Alembert imposed on the initial conditions, i.e., on the functions giving the initial position of the string and the initial distribution of velocities of the points of the string, severe restrictions as to their differentiability, essential, in his opinion, for an analytic solution of the problem to be possible. Euler, taking into account the special nature of the problem and bringing in geometric considerations, found d'Alembert's conditions unjustifiably restrictive, and to some extent anticipated the idea of solving the problem when the initial functions are non-smooth in some sense or other—an idea which was to be extensively elaborated upon and rigorously formulated in the 20th century in S. L. Sobolev's theory of generalized functions (1935), and then developed in detail by L. Schwartz as the "theory of distributions". Of course Euler's treatment of this theme cannot be considered satisfactory from the viewpoint of 20th century science, and perhaps even of his own time. All the same one cannot but admire the boldness and perspicacity of his mathematical thought. Note that a summation of the argument over the vibrating string, relative to the period of the 18th century, was given in an interesting essay by the French scholar Louis Arbogast on the nature of arbitrary functions featuring in solutions of partial differential equations, which earned a prize in a special competition organized by the Petersburg Academy of Sciences and was published in French in St. Petersburg in 1791.

A new method of mathematical physics consisting in the expansion of solutions of certain boundary-value problems in series of eigenfunctions, was proposed by Jean-Baptiste-Joseph Fourier at the beginning of the 19th century in his theory of heat, reviving an idea conceived in a particular case by Daniel Bernoulli, who believed it possible to represent every solution of the vibrating string problem as the sum of an infinite trigonometric series in sines and cosines. However in support of this idea Bernoulli could only offer physical analogies, which neither Euler nor d'Alembert considered proofs, and to which only Fourier managed to give analytic form, leading to an intensive development of the theory of trigonometric series, and in direct connection with this the theory of the integral and ultimately the general theory of functions of a real variable.

A little earlier I spoke of Euler's extraordinary boldness and keenness of vision as evinced by his mathematical ideas. A further instance of this is afforded by his thoroughgoing elaboration, in the middle of the 18th century, of the conception of the sum of an infinite series. The distinction between convergent and divergent series was at that time generally well known, and many mathematicians of the 17th century considered only the former to be of practical significance. However Euler, having obtained by means of certain divergent series several remarkable results—for example in the theory of the zeta-function—considered the exclusion of such series from use inexpedient, and worked out—to the extent that the contemporary analytic apparatus allowed—an original theory resting on a certain generalization of the concept of the sum of a series and a special method for transforming series, allowing under suitable conditions the calculation of such generalized sums, and also an improvement in the rate of convergence of a series. In the 19th century such applications of divergent series were subject to criticism and were rejected by several first-rate mathematicians, including N. H. Abel. It thus became the order of the day to formulate a general theory of convergence of series, hitherto non-existent. However later, when Cauchy and his followers had produced this theory—resulting in the intensive development of the theory of analytic functions—Euler's ideas were provided with the firm foundation not vouchsafed him to furnish, and became part of the general theory of summability of series, which has flourished for the last hundred years.

The vast accumulation of mathematical knowledge by the middle of the 18th century, most of all in analysis, was in need of systematization, the filling of gaps, the improvement in precision of basic concepts, and so on. Euler took upon himself the composition of a series of monographs providing an account of all the work in this area, with the aim of facilitating the efforts of others in furthering its development. This project had ripened in Euler's mind over many years, and even in St. Petersburg he had started work on it, although the actual publication of such a series of monographs began only during his Berlin period and was not completed till after his return to St. Petersburg.

On leaving St. Petersburg, Euler had undertaken to finish a certain essay, which in official documents was called "Higher Algebra". In Berlin Euler did in fact do work in algebra, however it is clear from various documents that the essay in question was really an extensive treatise on mathematical analysis. In 1744 Euler published in Switzerland *A method for finding curves possessing maximal and minimal properties*[3]. This work contained the first known method for finding extrema of certain classes of definite integrals (now called "functionals") whose value depends on the choice of unknown functions entering into them, by reducing the problem to the solution of a certain partial differential equation. It is remarkable that Euler achieved this reduction by means of a method nowadays termed "direct"; such so-called "direct" methods, which allow one to bypass integration of the corresponding differential equations, were systematically developed only in the 20th century. Several particular problems of this type had been posed and solved by Johann and Jakob Bernoulli, but both were very far from a providing general treatment of the question, this being due entirely to Euler.

In appendices to the above-mentioned book Euler provided solutions of several problems in mechanics, incidentally giving a precise formulation of the principle of least action, Maupertuis' version of which had been very special; it was here that he derived his formula for the critical loading of columns, which in the theory of resistance of materials bears his name. In the mid-1750s the young Lagrange, in introducing the concept and associated symbolism of the variational calculus, proposed a new, formally analytic presentation of the general problem independent of the geometric reasoning used in Euler's treatment, and extending readily to wider classes of functionals. Euler in turn wrote a more widely accessible exposition of Lagrange's method. Thus were laid the classical foundations of the calculus which, following Euler, came to be called variational.

At the time he was working on this book, Euler was also preparing a multi-volume work intended to encompass all areas of mathematical analysis. In the year of publication of *A method for finding curves*, he sent, again to Switzerland, the first part of this work, published in 1748 in two volumes under the title *Introduction to the analysis of the infinite*. The first volume contained a purely analytical treatment of the elements of analysis considered as a general study of functions, developed without recourse to the differential calculus as far as that was then feasible. The chief means for achieving this were infinite power series (as well as infinite products and sums of simple fractions). The unusually elegant and at the same time accessible presentation of the theory of the elementary functions (excluding the logarithm function, to which, apart from a few cursory remarks in Volume II of the *Introduction*, he devotes a separate work), for the first time considered as functions not only of a real variable but also of a complex variable, together with the beauty of the examples, make this one of the most distinguished works of the whole literature of mathematical analysis. Even today, after an interval of over two and a half centuries, one can recommend this book as absorbing reading to any novice lover of mathematics. The second volume of the *Introduction* is geometrical, for the most part devoted to the study of curves of degrees 2 and 3, and to the theory, expounded in detail for the first time, of surfaces of degree 2, with excursions into the realm of plane transcendental curves. A few years later there appeared his fundamental *Differential calculus*, whose publication in 1755 in Berlin was financed by the Petersburg Academy. In Berlin Euler also prepared a large portion of the manuscript of his three-volume *Integral calculus*[4], which was completed and published in St. Petersburg between 1768 and 1770. To describe briefly the riches contained in both of these works is

---

[3]The actual, Latin, title was *Methodus inveniendi lineas curvas maximi minimive proprietate gaudentes, sive solutio problematis isoperimetrici latissimo sensu accepti.*

[4]*Institutiones calculi differentialis* and *Institutiones calculi integralis.*

The house where L. Euler lived 1776–1783 (reconstitution)

impossible. It must be explained, however, that by the integral calculus Euler understood not only the computation of integrals in the narrow sense of that phrase—which in fact occupies only the first half of the first volume of the three-volume treatise of 1768–1770— but also the solution of both ordinary and partial differential equations; in this work he also provides a new, more complete exposition of the calculus of variations. Neither of these books contains geometric applications; these are to be found in his numerous articles.

This six-volume trilogy of Euler's played an extraordinarily large part in the development of analysis and of mathematics as a whole. Even by itself it fully justifies the words of Laplace: "Lisez Euler, lisez Euler, c'est notre maître à tous[5]". Euler's monographs have been widely used by authors of practically every mathematics textbook right up to modern times.

The circumstances surrounding Euler's personal life in Berlin with his family had turned out very favorably. In 1753 he acquired a farm at Charlottenburg, of which his mother, come from Basel, became manager. His son Johann-Albrecht was elected member of the Berlin Academy of Sciences, and his youngest son Christofor, born in Berlin, became an officer in the Prussian army. It was at this time also that Euler was awarded many prizes by the Paris Academy of Sciences, and in 1755 he was made a foreign member. The Royal Society of London had elected him much earlier, in 1746. Euler's relations with the Petersburg Academy were arranged to his satisfaction, and in the Berlin Academy he enjoyed very great scientific authority, in particular in connection with the running of that institution as Maupertuis' deputy during the years of the latter's absences. Even his relations with the king, who was for the most part preoccupied with war operations, were satisfactory.

However after Maupertuis' death the situation began to change for the worse. When d'Alembert refused the post of president of the Academy, the king took the directorship upon himself, and from 1762 onwards more and more frequently and to a greater and greater extent began to interfere in its affairs, in particular in the appointment of new mem-

---

[5]"Read Euler, read Euler, he is our master in everything".—*Trans.*

L. Euler's house (contempory view)
5 Lieutenant Schmidt Quay. 1956 photograph

bers, where he clearly showed his personal pro-French sympathies. At the end of the seven years' war, when the king took up permanent residence near Berlin, his relations with Euler steadily worsened. In 1763 Euler felt it prudent to sell the farm at Charlottenburg and to strengthen his contacts with the Petersburg Academy and with representatives of the Russian government. On examining the finances of the Berlin Academy, the management of which had fallen to Euler, the king expressed his dissatisfaction with their state and set up a special commission to audit them; in his opinion the revenue from the sale of the calendars should have been greater. For Euler these actions were oppressive and offensive.

By 1766 the discord between scholar and king had become such that Euler felt he had to resign. This decision was made easier by the ever more frequent invitations he was receiving to return to St. Petersburg. The king was reluctant to give up so useful a consultant and organizer, but with the weight of the Russian government behind Euler, had to release him. On June 9, 1766 Euler, together with all of his family except the youngest son, obliged to continue serving in the army, left for St. Petersburg, where they arrived on June 28. Some time later the king also released Christofor. Euler's position in the Berlin Academy was filled by Lagrange, who remained there till 1787, when he left permanently for Paris.

Thus Euler returned to Russia at the age of 60, with a rich experience of life behind him, a large number of unpublished and unfinished works in hand, and an abundant store of creative energy. He was received with the joy and esteem that his genius merited. Upon his arrival he was almost at once received by Empress Catherine II, and together with his son Johann-Albrecht appointed to the advisory council of the "director" of the Petersburg Academy of Sciences, Count V. G. Orlov, at that time deputizing for the official president count K. G. Razumovskiĭ, who had withdrawn from the imperial court. When a few years later both Euler and his son resigned their positions on Orlov's council over a disagreement with him, this did not at all reflect adversely on their situation as a whole; in particular Johann-Albrecht remained in the post of conference secretary assigned to him in 1769. In fact this time marks the beginning of a period of almost a hundred years during which the Euler family wielded great influence in the running of the Academy as a whole. Upon Johann's death in 1800, the post of conference secretary, i.e., the permanent secretary of the Academy, fell to Euler's grandson-in-law N. I. Fuss, and on Fuss' death, in turn, to *his* son, the mathematician P. N. Fuss, who filled the post till his death in 1855.

The Russian government was unchanging in the generosity of its financial support for Euler. To house his family, which now counted 16 people, a large house was built on the bank of the Neva, not far from the premises of the Academy of Sciences. When this house burnt down in the spring of 1771, it was rebuilt, and in a somewhat reconstructed form is preserved to this day. However two unhappy events clouded the last 17 years of Euler's life. In the autumn of 1766 he almost completely lost the vision of his hitherto healthy remaining left eye. From this time on he was only able to distinguish bulky objects and read large letters written in chalk on a blackboard. Although, fortunately, this did not affect his creative activity, it did change the way in which he worked. He now prepared his works with the help of a secretary, a position requiring a qualified specialist capable under his supervision of making necessary calculations and editing texts dictated to them. First Euler's son Johann-Albrecht worked in this capacity, then the physicist and academician L. Yu. Kraft, the son of his colleague from the 1730s, then the talented mathematician A. J. Lexell, and finally his students M. E. Golovin and N. I. Fuss, mentioned earlier, the latter having been invited from Basel as a youth on the recommendation of Daniel Bernoulli. These arrangements allowed Euler, with his constant freshness of intellect and his extraordinary memory, still fully intact, to continue working till the end of his days. The only change was a sudden falling off in the volume of his correspondence, in particular with Lagrange, since he was unable to re-read and verify the complicated arguments and computations contained in the latter's letters.

The second tragic event of these years was the death of his wife, with whom he had lived for almost 40 years. However his large family required a wife to manage it, and three years later Euler married Salome-Abigail Gsell, a sister of his first wife. Generally speaking, the family lived in complete contentment. The eldest son, a highly qualified, if not eminent, scholar, occupied, as they used to say, a responsible post at the Academy. The second son Karl became a very successful doctor, and the youngest, Christofor, enlisted in the Russian army, and for many years worked as director of an arms factory at Sestroretsk, near St. Petersburg, ending his career with the rank of general. Euler's sons became Russian citizens (although Euler himself remained a citizen of Basel for the whole of his life); to this day there are direct descendants of Euler living in Leningrad[6] and Moscow. For a long time Euler maintained excellent health and capacity for work. It was only a little before his death that he began to suffer from dizziness, and on September 18, 1783 died suddenly from a stroke. On September 22 at a general meeting of the Academy the oldest of the academicians, J. von Stählin, gave a funeral oration, and on November 3 N. I. Fuss delivered a eulogy, distinguished by its high seriousness. Finally, the great scientist was

---

[6]Now once more St. Petersburg.—*Trans.*

Bust of L. Euler by J.-D. Rachette
Installed in the building of the Presidium of the Academy of Sciences in Moscow.
There is also a copy by the sculptor in the State Hermitage.

accorded a special posthumous honor: On January 25, 1785 a bust in his likeness, prepared not long before by the well known French sculptor J.-D. Rachette, was solemnly placed on a pedestal in the large conference hall of the Academy. At the present time[7] this bust adorns the premises of the Presidium of the Academy of Sciences of the USSR[8] in Moscow.

Euler was buried in the Smolensk Lutheran cemetery, where in 1837 the Petersburg Academy of Sciences erected a huge monument with the inscription, in Latin: "To Leonhard Euler—Petersburg Academician". In 1957, in connection with celebrations of Euler's 250th birthday, both tomb and monument were moved to the Leningrad Mausoleum, where they were installed next to the burial site of M. V. Lomonosov, and a memorial marble plaque was affixed to Euler's house.

Over the second Petersburg period of Euler's life, from 1767 to 1783, he published around 250 works, including several lengthy books which had been partially composed in Berlin and were now completed and edited. However the publication of this output in the

[7] 1988

[8] Now once again the Russian Academy of Sciences.—*Trans.*

A group of academicians installing the bust of Leonhard Euler
From left to right: A. J. Lexell, J.-A. Euler, N. I. Fuss (holding
the amphora), I. I. Lepekhin, P. S. Pallas, and W. L. Kraft
Silhouettes by J. F. Anthing 1784

Commentarii of the Academy and other periodicals proved unmanageable. At Euler's death about 300 of his papers remained unpublished; of these about 200 eventually appeared in the *Commentarii* at the initiative of N. I. Fuss, and in 1862 P. N. Fuss had a further 50 or so published in the two-volume *Posthumous essays in mathematics and physics* of Euler. Many of these works opened fresh perspectives for research in mathematics and mechanics, receiving due attention and development only in the 19th century; some of them, long unnoticed by posterity, contained results rediscovered later. By way of example I note again Euler's purely analytic derivation of the so-called "Cauchy-Riemann equations", his calculation of many special integrals using functions of a complex variable (a method which Laplace arrived at at about the same time), and also his elementary derivation of the formulae for the so-called Fourier coefficients in the theory of trigonometric series. It is curious that Fourier derived them anew and in a different way, clearly unaware of Euler's work.

I now turn to some of the substantial monographs published by Euler in this last period of his life, apart from the three-volume *Integral calculus* and the three-volume *Dioptrics*, already discussed. Over the two years 1768-1769 there appeared in print a Russian translation of his two-volume *Introduction to algebra* (the German original was published later, in 1770); Euler had dictated the whole of this work to a young servant of German background. While on the one hand this highly original exposition of algebra to a very large extent defined—of course in abbreviated form—the content of all successive textbooks in algebra at the level of the gymnasium, on the other hand it contained many of Euler's own discoveries in the theory of Diophantine equations, reaching far beyond gymnasium level. The French edition of this work (translated by a member of the Bernoulli family), which appeared, with valuable additions by Lagrange, in 1774 in Lyon, represented a substantial leap forward in the development of Diophantine analysis. At about the same time, between 1768 and 1772, there appeared in French the three volumes of *Letters to a German princess on various questions of physics and philosophy*, published simultaneously in the Russian translation of S. Ya. Rumovskiĭ. These *Letters* became the most widely read of all of Euler's works, going through tens of editions in French, English, German, Russian, Dutch, Swedish, Danish, Spanish, and Italian. Written in the early 1760s, the *Letters* represented a popular exposition of the most fundamental questions of physics, philosophy, logic, ethics, theology, etc. On the scientific side this popularization answered to the highest standards of the time. It also reflected Euler's deep religious beliefs and his attitude to certain philosophical systems of the 17th and 18th centuries. As is well known, he was an opponent of Leibniz' monadology. In the *Letters* he expressed a negative attitude also towards subjective idealism and solipsism, and took up an intermediate position in the argument between the adherents of the rival natural philosophies of Newton and Descartes (in many respects, however, closer to that of Descartes). It cannot be doubted that the *Letters* exerted some influence on Kant in the first period of his philosophical creativity. I would be reluctant to say that Euler presented in his *Letters* some kind of complete philosophical system. Incidentally, the *Letters to a German princess* are to be discussed in two lectures of the present conference—by A. T. Grigor'ian and V. S. Kirsanov, and by K. Grau.

To the list of large monographs it is appropriate to add the *Theory of the motion of the moon, worked out using a new method*, published in 1772. This work was of even greater significance for the development of celestial mechanics than his book of 1753 devoted to that topic. It was prepared for the press under Euler's general supervision by three academicians: his eldest son Johann-Albrecht, and Kraft and Lexell, who were compelled to carry out very laborious calculations.

In conclusion I shall try to characterize Euler's output along general lines—an output which astounds anyone familiar with it by its volume, variety, and originality. Not for nothing did d'Alembert once call him a "man-devil". The Helvetic Society of Natural Scientists began to publish a complete works in 1911, and over the past 70 or so years there have appeared 69 thick volumes divided into three series: "Mathematical works", "Works in mechanics and astronomy", and "Works in physics, and others", and there remain four more volumes to be published[9]. Incidentally, two volumes from the first series were prepared under the editorship of A. M. Lyapunov. In 1975 the publication of a fourth series was begun with the participation of the Academy of Sciences of the USSR. This fourth series is subdivided into two subsets. The eight volumes of the first of these are devoted to Euler's scientific correspondence; of these three have appeared so far[10], a further one is in press, and the preparation of another for publication is almost complete. The second subset will consist of five or six volumes devoted to Euler's unpublished scientific manuscripts and fragments; this project is as yet in its initial stages and will take some years to complete.

While it is indeed unlikely that any mathematician or physicist ever published as many works as Euler, one is not any the less astonished by their extraordinary breadth. They

---

[9]As of 1988.
[10]As of 1988.

The memorial plaque affixed to Euler's house.
Marble bas-relief by Yu. Klyugge. 1957

embrace literally all areas of mathematics and the mathematical sciences, as well as a great many problems of technology, philosophy, and even theology. If one counts individual volumes, whose contents, incidentally, are often quite diverse in character, then almost 43% of the total are devoted to mathematics and the same percentage to mechanics, including astronomy, making a total of 86% of his published works[11]. However here it is essential to take into account the fact that Euler's works in mechanics and astronomy are replete with solutions of differential equations, expansions of functions in convergent or asymptotic series, etc., and very often contain completely original purely mathematical results not given a separate, self-contained treatment.

Thus notwithstanding the thematic variety of his researches, Euler was first and foremost a mathematician. The academician A. N. Krylov remarked a half-century ago that in essence Euler transformed mechanics from a physical science into a mathematical one. Following in Euler's footsteps, Lagrange announced in 1787 that mechanics had become a new branch of analysis, and then Fourier in 1822 that analysis was as vast as nature herself.

Naturally, in the majority of his mathematical works Euler's approach was that of an analyst. Of all his purely mathematical works, those in analysis account for around 60%, followed by geometry, chiefly differential, at 17%, then algebra, combinatorics and probability theory at 13%, and finally number theory, occupying 10%. These statistics—which are of course only approximate—indicate the organic connection of Euler's work with the investigation of the natural world. He was led to many of his new methods by the search for solutions to problems of the natural sciences amenable to mathematical formulation;

---

[11] The original has "of which 86% is published".—*Trans.*

however it was precisely in this that the special nature of Euler's mathematical genius consisted that he did not limit himself to solving individual concrete problems, whether applied or purely mathematical, but, returning constantly to a further deepening and generalizing investigation of the question to hand, was often able to create a new and fully independent mathematical theory. In this way not only did he extend enormously the boundaries of the analysis of Newton, Leibniz, and the elder Bernoulli brothers, but also created completely new branches of that subject: the variational calculus, the elements of complex function theory, the most important part of the theory of special functions, a whole system of approaches to the solution of differential equations, and so on. In terms of rigor, Euler's proofs remained within the bounds imposed by 18th century standards; however his rejection of excessive rigor, for which he was often reproached in the first half of the 19th century, turned out to be historically justified. His intuition shielded him from major error, and many of his bold ideas—such as for example the methods he invented for summing divergent series—became capable of being evaluated and developed on a new, more complete and rigorous basis only at the end of the 19th century and the beginning of the 20th. And it would be a great mistake to think that only applied problems served as the source of Euler's discoveries: a great many of his ideas arose in the course of pondering problems of the most purely mathematical character, beginning with his introduction of such important classes of functions as the beta-, gamma- and zeta-functions, and ending with all those problems of number theory with which he occupied himself unceasingly from the early 1730s till the end of his life.

It is perhaps Euler's attitude to number theory that is especially persuasive as evidence of the essentially mathematical style of his thinking. Following on the ingenious insights of Fermat, for many decades number theory suffered neglect. It failed to interest such outstanding mathematicians as D. Bernoulli, A. Clairaut, J. d'Alembert, and indeed the majority of the contemporaries of Euler, who, according to Chebyshev, was the first to turn number theory into a science in its own right. Of course number theory attracted Euler by its beauty together with the difficulty of many of its theorems, so easily formulated and yet requiring for their discovery acute discernment and for their proof means of extraordinary refinement. However here the chief consideration was Euler's awareness of the profound organic interdependence of all areas of mathematics; he understood mathematics as a single whole, of which number theory forms an integral part, and considered progress in that subject a precondition of the advance of mathematics over its entire front.

There are various approaches to defining a typology of mathematicians. On the one hand there have always been mathematicians of a definitely applied bent, such as Ch. Huygens and Newton, or D. Bernoulli, Clairaut, and d'Alembert, already noted, or J. B. Fourier and S. D. Poisson, M. V. Ostrogradskiĭ and V. A. Steklov, etc. On the other hand the "pure" mathematical tradition has existed for a very long time, some prominent representatives of which are: N. H. Abel and E. Galois—although the latter, it is true, died very young—, B. Bolzano, and somewhat later R. Dedekind and G. Cantor and L. Brouwer, E. I. Zolotarev and I. M. Vinogradov, N. N. Luzin and P. S. Aleksandrov. However Euler belongs to that category of mathematician combining these two tendencies organically, immanently mathematical and applied, of which some representatives were, in ancient times Archimedes, and in the last few hundred years J.-L. Lagrange, C. F. Gauss, A. Cauchy, B. Riemann, P. L. Chebyshev, H. Poincaré, D. Hilbert, etc. One should doubtless delineate also the cate-

Euler's tomb in the Leningrad mausoleum

gory of those mathematicians of a philosophical disposition, such as R. Descartes and G. W. Leibniz, as well as some of those already mentioned; I would include also among these N. I. Lobachevskiĭ.

It is appropriate to add a few words about Euler's relations with his contemporaries. As a rule, if one excludes a few heated discussions, his attitude to other scholars was always benevolent, and he never allowed any sense of his superiority to emerge. He was without envy towards those who preceded him in some discovery or other, and—to use the expression applied by B. Fontenelle to Leibniz—took pleasure in observing how plants springing from seeds that he had supplied, flourished in others' gardens. Always a student with the widest interests, throughout his life he was prepared to learn from others, and in his own works often expounded others' discoveries in more convenient and accessible form. However of course overall he gave to others immeasurably more than he took from them. He influenced the work of many generations of mathematicians; in particular the St. Petersburg mathematics school of the second half of the 19th century and the first quarter of the 20th, founded by P. L. Chebyshev, was very close in spirit to Euler. As the late B. N. Delone put it some years ago as a participant in the Euler Days of 1957, the guiding principle in the work of this school was that of Euler and Chebyshev: Confronted with a difficult problem—arising as often happens from a sister science or from technology—, construct a broad and deep mathematical theory, and once a solution has been obtained, convert it back into the numerical form required by the original practical problem.

I would like to end with the words pronounced 200 years ago by N. I. Fuss at the meeting of the Petersburg Academy of Sciences dedicated to the just-deceased Euler: "Such were the works of Mr. Euler, such is his right to immortality. His name will perish only with the end of science itself". To which I add: That will never be!

Leonhard Euler, active and honored member of the Petersburg Academy of Sciences

Yu. Kh. Kopelevich

It is perfectly natural that a solemn assembly to mark a jubilee of Leonhard Euler should take place not only in Moscow, whither the premises of the Academy of Sciences of the USSR were transferred in 1934, but also in Leningrad, a city with which he maintained a connection for all of the 56 years of his scientific life, from his first independent forays in science to his last breath. It was to the Petersburg Academy of Sciences that he came in 1727 as an almost unknown 20-year-old, and it was here that he became in the space of a few years the great Euler, renowned throughout the scientific world.

The theme "Euler and the Petersburg Academy of Sciences" recurs throughout the whole of Euler's biography, and in almost all publications on the history of our Academy in the 18th century. Over the past 25 years, since the celebration of Euler's 250th birthday, much new material has been discovered shedding light on just this theme. Through the efforts of Soviet archivists and historians of science, with the collaboration of researchers from the German Democratic Republic and Switzerland, more than a thousand documents and letters have been studied, described, and in part published, revealing to a greater or lesser extent Euler's fortunes as closely tied to the Petersburg Academy of Sciences at that time [1]. The early history of the Petersburg Academy—its emergence and establishment as one of the world's leading scientific collectives—has been even more thoroughly re-searched. These investigations allow us today to see more clearly than hitherto the relation between Euler's development into a great scientist and the conditions under which this development took place.

Euler's words in a letter sent to St. Petersburg from Berlin, dated 18th November 1749, i.e., 8 years after his move to Germany, are often quoted: "...I, together with all those who have had the good fortune to serve in the Russian Imperial Academy, must acknowledge that we owe what we have become to the favorable conditions prevailing there. As far as I personally am concerned, had this happy chance not occurred I would have had to devote myself to some other occupation, in which I would in all likelihood have become merely a drudge. When his royal eminence[1] recently asked me where I had gained my knowledge, I answered truthfully that I am in that respect wholly indebted to my years at the Petersburg Academy". [2] We shall here attempt to discover the real meaning of these words.

The history of the move to St. Petersburg in 1725 of Nicolaus II and Daniel Bernoulli— who soon engineered an invitation to join the Academy for their junior colleague and father's former pupil Leonhard Euler, languishing unemployed in his motherland in Basel— might perhaps be considered a series of accidents. However an analysis of the extensive correspondence around the time of the creation of the academy [3] shows that such was the scale of the process of selecting scholars to staff it—involving the co-opting of the most authoritative of Europe's scientists in the search for suitable candidates and aimed at filling the new institution with cadres of young, but very promising, scientists— that it was not in fact so accidental that the young sons of the renowned Johann I Bernoulli, lacking an outlet for their talents in Switzerland, should come within the ambit of this search.

May 1727, when Euler arrived in St. Petersburg, was perhaps the last month of that comparatively propitious initial period of the existence of the young Academy. The empress Catherine I had lavished favors of all kinds on the Academy, child of her late husband Peter the Great. Not for nothing did Christian Wolff, a patriarch of contemporary European science, say on hearing of Euler's imminent departure for Russia that he was going to "scientists' heaven" [4]. On the bank of the Neva opposite the Admiralty extensions were made to two large buildings assigned to the Academy: a medical museum with a library, anatomical theater and observatory, and an adjoining palace that had been built for the princess Praskov′ya Fedorovna for assemblies and other such functions. (Praskov′ya Fedorovna was the widow of Ivan V, full brother of Peter and till 1696 nominally joint ruler with him.) Scientific meetings and public lectures went full steam ahead. Public meetings of the Academy were conducted with great ceremony in the presence of high-ranking nobles, and the August meeting of 1726 was attended by the empress herself accompanied by her daughters; scarcely any other European academy of the time could boast of attracting so much attention. An academic printing office was set up and issued its first publications. The president of the Academy, Lavrentiĭ Blumentrost, doctor to the empress, exercised great influence at court and on more than one occasion used this influence in the interests of the Academy. A list of regulations for the Academy was drawn up with a view to guaranteeing the new institution and its members various privileges. Surviving letters written at that time by Petersburg academicians to former colleagues, are full of the rosiest optimism [5]. To summarize, in these first years all of the advantages of Peter the Great's conception of the Academy's organization were realized, without as yet any hint of the danger inherent in the excessive dependence of the scientific body on the intrigues and whim of the imperial court.

When Euler arrived in St. Petersburg the city was in mourning, and in a state of confusion. A week earlier the empress had died [6], and a struggle between rival court factions was intensifying around the child emperor Peter II. Six months later the court was moved to Moscow, and with it Blumentrost and not long afterwards also the conference-secretary of the Academy Christian Goldbach, who served as tutor to the emperor and his sister.

---

[1] Friedrich II

Thus the Academy was left without a head. The Academy's librarian, J. D. Schumacher, an adroit and energetic administrator, remained as plenipotentiary. Schumacher had earlier enjoyed the close confidence of Peter I and together with Blumentrost had actively assisted in the creation of the Academy, but now he artificially turned to his own advantage the circumstance that the academicians were unfamiliar with the Russian system of government and were thus dependent on him, long a resident of Russia and familiar with its ways. The Academy was still without a set of regulations—Catherine I had not had time to affirm them—, its position was shaky and Schumacher reckoned on securing its position at the expense of curtailing the extension of the auxiliary facilities—the printery, the instrumental and art workshops—, which, in view of the limited academic finances, added to the professors' difficulties and generated frustration—especially among the senior members. The growth of the Academy's administrative needs and the necessity of continually making application to some government department or other, led to the creation of the so-called Secretariat, an administrative office run exclusively by Schumacher. The professors were reluctant to acknowledge the power of the Secretariat and resisted the imposition of its regulations. This situation hardly changed with the accession to the throne in 1730 of Anna Ioannovna and the return of the court to St. Petersburg. After returning to the capital Blumentrost in practice no longer functioned as president of the Academy, and was soon thereafter removed from the presidency. The situation improved only in 1734 with the appointment of J. A. Korff as president. For five and a half years Korff fulfilled this role with great assiduity—he personally chaired the meetings of the Academic Conference and introduced a whole series of useful administrative changes—but even he was unable to secure official regularization of the affairs of the Academy or stabilization of its finances. Somehow or other he was able to make ends meet, in particular with the help of one-time-only subsidies from the empress. It proved impossible also to rein in the bureaucratic zeal of Schumacher and his Secretariat. Naturally, the flourishing favoritism of Anna Ioannovna's court, the continual waxing and waning of the power of minions, and political instability, did nothing to further the normal functioning of the scientific institution.

It might seem that this is all in conflict with Euler's high estimation of the conditions under which he began his scientific career. In fact it is precisely the way in which his career developed that shows that the above-described circumstances were not in fact decisive, being outweighed by the aspects of the organization of the Academy favorable to scientific creation, and the special situation of its members. Here most important were the general features of the Academy as a new form of scientific association, and then their realization in the innovative conceptions of Peter I and in the actual conditions created by his delegates.

The character of the faculty, i.e., of the scientific collective, constitutes perhaps the most significant of these features; the presence of such a group was the main factor distinguishing the research work of the Petersburg Academy and other academies from that of professors at western European universities, who for the most part worked in isolation. Although according to Peter I's proposed design the Academy was to embrace eleven disciplines, each represented by an academician or professor—the second of these titles being the more usual in the initial phase of the Academy's history—, in fact the majority of the first cohort of academicians, as scholars of wide knowledge and interests, formed together with the adjuncts a physico-mathematical community continually involved in collective discussions and investigations of problems of pure mathematics, astronomy, mechanics,

Jakob Hermann

and physics. The biweekly meetings of the Academic Conference provided a focus for
the steady interchange of opinions and ideas, a place for "interactive learning" of especial importance for the the younger members, among whom Euler. The stipulations as to
the distribution of ages among faculty were fortunate: more than half those joining the
Academy were under thirty, yet there were also "venerables" such as the mathematician
Jakob Hermann and the astronomer Joseph Nicholas Delisle. Moreover at the meetings no
account was taken of rank in the academic hierarchy—unlike the Paris Academy, where
members were seated according to rank, with the adjuncts immediately behind their respective supervisor-academicians. In contrast, the Petersburg Academy was characterized
by scientific democracy, and the degree of participation of each member in a meeting was
determined only by what he had to contribute to it. Although Euler was appointed to the
Academy as an "élève", i.e., literally as a "student"—the title of "adjunct" came into use
only later—, from the first he took the floor on an equal footing with the academicians.
Throughout his first Russian period, i.e., till the summer of 1741, Euler was the most active
participant in the meetings of the Academy Conference: he made research announcements
on an average of 10 times a year, whereas for the other members the figure was between 1
and 5. Of the 23 meetings of the so-called Mathematical Conference in the second half of
1735, 21 were taken up with reviews and talks by Euler.

The collective character of the work of the Academy is demonstrated not only by the
organization of the Conference meetings, but also by the way observations and experiments
were carried out jointly; such, for instance, were the trials in August and September 1727
set up to investigate the trajectory of a cannon ball, the combining of different kinds of expertise on questions posed by various government departments, the fulfilment of substantial

commissions, such as—to take an example from that time—the project of drawing up an official map of Russia, begun under the direction of J. N. Delisle, in which towards the end of the 1730s Euler played a leading role. The collective research in astronomy and the part Euler played in it are discussed in the article by N. I. Nevskaya and K. V. Kholshevnikov in the present collection [7].

Of crucial significance also was the prevailing great freedom in the choice of an area of research. None of the scientists was confined to the specialty which he officially represented, and even the adjuncts were not restricted to working on topics related to the work of their official supervisors. Thus for example Daniel Bernoulli, although appointed to the chair of physiology, occupied himself chiefly with mathematics and hydrodynamics, and from the very beginning Euler, invited to work in that same department, having demonstrated by his first talks his great mathematical gift, was given free rein to follow wheresoever his inclinations might lead him. In the Petersburg Academy, where, as in universities in the 18th century, scientists were hired on contract for a certain definite period, after which for various reasons they were sometimes let go, the younger members had a better chance of promotion than for instance in the Paris Academy with its appointments for life. Thus Euler's promotion to the professorship in physics in 1731 correlates with the return of J. Hermann to Switzerland and G. B. Bilfinger to Germany, and his appointment to the professorship in mathematics in 1733 with the departure of D. Bernoulli. However it must be supposed that even if such transfers and departures had not occurred, Euler would not have "stagnated" long amongst the adjuncts. To deal with the situation where an adjunct merited promotion but the appropriate vacancy in his speciality was lacking, the administration introduced the category of "extraordinary" professor, a departure from Peter I's conception of the structure of the faculty complement. Moreover in a few instances two professors were appointed to a single department, for example when one astronomer was needed for the work of the Academy and another for an expedition to Kamchatka. Yet another circumstance favorable to research work in St. Petersburg was the special status accorded to members of the Academy, who were materially provided for like university professors— salary, government (or paid) apartment, firewood, candles—, but whose teaching load (as we would put it now) was very light. For his part Euler never took advantage of the system of additional payments and rewards for the completion of supplementary projects. All the same the prevailing material conditions were such that he could start a family, acquire his own house, and live, if not luxuriously, then at least free from want.

The matters that occupied Euler in the Petersburg Academy in the 1730s are indicated by the minutes of meetings of the Academic Conference and the large number of administrative documents preserved in the Archive of the Academy of Sciences of the USSR[2], a description of which has been published [8]. Another source of illustrative material, affording a sort of "cross section" through the year 1737, i.e., the tenth anniversary of Euler's arrival in St. Petersburg, is furnished by the following report of Euler's, written on the 28th August 1737, which was discovered only recently and has not yet been catalogued[3]. At the request of the Academy's president J. A. Korff, the academicians had to report on what they were obliged to do under their contract, what in fact they had achieved over their years of service, and what they planned to do in the future. Euler wrote:

---

[2] Now once again the Russian Academy of Sciences.—*Trans.*
[3] As of 1988.—*Trans.*

"According to my conditions of service in the Imperial Academy of Sciences, I am obliged to fulfil the following:

1. To attend the meetings of the Conference, which I fulfil assiduously and always have in readiness articles to read there.

2. To give lectures to students on the higher branches of mathematics. This also, whenever students wanting to study that subject present themselves, I carry out according to their capabilities.

3. I have also been commissioned to participate in the work on the geography of Russia, and here I also work as far as my strength and my other duties allow.

As far as my other labors are concerned at present, and also in the future, I am now working on an arithmetic to be used in the Academy's gymnasium. Apart from that, I have the intention, if my other activities do not interfere, to bring to completion several works in hand, having to do with music, statics, the analysis of infinities, and the motion of bodies in water" [9].

The work of the Petersburg Academy favored Euler's research in applied mathematics, although in his basic interests he was more the theoretician. The duties of the academicians included the fulfilment of various commissions from government institutions, in particular the provision of expertise on technical literature from outside the Academy, and on various kinds of machines, inventions, and projects. It seems that the first such commission falling to Euler was the refereeing of a manuscript sent by the Naval Academy, of a book on navigation written by a lieutenant of the Russian fleet by the name of Stepan Malygin [10]. Euler announced the completion of this task in a letter to Schumacher dated September 14/25, 1731 [11]. His report was read at the meeting of the Conference held on October 19/30 [12]. The original has not been found, but on the basis of this report the Academy issued an "attestation" to Malygin, imprinted on the last page of his book, confirming that: "The Imperial Academy of Sciences handed over the aforementioned booklet to Leonhard Euler, member of the Academy and ordinary professor of physics, who, having faithfully examined the whole of the contents of the booklet, proposed to the Academy that it contains nothing contradicting the laws of the science of navigation, and that all problems are solved properly and correctly, and that the booklet is useful as a guide for training purposes...". This episode is characteristic of the early history of the Academy, when, striving to maintain the prestige of its "scientific collegium" among rival state institutions, it undertook commissions even when it had no expert of the requisite kind, gradually creating the necessary specialists from among its members. While it is not known whether this assignment fell to Euler by chance or on the strength of his essay on the rigging of ships, submitted to a competition of the Paris Academy when he was still in Basel (he carried out the experiments involved in this work in a bowl of water), what *is* clear is that although he had seen the sea for the first time as a passenger on a boat sailing from Lübeck to St. Petersburg, the last leg of his journey from Basel to Russia, nonetheless his report was considered an authoritative assessment of a book written by a sailor with some 15 years' experience at sea. However that may be, it is by no means ruled out that in composing the report in question Euler for the first time came into close contact with "the naval science" in which he was destined to achieve so much.

A great deal of consultation of this sort fell to Euler's lot from 1735 to 1740, during Korff's presidency. Among those of Euler's manuscripts preserved from that period there are about 20 reports and proposals of a technical character relating to the verification of various weights and measures, the testing of mechanisms and machines, and of drawing and measuring instruments, magnets, and fire-protection equipment. Several reports of this type have been published in *Materials relating to the history of the Imperial Academy of Sciences* [13], including one on a mechanism for raising a large bell in the Moscow Kremlin, and one concerning a "sawmill" built by the Academy's mechanic I. Bruckner for the admiralty. The technical contents of these reports still await assessment. In any case, it is noteworthy that their chronology indicates their connection with certain of Euler's theoretical works, for which these reports very likely served as cause or impulse. It would seem to be no accident that in February 1738, when Euler was occupied with the simultaneous testing of several sets of weights used by the Petersburg Customs, he gave a presentation to the Academic Conference entitled "Investigations into weights" [14], and just before he wrote the article "On the operation of machines, both simple and complex" [15], presented to the Conference in April of the same year, he had participated in expert consultations concerning various mechanisms designed by A. K. Nartov and I. Bruckner. These works were to form the basis for Euler's new theory of machines, the first attempt to apply the latest achievements of mathematical analysis to the working of machines. If Euler was led to consider questions concerning the navigation of ships through his refereeing of Malygin's book in 1731, then it is possible that two reports that he wrote in the years 1735 and 1736 on "An essay on the mechanism of motion of floating bodies", written by Delacroix, commissar-general of the French fleet [17], served as impetus to the composition of his substantial treatise on naval science, completed in Berlin and published in St. Petersburg in 1749 [16]. There is no doubt that the latest of the works Euler wrote on cartography, which laid the foundations for the future evolution of that subject, derived in the last analysis from his active involvement over the period 1735–1741 in the drawing up of an official map of Russia.

Yet another condition favoring the growth in the scientific capabilities of the members of the Petersburg Academy, was the official encouragement given to scientific correspondence. In contrast to its foreign analogues, the Academy arranged free postage not just for the Conference Secretary, responsible for the official correspondence of the Academy, but for all its members, provided the correspondence was scientific, not personal. Letters received and drafts of answers were presented and read at the meetings of the Conference, and copies made for the Archive. Contact with foreign colleagues by mail became especially extensive during Korff's presidency. This can be clearly seen from Euler's correspondence. Until 1735 it was almost entirely restricted to Basel, where his chief correspondents were his teacher Johann Bernoulli, and later Daniel Bernoulli and Jakob Hermann, his former Petersburg colleagues. Beginning in 1735 the circle of his correspondents quickly widened, to include, in particular, the astronomers G. Poleni in Padua and G. G. Marinoni in Vienna, the mathematician J. Stirling in London, the Danish naval officer F. Wegerslof, and the mathematicians Heinrich Kühn and Carl Leonhard Gottlieb Ehler in Gdańsk. Such correspondence allowed scientists to exchange unpublished discoveries and ideas, and to continually keep abreast of the latest developments in western European science. These aims were served also by regular contact between the Academy and scientific journals in Leipzig, Amsterdam, Nuremberg, and elsewhere, which were very efficient in their reviewing of the publications issuing from St. Petersburg, and even tried to outdo one another and

obtain information "first-hand" [18]. The Academy was concerned to disseminate its publications as rapidly as possible among interested European communities of scientists. This
is clear from the example of the distribution of Euler's *Mechanics* [19], published in 1736
and dispatched gratis to dozens of addresses: to big libraries, to individual scientific societies and academies, to publishers of journals, and to the envoys to European courts [20].
It is appropriate to note here that the time it took for scientific information to spread—for
example the time elapsed from the issue of a book from the press in St. Petersburg till the
receipt by the Petersburg Academy library of a journal from Amsterdam containing a review of the book—was often shorter than it is today. The Petersburg academicians were not
at all "in exile", but quite the contrary: they found themselves in the thick of the scientific
life of their time, as is clearly shown by the rapid spread of Euler's fame.

Of no small significance as a stimulus to scientific creativity was also the availability
at the Academy of a wide variety of outlets for publishing works of its members. No other
European academy of that time offered such opportunities. Here one cannot but give the
abovementioned Schumacher his due, since he was the chief organizer of the academic
publishing house and bookdealer's and conducted this business with great élan. From the
very beginning the publishing facility was not considered a commercial venture: its costs
were paid out of the general budget of the Academy, scientific publications were sold approximately at cost—sometimes even less—and losses were covered by the revenues from
the sale of popular multi-edition publications. It is true that the publication of the chief Academic journal, the *Commentarii*, was delayed in the 1730s through financial and other irregularities, but monographs given approval by the Conference were published without delay.
In any case Euler experienced no difficulties in getting his work published. Up to 1741 over
50 of his works appeared in St. Petersburg, and another 31, left behind in manuscript form,
were published after his removal to Berlin. In comparison with his subsequent published
output this may not seem such a lot, but scarcely any other scientist of his time published
more over such an interval.

Although it is well known which of Euler's monographs were published in St. Petersburg and which of his articles in the academic *Commentarii*, it remains unclear to what
extent he contributed to *Footnotes to the Bulletin*, the other journal of the Academy, published in German and Russian from 1728 to 1742 as a supplement to the newspaper *The
St. Petersburg Bulletin*. Until 1738 the articles in this journal were published anonymously,
and thereafter only with the authors' initials. The original manuscripts are lost. Hitherto
the only article definitely identified as Euler's is one entitled "On the external view of the
earth" (1738, Parts 27-32, 103, 104). Recently, on the evidence of a letter from G. F. Müller
to J. D. Schumacher, it has been ascertained [21] that the long article "On the recent exceptional display of the Northern Lights" (1730, Parts 14-17, 21, 25, 32, 35, 77, 78) was
written by Müller, G. W. Kraft, and Euler, who seems to have written the middle part. It
is almost completely certain that the article "How to observe ebbing and flowing tides"
(1740, Parts 9, 10) is also his; it was published (unlike the other articles printed in that year,
without identifying initials) at precisely the time when Euler was working on an essay on
the physical cause of ebbing and flowing tides, which he intended to enter in a competition
of the Paris Academy of Sciences [22], and at his request observations of this phenomenon
were made on the White Sea. The article is faithful to a manuscript of Euler's entitled "Essential comments which must be taken into consideration when carrying out observations

of ebbing and flowing marine tides" [23], and in it a sketch of measuring rods from the manuscript is reproduced together with part of the logbook of observations.

Towards the end of the 1730s Euler's highly varied output was on the increase, and, as is clear from his correspondence, he had no reason to be dissatisfied with his situation. However the political instability of the last years of the reign of Anna Ioannovna, which intensified after her unexpected death, and the very attractive conditions offered him by Friedrich II, inclined him to move with his family to Berlin, where he shortly became a central figure—Director of the Mathematics Class of the Royal Academy of Sciences and Philology—, while remaining an honored member of the Petersburg Academy. It is impossible to describe here the many threads connecting Euler with the Petersburg Academy throughout the 25 years of his sojourn in Berlin. Instead I shall confine myself to giving a short list of those areas of activity in which Euler, while in Germany, demonstrated that he remained an active member of the Petersburg Academy.

Throughout his years in Berlin Euler remained chief consultant of the Petersburg Academy on physico-mathematical questions. For example, he was appealed to as adjudicator in the disagreement which arose in 1745 between the academicians G. W. Richmann and J. Weitbrecht over the assessment of the scientific position of the Newtonian J. Jurin, and again in 1735 in the quarrel between the academicians Ch. G. Kratzenstein and N. Popov over the former's solution of the "marine problem" concerning the action of wind on water. In such cases Euler's opinion put an end to every argument. Euler was in practice the person responsible for maintaining the high standard of articles published in the journal of the mathematics department *Novi commentarii*, and he himself wrote editorial summaries of many of them. As chief referee of papers on physico-mathematical topics, he did Russian science a great service when he supported and encouraged M. V. Lomonosov's first steps in science, in particular by helping him secure his position in the Academy and maintaining a friendly correspondence with him over many years, until Lomonosov's death [24]. In his Berlin years Euler significantly widened the circle of his correspondence with European academicians and outstanding scientists, and thus incidentally provided the Petersburg Academy with an additional conduit for communicating with western Europe, and obtaining information about the latest scientific discoveries and events of scientific life.

When in 1749 the Academy began to announce annual competition problems, Euler was engaged as judge of entries on astronomy and physics. Many of the topics of the competition problems originated with him, and on more than one occasion he submitted an entire list of such topics [25]. Overall, in more than half of the competitions in physics and astronomy announced by the Academy between 1749 and 1783 the set topics had been proposed by Euler.

Although in all his years in St. Petersburg Euler had not "produced" any Russian mathematicians—Vassiliĭ Adodurov had shown exceptional promise, but switched to translation work and the composition of a Russian grammar—, he was more fortunate in this respect in Berlin, where from 1752 to 1756 there lived and studied in his home, sent to him from the Academy, S. Kotel'nikov, S. Rumovskiĭ, and M. Sofronov. Although the fortunes of the last turned out badly, the first two were to occupy positions of distinction in Russian science and mathematical education. Furthermore Euler was an active intermediary and consultant in connection with the choice of scientists to fill vacant positions in the Academy; in correspondence with the Academy dozens of candidates were discussed. In sum it can be said that more than a half of those taking up service in the Academy over Euler's Berlin period were accepted on his recommendation and through his intercession. These included the physicists Ch. G. Kratzenstein, J. A. Braun, and F. U. T. Aepinus, the mechanician J. E. Zeiher, the chemist J. G. Lehman, the botanist J. A. Güldenstädt, and the physiologist C. F. Wolff. The relevant correspondence bears witness to the high standards which Euler insisted on in choosing scientists for the Petersburg Academy.

It is impossible to describe briefly the multitudinous and varied commissions from the Academy fulfilled by Euler when in Berlin: subscriptions and the purchase of books, the ordering and dispatching of instruments, and so on. It is reasonable to suppose that Euler took on all these laborious and troublesome tasks for the Petersburg Academy [26] not only because of the salary he received as honored member (200 rubles per year) or out of gratitude to the country and academy where he matured as a scientist, but also because the role of "plenipotentiary representative" of the Petersburg Academy was important for reasons of personal prestige, and to his stature as one of the leading organizers of European scientific communications.

However, of the various forms of Euler's collaboration with the Petersburg Academy during his Berlin period, the most important was the sending of his works to St. Petersburg for publication. Over those 25 years the number of works he sent from Berlin to be published in St. Petersburg numbers around 100—roughly the same as the number published in Berlin in the same period. In this regard A. P. Yushkevich has with complete justice asserted that "His strength was more than enough for him to cope fully with his 'double appointment' at the two academies. In those years Euler published almost exactly half of his works at each academy; neither alone would have been able to manage and even together were unable to handle the unceasing flood of his works" [27].

It is interesting to consider not only the total number of Euler's papers published in the Petersburg journal in his Berlin period, but also the proportion these represented of all articles printed in that journal over that time interval. We give the composition of the mathematics section of the journal for the years 1744 to 1765, when the mathematics school was in a reduced state. For each volume indicated, the left-hand number denotes the number of Euler's papers appearing in that volume, and the right-hand number the total of all other authors': *Commentarii*, Vol. XIV: 4-1; *Novi commentarii*, Vol I: 4-2, II: 2-3, III: 7-2, IV: 4-1, V: 7-1, VI: 11-1, VII: 5-1, VIII: 10-6, IX: 10-0, X: 9-2, XI: 9-0. Moreover the few papers by others include some by Euler's former pupils Kotel'nikov and Rumovskiĭ. From this it is clear that throughout his Berlin years Euler continued as before to be the chief contributor to the mathematics section of the academic journal—in fact he almost filled it with his articles. Even had he lived in St. Petersburg all that time it is hardly likely that he could have published more there. It is thus not surprising that the Academy made no great effort to appoint a successor to Euler, his position remaining unfilled for the whole 25 years of his absence; in practice Euler himself occupied it in absentia. Thanks to him the academic journal kept its reputation as one of the leading physico-mathematical publications in Europe. However the obverse of this fact is also important: if Euler had not had this outlet for the publication of his works, then many of his discoveries would probably have remained unpublished and many papers unwritten.

When Euler returned to Russia, of those with whom he had served in the Academy in his younger days only a handful remained. However his former colleague J. von Stählin now

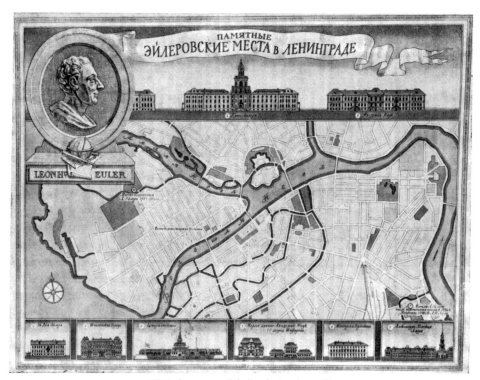

"Eulerian memorial sites in Leningrad"

filled the post of Conference Secretary, and his former pupils Kotel'nikov and Rumovskiĭ had become professors. Euler's eldest son Johann-Albrecht was appointed professor of physics in the Academy. At the first meeting following Euler's return, on August 7, 1766, he presented to the Assembly the manuscript of the first two volumes of his planned three-volume *Integral calculus*. From that time right up to his death his manuscripts came before the Assembly in a uninterrupted stream, sometimes five or six at a time, and once as many as 13 (March 18, 1772). Altogether more than 500 of Euler's works were either presented to the Academic Assembly over his second period in St. Petersburg or left in manuscript form after his death; the Academy continued publishing them for another 80 years, till 1862. It would seem that this level of productivity is unequaled in the history of science. There are among these works several monographs of fundamental importance. And all this was achieved in a condition of near-blindness, with his eldest son and pupils serving as eyes—N. I. Fuss, W. L. Kraft, M. E. Golovin, A. J. Lexell, P. B. Inokhodtsev. Fuss, who had married into the Euler family, and his sons N. and P. Fuss, Euler's great-grandsons, played a large part in the preservation and publication of Euler's literary legacy. Just as in the years of his Petersburg youth, in this later Petersburg period Euler acted as academic expert on a variety of practical questions, for example the organization of a widows' fund (1769), I. P. Kulibin's project to build a single-span bridge over the Neva (1776), and the determination of the fall and speed of the Neva's current (1780).

As senior academician, Euler now oversaw the work of the Academic Assembly. He was also a member of the Directorial Committee, a new administrative organ replacing the old Secretariat. Euler actively participated in the organization of large-scale academic expeditions, in particular the 1769 expedition to observe the transit of Venus across the disc of the sun. However not all of his ideas, as itemized in his memorandum "Plan for the reorganization of the Imperial Academy of Sciences" (1766?) [28], turned out to be feasible, in particular his proposal for increasing the revenues of the academic publishing house, or for revoking the regulation pertaining to the constancy of the number of members of the Academy and the fixed size of their salaries in favor of appointing only scientists of genius, attracted by high levels of remuneration. The idea favored by Euler of at least partial academic self-government was likewise realized only briefly—the despotism of count V. G. Orlov, director of the Academy from 1766 to 1774, was such that in 1774 both Euler and his son Johann-Albrecht resigned demonstratively from the directorial committee. Later Euler was compelled once again to struggle against a director's arbitrariness, namely that of S. G. Domashnev, director of the Academy from 1775 till 1782. It was only princess E. R. Dashkova, who took over the directorship in 1783 and occupied it till 1796, who showed Euler the deep respect due a senior academician—however by then he had but a few months of life remaining.

The whole of Euler's scientific fortunes, the fate of his literary legacy, the two-century-long labor of the Academy to publish all his works—all of this bears witness to the mutual good fortune of Euler's encounter with the Petersburg Academy of Sciences.

References

[1] *Manuscript materials of L. Euler in the Archive of the Academy of Sciences of the USSR.* Compiled by: Yu. Kh. Kopelevich, M. V. Krutikova, G. K. Mikhaĭlov, N. M. Raskin. M.: Published by Acad. Sci. USSR (Akad. nauk SSSR]), 1962. Vol.1: "Scientific description"; Euler L. *Correspondence.* With annotated index. Compiled by: T. N. Klado, Yu. Kh. Kopelevich, T. A. Lukina, I. G. Mel'nikov, V. I. Smirnov, A. P. Yushkevich, with the participation of K.-R. Biermann and F. G. Lange; Editors: V. I. Smirnov, A. P. Yushkevich. L.: Nauka, 1967; *Leonhardi Euleri Commercium epistolicum.* Eds. A. P. Juškevič, V. I. Smirnov, W. Habicht (*Opera* IVA-1); *Die Berliner und die Petersburger Akademie der Wissenschaften in Briefwechsel Leonhard Eulers.* Berlin: Akad.-Verl., 1959, 1961, 1976. Vols. 1–3. These volumes edited and with a preface by E. Winter and A. P. Yushkevich, with the participation of P. Hofmann, T. N. Klado, and Yu. Kh. Kopelevich. (Below this is referred to briefly as *Briefwechsel.*)

[2] *Briefwechsel.* Vol. 2. p. 182.

[3] See: Kopelevich, Yu. Kh. *The founding of the Petersburg Academy of Sciences.* L.: Science (Nauka), 1977. pp. 65–79.

[4] Letter from Ch. Wolff to L. Euler of 20th April 1727. LO Archive Acad. Sci. USSR, f. 136, op. 2, No. 6, l. 271.

[5] See: Kopelevich, Yu. Kh. "Letters of the first academicians from St. Petersburg". Communications Acad. Sci. USSR (Vesti Akad. nauk SSSR). 1973. No. 10. pp. 128–133.

[6] In his autobiography, written in 1767, Euler indicated mistakenly that he arrived in St. Petersburg on the very day of Catherine I's death. See: Kopelevich, Yu. Kh. "Materials for a biography of Leonhard Euler in Petersburg". Sources Math. Research (Ist.-mat. issled.). 1957. Issue 10. pp. 13–17; Mikhailov, G. K. "On Euler's move to Petersburg". News Acad. Sci. USSR. Technical sciences section (Izv. Akad. nauk SSSR. Otd. Tekhn. nauk). 1957. No. 3. pp. 1–37.

[7] See new material on this question in the book: Nevskaya, N. I. *The Petersburg school of astronomy of the 18th century.* L.: Science (Nauka), 1984.

[8]   In the book: *Manuscript materials of L. Euler...* . (See Reference 1 above.)

[9]   Leningrad Section Archive Acad. Sci. USSR, f. 3, op. 1, No. 860, l. 9. Document (in German) discovered by M. A. Alekseeva.

[10] Malygin, S. *Navigation (abridged) using a map "de réduction"*. SPb.: 1733. (Here the term "de réduction" means that the map is reduced to a region about a specified meridian[4].)

[11] *Briefwechsel*. Vol. 2. p. 52.

[12] *Minutes of the meetings of the Conference of the Imperial Academy of Sciences from 1725 to 1803*. SPb.: Published by Acad. Sci. (Izd-vo Akad. nauk), 1897. Vol. 1. p. 51.

[13] *Materials for the history of the Imperial Academy of Sciences: in 10 vols*. SPb., 1885–1900. Vol. 2. pp. 709–710; Vol. 3. pp. 29, 412–416, 631–634; Vol. 4. pp. 120-121, 173, 212, 301–302.

[14] Euler, L. *Disquisitio de bilancibus*. Opera II-16.

[15] ———. *De machinarum tam simplicium quam compositarum usu*. Opera II-16.

[16] ———. *Scientia navalis seu tractatus de construendis et dirigendis navibus*. Opera II-18, 19.

[17] de la Croix. *Extrait du mécanisme des mouvements des corps flottants*. Paris, 1735.

[18] Kopelevich, Yu. Kh. "J. A. Korff and the international relations of the Petersburg Academy of Sciences". *In: History of the natural science and technology of the Baltic region*. Riga: Zinatne, 1976. Issue 5. pp. 14–23.

[19] Euler, L. *Mechanica sive motus scientia analytice exposita*. Opera II-1,2.

[20] Leningrad Section Archive Acad. Sci. USSR, f. 3, op. 1, No. 436, l. 558–559 ob.

[21] Leningrad Section Archive Acad. Sci. USSR, f. 21, op. 3, No.308/30, l. 9-9 ob.

[22] Euler, L. "Inquisitio physica in causam fluxus et refluxus maris". *Rec. pièces remp. prix Acad. Sci. Paris (1738–1740)*. 1741. Vol. 4. pp. 235–350; Opera II-31.

[23] ———. "Nötige Erinnerungen, welche bei Beobachtungen der Ebbe und Fluth des Meeres in Acht zu nehmen". Archive Acad. Sci. USSR, f. 136. op. 1, No.121, l. 1–3.

[24] *Briefwechsel*. Vol. 3. pp. 2-4, 187–206; Lomonosov, M. V. *Complete collected works*. M.; L: Published by Acad. Sci. USSR, 1957. Vol. 10. pp. 435–438, 439–457, 464–467, 500–503, 515–518, 595–598.

[25] *Briefwechsel*. Vol. 2. pp. 21–22, 437–439; Bd. 3. pp. 236–238; Yushkevich, A. P., Winter, E. "On Leonhard Euler's correspondence with the Petersburg Academy of Sciences 1741–1757". Proc. Inst. History of Natural Science and Technology of Acad. Sci. USSR (Tr. In-ta istorii estestvozn. i tekhn. AN SSSR). 1960. Vol. 3 (34). pp. 428–491.

[26] *Briefwechsel*. Vols. 1–3.

[27] Yushkevich, A. P. "The life and mathematical works of Leonhard Euler". Results of math. sci. (Uspekhi mat. nauk). 1957. Vol. 12, Issue 4 (76). p. 12.

[28] Pekarskiĭ, P. P. *History of the Imperial Academy of Sciences in Petersburg*. SPb., 1870. Vol. 1. pp. 303–308.

---

[4] The original has "reduced to a specified meridian", but the French phrase "de réduction" means "on a small scale".—*Trans.*

The part played by the Petersburg Academy of Sciences (the Academy of Sciences of the USSR) in the publication of Euler's collected works[1]

E. P. Ozhigova

The idea of publishing the collected works of Leonhard Euler was a subject of discussion in the Petersburg Academy of Sciences for many years. It was first advanced by N. I. Fuss, Euler's pupil and assistant, and permanent secretary of the Petersburg Academy of Sciences from 1800 to 1826. M. V. Ostrogradskiĭ was insistent that the project be carried out. In 1843 Euler's great-grandson, P. N. Fuss, at that time permanent secretary of the Academy, with the help of his brother N. N. Fuss, published in two volumes and in the original languages a collection entitled *Correspondence on mathematics and physics of some renowned mathematicians of the 18th century*. When this publication appeared, P. N. Fuss proposed to the Academy that he tour western Europe on academic business. One of his aims was to acquaint eminent scientists with the *Correspondence*, to listen to their opinions of it, and to secure their collaboration in a new project—the publication of the complete works of L. Euler. Fuss had earlier discovered in the Archive of the Academy of Sciences and in the private archives of various of Euler's descendants—in particular his own—a large quantity of manuscript material of Euler's. He wished to include the most essential of these documents in the collected works.

P. N. Fuss left St. Petersburg on May 22, 1843 (old style). In his report on his journey he wrote: "The mathematical correspondence of famous geometers of the previous century: Euler, Goldbach and Bernoulli, published by the Academy, was received with delight by the scientists of every country". In Paris Fuss managed to acquire a valuable collection of unpublished letters from Euler to Lagrange, as well as correspondence between Lagrange

---

[1] A large portion of this article consists of quotations from letters from Switzerland to the Petersburg Academy of Sciences, which would have been written in French or German, but which in the original Russian article have been translated into Russian. In the present English translation, these translated excerpts have been subjected to yet another translation, so that the English version of the letters is twice removed from the original. It would of course be more scholarly to obtain copies of the original French or German letters and translate them directly into English. In view of the difficulties involved in obtaining copies of the original letters, this has not been done. However, it seems clear that the translation using the less scholarly approach nonetheless gives a good idea of the long struggle to publish the complete works of Euler. (This comment applies to other essays of the present collection, though to a lesser extent since fewer such translations-of-translations are involved.) —*Trans.*

and d'Alembert "having as its subject almost exclusively the works . . . of the immortal geometer, and in Basel additional correspondence between Euler and Nicholas Bernoulli of
the very greatest interest" [1]. Fuss also made a gift to the Academy of the manuscripts
of Euler that he personally had preserved. In an effort to demonstrate the necessity of
publishing Euler's *Nachlass*, Fuss asserted the following: "Before deciding the question of
whether to publish these manuscripts, I allow myself the comment that a new publication of
Euler's complete works is not only highly desirable and anticipated with great impatience
by the geometers of all nations, but is also considered as a kind of natural debt that Russia
owes to the memory of the great genius who found with us a second motherland—an un-
questionable, but as yet unfulfilled, obligation to the scientific world. A society of Belgian
geometers has already attempted to forestall us in this respect [2], however that enterprise,
badly planned and initiated, foundered after the publication of the fifth volume.

"During the tour which I undertook last year I did not meet a single geometer whose
first question to me was not: Isn't the Academy considering publishing Euler's works? Here
I mention in this regard only scientific luminaries: Messrs. Gauss, Bessel, Jacobi. In particular the last of these was so repeatedly insistent concerning this question that I gave him
my word that I would propose it to the Academy. Mr. Crelle, in the most recent issue of
whose journal the appearance of the *Mathematical correspondence. . .* was announced [3],
expresses the same opinion publicly" [4; 5, p. 154]. On Fuss' recommendation a committee
was struck to formulate a plan of action for publishing the complete works of Euler, including an estimate of the cost involved, and the preparation of a well-researched proposal to
present to the minister and president of the Academy S. S. Uvarov. The members of the
committee included P. N. Fuss, M. V. Ostrogradskiĭ, V. Ya. Struev, V. Ya. Bunyakovskiĭ,
and M. H. Jacobi. The committee fulfilled its mandate and its proposal was presented to
Uvarov; however the latter considered infeasible the requisition of the large subsidy required, and the project was set aside to await a more favorable time.

Having lost hope of publishing the complete works, the physico-mathematical section
of the Academy decided to publish at the Academy's expense the *Arithmetical works* of
Euler, to include certain manuscripts. At the meeting of October 23, 1846 (old style), it was
decided to publish such a collection in two volumes with an initial printing of 600 copies
in quarto. Two years were allowed for the printing to be completed. In his report for 1848,
Fuss stated that the publication of both volumes was nearing completion.

The preparation for publication of the *Correspondence* and the *Arithmetical works* of
Euler was ardently supported by C. G. J. Jacobi. On April 28, 1841 Jacobi wrote to Fuss:
"It would be a great favor done by your Academy to the mathematical world, an undertaking serving to enhance the greatness and honor of Russia, if the Academy were to publish
Euler's papers by subject. . . . It would be necessary to formulate a plan for the whole publication with regard to the distribution by subject and the contents of the individual volumes,
which would without doubt be difficult. The categories of the subdivision would need to be
determined from the actual contents, and it seems to me that this labor would be so great
that it would need to be divided among many people. Then its execution would be gradual
and if it were spread over several years the yearly publication costs would not be so large.
In the meantime you will have done something very useful if you republish and supplement the list of Euler's works which your father appended to the biography [of Euler]. . ."
[6, pp. 6–7]. Fuss replied that he had compiled such a list 20 years earlier for his own use,

P. N. Fuss
From a galvanograph made in the 1850s

furnished with an index showing where and when each work had been published and when presented to the Academy.

Jacobi also gave valuable advice concerning the systematic index that a complete works would need to be supplied with [6, pp. 50–73].

Jacobi expressed his pleasure on learning of the publication of the *Correspondence*, but added: "However I consider it more important that you return to the idea of publishing the complete works of Euler.... It seems to me that you must surely be incapable of giving up your large project. Think how closely connected to you personally it is and how it will never be realized if you yourself do not undertake it" [6, p. 18]. Jacobi began searching for Eulerian documents in the Archive of the Berlin Academy of Sciences, informed Fuss of his findings, and sent them to him. In particular he was able to establish from a perusal of minutes of meetings "the precise day (namely December 23, 1751)—of exceptional importance for the history of mathematics—when the Berlin Academy assigned Euler the task of refereeing a paper submitted to it by Fagnano.... With Euler's examination of this work the theory of elliptic functions was inaugurated" [6, p. 23]. Jacobi also gave advice as to how the material should be distributed across the separate volumes.

In a letter dated January 9, 1848 Fuss informed Jacobi that he is sending him the first volume of the *Arithmetical works* of Euler (for the time being without a table of contents or preface), and writes further that: "You have the right to receive a copy before everyone else; and send me your comments, which may be taken into account in connection with the printing of the second volume" [5, p. 45]. Judging by Jacobi's reply, this volume had been sent to him without the "systematic index" compiled by P. L. Chebyshev and V. Ya. Bunyakovskiĭ, so that what Jacobi—who had prepared his own system for classifying Euler's number-theoretic works—thought of it is unknown.

P. Fuss' labor in preparing Euler's works in number theory for publication was considerable: he chose the articles and ordered them chronologically; it remained only to annotate them and organize them according to their content—which was done by Bunyakovskiĭ and Chebyshev, who produced a "systematic index".

The young Chebyshev's labors over Euler's number-theoretical works were very useful to him personally. Simultaneously with the publication of the *Arithmetical works*, there appeared Chebyshev's doctoral dissertation *The theory of congruences*, containing a systematic exposition of number theory, with three appendices, one of which, "On the determination of the number of prime numbers not exceeding a given number" brought him wide renown.

Chebyshev had the greatest respect for Euler and constantly urged his pupils to study the latter's works. His student K. A. Posse wrote: "Chebyshev considered Euler the greatest mathematical genius after Newton. He often asserted that most questions occupying mathematicians at present originated with Euler and took pleasure in pointing out results attributed to some later mathematician which actually belong to Euler" [7, pp. 13–14].

P. N. Fuss did not confine himself to the publication of the two volumes of the *Arithmetical works*: he also prepared for publication, with the help of N. N. Fuss and P. L. Chebyshev, a further two volumes of Euler's works (E805)[2], which appeared only after his death, in 1862 [8]. This proved to be the last publication of Euler's works for many years.

In April 1907 celebrations of Leonhard Euler's 200th birthday were due. In connection with the approaching jubilee discussions were begun in the Petersburg Academy on the question of how exactly this date should be marked. At first it was debated whether a monument to Euler might be erected in St. Petersburg. However this proposal failed for lack of support from the General Assembly of the Academy. Then once again, as in the 19th century, the necessity of publishing the complete works of the great scientist began to be discussed.

In early 1902, A. M. Lyapunov had been elected a member of the Academy; he, like his teacher P. L. Chebyshev, nourished enormous respect for Euler. A. A. Markov's attitude to the memory and work of Euler was no less positive. At the suggestion of these two academicians the Academy acknowledged the publication of Euler's complete works as a desirable way of commemorating the 200th anniversary of his birth. In October 1902 a committee on the publication of the complete works was appointed, including among its members A. S. Famintsyn, C. G. H. Salemann, A. A. Markov, B. B. Golitsyn, and A. M. Lyapunov. At the same time Markov and Lyapunov declared that although the complexity of the enterprise was such that the Academy alone would not be able to complete the project, it must retain the initiative in the matter [9]. On instructions from the Academy, in January 1903 Famintsyn posed the following question to the permanent secretary of the Berlin Academy A. Auwers: Would the Berlin Academy of Sciences agree to participate in the publication of Euler's complete works? The history of this question has been investigated by K. R. Birman [10, pp. 489–500]. In the Berlin Academy mathematics was represented at that time by G. Frobenius, F. Schottky, and H. A. Schwarz. They welcomed the first efforts of the Petersburg Academy and recommended to their own Academy that it indeed take part in the project. However a final decision would be taken only after the required costs had been ascertained [10, pp. 496–497].

---

[2]This index refers to G. Eneström's classification of Euler's works; see later.—*Trans.*

The Petersburg Academy produced an estimate. According to academician Salemann, the total publication costs should be in the vicinity of 70,000 rubles, of which 52,000 would be for typesetting, paper, etc., 2,000 for the preparation of tables and engravings, and 16,000 for editorial work. The Petersburg Academy would assume 40,000 of the total if the Berlin Academy answered for the balance of 30,000 [11]. These figures were communicated to the permanent secretary of the Berlin Academy. In connection with this estimate Famintsyn noted that the costs might be reduced if the two volumes of posthumous works published in 1862 were included among the projected volumes [8], if only one volume were published prior to the impending jubilee, and if only papers that had appeared in journals were collected, so that books would not be reprinted as part of the collection.

Since the project involved publishing not just Euler's mathematical works, but also those in astronomy, mechanics, and physics, the Berlin Academy included in its committee, in addition to the above-mentioned mathematicians, the physicist M. Planck and the astronomer A. Auwers. On July 31, 1903 the Berlin Academy responded to the effect that in principle they were willing to collaborate in the publication of Euler's collected works, but that as no Berlin academician was prepared to undertake the editing of those works, additional funds would be needed to pay hired editors. Auwers estimated that the project would require a subsidy of 3500 marks per year over 12 years from the Ministry of Public Education (Kultusministerium), but the request was turned down and Auwers informed St. Petersburg of the setback. Reports of the Petersburg Academy for 1905 and 1906 note that it was not possible to begin the process of publishing Euler's complete works since the Berlin Academy of Sciences had hitherto been unable to persuade the ministry to allocate funds to the project.

On January 31, 1907 the Eulerian committee of the Berlin Academy of Sciences was convened. Planck submitted regrets for his absence; in his note he stated, among other things, that "As far as the situation under discussion is concerned I can provide the explanation that, in my opinion, the publication of Euler's papers on physics is not of interest to physical science and for my part I have therefore not been persistent in recommending it" [10, p. 497]. Of course the opinion of the famous Planck influenced the committee. On February 7, 1907 Auwers replied to Famintsyn that the publication of the collected works would indeed be a well-deserved memorial to Euler but that science would not derive benefits from it proportionate to the expense in labor and means. On receiving this response the Petersburg committee dissolved itself.

However in that same year 1907 the idea of publishing Euler's collected works was again broached, this time by the Helvetic Society of Scientists, and in 1908 it was given unanimous support by the 4th International Congress of Mathematicians in Rome. The academies of Berlin and Petersburg approved this initiative and decided to participate in the publication of Euler's complete works.

The Helvetic Society of Scientists sent a circular to the Petersburg Academy requesting that they support the proposed publication of the complete works through either donations or subscriptions. On May 2, 1909[3] a letter dated April 30, 1909, from professor F. Rudio to academician N. Ya. Sonin, was read to the General Assembly of the Academy; it contained

---

[3] Documents of the Petersburg Academy of Sciences and letters sent from Russia (before 1918) are dated according to the "old style" calendar; letters originating from outside Russia and documents from non-Russian institutions are dated according to the "new style".

the following message: "The day before yesterday I had the honor of welcoming Mr. Backlund to my home, and conveyed my request to him: that he send to our Eulerian committee those of Euler's manuscripts located in St. Petersburg. Thus the Academy is forewarned of this request and consequently I anticipate receiving this valuable parcel" [12].

At this meeting the academicians N. Ya. Sonin and O. A. Backlund, director of the Pulkovo astronomical observatory, spoke to this question. The General Assembly debated whether to support the Swiss initiative, and resolved on the following: 1) To allocate 5,000 franks by installments over 20 years (in accordance with Rudio's directions) towards the publication of the complete works of Euler; 2) To acknowledge as desirable that the Petersburg Academy of Sciences, on the example of the Paris Academy of Sciences, subscribe to 40 copies of the complete works of Euler; 3) To authorize academicians Salemann and Rykachev, as representatives of the Petersburg Academy at the impending meeting of the International Union of Academies [13], to communicate to that assembly the decisions taken by the Petersburg Academy and clarify various details of the question as to the kind of material assistance the Petersburg Academy of Sciences might lend the enterprise; 4) To strike a committee consisting of academicians Backlund, Salemann, Markov, Sonin, Golitsyn, and Lyapunov, instructed to look into the impending transfer to the Eulerian committee of materials preserved in the Archive of the Academy of Sciences relating to Euler's activities. In addition to these four resolutions, it was decided to enter into negotiations with universities and other higher educational institutions, libraries, etc., about taking out subscriptions to the complete works, while reserving to the Academy the organization of the whole project [14]. The General Assembly approved these proposals and confirmed the membership of the committee, with Backlund as chair. Backlund communicated these decisions to Rudio.

On May 20, 1909 Rudio wrote to Backlund: "Your message gave me great joy. Now we have indeed taken a large step forward in our enterprise! If, as can be expected, the Association [13] comes to a favorable decision and if your letter to Mr. Auwers, for which I give you many thanks, stimulates action in the Berlin Academy, then the Euler publication will be sufficiently secured. In any case, the Petersburg Academy has justified my hopes in the highest degree" [15].

In September 1909 the annual meeting of the Helvetic Society of Scientists in Lausanne confirmed the decision concerning the publication of the *Complete works of Euler* (*Leonhard Euleri opera omnia*), in order to "fulfil the oft-repeated desire of the whole mathematical world" [16].

Ensuing letters from Rudio to Backlund and Lyapunov contain many details concerning the further participation of Petersburg academicians in the preparation of the complete works.

Backlund was present at the aforementioned meeting of the Helvetic Society in Lausanne [16] where he apparently held discussions concerning the Euler project. On October 16, 1909 Rudio wrote to Backlund: "You have doubtless already returned from your long trip, so that it would appear that I may once again connect with you. I have the honor of sending you, on behalf of the Eulerian committee, copies of Stäckel's plan [17] (I can supply as many as you like), and I accompany these with a request that your committee arrange the Euler materials in the order indicated in the plan. This would greatly facilitate our work.... As soon as an editorial committee is formed it will address itself to the determination of the underlying distribution [of subject-matter] on which the Euler publication will be based. This plan will then be submitted to the three big academies. I have been able to prepare a great deal...together with Messrs. Stäckel and Krazer.

"I confirm somewhat belatedly receipt of the first Euler payment of 50 franks; I remain with deepest respect your devoted F. Rudio" [18].

We remark that subsequently professors A. Krazer, P. Stäckel, and F. Rudio participated in the editing of several volumes of the complete works of Euler as members of the editorial committee.

In St. Petersburg the Eulerian committee met five times over the year 1909. The first of these meetings was held on December 31, 1909 (old style), and was attended by Backlund, Salemann, Lyapunov, and Sonin. It was resolved that: 1) The Academy should be asked to leave to the committee all business relating to Euler, including the business concerning subscriptions to the new edition of Euler's works; 2) The Swiss Eulerian committeee should be asked to send to the Petersburg committee six offprints of the list of Euler's works compiled by Eneström when it is ready to be printed [19]; 3) Rudio, the Chair of the Swiss Eulerian committee, should be informed that in our library there are 17 volumes of Euler's papers, but told of the presence of manuscripts in the archive only after they have been fully sorted out; 4) The privy counsellor V. E. Fuss should be asked to inform the committee of all materials at his disposal. It was mentioned by Sonin at this meeting that apparently there existed a geometry textbook by Euler (in a translation by M. V. Lomonosov), and that there was a letter written by Euler in the Public Library [20].

In Rudio's letter to Backlund of November 23, 1909 he writes: "I express to you my deepest appreciation of your letter and the important information contained in it. I immediately requested Teubner [21] to send you six offprints of Eneström's list [19]. So far four pages have appeared, which you have probably already received. I await news of the progress in your work impatiently. Until it is completed it is scarcely possible for us to begin editing the papers.

"Apart from the necessity of reclassifying the unpublished materials expected from your Archive, we shall also need to collate papers that are already printed with the available manuscripts. Is the Academy prepared to furnish us with these? The collation is best carried out by the editors of the individual volumes. This, of course, will be possible only if the materials can be divided up [appropriately]. I would also like to ask the following question: Are there not other Russian scholars willing to participate in the preparation of these volumes? I would be grateful to you if you would find the occasion to give me two or three names. The final distribution of all the materials amongst the various collaborators can be undertaken only when their itemization is more or less complete. Hence it would be expedient to express your wishes in good time. You have kindly agreed to take on the preparation of the *Theoria motuum planetarum et cometarum* [22]. Will you take on the whole of the 14th volume of the second series (as Stäckel's plan has it)? I shall have several copies of that plan sent to you [17]. If required, more copies can be supplied in accordance with your instructions" [23].

This letter evoked a lively debate amongst the committee members. At the meeting of February 28, 1909 members voiced various ideas concerning the principles according to which the volumes should be edited. Sonin was of the opinion that each editor should not only re-derive the formulae but also verify all calculations, correcting all misprints and errors that might be discovered in the process. Salemann held the opposite point of view: no changes whatsoever should be made except for the correction of misprints, and even the original spelling should be preserved. Lyapunov held it desirable that the derivation of formulae should be redone, but that it would not be practicable to require of the editors that they check all the calculations. Markov and Golitsyn agreed. Backlund also thought it essential that the formulae be re-derived, but that as far as numerical applications were concerned the decision as to whether to carry out a verification or not might be left in each case to the discretion of the particular editor.

The committee finally decided to ask the Chair of the Swiss editorial committee what basic principles they had decided on in connection with this question [24].

Salemann instructed one of the librarian's assistants to draw up an inventory of those of Euler's works contained in the 17 volumes belonging to the Academy with a view to interceding with the General Assembly for payment of 50 rubles to this person for his labors. The head of the Archive of the Academy of Sciences, B. L. Modzalevskiĭ, was asked to inform the committee of the result of his search for a Russian version of Euler's *Geometry* [20]. They requested also that he produce an inventory of all of Euler's manuscripts stored in the Archive, and, with Sonin's assistance, to continue searching the Archive. Sonin informed the committee that he had sent G. Eneström a list of the misprints that he had found in the first four pageproofs of Eneström's "List" [19]. Lyapunov announced that he had found in J. Hagen's "List" [25] works not appearing in Eneström's list. It was decided to inform professor Rudio of this circumstance and also to send him four volumes of the Minutes of the Academy [26]. As far as the 17 volumes of Euler's papers were concerned it was also decided to request the Swiss committee not to split them up in the eventuality that they be sent to Switzerland.

On December 23, 1909, in reply to a communiqué on the upshot of the Petersburg committee's deliberations, Rudio wrote to Backlund:

"Many thanks for your so very precious letter. I anticipate the arrival of the promised index with great impatience.

"In the meantime we have introduced order into the organizational side [of the project]: the necessary committees have been formed, and the protocol governing their conduct and authority established. Of course these are auxiliary (äusserliche) things which, though requiring much labor and many meetings, still need to be put in order. Now at last we can begin the editorial work proper. Most importantly, the editorial committee has been formed (on Sunday last in Berlin): The general editor is Rudio and the co-editors Messrs. Stäckel and Krazer from Karlsruhe. The fact that two of the editors live in the same place is very convenient. Furthermore, Teubner has been chosen for the printing and publishing [21].

"The editorial committee is at present taking advantage of the Christmas break to formulate a precise editorial protocol to which every collaborator must adhere. At the same time of course we are looking into the questions you mention concerning information and so on.[4]

"In particular, I share your point of view that all algebraic formulae and calculations apart from astronomical ones—such as for instance lunar tables—should be verified.

"When we have drawn up our protocol we shall print it and submit it to the academies

---

[4] See the earlier summary of the meeting of the Eulerian committee of November 28, 1909.

and other large institutions for approval. Any suggestions emerging from this will be carefully examined and adapted.

"I did not know that Hagen has works absent from Eneström's list. In any case it would be useful if Mr. Lyapunov would name them" [27].

In his next letter to Backlund (January 13, 1910) Rudio notes that certain materials have arrived from St. Petersburg, including an "Inventory" compiled by B. L. Modzalevskiĭ [28]. He mentions also that he visited Karlsruhe to consult on the preparation of an editorial publication plan. He, Stäckel, and Krazer together settled the division of the future volumes into three series, and made a preliminary assignment of editors to the individual volumes. They decided to suggest to Backlund that he undertake the preparation of the 14th volume of the second series. Rudio noted that for this purpose they would send Backlund what was in essence a volume already prepared for publication, so that he had only to read the text critically and proofread. The volumes on astronomy (making up five volumes of the second series) were distributed approximately as follows: Volume 12 to the English astronomer Whittaker (whose address Rudio requested from Backlund); Volume 13 to Bauschinger; Volumes 15 and 16 to Bauschinger and Kobold.

Rudio repeated his request for names of members of the Petersburg Academy of Sciences or other prominent Russian mathematicians, physicists, or astronomers willing to take on the editorship of a volume from any of the series [29].

In a letter that has not come down to us, dated February 7, 1910, Backlund informed Rudio that academicians Lyapunov and Markov had agreed to his request, and on February 13 Rudio replied:

"...I would not wish to allow the chance to slip by of thanking you at once for your friendly letter of the 7th inst., and of telling you how pleased I am that you have finally decided to take on the whole of the 14th volume. At the same time I wish to thank you for the pleasant news concerning Messrs. Lyapunov and Markov. I immediately wrote to both co-editors Stäckel and Krazer, suggesting they indicate which volumes might be assigned to those gentlemen. As soon as I receive a reply I shall write to Messrs. Lyapunov and Markov and ask them to take on the assigned volumes. For the Euler publication it is highly desirable that your Academy be represented by three participants.... A draft of the editorial plan is now with the typographer and will shortly be sent to you" [30].

Rudio's next letter to Backlund, dated February 27, 1910, concerns a collection of Euler's works put together by P. N. Fuss (apparently the collection of published articles selected and bound according to his specifications): "Today I am writing to you specifically concerning Mr. Fuss' 'Conspectus' [31] of [his] collection of Euler's works. This collection is, of course, of great value for [our] publication, almost essential. If I understand correctly, it was put together by Fuss specifically for publication, and can now fulfil its intended role. For, if we ignore the copies marked 'Ms' and restrict attention to the printed texts, all of which are to be found in your library, then there is nothing to prevent us from using this collection in the capacity of an original printed text. And on the other hand it would hardly be possible for our editorial committee to re-compile such a collection without a disproportionately large sacrifice [of time and effort].

"Since your Academy has decided to place at the disposal of [our] Euler committee all those materials in its possession which might serve to enhance our edition of Euler's works, I can only make the following request of that Academy: To send us Fuss' collection in its original printed form. When the printing process is complete all such materials will of course be returned; naturally, minor damage may occur such as one might expect during typesetting. It is not always possible to avoid this, but I feel that once a successful printing has been achieved the originals will in any case be of no further use. Hoping that your Academy agrees to our request, which almost determines the question of life or death of the Euler edition, I remain with deepest respect your devoted F. Rudio" [32].

Apparently a copy of P. N. Fuss' collection of Euler's works was indeed sent to Zürich, since on June 22, 1910 Rudio wrote: "Regrettably I have not yet received official authorization from the Imperial Academy to disassemble the copy of Fuss' collection and use it as the original for printing. Until we receive such permission we cannot begin the work. Of course in using this copy as the original for printing purposes, it will inevitably be marked up, corrected, and so on. However the committee will ensure beforehand that this copy will then be returned to its former state.

"It becomes more and more urgent that the Academy send to Zürich the manuscripts itemized in the list you so kindly sent me. It would seem to be impossible to publish a volume of papers if one is unable to compare them with the original manuscripts. Four weeks ago I wrote about this to the permanent secretary of your Academy Mr. von Oldenburg, and now make bold to convey my wish in this regard to you personally with the hope that you will be able to expedite the matter. I have sent a similar request also to the Berlin Academy" [33].

In September 1910 Rudio, on behalf of the editorial committee, sent the following materials to the editors of the various volumes of the complete works: An editorial plan (in both German and French) [34]; Two versions of Stäckel's plan for organizing the material of the complete collected works into series and volumes 17]; A list of the editors and their addresses [35]; A sample page from Euler's *Mechanics* serving as an example showing mistakes in the script, the typesetting of formulae, the arrangement of the subject matter, etc., and examples of the table of contents and the cover design [36]; A reminder that the firm Teubner had been sent earlier the "List of Euler's works" compiled by G. Eneström (the proofs of the first version) [19].

Rudio wrote: "The enclosed materials will allow you to judge whether the volume you so kindly agreed to work on contains the articles appropriate to its assigned content, or whether further changes in the distribution of subject-matter are required (§26 of the editorial plan). It is in the very nature of things that many of the papers might be located in some other volume; however it is also clear that no such question should be decided in isolation, but rather in the final analysis with reference to the whole collection.

"The considered arrangement of materials decided upon in Stäckel's plan is the result of discussions that began long ago, and therefore, on behalf of the committee, I would ask you to confine yourselves to proposing only such changes as seem to you absolutely essential. I beg you to send me your communications by November 15 of this year. I beg you also to inform me as to how long you would like to keep your volume with you (§27 of the editorial plan) and how long you think your processing of it will take" [37].

(The materials itemized above were enclosed with this letter.)

In accordance with the editorial plan it was proposed that all of Euler's works be included, both those already published and those remaining in manuscript form, together with those of his letters having scientific content, and the works of J. A. Euler either written under his father's (L. Euler's) supervision, or actually written by Euler himself. It was planned also to include summaries of Euler's papers—to appear in small print at the beginning of each paper—, those works which had been published as individual books, and the essays awarded prizes by the Paris Academy. There would also be included the eulogies to Euler composed by N. Fuss and Condorcet, and a foreword relating the history of the publication of the complete works.

It was proposed that each article be printed from its first published version, with later appearances mentioned in the notes, while books were to be printed using the latest version edited by Euler himself. Individual books would be reprinted in their entirety. Collections of papers were to be broken up into their individual items and inserted into the appropriate volumes of the collected works.

Two types of notes were specified: footnotes referring to places in the text; and at the end of each article comments on the article as a whole. Such notes were to be confined to what was strictly necessary and not allowed to expand into historical excursions. At the same time important questions should not be passed over in silence—such as, for instance, the reciprocity law or the Riemann zeta-function. The papers were to be numbered according to Eneström's scheme.

In §27 of the editorial plan it was stated that the distribution of material by volumes would, as stipulated in Stäckel's outline, be carried out by the editorial committee. Each editor in charge of a particular volume would receive the materials for his volume pre-selected by the editorial committee. He would then need to check the text of the articles, make corrections or additions, and return the volume to the editorial committee. He might also make suggestions about changing the order of the articles or including other articles in that volume.

It was proposed that each work be published in its original language. Textual corrections were to be confined to mistakes visible at first glance (by entering omitted words between angle brackets, correcting references to other works and to diagrams, etc.). Comments on each paper were to be in the language of that paper, and the foreword to each volume in the language of choice of the author of that foreword.

In conformity with the decision of the Petersburg Academy, it was stated that the text and formulae should be checked and all formulae re-derived. It was also stated as desirable that all computations be verified, especially in works on pure mathematics. It was not necessary to check astronomical tables (§31).

In cases where the original was hand-written, the printed version should be checked against the original. In such cases mistakes and slips of the pen obvious at first glance should be corrected and these mistakes noted in the commentary. (The printed text should be based on the corrected version.)

Mistakes leading to incorrect results whose rectification would require substantial changes to the original were to be left alone. At the juncture where such a mistake first appeared, a comment should be entered under the line and if possible the correct result and the resulting conclusion indicated.

Apart from mistakes of the types mentioned in the preceding paragraphs, there were "incorrigible" errors, such as, for instance, those involved in the use of divergent series to yield valid conclusions. In such cases it was proposed that, in addition to interlinear comments, at the conclusion of the work there should be appended a general commentary in which others of Euler's works and relevant correct results that had appeared later would be referenced (§34). Generally speaking, it was stipulated as appropriate to indicate the most important works where the problems in question had been considered.

The conditions were also specified under which the editors of the individual volumes and the editorial committee were to work. The editorial committee entered into a contract with each editor of a volume. If a volume had more than one editor assigned to it then the contract was to be agreed to on their behalf by their representative, although the names of all of its editors would appear on the title page of that volume (§24).

Stäckel's plan was examined by the editorial committee, emended, and signed by Rudio, Krazer, and Stäckel.

It seems that Lyapunov and Markov requested some sort of clarification from Rudio, since in a letter from Rudio (from which the addressee's name is missing) preserved among the papers of the academician A. N. Krylov [38], one reads: "Since it was impossible to mention all of the finest details in the editorial plan, I allow myself the latitude of informing you that it is completely unnecessary to prepare a special manuscript. You can enter *directly in the volume*, in the margin or at the bottom of the page, those corrections and comments that you think necessary. When in paragraph 44 the great significance of the materials is mentioned, we only wanted to emphasize by this that *nothing should be lost*. The [material for the] volumes is intended *to be sacrificed*, so that comments, corrections, and notes, etc. can be entered in them by hand. Upon completion of the work on a volume, when you consider it ready for printing you should send it to the chief editor, i.e., to me. In accordance with paragraph 45, you will then *later* send the [corrected] proofs, which you are to correct, to Mr. Krazer or Mr. Stäckel, as well as the materials [originally] placed at your disposal.... I hope that your doubts have been dispelled. I beg you, my dear Sir and dear colleague, to accept the expression of my deep respect. F. Rudio" [38].

The materials from the Archive of the Academy of Sciences were selected, furnished with an inventory [28], and dispatched to Rudio in Zürich. They were accompanied by two plates engraved with portraits of Euler.

On October 29, 1910 Rudio wrote to Backlund: "Heartfelt thanks for your kind letter of the 23rd inst. and for your energetic efforts in connection with the manuscripts. I shall let you know, and of course also your Academy, as soon as they arrive. I am very glad that you find yourself able to prepare your volume quickly. I draw your attention to the fact that in the revised version of Stäckel's plan astronomy occupies a somewhat different place than in the original one.

"In this respect the decisive factor for us was C. G. J. Jacobi's proposal that all of Euler's works relating to the three-body problem be brought together. [See the correspondence between C. G. J. Jacobi and P. N. Fuss [5, p.55].] Thus Volumes 13 and 14 made their appearance. Then it began to seem desirable to undertake a narrower subdivision than before. The former Volume 16 has now become subdivided into three parts, and Volume 15 into two, so that Volume 15 contains the *Theoria motuum* [Berlin, 1744 (E66)], which you chose, with the result that you will now be responsible for Volume 15, although its content does not wholly coincide with that of the former Volume 14. I hope that you will agree to these changes, and that we can send you Volume 15 to work on. However this will be possible only after November 15 by which time we should have received all the responses to our circular 36" [39].

The Editorial Committee entered into contracts with the editor(s) of each of the individual volumes. Here, by way of example, is Lyapunov's contract [40]:

"Contract with the editor of an individual volume
of the Euler publication

Between the Swiss Society of Natural Scientists, represented by the Chair of the editorial committee professor F. Rudio in Zürich, on the one hand, and professor[5] A. Lyapunov on the other hand, there is today concluded the following agreement, formalized at Zürich.

§1. In connection with the Collected Works of Euler to be published by the Swiss Society of Natural Scientists, professor A. Lyapunov expresses his readiness to prepare the following parts: V. I–13 (Integrals) and undertakes to carry out the work in accordance with the editorial plan, and as necessary make such changes as may be desired by the editorial committee.

§2. Professor A. Lyapunov is in receipt of working materials from the editorial committee and agrees to present his work to this committee *in a form completely ready for printing*. The deadline for completion of the work is June 1913. All delivery of materials from either party to the other is to be free of charge.

§3. For its part, the Swiss Society of Natural Scientists agrees to pay our collaborator[6], on completion of the work, an honorarium of 30 f. (thirty franks) per printed sheet of 8 pages, and furnish him with 10 copies of the section prepared by him.

§4. In the case of death of our collaborator or the non-fulfillment of his obligations, the editorial committee has the right to conclude an agreement with another collaborator. In this case all claims of the first collaborator become null and void.

Zürich, October 10, 1911. Professor doctor F. Rudio.
St. Petersburg, October 10, 1911. A. Lyapunov" [40].

A. M. Lyapunov began work immediately on the Eulerian volume delivered to him (see below) and already by April 1912 had returned the volume to Rudio ready for the press. The latter responded on May 2, 1912 as follows: "My dear and highly respected colleague! I have the honor of confirming receipt of the materials relating to your volume, and hasten to repeat my congratulations and gratitude. However it would be futile to begin discussing your work by mail just at the moment since it will not be possible this year to deliver your materials to the printer. They are busy now coping with four volumes simultaneously. I may possibly be going to Cambridge [41] and shall then have the pleasure of seeing you" [42].

In 1913 Rudio appealed to the Petersburg Academy for approval of a new editorial plan. It seems that this approval was forthcoming. On November 20, 1913 Rudio wrote to Backlund: "I must first of all confirm receipt of the four payments of 50 f. I wanted to exploit that occasion in order to have the opportunity of thanking you very much for the energetic support of the Euler project that you show each time. The business is forging ahead. In a few days the tenth volume will appear [43]. We are fully supplied with prepared material for the time being, so that we can wait to submit the volume for which you so kindly took responsibility as long as need be. I can send you material as soon as you wish.[7]

---

[5]The original Russian has "Mr. professor Rudio", and later "Mr. professor doctor A. Lyapunov", indicating that the original contract was doubtless in German.—*Trans.*

[6]" Herr Mitarbeiter", perhaps.–*Trans.*

[7]More material?—*Trans.*

"At the same time as this letter I am sending you a circular concerning the newly formed Leonhard Euler Society [44] since you should have an idea of the state of the project's finances. I would rather not have had to send you this circular, taking into account the valuable contributions you have already made, but it has to be sent to all collaborators" [45].

Backlund agreed to become a member of this new society in order to lend support to the finances of the Euler publication project. In response to his letter (probably of December 19, 1913) Rudio wrote on January 23, 1914: "Many thanks for your kind letter of the 19th, and for your registration in the Euler Society. You can take as long as you like with your Euler volume. It would absolutely be soon enough if it arrived some time in 1915". [45]. However, apparently because of the state of his health and for other reasons (the First World War had begun), Backlund was unable to begin work on the Euler volume allotted to him. He died in 1916.

The proposal in the new editorial plan to change the distribution of material with respect to the volumes also affected the volume on which A. M. Lyapunov was working. On June 19, 1914 Rudio wrote to him as follows: "As you recall, the distribution agreed upon initially in connection with the publication of Euler's works had to be changed. According to the original plan the items on integrals were to be issued in two volumes—12 and 13. The new distribution, to which the Imperial St. Petersburg Academy of Sciences has agreed, provides for a greater number of volumes and as a result the items concerning integrals will appear in three volumes: Volumes 17, 18, 19. Volume 17 is to contain items 59 through 464 of Eneström's list [47]; Volume 18—items 475 through 640 [48]; Volume 19—items 651 through 819 [49].

"The volume edited by Mr. Gutzmer of Halle will appear in a few weeks: the other two volumes—18 and 19—should follow close behind. However in Volume 18, items numbered from 475 through 594 were prepared by Mr. Gutzmer, and from 606 through 640 by you. Thus we have no option but to indicate in the front matter of Volume 18 that it was: 'prepared by August Gutzmer and Aleksander Lyapunov', whereas Volume 17 was prepared by Mr. Gutzmer alone, and Volume 19 by you alone. It might also be specially noted in a short foreword that items 475 through 594 were examined by Mr. Gutzmer, and from 606 through 640 by you. Individual comments can be accompanied by the initials 'A. G.' in the case of the first items, and 'A. L.' in the case of the second, so that no confusion arise.

Gutzmer has declared his agreement with this course of action, which in any case would seem to be the only one possible, and I beg you to give your assent so that I can pass on to the printer Volume 19, on which you are working" [50].

On the same topic V. A. Steklov wrote to A. A. Markov: "In Cambridge I saw Rudio. He sends his respects to you. He says that Euler will not fit into 43 volumes and that the funds are insufficient to pay for the printing. Certain volumes will have to be divided in two since otherwise they will be too thick. He is afraid that the signatories will protest. He made an announcement concerning this at the meeting. He will be in Zürich in October and you'll get hold of him there. Please pay him my respects" [51].

It seems that Markov did indeed come to Zürich and spoke with Rudio about the volumes on which he would be working.

Unfortunately all of the volumes on which A. M. Lyapunov and A. A. Markov worked appeared only after their deaths. Of the volumes on which Lyapunov worked, Volume 18 appeared in 1920 and Volume 19 in 1932. In Volume 18 a note concerning Lyapunov's collaboration was included [52] in the form indicated by Rudio. In Volume 19, in addition to Lyapunov there appeared the names of two others: A. Krazer and G. Faber. In the foreword to that volume, written by Rudio [53], it is mentioned that the volume had undergone a difficult birth. To begin with, on November 3, 1918 Lyapunov died, and then on May 10, 1924 it was the turn of Gutzmer, who had managed together with Rudio and Krazer to prepare Volume 18 for the press. For Volume 19 the proofs [which, it is appropriate to mention, numbered more than a few!] had still to be corrected. Krazer drew up a survey of the articles contained in these three volumes, which was included in Volume 19. This was to be Krazer's last scientific endeavor. His place was taken by G. Faber, who completed the preparation of Volume 19, proofread it, and inserted several comments. It was he who completed Krazer's survey and included it in that volume. This explains the two additional signatories Krazer and Faber to Volume 19. The proofreading of Volume 18 was begun by Lyapunov and completed by Rudio, Gutzmer, and Krazer.

In his foreword to Volume 18, Rudio wrote: " The Swiss Society of Natural Scientists will remember the lively interest with which he [Lyapunov] collaborated in the Euler publication" [52, Vol. 18, p. VII].

Lyapunov's notes to the text, which have the initials "A. L." attached, are many and varied. In some cases these are merely references to relevant works of Euler in this or other volumes; such are for instance the notes on pages 263, 283, 293, 322, 375, 399, 466 of Volume 18. There are many textual corrections of earlier editions. For example on p. 288 of Volume 18, Lyapunov has the comment: "In the earlier edition No. XII is repeated incorrectly". In cases such as this the appropriate correction has been made in the text and the error mentioned in the notes. In cases where the text of the source edition has been changed, the original expressions are given in footnotes. There are some corrections of calculations. On p. 271 of Volume 18 there appears the more serious comment: "This formula makes sense only when $0 < \omega < 2\pi$, whence the condition $0 < \theta < 2\pi$ follows".

There is a lengthy and substantive comment on pp. 314, 315 of Volume 18, referring to Euler's paper entitled "On the use of imaginary quantities in analysis" (E621): "This series diverges", writes Lyapunov, "but the formula Euler gives in §37 can be proved as follows...". There follows a derivation of said formula. In his comment on p. 278 of Volume 18, Lyapunov states: "Although this is not fully proven, nonetheless one can in fact set $p = q\sqrt{-1}$ here, since when the value of p remains in modulus sufficiently small, the integral may be expanded in a power series in p. A. L.". On p. 383 he notes in a comment that "If the quantity f is negative then the integral does not exist", and on p. 401: "Obviously this condition is not necessary". In the comments on p. 318, in addition to giving references to works of Euler, Lyapunov refers to the book of L. Mascheroni published in Pavia in 1790—1792. On p. 318 of Volume 19 Lyapunov refers to Euler's *Differential calculus*, among others of his works, and on pp. 316, 317 to N. Bernoulli's work on summation of series of reciprocals of squares, and to letters from N. Bernoulli to Euler dated July 13 and October 24, 1742, published by P. N. Fuss in 1843.

It is thus clear that Lyapunov's work on his volume was very thorough, and that perhaps it would be appropriate to carry out a more particular investigation of his work on the Euler project. Some of his comments concern Euler's use of divergent series; it is probable that §§31 and 32 of the editorial plan originated from comments of Lyapunov that he had communicated to Rudio.

A. A. Markov participated in the preparation for publication of two volumes of the first series—Volumes 4 and 5 (*Arithmetical Essays*, Vols. 3 and 4). Both volumes were published after Markov's death: Vol. 4 in 1941, Vol. 5 in 1944. On the title page of Vol. 4, R. Fueter is given as editor; however, in the foreword Fueter states that that volume was first examined by F. Rudio, and then processed for publication by A. Markov using manuscripts of the essays. Unfortunately death intervened to prevent Markov's completion of this work. All notes are accompanied by their originators' initials [54, p. VII].

Markov's annotations, like those of Lyapunov, are various. He inserts references to other works of Euler. Several of his notes concern textual corrections of the original edition; these are indicated as follows: "Editio princips: R. Correxit A. M." (In the original edition: R. Correction by A[ndreĭ] M[arkov]) (Vol. 4, p. 131). He corrected many computational mistakes, including some used in compiling tables. For example on p. 47 of Vol. 5, the condition for integer substitutions in the formula $x^4 - mxxyy + y^4$ to yield a perfect square is corrected. In several places Markov corrected the values of the unknowns yielding solutions of equations, and supplied solutions missed by Euler. Certain formulae are corrected. On p. 269 of Vol. 4 Markov refers to a work of F. Grube in which a proof of Euler's is completed. He also makes more substantive comments.

Volume 5 is densely annotated by Markov; most of his comments arose from a comparison of the printed text with the original manuscript version.

As noted above, although on the title page of Vol. 5 only the name of R. Fueter appears, the latter states in his foreword that "A. Markov carried out the original examination of all articles in this volume", Markov's death preventing his completing the work. Volume 5 also contains a preface by A. Speiser to the sequence of volumes on number theory.

In summary, both A. M. Lyapunov and A. A. Markov put a great deal of effort into editing Euler's works, a fact not usually mentioned in biographies of them, and hitherto not investigated in detail.

Although, as mentioned earlier, O. A. Backlund did not live to do any actual editorial work, nonetheless it is clear from his correspondence with Rudio and from the minutes of the meetings of the Petersburg Euler committee that he deserves great credit for his help with the publication of the complete works in its initial stages, and also for arranging the participation of the Petersburg academicians A. M. Lyapunov and A. A. Markov in the project.

Further participation by Russian scholars in the publication of Euler's works was curtailed by the war. The initial stages of the renewal of relations between the Academy of Sciences and the Swiss Society of Natural Scientists after the First World War are reflected in the correspondence between the academicians A. N. Krylov and V. A. Steklov.

On May 2, 1921 A. N. Krylov, at that time on official leave abroad, wrote to V. A. Steklov, vice-president of the Russian Academy of Sciences (as it was called from 1917 through 1925), as follows:

... the rector of Zürich University, Prof[essor] Fueter (math[ematician]), who is now in charge of the publication of Euler's works, is very interested in knowing if the Petersb[urg] Academy will be continuing its participation in that publication and its subscription to 40 copies.

I spoke to rep[resentative]s of the Department of S[cience] and T[echnology], noting that the Academy had already received 40 cop[ies] of the first 11 volumes, and that therefore withdrawing from the subscription is not an option, and that, independently of appointments to the Mathem[atics] Office, I am already authorized to purchase for the Academy of Sciences 40 copies of each of the 5 volumes that appeared in 1914. Furthermore I have corresponded with Prof[essor] Fueter concerning the possibility of continuing with the publication, which is now running into financial difficulties because of an increase in costs and the Paris Academy's withdrawal not just of its subsidy but also of its subscription to the 30 cop[ies] that it was in the process of receiving, "because the publication is being carried out by a German firm".

I think that if our Academy explains to the Council of the People's Commissars the international character of a project such as the publication of Euler's works, its scientific import, and the necessity of supporting the publication in order to bring it to completion, then the CPC [Sovnarkom] will not refuse the necessary means. You know better than I how to couch the announcement appropriately, and I think that this matter is such that one might approach Vlad[imir] Il'ich[8] himself concerning it.

I have received an invitation from Prof[essor] Fueter to go to Zürich (he obtained the requisite sanctions with difficulty) in order to discuss this matter, and then report to the Academy accordingly. I shall inform you of all my discussions as soon as possible. (Of course I cannot myself presume to make any commitments since I do not have the Acad[emy]'s authority to do so.)

Prof[essor] Fueter has also sent me a [copy of a] letter of proposal from the Committee on the publication of Euler's works, addressed to the Petersb[urg] Acad[emy] of Sciences and dated Jan[uary] 12, 1921, containing the information that over the course of the war 5 volumes appeared, 10 copies of which are offered as a gift to the Academy, and that the task of delivering these has been assumed by Mr. Charles Moor. I heard nothing of this letter in the Academy prior to my departure from Petrograd; and have heard nothing from you concerning it, whence I infer that Ch. Moor has not yet arrived in Petrograd [56].

In Krylov's letter the following appeal from the Swiss Society of Natural Scientists was enclosed:

The Eulerian Committee of the Helvetic Society of Natural Scientists
Basel, 12 January 1921

To the Petrograd Academy of Sciences

Gentlemen, a large undertaking by the Helvetic Society of Natural Scientists—the publication of the complete collected works of Leonhard Euler—was subjected, like everything else, to the destructive action of the great war and its aftermath. Not only had we to overcome extraordinary difficulties in connection with the printing, but it also proved impossible for us to distribute the completed volumes to our subscribers in warring countries. In spite of difficulties of all kinds, the committee has managed to publish five volumes of Euler's works over the past few years.

The Petrograd Academy has been from the very beginning the most distinguished and powerful supporter of our enterprise, striving to immortalize the glory and great merit of one of its most famous members. And as a token of our profound gratitude we take the liberty of presenting you with 10 copies of those volumes. Mr. Charles Moor, a member of the Great Council of the city of Berne, has kindly agreed to deliver these volumes to Petrograd. Please receive him amicably. In the hope that our relations soon normalize, we beg you, gentlemen, not to refuse to continue your valuable endorsement of our enterprise, and to accept our expression of deepest respect.

Signed by: Fritz Sarazin, President of the Euler Committee of the Helvetic Society of Natural Scientists [57].

On May 12, 1921, V. A. Steklov replied to A. N. Krylov's letter and to the appeal of the Swiss Society of Natural Scientists as follows: "Dear Alekseĭ Nikolaevich, I have received your 'report' and the letter from the committee dealing with the Euler publication. First, the Academy had already decided when you were here to continue its collaboration in the publication of Euler's works, and the committee has chosen me to replace A[leksander] Mikh[aĭlovich] [58]. At the initiative of A. A. Markov the question of our relations with the

---

[8] V. I. Lenin.—*Trans.*

committee in Zürich was raised again at the meeting on May 9. I proposed that we prepare a detailed statement from our Department, indicate the members from our Academy, and request the Zürich committee to inform us about the progress of the work and its proposed further direction. I sent this statement to Revel Gutman, for him to forward to the correct quarters. I received your report opportunely the day after this decision was taken. It would be good if when you are in Zürich you personally delivered our statement and explained the details. Serg[eĭ] Fedor[ovich] [Ol′denburg] and I have again been ordered to Moscow by the General Assembly, and we leave on May 16; one item of business concerns the organization on behalf of the Academy of a foreign exchange of publications and books. Of course we shall take all steps to obtain the necessary credits towards the publication and purchase of 40 copies of Euler's works, and I think we shall succeed in this. I shall inform you [of the outcome] in good time. I only fear that you will already have left for Paris..." [59].

In a letter to Steklov dated June 15, 1921 A. N. Krylov described his stay in Zürich as follows: "I spent 10 days in Zürich: a meeting of the committee on the Euler publication was called, to which the president of the 'Naturforschende Gesellschaft' [the Swiss Society of Natural Scientists] Fr. Sarazin came from Basel. The committee was especially grateful to the Petrogr[ad] Academy for its active participation in the publication of Euler's works and for placing manuscript materials at their disposal. All of the manuscripts were shown me; they are kept in a special place at the university and are in perfectly good order and in a good state of preservation.

"I promised the committee that I would inform the Academy as to the state of affairs with the publication, and in the meantime for my part I shall try to negotiate with our Science and Technology Section in Berlin, which is responsible for the acquisition of foreign books and journals and for scientific publishing, and convince them that the publication of Euler's works represents a genuinely international enterprise in which Russia's interests are involved and that therefore the acquisition of 40 c[opies] of the volumes at the *nominal* price of 25 fr. = 250 Ger[man] marks per volume would show support for the publication and would correspond fully with the objectives of the S.T.S. [60]. The chairman of the S.T.S., Prof[essor] N. M. Federovskiĭ, and his closest co-workers A. A. Tretler and E. G. Lundberg [60] are in complete agreement with me[9], and have today sent off to the firm Teubner a cheque in the amount of 60,000 marks in payment for 40 copies of each of the 6 volumes issued. Teubner will send all of these to Berlin, to the S.T.S., and on my return journey I shall pick them up for delivery to the Academy" [61].

The volumes published during the war years were as follows: Vol. 2 (1915), Vol. 3 (1917), Vol. 13 (1914), Vol. 17 (1915), Vol. 18 (1920), all belonging to the first series; and in 1921 one further volume—Vol. 14 of the second series—was to appear.

A. N. Krylov also informed Steklov that, in view of the death of Professor Stäckel, the volume containing *The theory of motion of rigid bodies* was left without an editor, and that Rudio (who was still chief editor of the publication) would be glad to assign this task to some member of the Petrograd Academy of Sciences. "I promised to inform you of this, and if you yourself were to take on this task, or Yakov Viktorovich [62], I would willingly help you; in any case we will discuss this when we meet" [61].

---

[9]What at the beginning of this letter were expressed merely as intentions, seem to have been realized by the end.—*Trans.*

V. A. Steklov did indeed agree to undertake the preparation of the volume devoted to the motion of a rigid body. On learning of this, A. N. Krylov promised to obtain an original copy of this work from an antiquarian so that in addition to the copy that Rudio would be sending him, he would have a spare copy in case it prove necessary to engage young co-workers to help with the checking of calculations, proofreading, etc. [63]. In August, 1922, Krylov sent a telegram to Steklov informing him that on board the steamer "Red Profintern" there were 12 crates of Euler's works (40 copies of the 6 volumes) addressed to the Academy in charge of the courier Schute [64].

V. A. Steklov did not in fact manage to find time to work on the Euler volume assigned to him since his duties as vice-president of the Academy of Sciences kept him very busy and the times were hard. In 1974 a "Report" written by Krylov was published, containing a brief exposition of the history of the publication of the *Opera omnia*, and the participation of Russian scientists in that endeavor [65] (see also [66]).

References

[1] C. r. Acad. sci. St.-Pétersb. ann. 1844. Rec. actes séance publ. Acad. sci. Pétersb., 1845.

[2] *Oeuvres complètes en français de L. Euler.* Bruxelles, 1839. T. 1–5 (E786).

[3] Anzeige. J. für reine und angew. Math. 1843. Vol. 26. p. 368.

[4] LO Arkhiva AN SSSR[10], f. 1, op. 1-a, No. 70, l. 67–67 ob.

[5] Ozhigova, E. P. *Mathematics at the Petersburg Academy of Sciences from the end of the 18th century till the middle of the 19th.* L.: Nauka, 1980.

[6] *Briefwechsel zwischen C.-G.-J. Jacobi und P.-H. von Fuss.* Leipzig, 1908.

[7] Posse, K. A. "Chebyshev, Pafnutiǐ L'vovich". In: Vengerov, S. *A critical-biographical dictionary of Russian writers and scholars.* SPb., 1897–1904. Vol. 6. pp. 1–23.

[8] Euler, L. *Opera postuma.* Petropoli, 1862. Vols. 1, 2.

[9] "Report on the activity of the Imp. Academy of Sciences in its Physico-mathematical and Historico-philological departments for 1907". SPb., 1907, pp. 168–172.

[10] Bierman, K.-R. "Aus der Vorgeschichte der Euler-Ausgabe 1783–1907". In: *Leonhard Euler. 1707–1783.* Basel, 1983. pp. 489–500.

[11] *Report on the activity of the Imp. Academy of Sciences...for 1907.* pp. 170–171.

[12] LO Arkhiva AN SSSR, f. 707, op. 2, No. 29, l. 1.

[13] The International Union of Academies (IUA) (or the International Association of Academies) was founded in 1900. In 1901, under the presidency of G. Darboux, the first General Assembly of this organization met in Paris. Every three years the board of directors of the organization was chosen from a different country. The first delegates to the IUA from the Petersburg Academy of Sciences were C. G. H. Salemann and A. S. Famintsyn. From 1912 to 1914 the Union was administered from St. Petersburg, represented by O. A. Backlund. The directorate of the IUA was scheduled to be transferred to Germany in 1914, but in view of the outbreak of the First World War it was proposed that it should be moved instead to Amsterdam. However this transfer did not come to pass, and the Union ceased to exist. A new International Union of Academies was organized following the cessation of hostilities.

[14] LO Arkhiva AN SSSR, f. 707, op. 2, No. 29, l. 1.

[15] LO Arkhiva AN SSSR, f. 707, op. 3, No. 308, l. 1-1 ob.

---

[10]Leningrad Section of the Archive of the Academy of Sciences of the USSR; see also later references.—*Trans.*

[16] The Swiss (or Helvetic) Society of Natural Scientists (Schweizerische naturforschende Gesellschaft, Helvetische naturforschende Gesellschaft) was founded in Zürich in 1815. The Society meets annually in various towns in Switzerland. In September 1909 the annual meeting took place in Lausanne and it was there that the decision was made to publish the complete works of Euler.

[17] The first draft of the editorial plan for publishing Euler's works was sent in both French and German versions (see LO Arkhiva AN SSSR, f. 707, op. 2, No. 29, l. 14-29). In addition there were sent draft plans, drawn up by P. Stäckel, of the distribution by volume of the material to be published. The first of these was published in: *Vierteljahrsschr. Naturforsch. Ges.* Zürich. 1909. Vol. 54. pp. 1–28 ("Entwurf einer Einteilung der Sämtlichen Werke L. Euler, von P. Stäckel"); the second draft, vetted by the editorial committee, was published in: *Jahresber. Dtsch. Math. Ver.* 1910. Issues 5–6. pp. 104–116 (see: LO Arkhiva AN SSSR, f. 707, op. 2, No. 29, l. 192–207, 208–215). Other materials were also sent, including a list of the editors of the individual volumes (see: LO Arkhiva AN SSSR, f. 707, op. 2, No. 29, l. 12–13).

[18] LO Arkhiva AN SSSR, f. 707, op. 3, No. 308, l. 2–3.

[19] Eneström, G. "Verzeichnis der Schriften Leonhard Eulers". Jahresber. Dtsch. Math. Ver. 1910–1913. Erganzungsb. 4, Lief. 1–2. A copy of the proof pages of Eneström's list is preserved in: LO Arkhiva AN SSSR, f. 707, op. 2, No. 29, l. 80–191 ob. For the list's titles, see also loc. cit., l. 219–220.

[20] LO Arkhiva AN SSSR, f. 707, op. 2, No. 29, l. 5. This is a typewritten copy with some corrections (see loc. cit., l. 6). Appended to the minutes of the meeting of the committee is an announcement by B. L. Modzalevskiĭ (l. 10–10 ob.) to the effect that on instructions from the committee he has investigated the question as to whether in 1765 a translation from Latin of Euler's *Geometry* was published. Although Modzalevskiĭ reviewed a large quantity of material, including documents from the secretariat, the gymnasium, the bookstalls, and the printery, nowhere did he find any mention of a publication of Euler's *Geometry*. Between 1764 and 1765 there was published in the Academic printery the work *Practical geometry*, by S. K. Kotel′nikov, and in the printery of the Naval Corps N. Kurganov's *General geometry*. It would seem that by mistake the author attributed one of these books to Euler. It must surely be the case that N. Ya. Sonin was influenced by an assertion by V. V. Bobynin in the *Russian physico-mathematical bibliography* (M., 1889. Vol.2, Issue 1, No. 22, p. 13). For information concerning Euler's geometrical manuscripts, see the article: Belyĭ, Yu. A. "Concerning L. Euler's textbook on elementary geometry." Ist.-mat. issled. (Research in the History of Math.) 1961, Issue 14, pp. 123–284.

[21] The well-known Leipzig book-publisher B. G. Teubner.

[22] Euler, L. *Theoria motuum planetarum et cometarum....* Berolini, 1744 (E66). *Opera* II-28.

[23] LO Arkhiva AN SSSR, f. 707, op. 3, No. 308, l. 4–5 ob.

[24] LO Arkhiva AN SSSR, f. 707, op. 2, No. 29, l. 7–8.

[25] Hagen, J. G. *Index operum Leonhardi Euleri.* Berolini, 1896.

[26] Apparently, the printed *Minutes of the meetings of the Conference of the Imp. Academy of Sciences.* SPb., 1897–1911. Vols. 1–4.

[27] LO Arkhiva AN SSSR, f. 707, op. 3, No. 308, l. 6–7 ob.

[28] Inventory of the manuscripts of Leonhard Euler preserved in the Archive of the Conference of the Imp. Academy of Sciences. Compiled by B. Modzalevskiĭ. SPb., 1910. (Signed "For permanent secretary A. Karpinskiĭ".)

[29] LO Arkhiva AN SSSR, f. 707, op. 3, No. 308, l. 8–9 ob.

[30] LO Arkhiva AN SSSR, f. 707, op. 3, No. 308, l. 10–11.

[31] List of Euler's published works, compiled (and, apparently, bound) by P. N. Fuss for the proposed publication of Euler's collected works.

[32] LO Arkhiva AN SSSR, f. 707, op. 3, No. 308, l. 12–13. From the notation "Ms.", it appears that besides printed works the bound volumes contained certain works in manuscript form.

[33] LO Arkhiva AN SSSR, f. 707, op. 3, No. 308, l. 14–15 ob. This is S. F. Ol′denburg, permanent secretary of the Petersburg Academy of Sciences from 1904 through 1929.

[34] For the editorial plan, see: LO Arkhiva AN SSSR, f. 707, op. 2, No. 29, l. 14–21 ob. (German), l. 22–29 (French).

[35] For the list of editors of Euler's collected works, see: LO Arkhiva AN SSSR, f. 707, op. 2, No. 29, l. 12–13.

[36] For the sample page from Euler's *Mechanics*, see: LO Arkhiva AN SSSR, f. 707, op. 2, No. 29, l. 33–34 ob. For the cover and table of contents: l. 221–224.

[37] For the "Circular", signed by F. Rudio, and sent by the editorial committee to the editors of the individual volumes, see: LO Arkhiva AN SSSR, f. 707, op. 2, No. 29, l. 11–11 ob.

[38] LO Arkhiva AN SSSR, f. 759, op. 1-a, No. 14, l. 5–5 ob. (Signed "F. Rudio", but not in his hand.)

[39] LO Arkhiva AN SSSR, f. 707, op. 3, No. 308, l. 17–18. Apparently the distribution of materials by volume was altered once more, so that the volume assigned to Backlund as the 15th of the second series, ended up as Volume 28 in that series. It would seem that this was related to the circumstance that the volumes on mechanics were ready earlier that those on astronomy, and turned out to be significantly more numerous than expected.

[40] LO Arkhiva AN SSSR, f. 257, op. 2, No. 6. l. 3 ob.–4.

[41] The International Congress of Mathematicians at Cambridge took place August 22–28, 1912. The participants V. A. Steklov, A. M. Lyapunov, B. B. Golitsyn, and F. Rudio would have been able to discuss there all possible questions relating to the publication of Euler's works.

[42] LO Arkhiva AN SSSR, f. 257, op. 1, No. 52, l. 1–1 ob.

[43] Euler, L. *Opera omnia*. Sér. I. Vol. 10. Ed. G. Kowalewski. Leipzig, etc., 1913. Here Euler's *Differential calculus* appears.

[44] Circular from the Leonhard Euler Society: "Einladung zum Beitritt zu einer L. Euler-Gesellschaft". Signed by F. Rudio et al. See: LO Arkhiva AN SSSR, f. 257, op. 2, No. 6, l. 5–6.

[45] LO Arkhiva AN SSSR, f. 707, op. 3, No. 308, l. 20–21.

[46] LO Arkhiva AN SSSR, f. 707, op. 3, No. 308, l. 22–22 ob.

[47] Volume 17 of the first series of the *Opera omnia* contains the following papers of Euler, numbered in accordance with Eneström's list: E59, E60, E162, E168, E254, E321, E391, E421, E462–E464.

[48] Volume 18 of the first series of the *Opera omnia* contains the following papers of Euler: E475, E490, E500, E521, E539, E572, E587–E589, E594, E606, E620, E621, E629, E630, E635, E640, E651, E653.

[49] Volume 19 of the first series of the *Opera omnia* contains the following papers of Euler: E656, E657, E662, E668–E675, E688–E690, E694, E695, E701, E707, E721, E752, E807, E816, E819.

[50] LO Arkhiva AN SSSR, f. 257, op. 1, No. 52, l. 2–3 ob.

[51] LO Arkhiva AN SSSR, f. 173, op. 1, No. 20, l. 7 ob.

[52] F. Rudio's foreword to Vol. 18 of the first series of the *Opera omnia*. 1920. P. VII.

[53] F. Rudio's foreword to Vol. 19 of the first series of the *Opera omnia*. 1932. P. VII–VIII.

[54] Euler, L. *Opera omnia*. Sér. I. Vol. 4.

[55] Euler, L. *Opera omnia*. Sér. I. Vol. 5.

[56] LO Arkhiva AN SSSR, f. 162, op. 2, No. 214, l. 18–19.

[57] LO Arkhiva AN SSSR, f. 759, op. 3, No. 244, l. 9 (Copy in the hand of A. N. Krylov). Signed Fritz Sarazin.

[58] A. M. Lyapunov died on November 3, 1918.

[59] LO Arkhiva AN SSSR, f. 759, op. 3, No. 244, l. 7–7 ob.

[60] The Department of Science and Technology of the ACPE (All-union Council of the People's Economy) was created on August 16, 1918 by a decree of the Council of the People's Commissars. In August 1920 N. M. Fedorovskiĭ was sent to Berlin as director of the Bureau of Foreign Science and Technology of the Department of Science and Technology (BFST DST). The mandate of this bureau was the establishment of connections with foreign scientists, the gathering of information concerning the latest discoveries and achievements of world science and technology, the purchase and publication of foreign technical literature, and the publication of scientific literature. Evgeniĭ Germanovich Lundberg, a journalist who published material for the ACPE, was a co-worker at the BFST, as was A. A. Tretler.

[61] LO Arkhiva AN SSSR, f. 162, op. 2, No. 214, l. 22–24.

[62] Yakov Viktorovich Uspenskiĭ.

[63] LO Arkhiva AN SSSR, f. 162, op. 2, No. 214, l. 29.

[64] LO Arkhiva AN SSSR, f. 162, op. 2, No. 214, l. 40–40 ob.

[65] Academician A. N. Krylov. Publ. Yu. KH. Kopelevich. *Priroda* (Nature). 1974. No. 1. The original papers of A. N. Krylov are preserved in the LO Arkhiva AN SSSR, f. 759, op. 1-a, No. 14, l. 1–4. Reproduced with slight abridgments.

[66] At the present time (1987) there are scholars from Switzerland, the USSR, the GDR, France, as well as other countries, participating in the publication project.[11] The following are members of the editorial committee of the fourth series: From Switzerland, E. A. Fellmann, W. Habicht, Ch. Blanc, F. Fricker; from the USSR, A. T. Grigor′yan, A. P. Yushkevich, G. P. Matvievskaya, G. K. Mikhaĭlov. Three volumes of Series IV have appeared: Vols. 1, 5, and 6. See: Habicht, W. "Series I–III of the Euler publication of the Swiss Natural Science Society". In: Questions in the Hist. of Nat. Science and Technology. 1979. Vol. 51. pp. 78–86; Yushkevich, A. P., Grigor′yan, A. T. "A new series of Euler's works, and the Euler symposium in Basel". In: Questions in the Hist. of Nat. Science and Technology. 1973. Vol. 44. pp. 98–99; Burckhardt, J. J. "Die Euler-Kommission der Schweizerischen naturforschenden Gesellschaft: Ein Beitrag zur Editionsgeschichte". In: *Leonhard Euler. 1707–1783*. Basel, 1983. pp. 501–509.

---

[11] More recent information on the status of the *Opera omnia* may be found by visiting the website "The Euler Archive", which also gives access to over 800 of the originals of Euler's works.–*Trans.*

Leonhard Euler and the Berlin Academy of Sciences

K. Grau

Leonhard Euler spent 25 years of his 76-year-long life in Berlin. He arrived in Berlin in 1741 at the age of 34, and left the Prussian capital in 1766, in his 60th year. Around a third of his publications—or almost a half if one counts only those published in his lifetime— relate, as A. P. Yushkevich has shown, to this quarter-century. Euler's correspondence, if that portion of it that has been preserved is any indication, was especially intensive in his Berlin period. However, although over the period 1741–1766 Euler lived and worked in Berlin and its Academy of Sciences, the field of his activity was, then as later, the wide republic of scholars.

Let us glance at what was going on in the scientific world and in the lives of some of the scientists in Euler's sphere over that quarter-century.

A few months after Euler's departure from Russia, there returned to the Petersburg Academy from his time in Berlin as *stagiaire*[1] the outstanding Russian scholar M. V. Lomonosov, who was subsequently to nurture the Academy so solicitously in many scientific disciplines right up to his death. In 1743 A. L. Lavoisier, the inaugurator of the new quantitative chemistry, was born, and in 1749 the brilliant mathematician and astronomer P.-S. Laplace. In 1751 the Göttingen Society of Scientists was founded, in 1754 Columbia University in New York, and in 1755 Moscow University—now named after M. V. Lomonosov—began its activities.

In 1743 the Great Northern Expedition organized by the Petersburg Academy of Sciences took place, and in 1766 L.-A. de Bougainville set off on his voyage around the world, which so enriched geography by its many discoveries.

Certain important publications should also be noted. In 1749 G. Achenwall's *Abriss der neuesten Staatswissenschaft der vornehmsten europäischen Reiche und Republiken*[2] appeared, the first textbook on statistics. In the same year the publication of G. de Buffon's multi-volume *Histoire naturelle, générale et particulière* began. In 1751 Carolus Linnaeus'

---

[1]Or apprentice scientist.—*Trans.*

[2]*An outline of the latest political science of the leading European kingdoms and republics.*

Philosophia botanica was published, and in 1755 Immanuel Kant's *Allgemeine Natur-gesichte und Theorie des Himmels*[3]. Over the period 1757–1766 A. von Haller summed up in the eight volumes of his *Elementa physiologiae corporis humanis* the research in human physiology up to that time. In 1764 J. J. Winkelmann published his *Geschichte der Kunst des Altertums*[4] precisely at the time when the Berlin Academy desisted from its many attempts to attract him.

Here are a few more examples illustrating the progress of science over the Berlin period of Euler's life—although of course other choices are possible. In 1742 the 100-degree scale used in today's thermometers is introduced by A. Celsius. In 1745 E. G. von Kleist and P. van Musschenbroeck invent almost simultaneously the first electric condenser—the Leiden jar. In 1747 J. Bradley discovers the nutation[5] of the earth's axis, and A. S. Marggraf produces beet sugar. In 1748, in a letter to Euler, Lomonosov formulates the law of conservation of mass. In 1750 J. T. Mayer establishes the absence of a lunar atmosphere, and J. A. Segner builds a reactive water wheel. In 1758 J. Dollond constructs an achromatic telescope—the practicability of which Euler had indicated ten years earlier. In 1758 F. U. T. Epinus discovers electrical induction. In 1759 C. F. Wolff in his *Theoria generationis* describes his discovery of epigenesis. In 1760 Joseph Black distinguishes between temperature and quantity of heat, and J. H. Lambert lays the foundations of photometry. And to end, in 1766 Henry Cavendish demonstrates that carbon dioxide and hydrogen are different gases.

As evidenced by his published and annotated correspondence, Euler was acquainted with almost all of the above-mentioned scientists; some he had met in person, while with others he only corresponded. Unfortunately I cannot dwell here on the details of these scientific contacts.

And what research of import did Euler himself publish over that quarter-century?

Without entering into detail concerning his physico-mathematical works—in any case far from my area of expertise—I merely note that over that time period he published—for the most part in Berlin and St. Petersburg—around 250 articles and books, among the latter the following classic monographs, written in Latin: *A method for finding curves possessing maximum or minimum properties*[6] (1744); the two-volume *Introduction to the analysis of infinities*[7] (1748); the two-volume *Naval science, or a treatise on the construction and navigation of ships*[8] (1749); *The theory of the motion of the moon, exhibiting all of its irregularities*[9] (1753); *Differential calculus*[10] (1755), and *The theory of motion of rigid bodies*[11] (1765).

Euler's Berlin years were not such peaceful ones as might permit a scientist to remain indifferent to events occurring in the ambient society while devoting himself wholly to science. King Friedrich II, together with whom—as Voltaire put it—science and art ascended

---

[3] *General natural history and celestial theory.*

[4] *History of ancient culture.*

[5] Not the same as precession.—*Trans.*

[6] *Methodus inveniendi lineas curvas maximi minimive proprietate gaudentes, sive solutio problematis isoperimetrici latissimo sensu accepti.*

[7] *Introductio in analysin infinitorum.*

[8] *Scientia navalis seu tractatus de construendis ac dirigendis navibus.*

[9] *Theoria motus lunae exhibens omnes eius inaequalitates.*

[10] *Institutiones calculi differentialis cum eius usu in analysi finitorum ac doctrina serierum.*

[11] *Theoria motus corporum solidorum seu rigidorum.*

The old building of the Berlin Academy of Sciences on the avenue *Unter den Linden*
From an early engraving. Central Archive of the Academy of Sciences of the GDR

the throne, waged long wars with consequences ultimately felt in his capital. In 1760 Russian troops were advancing on Berlin. France, England, and Spain were fighting over their colonies in North America, and England was consolidating its supremacy in India. In 1751 *Die Vossische Zeitung* began appearing in Berlin, a newspaper acquainting its readership with current world events. Euler was made all the more attentive to the changing fortunes of the seven years' war by its direct impingement on his life and that of his family—as were others living in Prussia, and, in particular, Berlin. His concern with contemporary political villainy is evident, for example, in his letters to the Swiss K. Bettstein, who was living in England at that time. In 1758 Euler wrote the following fine words to him: "May God have mercy on us and send us an honorable and lasting peace". And although we here in Moscow, as well as our friends in Berlin[12] no longer expect an "honorable and lasting peace" from God, we nonetheless long for the triumph and preservation of peace no less fervently than did Euler in his time.

What was the Berlin of Euler like, 200 years ago?

The longest period of fixed residence of Euler and his family in Berlin was on Berenstrasse; there is a memorial plaque marking his residency, affixed to the building now occupying the site of his original abode. It is not far from this address to the avenue *Unter den Linden*, which in those days ran from the Royal Palace directly westwards to the *Tiergarten*. The Academy of Sciences was situated on *Unter den Linden*; to this day there are to be found there certain of the Academy's offices—not, however, in the building Euler knew. Euler came to the original observatory, built at the beginning of the 18th century, where the members of the Academy assembled until 1744. Subsequently, meetings were held in the Royal Palace, until in 1752 the Academy obtained new premises in a building

---

[12]The present essay dates from before 1988. The Berlin Wall fell in 1989.—*Trans.*

The Prussian king Friedrich II

on *Unter den Linden* not far from the observatory. This new edifice, in which the Academy remained until 1903, had been built on the site of the old one, destroyed in a fire on August 21, 1742. In this connection it may be appropriate to recall that in a letter written to G. F. Müller in July 1763, Euler gives his wife's fear of fires as one of their reasons for leaving St. Petersburg in 1741; and then very soon after arriving in Berlin they experienced just what they had quit Petersburg to avoid! However, at least Euler was there to witness the construction of the new Academic building, as well as that of the neighboring palace of prince Heinrich, brother to Friedrich II, now part of Humboldt University, named after the Humboldt brothers. On the square opposite, between 1741 and 1743 G. von Knobelsdorf built an opera house, now the German State Opera. In 1747, behind the opera house, the construction of the Catholic Church of St. Hedwig began, completed in 1773 when Euler had already returned to St. Petersburg. In walking from the observatory to the palace, one passed the arsenal, dating from before 1706, on which the architect A. Schlüter (who died in St. Petersburg in 1714) had worked. Today that building houses the German Historical Museum. The palace was destroyed during the Second World War, thus becoming a sacrifice to the imperialistic greed for power against which today more and more peace-loving people are voicing their protest. In its place the worker-peasant government of the GDR erected the Palace of the Republic.

Was Euler interested in the architectural achievements represented by the buildings amidst which he lived and worked? However that may be, several preserved or restored architectural details in the center of modern Berlin remind one of the time when Euler worked there, when Berlin was the capital of a state whose ruler, in 1741, summoned a Swiss *Bürger* from tsarist Russia to the Berlin Academy of Sciences.

The serfs, the soldiery oppressed by harsh service under the "soldier king" Friedrich Wilhelm I, the artisans, apprentices, and factory workers slaving under poor conditions—none of these held out great hopes for improvement in their situation with the accession of the new king Friedrich II in 1740. The nobility had nothing to fear from the enlightened despotism of the new king; their position as the ruling class remained unassailable once attempts to restore the former class rights had failed. Only the intelligentsia, busy absorbing the ideas of the enlightenment, felt able to breathe more freely, hoping that Berlin would be transformed from a Sparta into an Athens: it seemed that philosophy herself had ascended the throne with Friedrich.

Was the 33-year-old Bürger Euler in 1740 free of illusions with regard to the 28-year-old king Friedrich? It is clear from his letter to his friend G. F. Müller that he accepted the invitation to Berlin willingly. Although prior to leaving St. Petersburg he knew that "at present the king is busy subjugating Silesia", that he was conducting a war, and that he had commanded the muses to be silent, Euler set his hopes on the imminent reorganization of the Academy promised by Friedrich II. Did Euler have any idea then of what awaited him in Berlin, of how difficult a position he would eventually find himself in, regardless of his great scientific successes? Just before quitting St. Petersburg, Euler wrote to Müller: "His highness has appointed Mr. Wolff in Marburg privy counsellor and vice-chancellor of the university at Halle....However, apart from me, no foreigner has been invited to the Academy in Berlin, not counting Mr. Maupertuis of Paris, who is leading a French expedition to Lapland in order to establish the actual shape of the Earth". Friedrich II, Christian Wolff, and P. L. Moreau de Maupertuis exercised considerable influence—each in his own way—on Euler's situation in Berlin—more than any others of his contemporaries. And in 1766 Euler left the Prussian capital far less joyfully than when he arrived 25 years earlier at a king's summons. From this point of view it is historically justified in some sense that on the monument to Friedrich II standing in *Unter den Linden* neither Euler's likeness nor his name is reproduced, whereas Wolff and Maupertuis are immortalized there.

In 1741 Euler was invited to join the old Science Society, founded by G. W. Leibniz in 1700. On May 25, 1741, not long before Euler arrived in Berlin, the president of the Society, D. E. Jablonski, died; he, along with Leibniz and others, had been a founding member, and its president since 1733.

The Berlin Science Society, or, if you like, Academy, was made up of four schools: physics, mathematics, the school of philology and Germanistics (or history), and the school of philology and oriental studies. The mathematics school, which Euler joined, was directed by A. de Vignoles, whose chief interest was chronology. Euler was appointed director of the observatory immediately upon his arrival in Berlin and occupied this post until his departure in 1766.

The quarter-century during which Euler worked in Berlin can be divided into three periods in the development of the Academy: first, the period of reorganization, lasting from 1741 till 1746; second, that of Maupertuis' presidency, from 1746 to 1759; and third, the period 1759–1766 of especially intensive struggle for Euler over his position in Berlin.

The situation in which Euler found himself on his arrival in Berlin in 1741 was very complex. The king's promised reorganization of the Academy had to wait. Friedrich II was busy—as Euler put it in the above-quoted letter to Müller—with "the subjection of Silesia". Not only did the aggressive wars begun by the Prussian monarch delay by more than five

years the organization of the new Academy, they also cast a shadow over the whole of the
Berlin period of Euler's life. This circumstance did not, however, prevent Euler from setting
to work. In particular, concern for the progress of astronomy—and the circumstance that
astronomical data were used in compiling encyclopedic almanacs the profits from whose
sale constituted the most significant portion of the Academy's revenues—prompted Euler to
invite Johann Kies to fill the position of astronomer and member of the Academy. In 1743,
the seventh issue of the *Miscellanea Berolinesia* (the first journal of the Berlin Science
Society) contained in its 242 pages five articles by Euler, thus affirming his creative fertility
from the very beginning.

In 1743 a new "Société Littéraire" was founded in Berlin, in which Euler participated
along with other members of the Academy. The minutes of 21 of the meetings of this society from August 1743 through January 1744 have been preserved. Over this half-year Euler
gave five lectures[13]. In the Autumn of 1743 Friedrich II combined the literary and scholarly
societies into a single new Academy of Sciences, officially inaugurated in January 1744.
Euler now accepted the directorship of the mathematics school. The extent of his influence
on the organization of science in Berlin is indicated by his refusal of this directorship until
such time as the two scholarly societies should be united.

Negotiations with Maupertuis begun by Friedrich II in 1740, were concluded only in
1746; on February 1 of that year, Maupertuis was appointed permanent president of the
Academy. At that time the first issue of the new Academy's official publication appeared,
again containing many articles by Euler. On May 10, 1746 the Academy was furnished
with a new charter which, in keeping with his directorial role, Maupertuis had prepared.
Thus began the dozen-year interval during which Euler in concert with the president—and
often as his deputy—displayed his all-encompassing fecundity at the Berlin Academy.

As noted earlier, I shall not discuss Euler's creative achievements in the physico-
mathematical arena. I wish rather to consider certain philosophical and organizational
aspects of his scientific activity during his Berlin period, which I regard as of equal
importance with his narrowly physico-mathematical work; in my opinion it was in fact
these that played a decisive part in the ending of Euler's appointment in Berlin in 1766.

Twenty years ago the historian of science G. Kröber in his introduction to a selection
of philosophical excerpts from Euler's *Letters to a German princess on various topics from
physics and philosophy*[14], published over the period 1768–1772, came to the following interesting conclusion: "It is very often lost from view that Euler, through his activity in the
Academy, his opposition to Leibniz' monadology and Wolff's philosophy, his defense of
the mathematical principles of Descartes' physics, his attitude to the new materialism emerging from Newton's mechanics, and, finally, his influence on the youthful Kant, determined
quite decisively the direction in which philosophical thought in Germany developed in the
mid-18th century". In his article on Newton in the *Philosophen-Lexikon*[15], H.-J. Treder
writes about the struggle over the natural-scientific picture of the world during the Enlightenment: "On the basis of the glorious successes of Newton's mechanics in physics and
astronomy, followed by the acknowledgment, beginning with Voltaire, of Newton's natural

---

[13]Or research announcements.—*Trans.*

[14]*Briefe an eine deutsche Prinzessin über verschiedene Gegenstände aus der Physik und Philosophie.
Philosophische Auswahl.* Leipzig, 1964.

[15]Berlin, 1982.

Pierre Louis de Moreau de Maupertuis, president of the Berlin Academy 1746–1759

philosophy as providing the foundation for Enlightenment philosophy, Newton's mechanics was systematically identified in its content with natural science in general. Voltaire, Maupertuis, Euler, d'Alembert and others promoted Newtonianism as the prevailing world-picture counter to Cartesianism and the Leibniz–Wolff monadology". Except for Newton and Descartes the above-named were all members of the Berlin Academy.

Already in St. Petersburg Euler had become acquainted with Wolff's philosophy—a systematic elaboration of Leibniz' teachings—, and of course with Newtonian physics. G. B. Bilfinger, an adherent of Wolff's and since 1725 head of experimental and theoretical physics at the Petersburg Academy, had from 1726 onwards given public lectures there based on the *Foundations of Newtonian philosophy* (1723) of W. J. 'sGravesande, at a time when even in Cambridge the Cartesian system was in the ascendant. Since the content of Bilfinger's lectures is unknown, the question remains open as to whether his attitude to Newton was positive or negative, but whatever the case may be, this marked the beginning of the process whereby Newton was admitted to membership of the Petersburg Academy—just prior to Euler's arrival. At the official meeting of the Academy held on August 1, 1726 the mathematician J. Hermann had spoken on the history of geometry and on important mathematical discoveries. In this speech, which was published in 1728, he had talked of Newton's *Philosophiae naturalis principia mathematica* and the systematization of the differential calculus by Newton and Leibniz, and in this connection had expressed his objective opinion on the question of priority, allowing each his merits.

Euler was there to observe the intense dispute which broke out in the Petersburg Academy in 1729 between the proponents and opponents of Wolff, and, as E. Winter has correctly noted, it was at that time that he became a convinced opponent of Wolff, remaining so for the rest of his life.

In 1747 Euler published in German in Berlin a book entitled *Delivery of divine revelation from the objections of freethinkers*[16], written in opposition to deism. The appearance of

---

[16] *Rettung der göttlichen Offenbahrung gegen die Einwurf der Freygeister.*

this work coincided with the conclusion of a discussion in the Berlin Academy concerning the setting of the following as a competition problem: Perform an analysis of Leibnizian monadology. Euler, disregarding his duty as a member of the Academy to show restraint in such cases, became involved in a dispute—albeit anonymously, though the author's identity was known to all. He argued against Leibniz, and hence against Wolff, who as a result ascribed to Euler—erroneously—the desire "to domineer over all the sciences..., to the great detriment of his own renown—since only a few have any idea of his scientific merit—, and to the shame of the Berlin Academy of Sciences...". It was partly owing to Euler's support that at that time the opponents of the Leibniz–Wolff philosophy won recognition for their position from the Berlin Academy.

The 25-year-old Kant, who followed these Berlin conflicts closely, considered Euler's position from a viewpoint completely different from Wolff's. In 1749 he sent Euler his *Thoughts on a true evaluation of the life force*[17], which had appeared in 1746. In the accompanying letter Kant gives the following explanation: "That very boldness which prompted me to enquire into the true measure of the force of nature and the exertions of the defenders of Messrs. Leibniz and Descartes, was what allowed me to decide to send this essay to a person of such wisdom that he might elicit from this unworthy work the beginnings of an attempt at a final and complete settlement of the disagreement between such great scholars". Euler's reaction to this letter is unknown.

Somewhat later, between 1750 and 1753, Euler took part in another, even more important, exchange of polemics. This time the dispute concerned the principle of least action, which, as H.-J. Treder (1925) has shown, remained of great significance for the history of the Berlin Academy from Leibniz to Einstein. Maupertuis ascribed to himself the discovery and elaboration of the principle. However this was disputed by the Swiss mathematician S. König, who had been made a foreign member of the Berlin Academy in 1749. In 1751 König declared—as we now know, correctly—that Leibniz had formulated the principle earlier in one of his letters. Euler, though not completely satisfied with Maupertuis' exposition of the principle, took his side, at the same time producing the first mathematically correct formulation of it. On Euler's instructions, in 1752 the Berlin Academy of Sciences pronounced the above-mentioned letter of Leibniz (König being unable to produce the original) a forgery. König responded by turning in his diploma of membership to the Academy. Only in 1913 was another copy found of the letter of Leibniz whose authenticity had been put in question in 1752.

In connection with this quarrel, Voltaire published in 1752 a satirical essay directed against Maupertuis, entitled *Diatribe du Docteur Akakia*. G. H. Heinsius wrote from Leipzig to Schumacher in St. Petersburg that the *Diatribe* "deals very severely also with the esteemed Professor Euler". Euler himself considered the dispute to be based on personal enmity, in spite of his familiarity with its details. Concerning the *Diatribe*, he wrote to Schumacher that the latter would learn from the newspapers of the public burning of Voltaire's pamphlet by the executioner, and reminded him that Voltaire had attacked Maupertuis several times before. And although all of this bore very significantly on the scientific careers of Euler and Maupertuis, it seems that Euler was so sure of the correctness of his behavior that he even cooperated in the distribution of the pamphlet. He wrote to Heinsius in Leipzig

---

[17] *Gedanken von der wahren Schätzung der lebendigen Kräfte.*

that it would give Schumacher especial pleasure if he sent him some copies of the *Diatribe*, the sale of which in Berlin was completely prohibited, while as far as he knew it could be obtained in Leipzig.

While Maupertuis emerged from this period of strife in the early 1750's a completely broken man, almost fully retired from academic affairs, Euler's scientific reputation— certain hostile attacks notwithstanding—hardly suffered. Euler's recent biographer R. Thiele[18] suggests two basic reasons for Euler's behavior: "A duty to defend his president and the Academy" and "profound repugnance for the Leibniz–Wolff philosophy and the radical French Enlightenment", with the latter playing the decisive role. The fact that in the 1760s the Berlin Academy tended to develop more and more in both of these ideological directions played a decisive part in Euler's long anticipated return to St. Petersburg. This was reinforced by another circumstance which undermined Euler's position in the Berlin Academy.

Euler's extraordinary, multifaceted, and successful work of scientific organization in Berlin is documented in detail in the three volumes of *The Berlin and Petersburg Academies of Science in Euler's correspondence*[19], published by A. P. Yushkevich and E. Winter and their collaborators, and also in E. Winter's *Minutes of the Berlin Academy of Science. 1746–1766*[20], which I have consulted for my lecture. To economize on space and time I shall not touch on the material published and interpreted there. It seems to me that in the interests of arriving at a more complete idea of Euler's activities at the Berlin Academy it is essential to concentrate instead on one aspect of those activities hitherto insufficiently attended to, namely his participation in the administration of the Academy's finances.

It is well known that from its foundation the expenses of the Academy were defrayed chiefly by revenue from a monopoly on the sale of almanacs. A large portion of the work in producing the almanacs was carried out by the observatory, which was managed by Euler. Starting in 1744, the production of almanacs was organized as a single administrative unit, with Euler in charge. His most important colleague in this business was the almanac administrator D. Köhler. Euler managed to significantly increase the revenues from the sale of almanacs. Later he would propose that the Petersburg Academy begin publishing almanacs on the Berlin Academy's model.

As Maupertuis declared, Euler demonstrated over many years by "his honesty, ability, and zeal" that he was a most fitting manager of the business side of the Academy. When in 1753, after the scandal with König, an already incurably ill Maupertuis left Berlin, he sent a letter to Köhler confirming the king's approval of the transfer of the directorship of the Academy to Euler during his absence.

In connection with the economic reconstruction of Prussia after the seven years' war, a new budgetary system was introduced in the Academy. In practice this meant that the business of the production and sale of almanacs, of which Euler was the manager, was to be reorganized. In February 1765 Friedrich II appointed a commission to investigate the Academy's finances. Although Euler was a member of this commission, all the same as the only person having responsibility, he quite naturally understood this decision to be an

---

[18] Rüdiger Thiele. *Leonhard Euler*. Leipzig, 1982.

[19] *Die Berliner und die Petersburger Akademie der Wissenschaften im Briefwechsel Leonhard Eulers*. Berlin, 1959–1976.

[20] *Die Registres der Berliner Akademie der Wissenschaften. 1746–1766*. Berlin, 1957.

expression of mistrust in him and his colleague Köhler. (Here one should also recall that after Maupertuis' death in 1759, Euler had sufficient grounds for expecting that he would be promoted from his position as acting director of the Academy to the actual presidency.) Given the circumstance that he had been in fact already thinking about leaving Berlin, the appointment of the review commission represented another sharp jolt in that direction. Four months after the commission began its work, but without waiting for results, on June 16, 1765 the king commanded that the administration of the almanacs be reorganized as an office on lease from the government.

Two of Euler's proposals, contained in a single document dated October 3, 1765—that there be two Academic archives (for literature and for economics), and that the posts of Academic treasurer and archivist be combined so as to strengthen the influence of the latter—were not accepted. Instead a single archive with two departments was instituted. In November 1765 Euler resigned from the commission on Academic finances. The friction generated in the course of overseeing the Academy's finances, together with conflicts over ideology, strongly influenced his decision to quit the Academy. This has been convincingly shown in K. R. Birman's lecture "Was Leonhard Euler hounded from Berlin by J. H. Lambert?", delivered at the conference in Euler's honor held in Berlin on September 16, 1983[21].

This conference, which was organized by a special Euler committee of the Academy of Sciences of the GDR on the 200th anniversary of the great scholar's death, is the penultimate among those that I consider in connection with my theme "Leonhard Euler and the Berlin Academy of Sciences". Our own Academy has continually acknowledged its duty to honor the memory of its pre-eminent member Leonhard Euler—albeit with varying fervor.

After Euler's departure from Berlin, certain of his memoirs continued to be published in the Academy's journal. Euler's further communications with the Berlin Academy were influenced in particular by the fact that his son Johann-Albrecht, who was now conference–secretary of the Petersburg Academy, had become connected by marriage to Samuel Formey, the permanent secretary of the Berlin Academy: Johann-Albrecht's mother-in-law and Formey's wife were sisters. The extensive correspondence—little studied—between Johann-Albrecht and his uncle Samuel has been preserved. In the first half of the 19th century the German mathematician and academician C. G. J. Jacobi, through his examination of the minutes of the meetings of the Berlin Academy, was able to shed much light on Euler's activities in Berlin. The correspondence between the permanent secretary of the Petersburg Academy P. N. Fuss and Jacobi was published in Leipzig in 1908 in a volume entitled *On the publication of Euler's works*[22]. A project to publish Euler's collected works undertaken by the Berlin and Petersburg Academies in 1907 in connection with the 200th anniversary of his birth, lapsed because of financial difficulties. However the very same jubilee stimulated the Swiss Society of Natural Scientists in 1911 to launch the enterprise of publishing a multi-volume *Complete collected works of Leonhard Euler* (*Leonhardi Euleri opera omnia*), the publication of the first three series of which is now[23] nearing completion, while

---

[21] Reproduced in the present volume.

[22] *Der Briefwechsel zwischen C. G. J. Jacobi und P. N. von Fuss über die Herausgabe der Werke Leonhard Eulers. Hrsg., erläutert und durch einen Abdruck der Fusschen Liste Eulerschen Werke ergänzt von Paul Stäckel und Wilhelm Ahrens.* Leipzig. B. G. Teubner, 1908.

[23] 1988

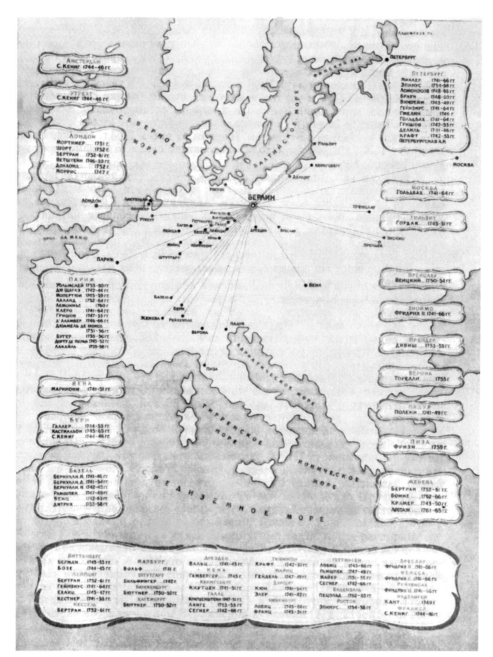

Schema of Euler's correspondence from Berlin

work on a fourth series, to contain Euler's correspondence and unpublished manuscripts, was begun a few years ago. As is well known, the volumes of the fourth series are being prepared for publication by the Swiss Euler Committee and the Institute for the History of Natural Science and Technology of the Academy of Sciences of the USSR, with the collaboration of scholars from the GDR and France[24].

---

[24] See also the essay by E. P. Ozhigova in the present volume. More up-to-date information on the status of the

When in 1925 the Academy of Sciences of the USSR celebrated the 200th anniversary of its founding, the Berlin Academy wrote in its congratulatory message: "Of almost the same age as your Academy, the Berlin Academy has in both good times and bad cooperated with the Petersburg Academy with unflaggingly amicable rivalry to produce many fruitful ideas which were then developed jointly, at the hands of such personages as Gottfried Wilhelm Leibniz and Leonhard Euler, whose names shine out to the world with undiminishing luster". Note also that in 1924 the Berlin Academy published Euler's correspondence with Lambert.

At present relations between the Academies of Sciences of the GDR and the USSR[25] are constantly developing and deepening, thereby creating for our own Academy favorable pre-conditions for exploiting the Eulerian tradition. In 1957, to mark the 250th anniversary of Euler's birth, official celebrations were organized in both St. Petersburg and Berlin, and in Berlin a conference on the theme "Euler and the conjunction of German and Russian science and culture in the 18th century". At that time also valuable publications appeared: a large two-volume anthology of articles commemorating Euler, the aforementioned three volumes of Euler's correspondence with the Petersburg Academy, and the minutes of meetings of the Berlin Academy during Euler's time there. This collaborative tradition was continued with the participation of a delegation from the Academy of Sciences of the USSR in a conference in September 1983 organized by the Academy of Sciences of the GDR, and that of a delegation of Academicians from the GDR in the present conference of Soviet Academicians. I would like to consider all of this as relevant to my theme "Euler and the Berlin Academy of Sciences". Euler felt himself closely tied to Berlin; in a letter of 1763 he tells his friend G. F. Müller of his "connection with this town Berlin for so many years".

I have already mentioned that in 1907 Berlin showed him due honor by installing a memorial plaque to mark his dwelling-place in the immediate vicinity of the Academy of Sciences. In 1980 the Academy of Sciences of the GDR established a Leonhard Euler medal for "outstanding merit in the fields of mathematics, cybernetics, and the theory of automatic information processing, as expressed by the solution of fundamental scientific problems leading to a significant acceleration in scientific and technological progress in the GDR". With this medal the Academy of Sciences of the GDR honored one of the greatest of its members, Swiss-born, but working from the age of 20 till the end of his life in Russia and Germany.

Opera omnia may be found by visiting the website "The Euler Archive", which also gives access to over 800 of the originals of Euler's works.—*Trans.*

[25] Acronyms become anachronisms since this was published in 1987–88; but the Academies remain.—*Trans.*

Was Leonhard Euler driven from Berlin by J. H. Lambert?[1]

K.-R. Biermann

On November 9, 1764 Leonhard Euler and Johann Heinrich Lambert, who had been in Berlin since January of that year, were dining with the visiting Russian chancellor count Mikhail Illarionovich Vorontsov as his guests. The theme of their discussion, which was the same as that of earlier talks with the Russian envoy prince Vladimir Sergeevich Dolgorukiĭ relating to the anticipated reorganization of the Petersburg Academy, concerned "the benefits that might accrue to the government from a well-organized Academy, and how its members might combine their individual strengths in the general interests of the nation". On November 10 Euler reported to St. Petersburg on this meeting, noting that Lambert, who "had drawn up the plan for the Bavarian Academy", "had made a great impression" on Dolgorukiĭ, and that Vorontsov "had evinced a great desire" to see Lambert enter into the service of Russia. He added that "such an association", i.e., where the members join forces in the service of general aims, "is lacking in almost all academies, where usually nothing more is achieved beyond what each member does by and for himself" [6, Vol. 1, p. 251]. What Euler is proclaiming here as both his and Lambert's conviction is just the thesis that teamwork is more effective than research in isolation. It would be most natural to conclude from this that two talents such as Euler and Lambert—who of course knew what they were talking about—would immediately rush to test the insight they had arrived at, i.e., to put their theory into practice; or, more explicitly, to collaborate. From the uniting of the intellectual capabilities of just these two—Euler, the most important mathematician of the 18th century, and Lambert, as deep a thinker as he was versatile in mathematics, natural science, and philosophy—in the pursuit of a single goal in the general interest, great benefits for science in general and for the Prussian Academy in particular might be expected. However things turned out completely otherwise. Not only did no collaboration occur, but quite the contrary: irreconcilable differences surfaced. Why this happened, why each began

---

[1] "Wurde Leonhard Euler durch J. H. Lambert aus Berlin vertrieben?" Abh. Akad. Wiss. DDR. Abt. Math. etc. 1985. No. 1. pp. 91–99. A talk delivered on September 16, 1983 at a scientific conference dedicated to Euler's memory. The present English translation is direct from the German original, with I. A. Golovinskiĭ's Russian translation used as a crib.

Jean-Henri Lambert

unswervingly to go his own way when they were under the king's command to cooperate, and why their paths ultimately diverged—all this I shall try to explain here, at the same time investigating the conjecture, already aired by Euler's and Lambert's contemporaries, that the real reason for Euler's quitting Berlin was Lambert [16, pp. 344–345].

When on April 9, 1761 Lambert, at that time in Augsburg, was unanimously elected to membership of the Berlin Academy [7, p. 214; 15, pp. 266–267], there can have been no doubt that this was at Euler's initiative. That the Prussian king, Friedrich II, withheld his assent to this election cannot be blamed on Euler. Although since the death of Maupertuis in 1759 Euler, as Director of the Mathematics School, had become in effect director of the Academy as a whole, the title of president was, to his great chagrin, not bestowed on him, and his authority limited. Thus all decisions regarding appointments were reserved to Friedrich II himself. Incidentally, it would seem that Lambert did not have a correct understanding of power relations in Berlin, since otherwise his statement in a letter of November 27, 1764 would be hard to interpret: "L'Académie m'avait élu membre depuis trois ans, sans qu'il y ait moyen de m'en expédier le diplôme[2]" [10, p. 228].

In any case it is certain that Euler had a high opinion of Lambert; letters from Euler to Lambert and others bear manifold witness to this. On the other hand we know that Lambert regarded Euler as the equal of d'Alembert in being the best mathematician of his time [14, pp. 31–32], and that this opinion was based on a close study of Euler's writings [7, pp. 212–214].

Thus when in January 1764, one year after the ending of the seven years' war, Lambert arrived in Berlin to try to obtain what had eluded him in the Bavarian Academy, namely a position in keeping with his abilities and inclinations, and also sustaining him, Euler was

---

[2]"The Academy elected me member three months ago [but] has not found it possible to send me the diploma [of membership]".—*Trans.*

glad to see him. No doubt Euler saw in this scholar, his junior by 21 years and as original as he was versatile, someone worthy of his patronage. Everything went as smoothly as could be wished. Four months after Lambert's arrival Euler reported that:

"We now have here another such person[3], at the extent of whose knowledge I have long marvelled. This is Mr. Lambert, who has already become famous on account of his *Photometry*[4] and other works, and has contributed more than anyone else to the establishment of a Bavarian Academy[5]. However, since he is a Swiss protestant from Mülhausen, he could not get along with the Jesuits and thus forfeited his sizeable Bavarian pension. He has come here on his own initiative, and although His Royal Highness spoke with him briefly, no indication has been given as to whether he will be hired here. However perhaps a decent post will soon be found for him. He is in all respects a person possessing enough talent for a whole academy" [6, Vol. 1, p. 245].

It is hard to imagine greater praise than this. It is absolutely clear that Euler was extremely well-disposed towards Lambert. This was not due only to his high regard for the depth and many-sidedness of Lambert's work: there was more behind such a high evaluation.

First of all, as a native of Basel, Euler considered Lambert, who as we have seen came from Mülhausen in Alsace, to be a Swiss compatriot—and on good grounds: at that time Mülhausen belonged as a "Donation" to the confederation of Swiss cantons. Thus Euler would have anticipated from Lambert's admission into the Academy a strengthening of the already sizeable Swiss contingent—however, not only this: it is reasonable to suppose that by advancing Lambert he hoped that his own position within the group of Swiss academicians would also be strengthened. The latter formed no monolithic bloc, and in fact some of them were definitely not in support of Euler, director of the Mathematics School, in his pretensions to the directorship of the whole Academy. In particular, Johann Georg Sulzer of Winterthur, an ardent Wolffite, and also the highly influential Swiss academician Johann Bernhard Merian of Liestal, near Basel, adopted a not uncritical stance towards Euler, and were not prepared to follow wherever he should lead. And with that we come to a further reason why Euler might welcome Lambert's admission into the Academy: With his belief in divine revelation, his strong aversion to Leibniz' philosophy, and his assumed duty to deliver "divine revelation from the objections of free-thinkers" [2], Euler played a significant role in the French Consistory, i.e., the French reformed church, and hence in the Huguenot "colony" in Berlin [5]. In this connection he was active on various councils, such as for example those dealing with allotments and with catechismal questions. As an Elder, since 1763, of the parish of Friedrichstadt[6], Euler exercised considerable influence in the granting of church offices, in matters relating to orphanages and other charitable enterprises, in the prosecution of measures for increasing church attendance, and on the instruction of children. We would not be mistaken in assuming that he saw in his *protégé* Lambert a potential ally in his struggle against the "spirit of indifference". And indeed, in accordance with Euler's expectations as it were, Lambert joined the Huguenot church in Berlin; it was discovered only a few years ago that in 1777 Lambert found his last resting-place in the

---

[3]That is, such as Johann III Bernoulli, whom Euler had—in the same breath as it were—described as having a "very nimble mind, which will bring honor to the Academy".

[4]See [9].

[5]See [8, 12].

[6]A suburb of Berlin on the left bank of the river Spree.—*I. A. Golovinskiĭ*

French-reformed cemetery of the parish of Dorotheenstadt[7] [1, p. 123]. The question must
be left unanswered as to whether this affiliation was due to Euler's influence, or sprang from
his own convictions, or whether other factors were involved (there was at that time active in
Berlin a preacher belonging to the "colony" named Jean Pierre Erman[8], whose ancestors—
known by the name of Ermatinger or Ermendinger—had settled in Mülhausen). However
there can be no doubt that Lambert's decision to join the Huguenot church strengthened
Euler's good opinion of him: as is obvious from the beginning of Euler's report of November 10, 1764 quoted above, there reigned between them complete mutual understanding.
And it was with satisfaction that soon afterwards, on January 10, 1765, Euler informed
the Academy of the king's decree that Lambert be admitted to that scholarly society [15,
p. 306]. Friedrich's change of heart, extremely irritating to Lambert's self-esteem, can be
traced to the threat of losing him to the Petersburg Academy, to which Euler had recommended him [6, Vol. 1, pp. 248, 249]. Euler could not have foreseen that these harmonious
relations were to end almost at once.

    In order to understand the tensions that arose, it is necessary to remind the reader of the
Academy's financial structure. In the 18th century the Academy's chief source of revenue
was its monopoly in calendars in the Prussian state; by comparison, the income generated
by other prerogatives—for instance the publication of lists of current ordinances or the supervision of geographical maps—was relatively paltry. The idea of a calendar monopoly
dates back as far as Leibniz' time[9]. The letters patent authorizing the calendar monopoly,
granted on May 10, 1700 by Friedrich III, Elector of Brandenburg[10], contained the following statement, among others:

    "Since we now…from our own lofty motives and so all the more graciously grant to the whole
    Society[11] publication rights throughout our electoral and other territories[12] to both the improved and
    standard calendars, in this regard holding it personally to be exclusively so privileged, in order to
    [put an end to the production of] calendars that hitherto were so often in part incorrect and in part
    vexatious, filled mostly with barbaric false histories, empty prophecies, and disgraceful discourses,
    and moreover produced by persons completely inexperienced in difficult and laborious astronomical
    computations, and furthermore so that in future the funds used to pay for them remain within our
    borders, we have deemed it necessary to make known…this our most gracious will and judgment"
    [4, Vol. 2, p. 88].

    The realization of an income from the calendar monopoly, on which the existence of the
Society—from 1744 the Academy—depended, initially encountered difficulties of various
kinds [4, Vol. 1, pp. 123–125]. At first the Academic calendars met with rejection by their

---

[7] A suburb of Berlin on the right bank of the Spree.–*I. A. Golovinskiĭ*

[8] J. P. Erman, who was later to become historiographer and director of the French gymnasium in Berlin, was
also destined to become the founder of a whole dynasty of scholars, of whom we note here only his son, the
physicist Paul Erman, his grandson, the geophysicist and circumnavigator of the globe Georg Erman, and his
great grandson the egyptologist Adolph Erman.

[9] Actual priority in conceiving this idea was claimed by E. Weigel; see [4, Vol. 1, pp. 64–66].

[10] From 1701 King Friedrich I of Prussia.

[11] The Berlin Academy of Sciences, founded in 1700 along lines proposed by Leibniz, was initially called
the "Scientific Society of the Elector of Brandenburg" (*Die Kurfürstlich–Brandenburgische Societät der Wissenschaften*), and then in succession the "Royal Prussian Scientific Society" and the "Royal Berlin Scientific
Society"; then in 1744, when it was integrated with the *Société Littéraire*, it became the "Royal Berlin Academy
of Sciences" (*Academia Regia Scientiarum Berolinensis*). See: Kopelevich, Yu. Kh. *The rise of the scientific
academies*. L.: Nauka, 1974.—*I. A. Golovinskiĭ*

[12] These included, in addition to the lands of the Brandenburg Electorate, East Prussia, part of Pomerania, and
certain territories along the Rhine.—*I. A. Golovinskiĭ*

readers, who missed the old calendars—until then obtained by subscription from abroad—with their weather forecasts and other predictions, and then there was embezzlement, which in spite of a steep increase in sales, restricted the Academy's profits. From 1738 the actual manager in charge of the sale of calendars was an accountant by the name of David Koehler, who significantly increased the sale of calendars, without, however, forgetting his own interests. He temporarily farmed out at a fixed fee the sale of four of the nine varieties of calendar published by the Academy [4, Vol. 1, p. 275]. How well the calendars sold can be gathered from the fact that the sole bookseller in Frankfurt-am-Oder ordered 6750 calendars for the period December 1740 to January 1741 [4, Vol. 1, p. 261]. That under Koehler's management the Academy received only a portion of the monies marked for the defrayal of the costs of personnel and materials was an open secret, and several attempts were made to get rid of this finagler, suspected of withholding from 25 to 50 percent of the Academy's revenues [4, Vol. 1, p. 265; 11, p. 176]. However Koehler had made himself so indispensable that he might be able to cling to his post for decades. The high-ranking officials and dignitaries of the Academy—patrons, curators, presidents—ordered no enquiries, with the result that Koehler was able to repeatedly slip out of the noose. Euler, who encountered Koehler when he arrived in Berlin in 1741, remained unaware of the extent of the losses to the Academy caused by the accountant. Koehler was astute enough to realize that Euler might be dangerous to him, so he adopted an ingratiating manner towards him and took care that his salary was paid punctually. Euler reciprocated by loyally acknowledging Koehler's undoubted service to the Academy.

Although during the seven years' war (1756–1763) the Academy's revenues were for obvious reasons significantly reduced, on the other hand economies had been made—or rather had had to be made—since the king had forbidden the redeployment of surplus resources as well as all innovations for the duration of the war [4, Vol. 1, p. 350]. Now, when peace had been concluded, the Academy's finances would have to be put in order, decisions made concerning savings and investments, and the administration reformed. From the members' ranks complaints about Koehler were once again to be heard. It was claimed that, although he had taken care that the calendar business did not fail, all the same he had not organized it as well as might have been, and that furthermore he continued to line his own pockets. However it was once again Euler who opposed any fundamental reform of the management and personnel of the Academy, and who maintained that the accountant was indispensable [4, Vol. 1, p. 363]. Then on February 21, 1765 Friedrich II appointed an Academic commission to investigate, reorganize, and monitor the administration of the Academy and its finances. To this economic body—most often called the "Royal Commission" to show that its authority derived from the monarch, not the Academy—, in addition to Euler, there were appointed Sulzer, the aforementioned opponent of Euler (it was actually he who had called for a commission), Merian, also mentioned earlier, the nonentities Louis-Isaac de Beausobre and Jean de Castillon, and finally Lambert, the youngest of the academicians. Euler took the very appointment of a commission as a slur. It had been set up against his will and its membership also represented a defeat—but that was not all. He sensed—justifiably, as it very soon emerged—the threat of a diminishment of his already limited authority as Director of Schools, and most of all of his own powers—those that he had wielded for years as president by default, as it were, in practice responsible for the running of the Academy, albeit within limits imposed by the king.

However gloomy Euler's forebodings may have been, he can scarcely have taken the full measure of what was to come. He was now to see an altogether different side of Lambert. While on the one hand Euler strove to use his plenipotentiary powers in their strictest possible interpretation in order to maintain the authority of the directorate of the Academy and thereby restrict the Economic Commission's activities as much as possible, on the other hand Lambert became extraordinarily active in the commission. It became clear that Lambert had good organizational skills which he at last felt free to display in the interests of the Academy. It was thus completely natural that within a very short time a conflict between Euler and Lambert developed in the commission: one wanted to go slower, the other faster. It is therefore certainly no coincidence that in Euler's correspondence dating from just after the formation of the commission his erstwhile praise for Lambert is altogether lacking, and that he attempted to distance himself from Lambert with the following characteristic statement: "When it is necessary for someone to direct the work of artists whilst occupying himself merely with invention, then there is no one more apt for this than our Mr. Lambert" [6, Vol. 3, p. 235]. Indeed Euler had by then ample evidence of Lambert's "talent for direction" from the proceedings of the "Royal Commission". It is therefore not surprising that now very quickly the differences in Euler's and Lambert's abilities and sympathies, in their characters and lifestyles, their social backgrounds, and finally their ages, began to outweigh the fact, mentioned above, that they were compatriots and of the same religious persuasion, and also their mutual respect as scientists. The son of a pastor from Basel, Euler was certainly not from a wealthy family, but all the same it would have to be called rich by comparison with the poverty prevailing in the dwelling of Lambert the tailor. Then there was the generation gap. When Lambert was born in 1728, Euler was already an adjunct of the Petersburg Academy, and was gaining experience in relations with the director of an academic administration. (Incidentally, the lessons he had learned in St. Petersburg from his acquaintance with J. D. Schumacher, also an Alsatian, and a bureaucrat as clever as he was ruthless, must have affected his relations later with Lambert: Schumacher had taught him how dangerous it could be for a scholar who dared oppose a chief administrator.) Then there were their different lifestyles. In Berlin Euler lived, one may say, in the grand style, while when he first arrived Lambert had had to bide his time on the verge of want. In Euler's patriarchally run household there lived around a dozen people made up of family members and guests, and he owned other houses, gardens, meadows, and farmlands; he had become wealthy and had a lot to lose. On the other hand we have the unmarried have-nothing Lambert, for whom a sister prepared meager meals; he had nothing to lose and everything to gain.

The algorithmist Euler was known throughout Europe as the great master of his field; Lambert, although known to some extent for his investigations into the origins of geometric representation and skill in logic, nevertheless stood on a significantly lower rung of the ladder of fame. Euler, the *de facto* director of the Academy and thus used to giving orders, stood opposed to an unsubmissive Lambert, who until his arrival in Berlin had worked as private tutor and independent scholar. Should we be surprised that, as Lambert's mentor, Euler felt he had a well-founded claim to respectful considerateness from his *protégé* Lambert? In this he was deceiving himself, however: his claim was not acknowledged. Moreover, the highly developed feelings of self-worth present in both Euler and Lambert were different in quality. Both were with justification imbued with a sense of the value of their

achievements without being vain. However, whereas Euler's conviction of his own significance was mixed with a salutory dash of modesty, Lambert's opinion of his own stature
was expressed with overpoweringly direct naiveté. This is supported by pronouncements
from both opponents—which is what they had indeed become following on the establishment of the Economic Commission—in answer to the king's standing question as to where
they had obtained their knowledge. Euler had answered in 1749 that he had his time with
the Petersburg Academy to thank for everything that he knew [6, Vol. 2, p. 182], while 15
years later Lambert claimed to be self-taught in mathematics [3, p. 15].

As already mentioned, all of these differences came to the fore when in the meetings
of the Commission conflicting opinions were voiced. This began as soon as the question
of the commission's authority was broached. Euler did not want to allow the directorate
to be deprived of its authority without a struggle. Thus he was insistent that payments
ordered by the commission should be made only with the consent of the directorate. The
other members of the "Royal Commission" shared Lambert's opinion that they had been
empowered by the king to manage their own finances [1, pp. 119, 120].

Tensions rose in the discussion as to how the sale of calendars should be organized
in order to increase revenues. The commission split into three factions. One of these recommended leasing out the marketing side of the calendar business, Euler wanted Koehler
to be kept on under altered conditions, and Lambert favored the management of the sale
of calendars by the commission independently of the directorate. Euler now attempted to
have his way by twice petitioning Friedrich II directly, receiving however only the well-
known crude rebuff: "It's true that I don't understand how to calculate any curves, but I
do know that 16,000 talers is more than 13,000" [4, p. 364]. It was ordered that the management of the sale of the calendars be leased out, and as a result the Academy's revenues
increased noticeably in spite of Euler's gloomy prognostications. However this did not end
the friction; on the contrary at the meetings of the commission further inflammatory topics
continually arose. When somewhere near the beginning of 1766 the commission member
Castillon petitioned the Schools of Mathematics and Physics for funds for experiments and
the purchase of instruments, Euler observed frostily that the Mathematics School would
be acting improperly in considering such a petition, since according to Castillon himself
the distribution of funds was the responsibility of the Economic Commission alone, and
to this, as Euler's son pointedly remarked, the petitioner himself belonged. He would thus
be doing those Schools too great an honor in requesting their approval when this was perfectly meaningless. In contrast, Lambert expressed calmly and objectively his confidence
that after submission of a report on said experiments and an account of the expenses in-
curred, the Economic Commission would arrange appropriate reimbursement. The records
contain a great many instances of such friction between Euler and Lambert over the period 1765–1766. Often dissatisfaction can be discerned only by reading between the lines,
but sometimes it becomes explicit. A certain document dealing with the botanical garden
contains such an example. It was determined that certain specific expenses were essential.
Although Euler had gone along with this decision, this did not prevent his objecting to the
payment of these some days later. In his director's hat, so to speak, he rejected what he had
agreed to as member of the commission, thus taking away with one hand what he had just
given with the other. On October 19, 1765 Lambert made the following dry observations
concerning such conduct:

"Il seroit un ouvrage trop long, si on vouloit rendre raison de toutes les anomalies de l'esprit humain. Il s'embrouille d'avantage à mesure qu'on se met à le débrouiller. Les phénomènes, qui s'y offrent, pourroient servir d'amusement à ceux, qui les voient de près, s'il ne s'agissoit pas des choses, dont il faut pourvoir ensuite rendre compte. Mr. Euler se fait du tort, en se repentant le lendemain des résolutions qu'il avait prises, approuvées, signées le jour précédant, comme p. ex. lorsqu'il s'agissoit des archives et à l'heure qu'il est, où il s'agit du rétablissement du jardin. Sur ce dernier Sujet nous avons sa propre Signature, ainsi je crois qu'il doit être naturellement intéressé à la faire valoir[13]" [17].

This is strong and clear, and would not, one may suppose, have been kept secret from Euler, gossip being essential for lending flavor to the daily round of the Academy. Events came to a head as they had to: as soon as one month later, on November 25, 1765 Euler, in a formal "Déclaration à la Commission de l'Académie", in blunt terms incapable of misinterpretation, declined to participate further in the commission under the prevailing conditions [18].

On February 2, 1766 Euler petitioned the king for permission to resign, this being granted very ungraciously only on May 2, after many attempts to dissuade him and only after insistent renewals of his request. On June 1, 1766 Euler left Berlin with his family, destined once more for St. Petersburg, whence he had come 25 years earlier. Just prior to leaving he paid a farewell visit to Lambert, who reported on it in the following noteworthy fashion:

"Je n'ai reçu la chère vôtre que deux jours après Mr. Euler était venu me rendre la visite de congé. Vous ne sauriez croire, Monsieur, avec combien de joie il va à Pétersbourg, où il vaut tout ce qu'il peut désirer. L'Académie de Pétersbourg va être mise sur un nouveau pied, et à force d'argent on y attirera tout ce qu'il y a de plus savant en Europe.... De la sorte il est très sûr que Mr. Euler aurait pris la résolution qu'il a prise, encore qu'il n'eût aucun différend avec le Comité économique de l'Académie, dont il était pareillement membre. Ce qu'il pouvait attendre de devenir ici, il vit, qu'il n'en serait jamais rien, et il n'aurait non plus été spectateur tranquille, si tôt ou tard un autre l'avait devancé. Enfin quoiqu'il était lui-même membre du Comité économique, il ne laissa pas d'être intimement persuadé que ce Comité avait été nommé, si non pour redresser du moins pour perfectionner ce qu'il avait fait en qualité de Directeur de l'Académie[14]" [10, p. 230].

Not everything that Lambert writes here is accurate. It is possible to maintain [1, pp. 119–120] that even before Lambert's arrival in Berlin Euler had begun preparations to return to St. Petersburg. Thus as early as June 1763 he had begun to sell off his acquisitions in real estate in order not to be tied down. In fact there were many reasons why he might wish

---

[13]"It would be too great a task to provide reasons for all the anomalies of the human spirit. The more one tries to sort it out the more complicated it becomes. The phenomena that occur in this regard might serve to amuse those observing them from close at hand if it were not a question of their effect. Mr. Euler is wrong to reject on one day resolutions that he accepted [for consideration], approved, and signed on the previous day, as when, for instance, the archives were being discussed, and the current topic concerning the refurbishing of the garden. On the latter subject we have his personal signature, so that I think he should naturally have an interest in honoring it".

[14]I only received your cherished [letter] two days after Mr. Euler paid me a farewell visit. You would find it hard to believe how happy he is about going to St. Petersburg, where he will obtain everything he might wish for. The Petersburg Academy will be placed on a new footing, and by financial means they will attract all of the most erudite of Europe.... Thus it is quite certain that Mr. Euler would have come to this decision in any case, since he had actually made his mind up before any disagreements had arisen in the Economic Commission of the Academy, in which he nonetheless continued as a member. He saw that here he could never attain to the position he wished for, and furthermore he could not have remained a passive spectator if sooner or later someone were promoted ahead of him. Finally, although he himself was a member of the Economic Commission, he could not rid himself of the conviction that the commission had been established if not to mend then at least to improve on what he had done as [acting] Director of the Academy".

to return to his former workplace. In the first place one might point to the dominance of free-thinkers and critics of religion in the Berlin Academy; and added to this there was the preference for French candidates in the appointment of new members by the king without input from the Academy. Furthermore Euler took offence at Friedrich II's continued withholding from him the post of president of the Academy—to which he had a well-founded claim and which would have meant a salary increase of 50%—an offence exacerbated by the influence wielded from Paris by d'Alembert—the "secret president" of the Academy as he was called—with whom the king was consulting. Finally he felt his Swiss sense of freedom violated by the despotic ways of the Prussian king. Thus the affair of the calendars and the clash with the Economic Commission were no more—but also no less—than the straws that broke the camel's back, as it were [13, p. 24]; but this is far from saying that Lambert treated Euler brusquely and drove him from Berlin. Only insofar as it was completely foreign to Lambert to allow personal considerations to interfere with matters of a purely impersonal nature, had he unintentionally put a fullstop to the Berlin chapter of Euler's life. It is likely that Euler himself also regarded the matter in this light, since in the covering letter written by Euler's son Johann Albrecht accompanying the scientific reply prepared by Lexell to a letter that Lambert sent to Euler in October 1771, there is no acknowledgement of Lambert's expression of concern for Euler's health and eye-operation [1, p. 121]. On the other hand, while in Petersburg Euler continued to show concern for the state of Koehler's health [13, pp. 44–45].

Thus in the end what estranged Euler and Lambert from one another proved stronger than that which united them.

References

[1] Biermann, K.-R. "J.-H. Lambert und die Berliner Akademie der Wissenschaften". In: *Colloque international et interdisciplinaire Jean-Henri Lambert*. Mulhouse, 26–30 septembre 1977, Éditions Ophrys, Paris 1979, pp. 115–126.

[2] Euler, L. *Rettung der göttlichen Offenbahrung gegen die Einwürfe der Freygeister. Opera* III-12.

[3] Graf, M. "Lamberts Leben". In: Huber, Daniel: *Johann Heinrich Lambert*. Basel 1829.

[4] Harnack, A. *Geschichte der Königlich Preußischen Akademie der Wissenschaften zu Berlin*. Vols. I/1 and II, Berlin 1900.

[5] Hartweg, F. G. "Leonhard Eulers Tätigkeit in der französisch-reformierten Kirche von Berlin". Die Hugenottenkirche **32** (1979), No. 4, pp. 14–15; No. 5, pp. 17–18.

[6] *Die Berliner und die Petersburger Akademie der Wissenschaften im Briefwechsel Leonhard Eulers*. A. P. Yushkevich and E. Winter (eds.). Berlin: Akad.-Verl., 1959–1976. Vols. 1–3.

[7] Yushkevich, A. P. "Lambert et Léonhard Euler". In: *Colloque international et interdisciplinaire Jean-Henri Lambert*. Mulhouse, 26–30 septembre 1977, Éditions Ophrys, Paris 1979, pp. 211–223.

[8] Kraus, A. "Lambert und die Bayerische Akademie der Wissenschaften". In: *Colloque international et interdisciplinaire Jean-Henri Lambert*. Mulhouse, 26–30 septembre 1977, Éditions Ophrys, Paris 1979, pp. 105–113.

[9] Lambert, J. H. *Photometria sive de mensura et gradibus luminis, colorum et umbrae*. Augustae vindelicorum (= Augsburg) 1760.

[10] Speziali, P. "Lambert et le Sage". In: *Colloque international et interdisciplinaire Jean-Henri Lambert*. Mulhouse, 26–30 septembre 1977, Éditions Ophrys, Paris 1979, pp. 225–234.

[11] Spiess, O. *Leonhard Euler.* Verlag Hubner & Co., Frauenfeld und Leipzig 1929.

[12] Spindler, M. *Electoralis academiae scientiarum boicae primordia.* München 1959.

[13] Stieda, W. "Die Übersiedlung Leonhard Eulers von Berlin nach St. Petersburg". Berichte über die Verhandlungen der Sächsischen Akademie der Wissenschaften zu Leipzig, Phil.-hist. Klasse **83** (1931), Issue 3.

[14] Thiébault, Dieudonné. *Mes souvenirs de vingt ans de séjour à Berlin.* Paris 1804.

[15] *Die Registres der Berliner Akademie der Wissenschaften. 1746–1766.* E. Winter and M. Winter (eds.). Berlin: Akad.-Verl., 1957.

[16] Wolf, R. *Biographien zur Kulturgeschichte der Schweiz.* Vol. 3, Zürich 1860.

[17] Zentrales Archiv der Akademie der Wissenschaften der DDR. I–XIV–37, Bl. 30–31.

[18] *Ibid.* I–IV–16, St. 78.

Euler's Mathematical Notebooks[1]

E. Knobloch

The goal of the present paper is to convey the gist of Euler's mathematical notebooks. Of the works up till the present[2] devoted to the notebooks, mention must be made above all of the articles by G. K. Mikhaĭlov on Euler's research in mechanics [1] and by G. P. Matvievskaya on his investigations in number theory [2]. And the systematic classification of Euler's scientific works given by the catalogue [3] is of course crucial to such a survey as this. There the notebooks are indexed as f. 136, op. 1, Nos. 129–140. Here I shall refer to each of them only by its number (between 129 and 140) followed by a colon and the numbers of the relevant pages.

In attempting a systematic sorting of the notes one encounters problems to do with their classification. Thus I have assigned navigation and chronology to astronomy, and meteorology to physics. And a need arose for additional areas: philosophy and philology, and also medicine, chemistry, and geography. In connection with the mathematical subdisciplines, difficulties in delimitation occurred; thus while continued fractions belongs essentially to the group 1.4[3], these also occur in problems and solutions in 1.5, 1.6, and 1.8, and although differential equations comprises the group 1.6, these play an important role also in series expansions and in problems of differential geometry. Similarly, research on the ζ-function in essence belongs to the group 1.4, but is important in connection with the number–theoretical investigations assigned to the group 1.1. Thus the general data below, as well as the systematic index of notes given at the end of the paper[4], are based on a classification that in certain cases might have been done differently.

The notes frequently switch from one area to another. Often they were occasioned by perusal of related papers from scientific journals, or by correspondence with scholarly friends, and this makes it sometimes possible to date them quite accurately. Often Euler

---

[1] The Russian version of this essay is a translation by I. A. Golovinskiĭ of the German original, as in the case of the preceding essay. The original reference is: Eberhard Knobloch. "Leonhard Eulers Mathematische Notizbücher". Annals of Science, 1989. Vol. 46. pp. 277–302. The present English translation has been made directly from the Russian version. The translation has, however, been checked for accuracy by Professor Abe Shenitzer in consultation with the author.—*Trans.*

[2] That is, 1987.

[3] These numerals refer to the table below.

[4] This index was omitted from the Russian version of the essay, and is likewise omitted from the present version.—*Trans.*

Scientific fields	Number of pages		
	octavo	quarto	duplo
1. Mathematics			
1.1. Number theory	152	346	83
1.2. General questions of algebra and algebraic equations	$\frac{1}{2}$	$112\frac{1}{2}$	10
1.3. Probability theory, combinatorics, and mathematical games	—	86	—
1.4. Series, products, and continued fractions	18	337	146
1.5. Differential and integral calculus	$5\frac{1}{2}$	270	$95\frac{1}{2}$
1.6. Differential equations	13	250	$92\frac{1}{2}$
1.7. Geometry	$18\frac{1}{2}$	$160\frac{1}{2}$	23
1.8. Differential geometry and the calculus of variations	$5\frac{1}{2}$	$219\frac{1}{2}$	$32\frac{1}{2}$
2. Mechanics	$21\frac{1}{2}$	554	46
3. Astronomy	12	103	46
4. Physics	2	$102\frac{1}{2}$	9
5. Geography, medicine, chemistry	—	3	—
6. Philosophy, philology	1	68	—

gives a reference, with the relevant name included, for example: Clairaut in connection with differential geometry; Fontaine[5], Lambert, and Lagrange in connection with the calculus of variations; Goldbach in connection with infinite series and number theory, and so on. Euler proposes particular problems and formulates and proves theorems. Sometimes he notes that a certain work has already been presented to the Academy: "All of this, as well as that further pertaining to the present topic, has been communicated to the Academy" (**140**:35).

Notebook No. 137 occupies a special place. A significant portion of it consists of printed excerpts from texts by other authors or copies of such: from John Wallis' *The arithmetic of infinities*, from works on algebra, geometry, astronomy, and optics (the last by Aegidius Franciscus de Gottignies[6]), and from a work of Francis Linus[7] on the construction of sundials. On many of these pages Euler inserted comments where space permitted.

During my investigations I received great encouragement and support from A. P. Yushkevich, E. P. Ozhigova, and Yu. Kh. Kopelevich. I would like to extend to them my heartfelt thanks.

1. Mathematics

1.1 Number theory

By taking into account the article of Matvievskaya published in 1960, and E. P. Ozhigova's work in preparing material for the corresponding volume of Series IVB of *The complete*

---

[5] Alexis Fontaine des Bertins (1704–1771).—*Trans.*
[6] Or Gilles François de Gottignies.
[7] A 17th century English Jesuit.—*Trans.*

collected works (Opera omnia) of Euler, I can here be brief. Much of the number-theoretic material of notebooks 138, 139, and 140 is published as item 111 entitled: "Fragments extracted from the mathematical notebooks" ("Fragmenta ex Adversariis mathematicis deprompta") of the two-volume *Leonhardi Euleri Opera postuma...quae ediderunt...P. N. Fuss et N. Fuss* (Pétersbourg, 1862). These will not be considered in the present survey.

Euler investigates the distribution of the primes using a graphical representation appearing in Leibniz' manuscripts (**129**:43; **134**:205).

He discusses questions relating to divisibility in the domain of natural numbers, such as remainders after division (**131**:232), amicable numbers (**133**:45, 56; **134**:32), perfect numbers in the sense of Euclid (**130**:35, 58; **131**:56), and expresses doubt as to the existence of odd perfect numbers (**134**:115). He considers division by arbitrary natural numbers (**134**:64), the situation of two such numbers and their remainders (**138**:105), the problem of finding a number of the form $a^2 + 1$ with prescribed divisors (**132**:258), and certain of Fermat's theorems (**131**:22; **132**:77).

Euler studies equations involving exponents, such as $2^x = x^2$, $3^x = x^3$, $x^y = y^x$ (**130**:37, 39). He works with "dyadic arithmetic" ("arithmetica dyadica") (**131**:208; **133**:52). "Partitio numerorum" is an especially important theme (**132**:17, 18, 105, 106, 196); here he uses both analytic and non-analytic methods. Schooten's[8] "weighing method" ("praxis ponderandi") receives especial attention (**132**:94, 97). In analytic number theory he considers the problem of the frequency of occurrence of a given integer in certain sequences (**131**:12). His attention is taken up by the Bernoulli numbers on many occasions (**134**:117; **139**:23, 106). The note headed "Exposition of the science of (whole) numbers" would appear to represent the beginnings of an article (**134**:117).

There are a great many problems of Diophantine analysis, some of which Euler attributes to Ozanam[9] (**131**:123; **133**:11, 136, 143), Leibniz (**131**:60; **132**:76), Pell (**132**:247), and Goldbach (**132**:90, 220). In fact, in asserting that certain problems originated with Leibniz, Euler commits an historical error. The problems in question involve polygonal numbers (**131**:21, 152; **134**:130), the extraction of rational roots by an appropriate assignment of values to variables appearing in expressions under root signs (**131**:17, 53, 54), and rational or right-angled triangles (**129**:46; **132**:167). Euler shows that $\sqrt{a - x^2}$ is never a perfect square if a is any number from the sequence $3, 6, 7, 11, 12, 15, \ldots$ (**130**:23), and that $2n^2 + 1$ is never the fourth power of an integer (**130**:112). He considers decompositions into sums of squares (**131**:61, 62, 116, 117; **132**:82), a problem concerning six squares (**132**:147), linear Diophantine equations (**131**:207), and also the following Diophantine equations:

$$x + y + z = xyz = x^2 + y^2 + z^2 \quad (\mathbf{130} : 60);$$
$$x^2 + y^2 + z^2 = x^2 y^2 z^2 \quad (\mathbf{135} : 3);$$
$$ax^2 + 2bxy + cy^2 = p^2 \quad (\mathbf{136} : 8).$$

He also investigates the conditions under which $\alpha x^2 + \beta x + \gamma$ is a perfect square (**132**:109, 112).

---

[8] Frans van Schooten (1615–1660).—*Trans.*
[9] Jacques Ozanam (1640–1717).—*Trans.*

Among the particular problems of Diophantine analysis he considers, the following are also noteworthy:

Find four natural numbers whose pairwise differences are all squares (**129**:50).

Find two natural numbers with sum a square, the sum of their squares a cube, and the sum of their cubes a fourth power (**131**:80).

"The most difficult Diophantine problem: find three numbers ka, kb, kc such that I) their sum, II) the sum of their pairwise products, and III) the product of all three are square numbers" (**134**:58).

1.2 General questions of algebra and algebraic equations

In his notebooks Euler devotes much space to solving equations of the second, third, fourth, fifth, sixth, and nth degrees (**129**:1, 8, 67, 158;**131**:25, 44, 70/71, 77/78, 193/194; **132**:46, 50, 67–69, 77, 85, 193, 237; **133**:33, 65, 103, 174 ("the usual criteria of Newton"—"criteria vulgaria Neutoni"); **135**:68–70; **138**:56), occasionally by way of completing work of others (**133**:44), and particularly equations of even degree (**132**:123; **133**:61). He also investigates equations of the form

$$ay^m + bx^n = cx^p y^q \quad (\mathbf{132} : 191)$$

and of the form

$$1 = \frac{A}{x^\alpha} + \frac{B}{x^\beta} + \frac{C}{x^\gamma} \quad (\mathbf{139} : 26).$$

He considers the following problem: Given an equation of degree four, find one of degree six whose roots are the the pairwise sums of the roots of the given equation (**133**:167). He attempts to eliminate the terms of degrees $n - 1$ and $n - 2$ from the general equation of degree n (**132**:37), and discusses the rule of false position (**131**:193). The fragment (**133**:62) also contains material on elimination, among other things.

Euler also discusses "the theorem on the alternation and permutation of signs" proved by Segner and going back to Harriot[10] (**134**:171), the result that every [real] equation of odd degree has at least one real root (**131**:91), the problem of "finding a bound for the roots of a given equation" (**131**:139), and other problems related to estimation (**131**:138; **136**:82). He also investigates the fourth and fifth roots of unity (**129**:47; **130**:47; **132**:43).

The notes show him to have been much occupied with inventing equations (**131**:135; **135**:10, 63), with the elementary symmetric functions and sums of powers (**131**:77/78, 169; **132**:45/46), and with questions concerning "the application of Cardano's rule to rational roots" (**138**:145).

Following Nicholas Bernoulli he investigates the factorization of equations and fractional (rational) expressions (**132**:106, 179), and algebraic factors and divisors (**131**:195).

Many of the notes have to do with extraction of roots (**129**:66; **131**:256), especially of binomials (**129**:38; **131**:246/247; **133**:59), or iteration of radical expressions such as for example

$$x = \sqrt{1 + \sqrt{2 + \sqrt{3 + \sqrt{4 + \sqrt{5 + \ldots}}}}} = 1.757829\ldots \quad (\mathbf{133} : 12; \mathbf{135} : 51).$$

---

[10]Thomas Harriot (ca. 1560–1621).—*Trans.*

Euler analyses "the calculus of irrationalities" ("Calculus irrationalium") (**132**:246; **138**: 138), the rationality of a certain product of Segner's (**136**:90), reduction to rational quantities (**129**:157), incommensurable quantities (**138**: 128), and irrational trinomials (**132**:103).

He considers algebraic, especially irrational, identities, such as, for example

$$\sqrt{a + \sqrt{b}} \equiv \sqrt{\frac{a + \sqrt{a^2 - b}}{2}} - \sqrt{\frac{a - \sqrt{a^2 - b}}{2}} \quad (\textbf{131} : 123),$$

or Goldbach's identity

$$a^2 + b^2 + c^2 = \frac{1}{6}\{(2a + b + c)^2 + 2(a - b - c)^2 + 3(b - c)^2\} \quad (\textbf{133} : 121),$$

and also certain results concerning sums of ratios (**131**:121).

Notes forming a somewhat distinct subset are those devoted to particular problems in part relating to Pappus, such as, for example, angle trisection (**135**:6, 54; **136**:63), particular problems concerning commerce (**129**:151; **134**:228), problems involving percentages (**134**:39), and facetious puzzles such as the following: "How much time would be needed to write down all numerals from 1 through $90, 000, 000, 000$?" (**134**:24/25).

1.3. Probability theory, combinatorics, and the theory of mathematical games

Euler's notes on these topics are concentrated mainly in seven notebooks dating from his early years in Basel, his first Petersburg period and his Berlin period. Of the other notebooks, there are six[11] (**130**, **136**–**140**) containing nothing on these topics. Those relevant to the present section contain notes on lotteries of many different types or involving a predetermined number of tickets (**129**:38, 61/62; **134**:23, 83/84; **135**:3–5; **137**:351, 366), and on the chances of winning certain games, especially those involving dice or cards (**131**:97/98, 141; **132**:92/93, 104/105, 250/251 (concerning an n-sided die); **135**:85/86, 144–148, 227). Certain of the notes already published in Volume I–7 of the *Opera omnia* also pertain to this section. Euler calculates the probability of obtaining or drawing particular cards (**132**:16/17; **133**:24/25, 51 (drawing n slips from a set of s), 111 (two players); **135**:145). He solves the combinatorial problem of counting the number of ways a given polygon can be subdivided into triangles by means of diagonals (**133**:81). It is appropriate to include under the heading of combinatorics also Euler's detailed investigations into magic squares and rectangles (**129**:147–151; **134**:1, 52–55, 229–239, 241). He also investigates questions concerning insurance, for instance life premiums and mortality (**134**:17/18, 164–166), and problems in commercial insurance (**131**:108/109). There are many problems concerned with commerce, capital, and interest (**129**:106, 174–176; **134**:18/19, 24/25, 30, 35, 43/44, 54/55, 84).

1.4. Series, products, and continued fractions

Although Euler did consider infinite series from a general standpoint, he was primarily concerned with their application to particular problems, such as the theory of the ζ-function, or

---

[11] This contradicts the earlier statement that in all there are just 12 notebooks.—*Trans.*

continued fractions, or the computation of integrals and solutions of differential equations. Among his notes there are those entitled "On the summation of series" (**129**:33), "My general method for summing a series" (**131**:43), "Transformation of series composed of sums of terms into series composed of products" (**131**:4), "Finding series summing to infinity but not having an indeterminate sum" (**131**:17), "Transformation of series" (**132**:31), and "To evaluate a sum of hyperbolic logarithms" (**132**:233; compare **134**:204).

The ζ-function is investigated in its general form and for particular values of the exponent [argument] (**131**:92–95, 166; **132**:194–196). Some special cases and variations on these are as follows:

$$\text{"The Basel problem"} \quad S_2 = \sum_{k=1}^{\infty} \frac{1}{k^2} \quad (\mathbf{132} : 92);$$

$$\text{"Conjectures concerning the sum} \quad \sum \frac{x^n}{n^2}\text{"} \quad (\mathbf{139} : 24);$$

the series:

$$1 + \frac{1}{9} + \frac{1}{25} + \cdots \quad (\mathbf{131} : 113);$$

$$\frac{1}{1 - a^2} - \frac{1}{4 - a^2} + \frac{1}{9 - a^2} - \frac{1}{16 - a^2} + \frac{1}{25 - a^2} - \cdots \quad (\mathbf{134} : 74);$$

$$\frac{1}{a + p} + \frac{1}{b + p} + \frac{1}{c + p} + \cdots \quad (\mathbf{132} : 60);$$

$$\frac{1}{1 + p} + \frac{1}{4^2 + p} + \frac{1}{9^2 + p} + \frac{1}{16^2 + p} + \cdots \quad (\mathbf{131} : 249);$$

$$s = \frac{x}{1} - \frac{x^2}{4} + \frac{x^3}{9} - \frac{x^4}{16} + \cdots \quad (\mathbf{132} : 152).$$

Euler makes special mention of "the theorem communicated to me by Goldbach:

$$1 + \frac{1}{2^n} + \frac{1}{3^n} + \cdots = \alpha \pi^n \text{"} \quad (\mathbf{131} : 251).$$

There are a great many results of a similar type giving series with sums $\frac{\pi}{4}$, $\frac{\pi}{8}$ (**131**:90, 113; **132**:242), $\frac{\pi}{6}$ (**132**:153), $\frac{\pi}{2}$, $\frac{\pi}{3}$, $\frac{3\pi}{8}$, $\frac{\pi}{12}$ (**133**:174), $\frac{1}{\pi}$, π^2 (**132**:139), $\frac{\pi^2}{6}$ (**135**:45), $\frac{\pi^2}{6} - (\ln 2)^2$ (**133**:152), $\frac{\pi^2}{16}$ (**131**:156), $\frac{\pi^2}{32}$, $\frac{\pi^2}{72}$ (**131**:236), "Goldbach's theorem":

$$\frac{\pi^4}{72} = 1 + \frac{1}{2^3}\left(1 + \frac{1}{2}\right) + \frac{1}{3^3}\left(1 + \frac{1}{2} + \frac{1}{3}\right) + \cdots \quad (\mathbf{132} : 190),$$

and also

$$\frac{\pi^4}{32} - \frac{\pi}{16} = \frac{1}{1^2 \cdot 3^2} + \frac{1}{5^2 \cdot 7^2} + \frac{1}{9^2 \cdot 11^2} + \cdots.$$

Euler records interesting new problems and methods concerning series, communicated to him by Goldbach:

"Find the value of the expression

$$\frac{\pi^2}{6n(n-1)} - \frac{2(n-1)(1 + \frac{1}{2} + \cdots + \frac{1}{n})}{n^2((n-1)^2} + \frac{1}{n(n-1)^2}$$

for $n = 1$" (**132**:203);

"The Goldbach summation of the series with general term $\frac{1}{a^n - 1}$" (**131**:5);

"The method of the renowned Goldbach for expressing double rectangles of series by a single infinite series" (**132**:245);

"A theorem discovered by Goldbach: If A is the square root of the length of an arc of the circle of radius 1, then

$$nA = 1 - \frac{1}{2}(1 - n^2) - \frac{1}{2 \cdot 4}(1 - 2n^2 + 2n^4) - \cdots"\ ^{12} \qquad (\mathbf{131} : 87);$$

the series

$$1 - \frac{1}{2^n} - \frac{1}{3^n} + \frac{1}{4^n} - \frac{1}{5^n} + \frac{1}{6^n} - \frac{1}{7^n} - \frac{1}{8^n} + \frac{1}{9^n} + \frac{1}{10^n} - \frac{1}{11^n} - \cdots$$

"where the rule determining the signs is that primes [under the exponent n in the denominators] get a minus sign, numbers which are products of two primes get a plus sign, products of three primes a minus sign, and so on" (**131**:244).

Euler computes a great number of interesting series expansions, for example, of x^x and $(1 + z)^{1+z}$ (**129**:61), c^x (**129**:205), $\int \sqrt{a^2 - x^2} dx$ (**129**:44), $x^{\frac{1}{x}}, e^{\frac{1}{e}}$ (**131**:263/264), $\cos e^x z$, $\sin e^x z$ (**132**:2), $x/(x^2 - 1)$ (**131**:141), $(1 + x + x^2)^n$ (**133**:74),

$$\Phi = \frac{1}{(1 - n \cos \phi)^\lambda} \qquad (\mathbf{139}:76),$$

$\int n^m$ ^{13}—an example due to Christian Wolff (**131**:86). He considers "Certain inexpressible^{14} functions worthy of note". (**140**:49). He also investigates the iterated exponential function $x^{x^{\cdot^{\cdot^{\cdot x}}}}$ (**130**:70). Binomial expansions such as $(1 + p)^{\frac{1}{p}}$ receive special attention (**131**:38, 99, 265; **132**:248; **138**:21).

Euler notes that the continued fraction expansion for $(1 + x)^n$ was found by Lagrange. He often turns his attention to the interconnection between continued fractions and infinite series (**131**:159; **132**:77; **140**:46), for instance in his note "To define the series arising from division of a fraction" (**132**:51). He often investigates continued fractions purely for their own sake (**131**:27/28; **140**:23), for instance in the note "Arguments concerning the formation of continued fractions" (**138**:14).

He studies the expansions of various rational functions as continued fractions (**131**: 223/224), and of the expressions $e^{\frac{1}{a}}$ (**131**:51), $e - 1$, $\frac{e+1}{e-1}$ (**134**:131), $\sqrt{\pi}$ (**131**:137), and

---

^{12}This statement seems incomplete as it stands.—*Trans.*

^{13}Here m is fixed and the integral sign denotes summation over n—*Author*

^{14}It is not clear what "inexpressible" means here.—*Trans.*

also considers

$$s = 1 + \cfrac{a}{1 + \cfrac{a}{1 + \cfrac{2a}{1 + \cfrac{2a}{1 + \cfrac{3a}{1 + 3a}}}}} \qquad (\mathbf{140} : 21).$$

$\ddots$

Euler considers divisors of infinite series ($\mathbf{132}$:125). He also formulates approximations to various expressions, for example $\sqrt{(x^2 + y)}$ ($\mathbf{131}$:443), and $(1 + \frac{x}{n})^n$ for large n ($\mathbf{131}$:31); he calculates e to 12 ($\mathbf{130}$:35) and 14 ($\mathbf{131}$:31) decimal places.

Series given by recurrence relations are prominent in Euler's notes ($\mathbf{129}$:79: $\mathbf{131}$:21); he investigates these in particular in connection with James Stirling's "difference method" of 1730. Other frequently appearing types of series are: geometrical series ($\mathbf{129}$:66); hypergeometric series ($\mathbf{140}$:41); series arising from the squaring of the circle, of the form

$$\frac{n}{1} + \frac{n^2}{3} + \frac{n^3}{5} + \cdots \qquad (\mathbf{131} : 143);$$

the series for the arctangent ($\mathbf{131}$:146), and Wallis' product ($\mathbf{140}$:30).

Infinite products form a special, frequently recurring theme ($\mathbf{133}$:19, 49, 130), on a par with series solutions of differential equations, for instance in the note "On the use of recurrent series in the integration of differential equations of any order" ($\mathbf{132}$:172).

Euler studies a multitude of infinite sequences, such as $1, 6, 35, 204, \ldots$ ($\mathbf{131}$:64), and $1, 1, 2, 4, 14, 28, 216, \ldots$ ($\mathbf{131}$:122), and applies Brouncker's[15] interpolation method to the sequence $1, \frac{1}{2}, \frac{1 \cdot 3}{2 \cdot 4}, \ldots$ ($\mathbf{133}$:56). He studies the sequence $1, 2, 6, 24, 120, \ldots$ ($\mathbf{129}$:36, 169), and the related series $1 - 2 + 6 - 24 + 120 - \cdots$ ($\mathbf{131}$:197) and

$$s = 1 - x + \frac{x^2}{2} - \frac{x^3}{6} + \frac{x^4}{24} - \frac{x^5}{120} + \cdots \quad (\mathbf{129} : 55).$$

Of the great number of interesting special series appearing in the notebooks, we provide here the following selection:[16]

$$\sum \frac{1}{x^2 + nx + \frac{(n^2-1)}{4}} \qquad (\mathbf{130} : 19);$$

$$\ell(1 + x) = \ell x + \frac{1}{x} - \frac{1}{2x^2} + \frac{1}{3x^3} - \frac{1}{4x^4} + \cdots \qquad (\mathbf{130} : 28);$$

$$z = \frac{x}{1} - \frac{x^3}{3a^2} + \frac{x^5}{5a^4} - \frac{x^7}{7a^6} + \cdots \qquad (\mathbf{130} : 36);$$

---

[15] William Brouncker (1620–1684).–*Trans.*

[16] The first and sixth of the following series appear in the original without summation signs. In the second series (and later) ℓx denotes the logarithm function to any base.—*Author*

$$s = \frac{2}{1 \cdot 3} + \frac{2}{5 \cdot 7} + \frac{2}{9 \cdot 11} + \cdots \qquad (\mathbf{130} : 47);$$

$$s = \frac{m-1}{a-1} - \frac{(m-1)(m-a)}{a^3 - a} + \frac{(m-1)(m-a)(m-a^2)}{a^6 - a^3} - \cdots \qquad (\mathbf{131} : 109);$$

$$\sum \frac{1}{n^2 + (a+b)n + ab} \qquad (\mathbf{131} : 191);$$

$$\frac{1}{1} - \frac{1}{1} + \frac{1}{3} - \frac{1}{3} + \frac{1}{5} - \frac{1}{5} + \frac{1}{7} - \cdots \qquad (\mathbf{131} : 225);$$

$$\frac{n^2}{1} - \frac{(n+1)n(n-1)}{1 \cdot 2} + \frac{(n+1)^2 n(n-1)(n-2)^2}{1 \cdot 2 \cdot 3} - \cdots \qquad (\mathbf{132} : 34);$$

$$\frac{1}{a} + \frac{2b}{n^2 - b^2} - \frac{2a}{4n^2 - a^2} + \frac{2b}{9n^2 - b^2} - \frac{2a}{16n^2 - a^2} + \cdots \qquad (\mathbf{132} : 78);$$

$$\frac{1}{x^m y^n} \text{ as the sum of a series} \qquad (\mathbf{137} : 293);$$

$$1 + \frac{A}{x} + \frac{B}{x^2} + \frac{C}{x^3} + \frac{D}{x^4} + \cdots \qquad (\mathbf{138} : 30);$$

$$s = \frac{1}{1} - \frac{1}{1+a} + \frac{1}{1+2a} - \frac{1}{1+3a} + \frac{1}{1+4a} - \cdots \qquad (\mathbf{138} : 81);$$

Lambert's series :

$$1 + \frac{n}{\alpha}\gamma + \frac{n}{\alpha}\frac{n+\alpha-2\beta}{2\alpha}\gamma^2 + \frac{n}{\alpha}\frac{n+\alpha-3\beta}{2\alpha}\frac{n+2\alpha-3\beta}{3\alpha}\gamma^3 + \cdots \qquad (\mathbf{139} : 28/29);$$

$$\frac{a}{b} + \frac{a}{b}\frac{a+\vartheta}{b+\vartheta} + \frac{a}{b}\frac{a+\vartheta}{b+\vartheta}\frac{a+a\vartheta}{b+b\vartheta} + \cdots \qquad (\mathbf{139} : 60);$$

$$\frac{1 \cdot 2 \cdots p}{(q+1)(q+2)\cdots(q+p)}x + \frac{1 \cdot 2 \cdots (p+1)}{(q+1)(q+2)\cdots(q+p+1)}x^2 + \cdots \qquad (\mathbf{139} : 75).$$

1.5. Differential and integral calculus

In the theory of real functions Euler's interest—insofar as it is indicated by his notebooks—is taken up mainly by: various functions involving natural, base-ten, or hyperbolic logarithms (**129**:35, 53; **130**:14; **131**:35, 99; **132**:129; **133**:59; **134**:23); homogeneous functions of x, y, z (**132**:33); equations of the form $x^{y^m} = y^{x^n}$ (**130**:33); equations of algebraic curves and their integration[17] (**129**:105; **131**:83); topics indicated by the headings "On hyperbolic curves" (**139**:57), "On Lambert's hyperbolic sines and cosines" (**140**:47), and "Goldbach's investigation of two unrectifiable algebraic curves with the property that the sum of their arcs corresponding to one and the same [interval of the] abscissa, can be expressed algebraically" (**131**:124).

He examines specific ways of constructing curves (**129**:160, 207/208), "lines" of the first, second, third, and fourth degree (**132**:182), results on sines and cosines (**132**:156–158), and "Mr. Lagrange's most universal theorem: If the quantity t is determined by x via

---

[17]This might mean the graphs of algebraic functions, and their antiderivatives—and then again might not.—*Trans.*

$t = x - n\varphi x$, where φx denotes an arbitrary function of x, then not only the quantity x, but any function ψx, can in turn be defined in terms of t in the following way. It is assumed that φt and ψt are the same functions of t as φx and ψx are of x" [18] (**138**:39).

Other headings of items in the notebooks are: "On the branches of curves going to infinity" (**132**:180); "On rectifiable curves (**134**:45); "Various theorems on differential equations, divided by ℓx" (**139**:40); and "Maclaurin's theorem" (from his *Treatise of fluxions*) (**132**:212). Euler discusses problems concerning interpolation (**134**:166), such as that of finding a curve passing through prescribed points, bounding a squarable region (**129**:171/172). The notes concerned with circles or arcs thereof form a distinct group (**129**:127, 173; **130**:27/28); one finds headings such as "An arc equal to the radius" (**131**:26), and "Finding an arc of a circle equal to the hyperbolic logarithm of its chord, assuming the radius $= 1$" (**131**:241). He mentions the calculation of π to 100 decimal places by de Lagny[19] (**131**:36).

On complex analysis one finds: an investigation of imaginary numbers (**133**:59); "de Moivre's theorem" and "Cotes' theorem" (**131**:36, 106); and complex-valued functions (**139**:64, 74/75; **140**:48). The entries under the following heads belong to this group: "Differential irrational expressions that seem to be insoluble without imaginary substitutions" (**139**:66), and "Investigate the integral

$$s = \int x^{n-1} dx (x^{\sqrt{-1}} + x^{-\sqrt{-1}})\text{ "}\quad (\mathbf{140} : 48).$$

Of the notes concerned with the differential and integral calculus, problems of integration are very much in preponderance. Here one finds discussions of a connection between expansions in continued fractions and computation of integrals (**134**:119), considered also in the note "Generation of continued fractions from integrable expressions" (**131**:188–190), in notes on squaring the circle (**131**:105/106; **132**:135), and in connection with Descartes (**134**:114), and the quadratrix (**135**:50).

There are brief remarks and phrases, such as: "A type of integral calculus" (**134**:59); "New discoveries in the calculus of exponents" (**129**:118); "Theorems on the integration of certain expressions" (in connection with Lagrange) (**139**:2); and the "Theorem: If P is any function of $\sin\varphi$ and $\cos\varphi$, then the integral $\int P\,d\varphi$ always has the form $\alpha\varphi + Q$, where Q is [again] a function of $\sin\varphi$ and $\cos\varphi$" (**133**:166). Often Euler gives lists of integrals (**140**:52), for example: $\int (x^i\,dx)/(1 - x^2)$, for $i = 1, 3, 5, 7, 9$ (**136**:28); "Formulae for the [result of] integration of $\int x^n\,dx/(1 - x)$" (**131**:34); "Remarkable reductions of integral expressions[20]" (**139**:123); "Expressions integrable using

$$\int du \sqrt{\frac{\alpha + \beta u^2}{\gamma + \delta u^2}}\text{ ,"}\quad (\mathbf{134} : 125);$$

and "List of 12 cases of the integral expression

$$\int dz \sqrt{\frac{f + gz^2}{h + kz^2}}\text{ ,"}\quad (\mathbf{134} : 180).$$

---

[18] This would seem to require amplification.—*Trans.*

[19] In 1719 de Lagny calculated π to 127 decimal places, of which 112 were correct.—*Trans.*

[20] That is, of integrals.—*Trans.*

He investigates integrals suggested by John Wallis' *Arithmetic of infinities* (**131**:13). A dominant theme is the reduction of integrals to algebraic form (**131**:14–16), and the search for algebraic means of integrating: "Reduction of transcendental integrals to algebraic integrations in the case $x = 1$", for example $\int \frac{dx}{1-x}$ (**131**:221); "Find x and y such that the expression $\int \left(\frac{ydx}{x} + \frac{xdy}{y} \right)$ is algebraic" (**133**:52); "Integration of a differential transcendental expression" (**131**:213/214); "Find an approximate algebraic quantity [for] $\int dr \sqrt{1 + r^2} = R$" (**133**:118).

Euler considers the following problem: "[Expressions] x and y are sought such that dx^2/dy is integrable" (**131**:261). He calculates integrals for particular values of the variables, for example: "[Evaluate] $\int z^{m-1} dz (1 - z^n)^{p/q} R = S$, taking $z = 1$ after integrating" (**131**:261); "Find the quantity $\int \frac{x^m dx}{1+x^n}$, if after integrating we set $x = 1$" (**133**:144).

Euler also investigates series expansions as an approach to solving problems in integration, such as: "Expand $\int ydx$ in an infinite series" (**129**:144/145), and problems in integration related to this method, such as in the note "On a special method for differentiating and integrating terms of series[21]" (**139**:20).

We conclude this section with a list of some of the great many problems in integration appearing in the notebooks:[22]

$$\int dy(a + by^{-1/k})^{k-1-i} \qquad (\mathbf{131} : 82);$$

$$\int x^m dx (a + bx^n)^k \qquad (\mathbf{131} : 95);$$

$$\int (b + x)^n dx \sqrt{a^2 - x^2} \qquad (\mathbf{132} : 88);$$

"Integrate $\dfrac{dp}{ap + \sqrt{bp^2 + c}}$" $\qquad (\mathbf{132} : 110);$

$$\int x^{2m} dx \sqrt{a^2 - x^2} \qquad (\mathbf{131} : 108);$$

$$\int \frac{dx}{\sqrt{1 - x^2}}, \int \frac{x^{m+2}dx}{\sqrt{1 - x^2}} \qquad (\mathbf{131} : 111, 205);$$

$$\int e^{x \int \frac{dx}{\sqrt{1-x^2}}} \qquad (\mathbf{131} : 119);$$

$$\int \frac{dx}{1 + \alpha x + x^2} \qquad (\mathbf{131} : 443);$$

$$y = \int \frac{(c^2 - a^2 - x^2)dx}{\sqrt{(c^2 - x^2)(2a^2 - c^2 + x^2)}} \qquad (\mathbf{132} : 185);$$

$$\int x^m dx (1 - x^n)^l \qquad (\mathbf{131} : 186);$$

---

[21] Term by term?—*Trans.*

[22] In the fourth of the following items the original has dp rather than ap in the denominator. In the third last, the integral is to be evaluated from $\varphi = 0$ to $\varphi = 180°$.—*Trans.*

$$\int \frac{x^{2m-1}dx}{(1-x^n)^{2m/n}} \qquad (\mathbf{131}:221);$$

$$\int c^x dx = c^x \qquad (\mathbf{129}:205);$$

"Integration of the function $y = x^x$" $(\mathbf{132}:32);$

$$\int \frac{d\varphi}{(c+d\cos\varphi)^{n+1}} \qquad (\mathbf{134}:74);$$

$$\int \frac{nP^{n-1}}{Q^{n+1/2}}dy \qquad (\mathbf{134}:81);$$

$$\int x^{p-1}y^{q-n}dx \qquad (\mathbf{134}:169);$$

$$\int \frac{dx\sqrt{x}}{(f+x)(g+x)} \qquad (\mathbf{134}:176);$$

"Expansion of the expression $\displaystyle\int \frac{Bzzdz}{(f+gz^2)(h+kz^2)} = Q$" $(\mathbf{134}:177);$

"Find the integral $\displaystyle\int \frac{d\varphi}{\sqrt{\cos\varphi}} = P\sqrt{\cos\varphi}$" $(\mathbf{133}:80);$

"D'Alembert's difficult case $\displaystyle\int \frac{zdz}{\sqrt{A+Bz+Cz^2+Dz^3+Ez^4}}$" $(\mathbf{134}:189);$

"Find the integral $\displaystyle\int \frac{d\varphi(a\cos\varphi+b)}{\sqrt{a^2+2ab\cos\varphi+b^2}}$" $(\mathbf{134}:220);$

"Reduction of a particular integration" $(\mathbf{132}:7);$

"A particular reduction of the form $v = e^x\displaystyle\int e^{-x}ydx - e^{-x}\int e^x ydx$" $(\mathbf{135}:15);$

"A special reduction of the integrals $\displaystyle\int e^{m\varphi}d\varphi\sin\varphi^n$" $(\mathbf{136}:19);$

"Reduction of the expression

$$\int \frac{(A+Bx)dx}{(a^2-2abx\cos x+b^2x^2)^n} \text{ to a simpler integral}$$ $(\mathbf{136}:22);$

"Reduction of the integral $\displaystyle\int \frac{(ax^{\lambda-1}+bx^\lambda)dx}{\sqrt{1-x^2}}$ to two others" $(\mathbf{136}:27);$

"Form of the most difficult integrations" $(\mathbf{139}:65);$

$$\int \frac{dz\sqrt{1+z^2}}{1-z^4} \qquad (\mathbf{139}:44);$$

"A special integration of the expression

$$\int \frac{d\varphi}{(1-2b\cos\varphi+b^2)^{n+1}} \begin{bmatrix} \text{ab } \varphi = 0 \\ \text{ad } \varphi = 180° \end{bmatrix} = 0$$ $(\mathbf{139}:112);$

$$z = \int d\varphi \sin \varphi^{\lambda-1}(1 + t \cos \varphi)^{n+1} \qquad (\mathbf{140} : 130);$$

$$\text{``}s = -\int \frac{x^{m-1}dx\ell x}{\sqrt[n]{(1 - x^n)^m}} \text{ ab } x = 0 \text{ ad } x = 1\text{''} \qquad (\mathbf{140} : 55).$$

1.6. Differential equations

The notebooks contain on the one hand general investigations of differential equations and on the other discussions of a multitude of special equations.

Thus one encounters propositions on "difference-differential equations" ($\mathbf{111}$:111), a note entitled "Finding differential equations admitting integration" ($\mathbf{132}$:34–36), headings such as "Analytic theorems" ($\mathbf{139}$:62), and "On functions of three variables" ($\mathbf{140}$:9), differential equations that are soluble by means of infinite series ($\mathbf{138}$:28/29; $\mathbf{140}$:31)), and those that are not ($\mathbf{134}$:134).

An important[23] proposition: "If

$$\int \frac{dx(A + Bx^2)}{\sqrt{\gamma + (\beta\beta + \alpha\gamma - 1)x^2 - \alpha x^4}} - \int \frac{dy(A + By^2)}{\sqrt{\gamma + (\beta\beta + \alpha\gamma - 1)y^2 - \alpha y^4}} = C - Bxy,$$

then $x^2 + y^2 + \alpha x^2 y^2 = 2\beta xy + \gamma$" ($\mathbf{133}$:101).

Euler considers the following problem: "Suppose we have an arbitrary equation $Y = 0$, where Y is a function of an unknown y and has several roots relative to y.[24] Find conditions under which one root is greater than another root on some set a" ($\mathbf{132}$:3). He also considers differentiation of the equation $x^x = y^y$ ($\mathbf{130}$:39).

Riccati's equation $y' = f_0(x) + f_1(x)y + f_2(x)y^2$ is of especial interest to Euler ($\mathbf{130}$:16, 32; $\mathbf{131}$:100; $\mathbf{134}$:47, 56; $\mathbf{139}$:12, 52/53; $\mathbf{140}$:7), as are equations that can be solved using integration factors ($\mathbf{133}$:170; $\mathbf{135}$:72; $\mathbf{136}$:48). He also works on homogeneous equations ($\mathbf{136}$:20–22), and transformations of differential equations ($\mathbf{133}$:161; $\mathbf{135}$:91; $\mathbf{140}$:39), and investigates the connections with complex analysis ($\mathbf{136}$:56; $\mathbf{140}$:39).

Of the particular differential equations examined in the notebooks, we list, in increasing order, the following ones.

1. First-order differential equations:

$$axdy + dx\sqrt{c^2 + y^2} = 0 \qquad (\mathbf{129} : 80; \text{compare } \mathbf{129} : 102/103);$$
$$x^m dx + y^2 dx = dy \qquad (\mathbf{130} : 37/38);$$
$$dy + Pydx + Qy^2 dx = Rdx + QSSdy \qquad (\mathbf{132} : 133);$$
$$ax = y - \frac{sdy}{ds} \qquad (\mathbf{130} : 52);$$
$$adq = q^2 dp - dp \qquad (\mathbf{131} : 70);$$
$$Pdx + Qdy = 0 \qquad (\mathbf{131} : 98/99);$$
$$ccdz - zzdz = xzdx + cdx\sqrt{x^2 + z^2 - c^2} \qquad (\mathbf{131} : 100);$$

---

[23] Important to Euler, or to the author? And why?—*Trans.*
[24] That is, values of y at which $Y = 0$ (*zeros* of Y).—*Trans.*

$$ax^m dx + bx^n y dy + cx^{n-m-1} y^2 dx + dy = 0 \qquad (\mathbf{131} : 204);$$

$$v dv + vP dz = Q dz \qquad (\mathbf{131} : 218);$$

$$dy + yy dx = ax^n dx + b^2 x^{2n} dx \qquad (\mathbf{131} : 223);$$

$$x^m dx = x^{q+1} dz + (p - m)x^q z dx + z dx \qquad (\mathbf{133} : 125);$$

$$dy + P y dx = Q yy dx + R dx \qquad (\mathbf{134} : 7);$$

$$dy + cy^2 dx + bx^{m+1} y dx = ax^m dx \qquad (\mathbf{134} : 119);$$

"Given $dz = p dx + pz dy$, find the relationship between x, y, z" $(\mathbf{136} : 34);$

Integrability conditions for the following differential expression:

$$(M + V) dx + (N + V) dy \qquad (\mathbf{136} : 58);$$

$$\frac{dp(p - u)(1 + pu)}{(1 + p^2)^2} = n du \qquad (\mathbf{139} : 58);$$

$$\frac{dx}{\sqrt{x}} = \frac{dy}{\sqrt{y}} \text{ (in connection with which}$$

   Euler mentions "Lagrangian analysis") $(\mathbf{139} : 86 - 90, 104);$

$$\frac{dx}{\sqrt[n]{a + bx^n + cx^{2n}}} = ds \qquad (\mathbf{140} : 34);$$

Eliminate u from $du + u^2 dx + Pu dx + Q dx = 0$ $(\mathbf{134} : 134);$

"A remarkable integration of the equation

$$a dp(1 - p)(bc - z^2) + dz(bc + a^2 p^3 + apz)(1 + p) = 0" \qquad (\mathbf{140} : 73).$$

2. Second-order differential equations:

$$x^m dx^2 = y^n d dy \qquad (\mathbf{134} : 47);$$

$$d dy + P dx dy + Q y dx^2 = 0 \qquad (\mathbf{132} : 132; \text{compare } \mathbf{131} : 215);$$

$$d dx + mm d\varphi^2 = \alpha d\varphi^2 \sin \varphi \qquad (\mathbf{133} : 48/49);$$

$$\frac{d dr}{r} + \frac{A du^2}{(Cr^2 + F + 2Gu + Hu^2)^2} = 0 \qquad (\mathbf{133} : 158);$$

$$(\alpha x^3 + \beta x^2) d ds + (\gamma x^2 + \sqrt{x}) dx ds + (\varepsilon x + \zeta) s dx^2 = 0 \qquad (\mathbf{135} : 57);$$

$$d dz + Xz dx^2 = 0 \qquad (\mathbf{134} : 202);$$

$$\frac{d dz}{dy^2} = \frac{d dz}{dx^2} + Xz \qquad (\mathbf{135} : 14);$$

$$P ds^2 + 2Q s dy ds + R ss dy^2 = A dx^2 \qquad (\mathbf{135} : 56);$$

"A new reduction of the equation $\dfrac{d dv}{v} + \dfrac{T dv dx}{v} + \gamma dx^2 = 0$" $(\mathbf{135} : 85).$

3. Differential equations of order higher than two[25]:

$$ydx^4 + a^4d^4y = 0 \qquad (\mathbf{131}:227);$$

Look for y such that $\alpha\left(\dfrac{ddy}{dt^2}\right) = \beta\left(\dfrac{d^4y}{dx^4}\right)$ $\qquad$ (**134**:88);

$$x = Ay + \frac{Bdy}{dx} + \frac{Cddy}{dx^2} + \frac{Dd^3y}{dx^3} + \frac{Ed^4y}{dx^4} + \frac{Fd^5y}{dx^5} + \cdots \quad (\mathbf{133}:57).$$

1.7. Geometry

A large proportion of Euler's notes on elementary geometry concern conic sections (**131**: 107; **132**:64), especially ellipses (**131**:111) and parabolas (**131**:20, 99; **136**:70). Lambert communicated theorems and conjectures on these to Euler. Conjugate diameters form a frequently recurring topic (**132**:125/126). There are many problems concerning circles: for instance, concentric circles (**129**:152) or a pair of intersecting circles (**131**:133/134). There is one problem related to Apollonius' tangent problem: Find a circle tangent to three given circles (**132**:7).

Euler also considers arcs of circles, in particular their division[26] (**131**:154, 263/264; **132**:88), chords (**131**:154), semicircles and their chords and arcs (**129**:1a; **131**:139; **132**:73). Following Frenicle[27] he considers a semicircle to which a rectangle has been adjoined (**132**:111).

There are many notes dealing with triangles (**129**:29, 32, 69, 86; **130**:44, 78/79; **131**:6, 76; **132**:56, 73), quadrilaterals (**131**:7, 8, 36, 51, 65/66, 71), and arbitrary polygons (**129**: 32/33). For example under the heading "A geometrical problem" one finds the following: "Given the four sides of a quadrilateral $ABCD$ and the angle between the two opposite sides AB and CD, construct the quadrilateral" (**132**:55).

The many notes on regular polygons form a distinct subset (**129**:48, 79; **130**:77, 80; **131**:14, 36/37, 140/141, 143, 154, 212), including "Renaldini's rule" for inscribing a regular polygon in a circle[28], and "a proof of the theorem of the renowned Segner on the number of subdivisions of an arbitrary polygon into triangles" (**134**:96).

Euler often considers regular polyhedra (**131**:22; **134**:39). In the note (**133**:65/66) he works out his theorem on polyhedra:

$$E + F = K + 2.$$

He considers questions concerning parallelograms (**131**:3), trapezia (**131**:216), circular cylinders (**129**:140), and pyramids (**133**:59). Again under the heading "A geometrical problem" one finds the following: "Given four straight lines construct a straight line such that the intervals of this line [determined by the points of intersection] are in a given relation to one another" (**132**:232).

---

[25] Although in the second of these differential equations it may seem that t should be replaced by x, the original was as here.—*Author*

[26] Cyclotomy, perhaps.—*Trans.*

[27] Bernard Frenicle de Bessy (1605–1675).—*Trans.*

[28] Count Carlo Renaldini (1615–1698).—*Trans.*

There are many investigations of plane (**131**:22, 197/198, 222, 243/244; **132**:25) and spherical trigonometry (**129**:210/211: **131**:22, 24, 55 (on spherical 2-gons)). He considers the problem of finding three angles in arithmetic progression[29] (**129**:120), and the following one: "For a given angle φ find an angle ω such that $\cos \omega = \frac{\cos \varphi + n}{1 + n \cos \varphi}$, where $n < 1$" (**139**:106).

The notes concerned with transformations of the plane also form a distinctive group: transforming a circle into a square (**131**:137); a triangle into a circular sector (**131**:142); and various figures into triangles (**131**:86); and tessellations of the plane. Following Saint Vincent[30] Euler considers also subdivisions of a quadrant of a circle (**131**:206; **136**:92), subdivisions of a circle into three regions of equal area by means of two straight lines (**131**:140), and the following problem: "On the diameter AB of a semicircle ADB construct a segment APB of another circle dividing the semicircle into two equal parts" (**136**:92). Euler also works on Alhazen's[31] problem: "Given a point-source of light A from which a ray reaches another point B [after reflection by a spherical mirror], find the point on the circle [i.e., on the mirror] where the ray is reflected" (**192**:42), on representing the surface of a sphere on the plane (**139**:10), and on an instrument for solving geometrical problems (**135**:6).

1.8. Differential geometry and the calculus of variations

Of the notes devoted to differential geometry, a large proportion are concerned with finding curves or families of curves satisfying certain specified conditions (**139**:16). Here is a selection of these:

"Find a curve AM such that if one draws the tangent MT and the normal MN, then $AT = PN$"[32] (**129**:87).

"Find a curve AM equal to its subnormal[33] PN" (**129**:88–90).

"Find a curve AM whose radius of curvature $MR = \tan g\ MT$" (**129**:92–99).

"Find a curve whose evolute is itself" (**129**:131).

"Find a curve AM such that the arc AM is equal to the normal TM" (**129**:137–140).

"Find a curve DM such that all rays issuing from a point A are reflected to a point B" (**133**:25).

"Find an infinite family of curves all equal in length, passing through two given points" (**133**:51).

Euler looks for algebraic curves whose quadrature depends on the quadrature of a given curve (**131**:30). He seeks curves satisfying conditions on their curvature (**140**:50), and a general method for constructing the radius of curvature (**129**:155–157). He considers the following problem, posed by Clairaut in 1740:

"When a circle rolls around another circle, a peg attached at a point M of the first circle describes a curve EM, whose nature it is required to discover" (**133**:139).

---

[29] The intended context is missing here.—*Trans.*

[30] Gregory Saint Vincent (1584–1667).–*Trans.*

[31] ibn-al-Haitham (ca. 965–1039), known in the West as Alhazen.–*Trans.*

[32] The points T and N are where the tangent and normal through M meet the x-axis, and P is the orthogonal projection of the point M onto TN.—*Author*

[33] The *subnormal* is the projection on the x-axis of the segment of the normal from the point of interest to the point where it intersects the x-axis.—*Trans.*

Euler evinces particular interest in questions relating to rectification or rectifiability of a curve. Thus he looks for a general equation for a rectifiable algebraic curve (**131**:112; **133**:57; **132**:32–34), and investigates algebraic curves whose rectification depends on their own quadratures or on given quadratures (**131**:128, 130), for example, algebraic curves whose length has the form $\alpha \int \frac{dz}{\sqrt{1-z^4}}$ (**133**:161).

Euler investigates curves introduced by Varignon[34] and Leibniz (**132**:169–170), tautochrones[35] both in a vacuum and in a resisting medium (**130**:31, 34, 40; **131**:195; **133**:139; **138**;94, 99), influenced in particular by Fontaine (**132**:137, 139). Euler examines Daniel Bernoulli's method (communicated by letter on October 20, 1742) for finding an "elastic curve"[36] (**132**:173, 183–185). There are notes on the logarithmic (**139**:60) and Archimedean spirals (**133**:22–24, 123), Archimedes' snail[37] and the lemniscate (**133**:159–160), the tractrix (after Clairaut) (**132**:108, 164), and conic sections (**132**:134), in particular quadrants and perimeters of ellipses (**138**:38).

Euler investigates also the conchoid[38] (**133**:65), caustic curves[39] (**129**:45; **134**:160), the cycloid (**129**:131; **131**:82, 131; **133**:64; **139**:60), and various trajectories (**132**:200; **134**:64; **137**:1–2), especially orthogonal families of these (**133**:131; **134**:125; **137**:271–273, 306, 358–360). He studies evolutes of various curves (**129**:131, 155–157; **131**:119; **133**:67; **134**:64), in particular in the note "On the continuous development, after Bernoulli, of a curve with orthogonal ends" (**133**:70). He mentions paradoxes involving curves (**139**:53), and gives a general explanation "Concerning the nature of curved lines" (**139**:59).

Those notes devoted to investigations of the theory of surfaces (**138**:1–2) and space curves, for example spherical epicycloids (**132**:222), form an important subset. Of these the following are worthy of note: "Reason about the curvature of an arbitrary surface" (**135**:29); "Find all geometrical curves on a spherical surface" (**138**:16–18); "Define the radius of osculation of curves arbitrarily drawn on a spherical surface" (**140**:40); the problem solved by Huygens concerning the situation where "the sum of the surface areas of elliptic and hyperbolic conoids are equal to [the area of] a circle" (**133**:108). The following problems also belong here: "Of any surface constructed above a horizontal plane, determine that portion covering a given region…" (**131**:314); "the Florentine problem: On a spherical algebraic surface draw a curve AM which together with a line of latitude AP and a meridian PH contains a squarable region APM"[40] (**131**:127; **133**:162/163).

In the variational calculus Euler shows particular interest in isoperimetric problems (for regions and curves bounding them) (**129**:213/214; **131**:49/50; **132**:174), and considers "Jakob Bernoulli's second problem on isoperimetries, which he was the first to propose to his brother" (**131**:109).

---

[34]Pierre Varignon (1654–1722).–*Trans.*

[35]A *tautochrone* is a curve in a vertical plane with the property that at whatever point of it a bead is released, it will always slide to the bottom in the same amount of time.—*Trans.*

[36]Daniel Bernoulli and Euler derived the "equilibrium curves" for flexible and elastic bodies in 1728, so this item belongs more properly, perhaps, to the next section, on mechanics.—*Trans.*

[37]A tube wrapped helically around a pole, which when rotated acts as a pump. (Thus this item also should perhaps be relegated to the mechanics section.)—*Trans.*

[38]"The conchoid of Nicomedes" (ca. 200 BC) has equation $r = a + b \sec \theta$ in polar coordinates.—*Trans.*

[39]A *caustic curve* is the envelope of rays of light emitted by a point-source of light and reflected (or refracted) by a given curve.—*Trans.*

[40]The phrase "spherical algebraic" needs clarification. There is a problem known as "the Florentine problem" attributed to a Florentine by the name of Viviani, concerning the intersection of a sphere with a cylinder (see: D. J. Struik's *Lectures on classical differential geometry*), but this would seem to be a different problem.—*Trans.*

He poses the problem of finding among all curves one for which integrals of the forms $\int Q\,dx$ (**131**:49, 82), $\int R^m\,ds$ (**131**:18), and $\int z\,dx$ (**131**:183, 269) attain a maximum or minimum. He works on the brachistochrone (**132**:206; **131**:139; **140**:88), on the problem of "determining the line of quickest descent on a spherical surface" (**134**:141), and on that of "drawing the shortest line on an arbitrary surface" (**139**:119), and also on minimal surfaces (**134**:161).

Euler discusses "problems concerned with finding curves enjoying the property of maximality or minimality absolutely or relative to certain conditions" (**132**:15/16).

He also occupies himself with "an extension of the method of maxima and minima" of Lagrange, in connection with a letter from Lagrange of August 12, 1755 (**134**:139/140; compare **136**:53/54). He gives various expositions of the variational calculus, under headings such as "Elements of the calculus of variations" (**132**:2–5), and "Treatise on maxima and minima" (**139**:61), and remarks that "Everything that one investigates in the calculus of variations reduces to that area of analysis where functions of two variables are studied" (**138**:2/3).

2. Mechanics

In Euler's notes one very often encounters headings or drafts of notes on mechanics (**129**:134–136; **130**:52), lists of problems in mechanics (**134**;40–41), and paradoxes (**135**:22).

1. Euler considers, in particular, friction ("The laws of friction") (**131**:26), and the laws governing collision of objects (**131**:73, 84, 85; **132**:215; **138**:360/361).

2. A substantial number of notes are concerned with the mechanics of material points and rigid bodies. Thus there are many investigations of attraction between bodies, especially of the three-body problem (**129**:127–131; **130**:24–27; **132**:22, 149; **133**:2, 76; **134**:49, 50, 78, 92, 93; **135**:52), of ballistics (**129**: 64, 87; **132**:187), of the motion of several bodies in a tube that is itself in motion (**132**:205, 207), or on inclined planes (**132**:60; **131**:1), or in a resisting medium (**129**:177), and of isochronism[41] (**129**:129). Euler investigates pendula (**129**:70, 71, 119, 120; **132**:127/128, 192/193; **133**:43), and considers a great many problems on vibrational or oscillatory motion, such as "On the oscillation of rigid bodies" (**132**:22), and "To determine the center of oscillation of an arbitrary figure hanging from a given axis" (**131**:135).

A few other, more specialized, problems:

"On the action of attractive forces on bodies connected by strings" (**131**:63); "Finding four weights that can be attached to the corners A, B, C, D of a uniform trapezoid $ABCD$"[42] (**131**:165); "On the periods of revolution of bodies in motion about an arbitrary center under the influence of an arbitrary force" (**129**:30); "On a new method for determining the motion of rigid bodies" (**139**:13).

There is a note on elliptic orbits pertaining more to celestial mechanics (**136**:81). Here and there one finds references to "the life force" (**132**:202).

---

[41] The property of an oscillating mechanical system of having the same period of oscillation regardless of amplitude, or more generally, the occurrence of similar actions over time intervals of equal length.–*Trans.*

[42] So that the center of gravity remains unchanged?–*Trans.*

3. Notes on the mechanics of flexible and elastic bodies deal with the catenary (**133**:44), and "a perfectly flexible string" (**131**:182; **132**:26; **133**:133/134 ("Hermann's method); **137**:167; **139**:8). The problem of "finding the acceleration of a vibrating string AQC" is proposed (**129**:68/69).

4. Most of the notes on mechanics are concerned with the mechanics of liquids and gases (**129**:36, 107, 207; **136**:66), in particular hydrostatics (**131**:38) and the hydrodynamics of Daniel Bernoulli (**131**:172).

Of the great many problems of this type that Euler considers, we mention only the following few:

"On the condensation of liquids produced by an internal force" (**129**:34);

"On the motion of fluids" (**129**:38);

"On the motion of a liquid along a canal of arbitrary shape" (**131**:211);

"An investigation of the motion of currents" (**133**:25);

"An experiment in the theory of pulsations transmitted through the air" (**136**:40);

"To find the resistance to the motion of a body in perfectly elastic media" (**129**: 67);

"To determine the oscillation and period of oscillation of vibrating air" (**129**:67, 72);

How triangular, quadrilateral figures, and other bodies float when immersed in water (**131**:68, 72, 77);

Note also the heading: "Remarks on Mr. d'Alembert's treatise on the equilibrium and motion of fluids" (**134**:14).

5. The theory of mechanisms, machines, and instruments of various kinds. Euler writes for example on pumps (**133**:26/27), machines (**133**:34/35), saws (**134**:34/35), toothed wheels (cogs) (**136**:32), and Segner's "water wheel" (**133**:58).

6. There is a series of notes relating to ships (**129**:56, 61, 170), in particular concerning a boat on a river (**131**:47–49), and the initial motion of a ship (**133**:105). He appends an "excerpt from the memoir of Mr. Shosho" (**134**:102).

3. Astronomy and geodesy

All 12 of the notebooks contain notes on astronomy and geodesy. The following fragments are concerned with celestial mechanics and the motion of the planets: (**131**:64, 233, 266; **132**:133, 135/136, 154, 199, 251/252; **133**:2, 7, 8; **134**:172; **135**:53). In particular, Euler discusses the earth's orbit (**129**:121/122, 146; **131**:65, 269; **132**:182; **133**:74; **135**:34), the earth's motion under the influence of the sun's attraction (**133**:39, 41), the annual periodic motion of the poles (**133**:40), and the distance from the earth to the sun (**130**:34). He attempts to determine the temperature at every point on the earth at every moment in time (**131**:120–122).

Euler considers the motion of a single body under attraction from a fixed center (**133**:2, 3, 5), the same for two bodies (**133**:35, 52), and the three-body problem (**133**:76). (See also the mechanics section above.) There is a thoroughgoing discussion of the distances from the sun to the planets (**132**:214, 217), the angular velocities of the planets (**131**:32), the computation of the perturbation arising in the case of two planets (**131**:172/173), and from all planets combined (**130**:40/41). Euler often considers the planets in conjunction with the moon or comets (**130**:50/51; **131**:266/267; **132**:133, 244; **133**:2, 7, 8, 18).

Several notes are concerned with practical, observational astronomy. Euler cites the astronomical observations of various astronomers, for instance Ptolemy and Hipparchus (**130**:50; **132**:128; **134**:182–186), including observations of the planets (**131**:268), in particular Mars (**131**:86). He describes a method for taking the height[43] of the pole star and the inclination of the other "fixed" stars (**132**:212). He studies the theory of refraction in connection with the observation of the stars (**131**:201/202), discusses various instruments, such as a parabolic mirror for observing the sun (**133**:79), the astronomical quadrant (**132**:66), the naval quadrant for measuring the heights of stars (**132**:15), and sundials (**131**:107; **134**:51).

Notes on lunar theory include: theory of the lunar orbit (**131**:99; **133**:2, 7, 8; **134**: 97–99; **135**:51, 70–72; **138**: 29–37), determination of the moon's position (**133**:73), equations for the moon's motion (**134**:36–38), lunar eclipses (**134**:50), and the period of revolution of the moon (**130**:38). In connection with sun, moon, and earth, Euler considers the problem of "Finding, at a given instant when the luminous bodies are in conjunction, the centers[44] on the surface of the earth" (**133**:13–16). There are fragments exclusively concerned with comets (**132**:155, 183), in particular the orbit of the 1759 comet (**136**:43).

A few fragments are devoted to the fixed stars, in particular the sun: the declination of a star (**132**:238; **139**:6), the obliquity of the ecliptic[45] (**131**:24), the height of the pole star and the ecliptic (**134**:92), the sun's motion (**133**:119), the sun's radius (**134**:95), solar parallax (**139**:54/55), and the determination of the length of twilight in terms of the height of the pole star and the sun's declination (**131**:45/46, 49).

Euler also considers certain questions relating to chronology (**131**:267), in particular the length of the year (**133**:128), the uniformity of time (**132**:141), lunar and solar cycles (**131**:134, 263), and a method for determining the date by the Julian calendar from the date by the Gregorian calendar (**130**:29/30).

Finally, there are discussions of problems of geodesy (**140**:82/83), and cartography (**131**:260; **137**:317—Delisle's projection), in particular that of determining the shape of the earth using a given value for lunar parallax[46] (**133**:122/123), loxodromes[47] (**139**:19), navigation and a ship's course (**131**:8, 257–259, 261), for example the problem: "Find increasing latitudes in Mercator's table", and hydrographical maps (**140**:46/47).

4. Physics

Euler's notes contain occasional lists of problems in physics (**134**:22), discussions of the specific gravity of various materials (**129**:169, 173), in particular the earth (**132**:221), of the efflux of water under pressure ("On the ejection of water") (**134**:71), and of the motion of air in pipes (**137**:130/131).

The majority of the notes on physics are devoted to optics (**133**:51, 53, 86–88, 106–108, 150/151; **134**:50, 67, 70, 96–107), sometimes with reference to Newton and [his theory of]

---

[43]That is, the angular height measured from the horizon.—*Trans.*

[44]Of attraction? —*Trans.*

[45]The *ecliptic* is (the plane of) the apparent path of the sun among the stars; it is inclined to the celestial equatorial plane.—*Trans.*

[46]*Lunar parallax* is the difference between the moon's apparent celestial position as measured from the center of the earth and its position when on the horizon as seen by an observer on the surface of the earth.—*Trans.*

[47]A *loxodrome* is a line of constant (compass) bearing on the earth's surface.—*Trans.*

refraction (**135**:11, 47; **138**:153/154). Euler considers combinations of lenses (**133**:151), refraction in flint glass (**137**:1) and in convex lenses in connection with Dollond's experiments (**136**:67–80), and the *camera obscura* (**134**:31). He examines how shadows are thrown by objects (**138**:12/13), the theory of optical instruments, such as telescopes (**136**:6, 17), spherical mirrors (**136**:54/55), and igniting glass ("vitrum causticum") (**129**:152, 155).

He discusses problems in magnetism (**133**:174), for example that of defining the direction of a magnetic needle on the earth's surface (**136**:57/58), the theory of magnetic force (**134**:137–139, 143, 217/218), and investigations in electricity (**134**:39, 206–208).

Euler also touches on metereological questions, for example concerning clouds (**130**:23), barometers (**136**:52), and the density of air (**129**:62).

In terms of the number of notes, acoustics occupies a secondary place. Here musical themes predominate (**129**:35–37, 39–40, 51–53, 80/81, 132; **130**:45/46; **135**:62/63; **138**:129), and musical instruments, for example the flute (**129**:90/91; **135**:27/28; **138**:157/158).

5. Geography, medicine, chemistry

At some places in the notes Euler cites geographical latitudes and longitudes of certain places (**130**:56, 72), and at others he discusses the movement of muscles (**134**:51). With chemistry in its wider sense the following notes are concerned: the dissolving of metals—in connection with which he refers to Lomonosov (**134**:186); the durability of wood (**134**:259); on recipes for paint (**135**:90); and the salinity of the sea (**136**:22). He also considers how to obtain fresh water from sea water (**134**:56).

6. Philosophy, philology

Notebook No. 134 contains extensive philosophical discussions. Of these the "Ontological elucidations" are concerned with Wolffian philosophy (**134**:4–7), the natural philosophical theses (**134**:8–14) and logical theses (**134**:15–17), and with the definition of "the art of discovery" (" ars inveniendi"): "The art of discovery is a method of determining from one or several given properties or relations pertaining to any sorts of entities, other relations or the nature of the entities themselves" [48]

Euler discusses the concepts of time and space, and also the resultant of motions when bodies collide (**134**:19–21). To this there are appended further notes on logic (**134**:42/43). There is a note on philosophy also in (**136**:65). There is also a plan for a dictionary (**130**:11/12), a list of arithmetical, geometrical, and algebraic subdisciplines, belonging to a "universal mathematics" ("mathesis universalis").

To philology belong remarks on Latin and German (**134**:68, 150–165), on phonetics ("the elements of speech") ("elementa sermonis"), on the frequency of letters in German sentences (**134**:253), and also various aphorisms and maxims, such as the following:

> Voltaire a de l'esprit, il est vrai, mais, mais, mais,...
> Les mais à son égard ne finissent jamais.[49]

---

[48]"Ars inveniendi est methodus ex una vel aliquot datis alicuius rei proprietatibus seu affectionibus, alias affectiones, vel naturam ipsius rei definiendi".

[49]Voltaire is clever it is true, however, however, however,... In his case the "howevers" go on for ever.

7. Miscellaneous

The notebooks contain various notes whose content is either of a personal nature (**130**:0–9), or in any case unrelated to science, which are nonetheless of particular interest for a biography of Euler. There are Russian exercises (**130**:11, 17, 18), lists of expenses and registers of purchases and payments (**130**:82/83; **134**:244–251; **135**:3–5), specifications for a certain geographical map (**130**:61–72), medicinal recipes (**130**:2/3), prices of the noble metals (**134**:30, 47), book indices (**134**:1), especially of borrowed books (**134**:257/258), a list of 509 book titles (**134**:192–201), and also a list of Euler's unpublished articles: "Catalogue of works presented to the Academy but not yet published, 1773" ("Catalogus dissertationum Academiae traditarum quae nondum impressae sunt 1773") (**139**:124–131). Of the 312 titles, many number-theoretical, only 98 are crossed out.

References

[1] Mikhaïlov, G. K. "Euler's notebooks in the Archive of the Academy of Sciences of the USSR". Research in the history of mathematics (Ist.-mat. issled.), 1957. Issue 10. pp. 67–94.

[2] Matvievskaya, G. P. "On the unpublished manuscripts of L. Euler on Diophantine analysis". Research in the history of mathematics (Ist.-mat. issled.), 1960. Issue 13. pp. 107–186.

[3] *Manuscript materials of L. Euler in the Archive of the Academy of Sciences of the USSR.* M.; L.: Published by Acad. Sci. USSR, 1962. Vol. 1. pp. 6–7.

On Euler's Surviving
Manuscripts and Notebooks

G. P. Matvievskaya

During the two centuries that have passed since the death of Leonhard Euler[1], a plethora of articles and books have been written about various aspects of his life and work. There have also been more than a few serious attempts at composing a complete scientific biography of the great scholar by extrapolating from the known facts. However all such endeavors have proved of limited success since their authors invariably encountered difficulties not within their competence to overcome.

Although Euler's life was not rich in outward events, and it might appear that tracing the path of his creative development presents no great difficulties, in fact the bare facts of his biography give scarcely an inkling of his personality. Even Euler's contemporaries, who, wishing to do him justice, strove to throw light on his multifaceted scientific activity and preserve for posterity a vivid portrait of their colleague and teacher, understood the complexity of this undertaking. This is apparent from the texts of the orations of N. I. Fuss and J. von Stählin on the occasion of Euler's death. Stählin, noting "the exceptional qualities that evoked astonishment by their rarity, and singled him out from among the millions of other human beings" [1, p. 37], asserted that a biography of this scientific genius "would fill a massive tome" [*loc. cit.*, p. 40]. Fuss, who had engaged in joint work with Euler over many years, spoke of Euler's life in the eulogy he delivered on October 23, 1783. This speech still impresses by its sincerity and the profound understanding it conveys of Euler's enormous scientific merit[2]. However even Fuss considered an "accurate laudatory assessment" beyond his competence, and confined himself to expressing his readiness to hand over the necessary materials to someone prepared to undertake the task [2]. As the future was to show, later biographers of Euler found the challenge no less great, so that even a century later the situation in this regard remained essentially unchanged.

Over and above bare biographical facts and analyses of concrete results of the scholar's

---

[1]This was more accurate at the time this essay was written, in 1986. Euler died in 1783.—*Trans.*

[2]N. I. Fuss' eulogy, which appeared in a Russian translation in 1801, is reproduced in the present volume.— *Eds.*

activity, a full-fledged scientific biography should reveal to the reader its subject's interior world and provide a glimpse into the laboratory of his creations, as the well-used phrase has it. In Euler's case this task certainly exceeds the capabilities of the individual investigator. The difficulties involved arise in the first place from the colossal amount of material to be studied. In terms of breadth of interests, number of discoveries of key importance in every area of mathematics he turned his attention to, and wealth and variety of original ideas, Euler had no peer, and this fact was of crucial import for his scientific heritage; even a simple survey of the results he obtained becomes in his case a complex problem occupying more than one generation of scholars. Euler's legacy was so vast that just compiling a list of his works turned out to be no easy assignment for his biographers.

In 1784 the list of Euler's published works contained 562 titles; however there remained many completed but unpublished items. The last publication in which he himself participated was Volume 1 (1783) of a collection entitled *Opuscula analytica*, of which a second volume appeared in 1785. By 1826, the list of Euler's published works had risen to 771 titles, and there remained only 14 unpublished treatises. These were included in the volume of the "Mémoires" of the Academy published in 1830 as a supplement to the series with this title that had been discontinued in 1826.

Once the publication of those of Euler's works in finished form was complete, attention turned to his enormous archive, which contained—as had been correctly surmised—important materials for his scientific biography. This was true above all of his scientific correspondence with prominent contemporary scientists, which contained a great many ideas of extraordinary interest.

On examining the manuscript materials in the Academy's archive and that of his own family, it became clear to P. N. Fuss that Euler's scientific *Nachlass* was far from exhausted. In 1844 he found many (61) manuscript works of Euler's on mathematics and mechanics ready for publication but unknown to the publishers of his surviving manuscripts. This remarkable find served as a stimulus to the inauguration of a project to publish the complete works of the great scholar. The idea of such a publication was warmly endorsed by M. V. Ostrogradskiĭ, V. Ya. Bunyakovskiĭ, and P. L. Chebyshev [3], and C. G. J. Jacobi participated actively in the planning of the project [4]. Initially it was proposed that Euler's complete works be published in 25 volumes but then in view of financial constraints it was decided to refrain from reprinting the works already published independently, and also to restrict publication to articles. Of this projected publication (*Leonhardi Euleri Opera minora collecta*), only the first two—number-theoretical—volumes (*Commentationes arithmeticae collectae, 1849*) actually appeared.

By 1862 Euler's strictly scientific legacy had largely been realized in printed form but the idea of the necessity of publishing his complete works continued to exercise mathematicians. There was likewise no decrease of interest in examining further Euler's correspondence and other materials from his archive relevant to his scientific biography.

Towards the end of the 19th century it became clear that such a biography could be written only when Euler's published works, surviving manuscripts, letters, and also relevant official documents, had all been assembled in one place. The attention of the scientific world was drawn to this need first by the official meeting of the Basel Society of Natural Scientists held on November 17, 1883 to mark the hundredth anniversary of Euler's death, and then by the preparations for celebrating his two-hundredth jubilee in 1907. At this time there appeared a great many publications on his life and work and there was consistent advocacy of the idea of publishing the complete works. At the initial planning stage it was proposed that scientists from the three countries with which Euler's life had been connected collaborate on the project.

In 1907 the Swiss Eulerian Commission was created. It passed a resolution to begin preparatory work on the publication of the complete works and addressed an appeal for material support to the scientific community at large—to scientists and scientific amateurs of all nations. This initiative received an enthusiastic response and as early as 1911 the first volume of a unique publishing enterprise issued from the press, supported by international subscriptions and continuing right up to the present. In his foreword, F. Rudio, the chief editor—and one of the enthusiasts of the project to publish of *Leonhardi Euleri Opera omnia*—related the history of the investigation and publication of Euler's scientific *Nachlass* up to that time [6].

The Petersburg Academy of Sciences participated very actively in the project[3]. At the General Meeting of the Academy of May 2, 1909 a special committee was formed "to look into the imminent transfer to the Eulerian Commission of materials relating to Euler's activities, preserved in the Academic Archive" [7]. In December 1910 the Euler material held in the Archive of the Academy of Sciences, together with an inventory compiled by B. L. Modzalevskiĭ [8], was dispatched to the Swiss Society in Zürich "subject to an undertaking to return them within a certain time" [9, 10]. The investigation of these materials was undertaken by G. Eneström, the well-known Swedish historian of mathematics and compiler of a basic list of Euler's works [11].

In 1913 Eneström made a formal announcement of the result of his labors to the Eulerian Commission [12]. First of all he had placed on one side those of Euler's manuscripts relating to works already in print, admittedly without making an accurate comparison of these with the corresponding published versions. He had then sifted out the materials not belonging to Euler and classified the remainder under the following three heads: a) general and miscellaneous; b) scientific works, lectures, and reports; and c) biographical and bibliographical materials.

Eneström paid particular attention to the manuscripts, which evoked "great astonishment", and to the notebooks that had turned up "in which Euler casually noted those things that he was busy with at the time" [11, p. 193]. According to him, "two of the oldest of the notebooks contain essentially just exercises and are therefore of subordinate significance, but with the aid of the other notebooks one can get a very good idea of the history of Euler's discoveries" [*loc. cit.*]. Expressing a wish that these manuscripts be thoroughly examined, Eneström suggested that for each mathematical item a note be written conveying its content, although, in his words, "since the notebooks in all come to about 3000 closely written pages and the individual items are often very short, this work would require many months to complete" [11, p. 194]. It should be mentioned that in due course it became clear that Eneström had seriously underestimated the amount of labor involved. As he says, he himself was only able to flip through the pages of the notebooks in order to establish their chronological order. In connection with the publication of the *Opera omnia*, of the archival documents only the manuscripts of works already in print were used, while it was

---

[3]For details see E. P. Ozhigova's essay "The part played by the Petersburg Academy of Sciences in the publication of Euler's collected works" in the present collection.—*Eds.*

proposed that the notebooks be published separately in a series devoted to biographical materials. Along with Euler's other manuscripts, the notebooks remained in Switzerland for a very long time. They were returned to Russia only after the second world war, in 1947–1948.

The absence of the Euler archive was sorely felt by Soviet historians of science pursuing the study of Euler's works [13] and those of his contemporaries—especially M. V. Lomonosov. Lomonosov's letters to Euler, which formed part of that archive, were now inaccessible to them [14, pp. 22–23].

The return of the manuscripts to the Archive of the Academy of Sciences of the USSR coincided with the beginning of preparations to mark the 250th anniversary of Euler's birth, so that once again his scientific *Nachlass* became the center of attention of many scholars, some of whom were mobilized at this new stage to collaborate in the prodigious collective labor of publishing the scientific biography of the great man.

The publication of Euler's complete works greatly simplified work on his scientific legacy. However the question of the genesis of his ideas, so important for the history of modern mathematics, is still far from being settled; the answer requires a knowledge not only of the facts, but also of the path by which he arrived at each of his discoveries. Euler was unique in seeking to show the reader this path by tracing his train of thought. This splendid feature of his style, present in all of his work, was remarked upon by his earliest biographers. At the same time his results in various areas of natural science are so deep and so closely interconnected that, in spite of the extreme clarity of his exposition, reconstruction of a picture of his scientific *oeuvre* is very difficult. For this reason it was necessary to bring to the investigative process, in addition to his published works, the further resource of Euler's invaluable manuscript documents: his notebooks and scientific correspondence.

As soon as they were returned to the Archive of the Academy of Sciences of the USSR, these manuscript materials attracted the attention of Soviet scholars [15–17]—above all the late V. I. Smirnov, and A. P. Yushkevich—who has with great dedication appraised Euler's legacy for almost 40 years—, and also G. K. Mikhaĭlov. Gradually the circle of investigators, researching along lines suggested by these three, has widened, and at the present time the literature on Euler based on analyses of his manuscripts is substantial.

In collaboration with foreign[4] historians of science much has been achieved in connection with the publication of Euler's scientific correspondence [18]. In Euler's time correspondence played an enormous role in the life of scholars, allowing the promulgation of scientific discoveries well before they were published. Euler wrote more than 4000 letters on scientific matters in which he communicated to his correspondents problems of interest, methods of solution, etc. It is difficult to overestimate the importance of these letters for the realization of his scientific biography.

As already mentioned, the most important material for the history of Euler's scientific discoveries is that contained in his notebooks: 12 bound volumes of varying size, preserved in the Archive of the Academy of Sciences of the USSR [19]. The number of pages varies from 152 (in Notebook No. 130) to 544 (in No. 131) and altogether they comprise over 3000 pages. The study of these notebooks presents special difficulties: the investigator encounters here a truly immense chaos of the most variegated of scientific notes, drafts of

---

[4] That is, non-Russian.—*Trans.*

letters and articles, formulations and proofs of theorems, with drawings, problems, and computations sometimes lacking full specification, or not carried through to completion. On the other hand, each item corresponds to a definite stage in Euler's work on some problem or other and without doubt there is also an internal connection between successive items. It may confidently be maintained that the deciphering of such connections would result in the elucidation of many important aspects of Euler's mathematical work, not attainable from an examination of his published works alone. Moreover a close study of the notebooks may yield material affording interesting conclusions concerning the psychology of scientific creation in general.

In Euler's notebooks the sequence of steps stands clearly revealed in his progress towards a solution of many a scientific problem whose ultimate solution is known in published form. There are also more than a few notes concerning questions not appearing in any of his published works; such notes are naturally of the greatest interest.

There are many entries concerning solutions of problems discussed by Euler in his correspondence with other scientists. In such cases it turns out to be possible to date these and neighboring notes with reasonable accuracy and furthermore, depending on the content of the letter in question, determine when and in what connection one or another mathematical problem came to Euler's attention. The correlation of the notes with the corresponding letters affords abundant possibilities for the study of Euler's creative approach. Occasionally it is possible to trace how, in developing an idea that has come to him, he arrives at a completely new way of posing the question, far from its initial form, and yielding answers apparently unexpected even by him.

It seems that in his notebooks Euler wrote down results as soon as they occurred to him in order to "fix" them. Some of them, perhaps those he felt were most topical from his point of view and about which he harbored no doubts, he would at once communicate to an appropriate addressee. Others he would verify at length: these are repeated over and over in the notebooks (sometimes in a different form) and emerge into the correspondence only significantly later. In this sense the notes may be considered in essence rough drafts of letters.

From the notebooks one derives a vivid idea of the style of Euler's research activity, of which the most striking feature is its purposefulness: the retention of a fundamental idea over many years and the persistent search for a solution to a problem, in the face of false diversions and mishaps along the way. It is likewise clear from the notebooks just how important a combination of theory and scientific experiment was for Euler: work on applied problems provided him with the stimulus to development of theory.

The first stage in a systematic investigation of the notebooks consists in dating the entries and providing a general survey of each of them. As already mentioned, the first efforts in this direction were made by G. Eneström in 1913 [12]; subsequently the accuracy of his dating was improved and the content of each notebook elucidated more fully (see [15, 20–23] *et al.*)

The next, more complicated, stage in this endeavor consists in the classification of the items in the notebooks by subdiscipline, together with a process of familiarization with the content of the respective notes. This is an extremely complex task in view of the enormous number of notes and their variety, simultaneously rendering a complete survey and thorough scientific analysis difficult. For this reason a twofold approach to the problem is appropriate: On the one hand one could examine the notebooks as a totality—without dwelling on particulars or analyzing individual problems in any detail—allowing the over-all course of the creative process to be traced. On the other hand one might separate off from the rest all notes relating to some particular question or other and then carry out a detailed examination of each of these, in this way illuminating the progress of Euler's creativity in a special narrowly defined area.

Our research, begun in 1954 at the initiative and under the supervision of V. I. Smirnov, who had brought to people's attention the importance of a close study of Euler's note-books, has been devoted chiefly to those of the notes concerned with the theory of numbers [23–29], in particular Diophantine analysis. These notes comprise a substantial proportion (around 800 pages of text) of the whole. A general survey of the relevant material was car-ried out, a proposal for classifying it made, and on the basis of analyses of the notes conclu-sions were drawn concerning those of Euler's unpublished manuscripts having arithmetical content [23–25]. However the text of these notes is for the most part still not published—neither in the original Latin nor in translation. (The exceptions are the notes on perfect numbers [26], "Bertrand's postulate" [27], "partitio numerorum" [28], polygonal numbers [29], the Euler function [32], and amicable numbers [31].) At the same time it cannot be doubted that the publication not only of the number-theoretical items from the notebooks, but also of a Russian translation of the whole of Euler's published *oeuvre* in number theory, would be extraordinarily important for the history of mathematics. This aim is being kept in mind in connection with continuing investigations of the number-theoretical notes [30–32] and their preparation for publication[5].

In the notebooks one also finds notes that are not—at least at first glance—of a purely scientific character, but contain valuable biographical information. An example of this is an item from the notebook spanning, approximately, the period 1749–1755 [32]; this note-book contains a catalogue of the books from Euler's library, with the heading "Catalogue of my books" ("Catalogue Librorum meorum"). This catalogue is mentioned cursorily in the article [20, see p. 88] by G. K. Mikhaĭlov; however it merits much closer study since the collection of books in a library bears very eloquent witness to the interests and inclinations of its owner [33]. This list occupies 20 pages of the notebook (1. 192–201 ob.) and contains 539 titles in various languages: Latin, German, French, English, Russian, and Greek. The books are listed according to no definite system, except for a subdivision into volumes *in quarto* or *in folio*. Even a cursory perusal of the catologue suffices to show that around 1750 Euler was in possession of a superb library. Not only the natural sciences but also the hu-manities, and even religion and fine literature—predominantly classical—were represented on its shelves.

At the present time[6] the notebooks are the subject of further intensive investigations. To some extent efforts have been made in connection with the notes on mechanics, number theory, algebra, geometry, physics, and astronomy (Yu. A. Belyĭ, R. I. Galchenkova, A. A. Kisilev, G. P. Matvievskaya, I. G. Mel′nikov, L. S. Minchenko, G. K. Mikhaĭlov, N. I. Nevskaya, and E. P. Ozhigova), but fundamental work remains to be done.

At present the possibilities for using the notebooks as material for a scientific biog-raphy of Euler are much greater than they were, say, 25 years ago. Over the intervening

---

[5] By the present author and E. P. Ozhigova.
[6] 1986

years the history of the Petersburg Academy of Sciences in the 18th century has been studied in depth, scientific biographies of many of the scientists who worked with Euler have been written, and their correspondence published, fresh conclusions have been drawn via archival documents as to the conduct of research in the Academy, and so on. Correlation of such new data with notes from Euler's notebooks will facilitate the deciphering and dating of those notes, provide reasons for Euler's having written them, and lead to a better understanding of the connections between the various notes.

Without doubt the further study of Euler's notebooks will yield valuable material for his scientific biography.

References

[1] Kopelevich, Yu. Kh. "Materials for a biography of Leonhard Euler". Research in the history of mathematics (Ist.-mat. issled.). 1957. Issue 10. pp. 9–65.

[2] Fuss, N. "Éloge de Monsieur Euler, lu à l'Académie Imp. des sciences, dans son assemblée du 23 octobre 1783 par Nicolas Fuss. Avec une liste complète des ouvrages de M. Euler". SPb., 1783.

[3] Kostryukov, K. I. "Concerning an attempt to publish the works of L. Euler". Research hist. math. (Ist.-mat. issled.). 1954. Issue 7. pp. 630–640.

[4] Stäckel, P., Ahrens, W. *Briefwechsel zwischen C. G. Jacobi und P. N. v. Fuss über die Herausgabe der Werke Leonhard Eulers*. Bibl. math. 1907. Vol. 8. pp. 233–306.

[5] Euler, L. *Commentationes arithmeticae collectae*. Petropoli, 1849. Vols. 1–2.

[6] Rudio, F. Foreword to: Euler, L. *Opera omnia*. Ser I. Vol. 1.

[7] Izv. Akad. nauk. Ser. 7. 1909. Vol. 3, No. 14. pp. 929–930.

[8] Modzalevskiĭ, B. L. *Inventory of the manuscripts of Euler preserved in the Archive of the Conference of the Imperial Academy of Sciences*. Academy of Sciences. 1910. (Publication rights reserved.)

[9] *Survey of archival materials (Archive of Acad. Sci. USSR)*. L.: Publ. Acad. Sci. USSR, 1933. Issue 1. pp. 73–74.

[10] Rudio, F., Schröter, C. "Die Eulerausgabe". Vierteljahrsschr. Naturforsch. Ges. Zürich. 1908. Vol. 53. N. 24.

[11] Eneström, G. "Verzeichnis der Schriften Leonhard Eulers". Jahresber. Dtsch. Math. Ver. 1910–1913. Erganzungsb. 4, Lief. 1–2.

[12] ———. "Bericht an die Eulerkommission der Schweizerischen naturforschenden Gesellschaft über die Eulerschen Manuskripte der Petersburger Akademie". Jahresber. Dtsch. Math. Ver. 1913. Vol. 22, H. 1–2. pp. 191–205.

[13] *Leonhard Euler: Collection of articles and other materials on the occasion of the 150th anniversary of his death*. M.; L.: Publ. Acad. Sci. USSR, 1935.

[14] Modzalevskiĭ, B. L. "From the compiler". In: *Manuscripts of M. V. Lomonosov in the Archive of Acad. Sci. USSR: Scientific description*. M.; L.: Publ. Acad. Sci. USSR, 1937.

[15] *Manuscript materials of L. Euler in the Archive of the Academy of Sciences of the USSR*. M.; L.: Publ. Acad. Sci. USSR, 1962. Vol. 1: "Scientific description."

[16] Euler, L. *Letters to scientists*. M.; L.: Nauka, 1963.

[17] ——. *Correspondence. Annotated index*. V. I. Smirnov, A. P. Yushkevich, eds. M.: Nauka, 1967. (More complete version: *Opera* IVA–1).

[18] Yushkevich, A. P. "On the archival legacy of Leonhard Euler". Questions in the hist. of natural science and technology (Vopr. istorii estestvozn. i tekhn.). 1982. No. 3. pp. 137–139.

[19] Archive Acad. Sci. USSR, f. 136, op. 1, Nos. 128–140.

[20] Mikhaĭlov, G. K. "The notebooks of Leonhard Euler in the Archive of Acad. Sci. USSR: General description and notes on mechanics". Research hist. math. (Ist.-mat. issled.). 1957. Issue 10. pp. 67–94.

[21] Mikhaĭlov, G. K., Smirnov, V. I. "Unpublished materials of Leonhard Euler in the Archive Acad. Sci. USSR". In: *Leonhard Euler*. M.: Publ. Acad. Sci. USSR, 1958. pp. 47–79.

[22] Mikhaĭlov, G. K. "Notizen über die unveröffentlichen Manuskripte von Leonhard Euler". In: *Leonhard Euler. Sammelband der zu Ehren des 250. Geburtstages Leonhard Eulers der Deutschen Akademie der Wissenschaften zu Berlin vorgelegten Abhandlungen*. Berlin: Akad.-Verl., 1959. pp. 256–280.

[23] Matvievskaya, G. P. "On the unpublished manuscripts of Leonhard Euler on Diophantine analysis". Research hist. math. (Ist.-mat. issled.). 1960. Issue 13. pp. 107–186.

[24] ——. "Unpublished manuscripts of Leonhard Euler on number theory". Author's summary of dissertation submitted for degree of "Candidate" in physical and math. sci. L., 1958.

[25] ——. "Unpublished manuscripts of Leonhard Euler on Diophantine analysis". Proc. inst. hist. nat. sci. and technology Acad. Sci. USSR (Tr. In-ta istorii estestvozn. i tekhn. AN SSSR). 1959. Vol. 22. pp. 240–250.

[26] ——. "The notes on perfect numbers in Euler's notebooks". Proc. inst. hist. nat. sci. and tech. Acad. Sci. USSR (Tr. In-ta istorii estestvozn. i tekhn. AN SSSR). 1960. Vol. 34. pp. 415–427.

[27] ——. "Bertrand's postulate in Euler's notes". Research hist. math. (Ist.-mat. issled.). 1961. Issue 14. pp. 285–288.

[28] Kiselev, A. A., Matvievskaya, G. P. "Unpublished notes of Euler's on partitio numerorum". Research hist. math. (Ist.-mat. issled.). 1965. Issue 16. pp. 145–180.

[29] Matvievskaya, G. P. "Notes on polygonal numbers in Euler's notebooks". Research hist. math. (Ist.-mat. issled.) 1983. Issue 27. pp. 27–50.

[30] Matvievskaja, G. P., Ožigova, E. P. "Leonhard Eulers handschriftlicher Nachlass zur Zahlentheorie". In: *Leonhard Euler. 1707–1783: Beiträge zu Leben und Werk*. Basel, 1983. pp. 151–160.

[31] Mel′nikov, I. G. "Amicable numbers in the manuscript legacy of Euler". Research hist. math. (Ist.-mat. issled.). 1983. Issue 27. pp. 10–24. With an addendum by E. P. Ozhigova. *loc. cit.* pp. 24–27.

[32] Ozhigova, E. P. "The Eulerian function in Euler's notebooks". Research hist. math. (Ist.-mat. issled.). Issue 27. pp. 50–63.

[33] Archive of Acad. Sci. USSR, f. 136, op. 1, No. 134.

[34] Matvievskaya, G. P. "Euler's library". Questions in hist. nat. sci. and tech. (Vopr. istorii estestvozn. i tekhn.). 1982. No. 3. pp. 139–140.

The Manuscript Materials of Euler on Number Theory

G. P. Matvievskaya and E. P. Ozhigova

The theory of numbers accounts for a large proportion of Euler's work: about one-sixth of his published works contain results, methods, or applications of number theory, and the same holds for those of his manuscripts and letters preserved in Russian archives.

1. The publication of Euler's surviving number-theoretical manuscripts

Euler's unpublished manuscripts attracted the attention of scholars as early as the 18th century. A large number of these remained after his death. Between 1783 and 1785 N. I. Fuss published two volumes of articles by Euler that had not earlier appeared in print. Eleven of these articles dealt with number theory [1]. Fuss went on to prepare eight further works on number theory for publication, which appeared only posthumously, in 1830 [2].

Number theory is very prominent in the *Correspondence* of Euler with certain leading 18th century mathematicians—especially with Gol'dbach—, published by P. N. Fuss in 1843 [3]. In 1849 P. N. Fuss, V. Ya. Bunyakovskiĭ, and P. L. Chebyshev published the two-volume *Arithmetical works* of Euler [4] in which they gathered together both published and unpublished works, including a large 16-chapter treatise on number theory. The first volume contains a very valuable systematic table of contents, in French, listing all of the works contained in the two volumes [4, pp. LI–LXXX]. Finally, in 1862 P. N. Fuss and N. N. Fuss published a collection entitled *Posthumous works*, likewise in two volumes, including fragments of manuscripts dealing with number theory [5].

Interest in Euler's manuscript *Nachlass* was aroused at the turn of the 19th century in anticipation of the impending jubilee. At the International Congress of Mathematicians in Zürich in 1897, and again in Rome in 1908, the question of the publication of the complete works of Euler was broached. In connection with the preparation and celebration of the 200th anniversary of the birth of the great scholar, committees were formed in Switzerland, Germany, and Russia to look into the possibility of such an enterprise. Russian scholars (D. K. Bobylev, A. M. Lyapunov, O. A. Backlund, A. A. Markov, N. Ya. Sonin, and others) evinced great interest in Euler's *oeuvre*, and stressed the importance of the publication of the complete works [6, p. 243]. However in the end the responsibility for the publication of *The complete collected works* (*Leonhard Euleri Opera omnia*) was assumed by the Swiss Society of Natural Scientists [7]. The Swedish bibliographer and historian of mathematics G. Enestrőm, compiler of a complete list of Euler's works [8] which became, in the words of F. Rudio, "the backbone of the whole publication" [7, I-1, pp. XXXV–XXXVI], undertook to examine those of Euler's manuscript materials temporarily placed at the disposal of the Swiss Society of Natural Scientists by the Petersburg Academy of Sciences. In 1913 he informed that Society of the results of this examination.

It was proposed that the complete collected works, which began to appear in 1911, should be published in three series, the first to contain Euler's mathematical works, the second those on mechanics and astronomy, and the third those on physics and other subjects. In preparing material for publication in these series, almost exclusive use was made of the manuscripts of works already published—for the sake of comparison with the originals and greater accuracy. Although the examination of the Euler materials surviving only in manuscript form continued to be of pressing urgency in respect of the goal of publishing the complete works, this part of the project was postponed for what turned out to be a very long time.

Over the period 1947–1948 the Euler manuscripts were returned from Switzerland to the Archive of the Academy of Sciences of the USSR in Leningrad (now[1] the Leningrad Section of that archive: LS AS USSR). At the initiative of V. I. Smirnov a systematic investigation of Euler's manuscript *Nachlass* was begun. The first results of this research were announced by Smirnov at the jubilee meeting of the Academy of Sciences of the USSR held on April 16, 1957 and published in articles by Smirnov and G. K. Mikhaĭlov [10, 11] and others [12–14]. A. P. Yushkevich has played—and continues to play—a leading role in the further study and publication of the Euler manuscript materials

The process begun in 1911 of publishing the vast conglomeration of Euler's works continues to this day and provides an example of fruitful international cooperation in a scientific endeavour. Over the last decade[2] the most active participants in this work have been Soviet scholars:–on the organizational side: V. I. Smirnov, A. T. Grigor′yan, G. K. Mikhaĭlov, and A. P. Yushkevich; and in direct investigation of the archival materials: Yu. A. Belyĭ, R. I. Galchenkova, A. A. Kiselev, T. N. Klado, G. A. Knyazev, Yu. Kh. Kopelevich, M. V. Krutikova, T. A. Lukina, G. P. Matvievskaya, I. G. Mel′nikov, L. S. Minchenko, G. K. Mikhaĭlov, N. I. Nevskaya, E. P. Ozhigova, N. M. Raskin, V. I. Smirnov, and A. P. Yushkevich. In particular, over the period 1957–1960 a start was made on the detailed examination of those of Euler's manuscripts concerning number theory (A. A. Kiselev, G. P. Matvievskaya, and I. G. Mel′nikov [15–24]).

In 1965 a new edition of *Euler's correspondence with Gol′dbach* was published in Berlin by E. Winter and A. P. Yushkevich [25], jointly sponsored by the academies of the USSR and the GDR. This publication includes detailed commentaries on number-theoretical questions—among others—touched on in the correspondence (A. A. Kiselev and I. G. Mel′nikov), in connection with which certain manuscript material from Euler's notebooks was used.

In more recent times, in view of the imminent completion of the publication of the first

---

[1] 1986
[2] From 1976 to 1986 approx.—*Trans.*

three series of the complete collected works of Euler, the process of publishing Series IV has begun, to include the scientific manuscript and epistolary *Nachlass* of Euler. It was within the framework of this project that G. P. Matvievskaya, I. G. Mel'nikov, and E. P. Ozhigova undertook to investigate the Euler manuscripts relating to number theory.

2. A general survey of the Euler manuscript materials on number theory

The manuscript materials on number theory—indeed all of Euler's manuscripts—may be subdivided into three subsets: drafts of published works; fragments whose contents are already partially or wholly published; and fragments remaining unpublished hitherto and therefore of especial interest.

Research on these materials began with an examination of all of Euler's notes in order to sift out the fragments with number-theoretic content. These fragments were then gradually deciphered, translated from Latin into Russian, and their mathematical content elucidated. Next the material was classified by theme, and a comparison of the fragments made with corresponding published works of Euler in order to establish whether in fact the fragments had been published, and if so whether they had been reproduced accurately in print. Finally, commentaries were written on individual fragments or groups of notes united by a common theme.

Euler's "notebooks" ("Adversaria mathematica") form the bulk of Euler's manuscript materials [31]. These are bound volumes of varying format and thickness, with the numbers of pages varying from 152 (No. 130) to 544 (No. 131). In all they come to around 3000 pages, for the most part in Latin, more rarely German, and very occasionally French. These notebooks trace in chronological order the stages in Euler's progress in his work on problems occupying him at that juncture, and bear witness to the breadth of his interests and the ease with which he transferred his attention from one topic to another. Questions in mechanics give way to geometrical ones, algebraic and number-theoretical problems to questions in mathematical analysis, and notes on musical theory are followed by entries on differential equations and physics. These materials are thus of extreme importance for the study of the history of Euler's discoveries since they allow a glimpse into the laboratory of his creation.

More than 1000 pages of the notebooks contain entries on the theory of numbers. Adjacent entries are usually not at all concerned with number theory, but there are pages devoted exclusively to number theory. Close examination of the notebooks reveals that although the intervals when Euler was working most intensively on number theory (1736–1741 and 1767–1783 [20]) fall within his two St. Petersburg periods, he nonetheless continued to work unremittingly on that topic during those years when he published scarcely anything in number theory. Some of his well known number-theoretic results had been obtained long before they were published.

Apart from the notebooks, there are other manuscripts containing material on number theory—assorted drafts of notes and the myriad letters.

Analysis of all of the above-mentioned materials often reveals the path by which Euler came to his results. Sometimes his research on a particular problem extended over many years. Often he drew conclusions on the basis of the behavior of numbers as revealed by long calculations and the compilation of tables. He would register a mathematical fact he had observed and then subsequently return to the question it raised; over and over again he would attempt to prove a result he had formulated on the basis of empirical evidence until at last he had obtained a strict proof, only then considering the question settled. However it remains true that several of his theorems were proved only after his death.

While this peculiarity of Euler's creative method can be discerned also in his published work, it is especially striking in his notes. Euler often solves one and the same problem in a variety of ways and applies one and the same method to the solution of a variety of problems. For example, he used mathematical analysis as a tool for solving number-theoretical problems, and number theory in his research on the differential and integral calculus. His notes not only reveal the progress of his investigations but also various doubts he was subject to along the way, and how he settled them. In this respect examples play a leading role.

3. Methodology and results

Much as in the 1849 publication [4], we classified all the manuscript materials on number theory according to the following four categories: 1. Divisibility; 2. Representation of a number as a sum of some kind; 3. Diophantine analysis; 4. Applications of number theory to mathematical analysis and of mathematical analysis to number theory; and 5. Other questions.

Each of these five categories is further subdivided. Thus for example the category "Divisibility" consists of the following topics: Divisors; Fermat's little theorem and Euler's theorem; The Euler function; Residues; and Prime numbers. Each of these subcategories is then subject to yet further subdivision; for instance the subcategory concerned with residues includes quadratic residues, residues of powers, primitive roots, and the law of reciprocity.

The entries relating to Diophantine analysis (Category 3), which comprise the bulk of Euler's number-theoretic notes, are subdivided as follows (as in the book [32] by L. E. Dickson): 1) Polygonal numbers; 2) The equation $ax^2 + bx + c = y^2$ and Pell's (or Fermat's) equation; 3) Problems involving triangles; 4) Various quadratic equations; 5) Systems of indeterminate [i.e., Diophantine] quadratic equations; 6) Indeterminate equations and systems of cubic equations; 7) Squares occurring in an arithmetic progression; 8) Fermat's last theorem; 9) Integer solutions of the equation $x^y = y^x$. Each of these subcategories is subject to further subdivision.

We now give a few specific examples from the various categories defined above.

Page 18 of notebook No. 131 is concerned exclusively with number theory. It begins with the formulation of the following general problem: Determine whether a given number n is prime. With this aim in mind he considers the number $2^{n-1} - 1$ and divides it by n. Euler reasons that if the remainder is zero, then the number n should be prime, and in the contrary case composite. It would seem that this method of testing a number for primality (Is it prime or not?) occurred to Euler after he had learned of Fermat's little theorem: If p is a prime and a any number not divisible by p, then the difference $a^{p-1} - 1$ is divisible by p. Euler mentions this theorem in his first published work in number theory: "Remarks concerning Fermat's theorem and others, relating to the investigation of prime numbers", numbered 26 (E26) in Eneström's list [8].

Subsequently Euler gave four proofs of this theorem[3] (E54, E134, E262, E271), and proved the theorem generalizing Fermat's little theorem which now bears his name (E271). Various authors have written on Euler's published works around this question: L. E. Dickson [32]; the editors of the arithmetical volumes [7]; A. A. Kiselev and I. G. Mel'nikov in their comments on Euler's correspondence with Gol'dbach [25]; and others.

The above-mentioned formulation (on p. 18 of notebook No. 131) of a test for primality is left without proof. Euler seeks to verify it through examples, but the question immediately arises as to how to do this if the numbers n one is testing for primality are very large. He proposes finding the remainder on dividing such numbers n into 2^{n-1}; the remainder will be 1 if n is prime. At this point Euler formulates a different theorem: If 2^m yields remainder p on division by n, then 2^{m+1} will yield remainder $2p$ after division by n, and 2^{2m} will yield remainder p^2 after division by n.

Euler now verifies the putative primality test in some particular cases ($n = 61$ and $n = 34$) assuming the latter statement, formulated as Theorem 2. (If we express the divisibility of the number $2^m - p$ by n in the more modern congruence notation $2^m \equiv p \,(\text{mod } n)$, then Theorem 2 can be restated as follows: If $2^m \equiv p \,(\text{mod } n)$, then $2^{m+1} \equiv 2p \,(\text{mod } n)$ and $2^{2m} \equiv p^2 \,(\text{mod } n)$.)

As is well known, the theory of congruences was given its final form by C. F. Gauss in 1801; however we may infer that the basic properties of congruences—not, of course, in modern notation—were known to Euler at the very beginning of his work on number theory from the fact that notebook No. 131 spans the years 1736–1740, so that he would have filled its early pages in 1736. Euler published Theorem 2 in 1761 in the article "Theorems on the remainders after division of powers" (E262) in the following more general form: "If a^μ yields remainder r on division by p, then $a^{2\mu}$ yields remainder r^2 on division by p, $a^{3\mu}$ yields remainder r^3 on division by p, and so on". Euler infers from this result that if a^μ leaves remainder 1 on division by p, then so do $a^{2\mu}, a^{3\mu}, a^{4\mu}, \ldots$ (E262) [7, I-2, p. 496].

However let us return to p. 18 of notebook No. 131. Euler has doubts as to the truth of the theorem he has formulated (an exact criterion for primality). On the next page (18 ob.) he notes that the rule does not always hold. If, for example, one takes $n = 2^{32} + 1$, then since this number is divisible by 641, it is composite, yet the difference $2^{n-1} - 1 = 2^{2^{32}} - 1$ is divisible by $2^{32} + 1$ (without remainder). Hence the proposed rule for establishing primality is invalid. Euler concludes that all one can say in general is that if the remainder after division of $2^{n-1} - 1$ by n is not zero, then n cannot be prime, while if the remainder *is* zero, i.e., if n divides $2^{n-1} - 1$, then no conclusion as to whether n is prime or composite follows.

In fact similar questions interested Euler before he began notebook No. 131. In letters written to Ch. Gol'dbach dated June 4 and 25, 1730 (and also much later, on October 28, 1752) he mentions divisibility questions concerning numbers of the form $2^n - 1$ and $2^n + 1$, in particular the fact that $2^{2^5} + 1$ is divisible by 641 [25, pp. 30, 34, 357]. On divisibility problems concerning $2^n - 1, 2^n + 1$, he wrote a series of articles (E26, E134, E271).

The notebooks show that from the very beginning of his number-theoretic investigations Euler was studying the series $\sum\limits_{m=1}^{\infty} \dfrac{1}{m^s}$, for natural numbers s, now denoted by $\zeta(s)$ and called the "Riemann Zeta-function." In notebook No. 131 there is an entry relevant to the

---

[3]Fermat's little theorem, presumably.—*Trans.*

prehistory of Euler's investigations of this function; see [33, pp. 52–56] for references to the literature concerning this area of Euler's research.

The very earliest number-theoretical problems considered by Euler concerned—as is shown by the notebooks—Diophantine analysis. Among the very first entries in notebook No. 131 there is one headed "Determine how many times a given number occurs among all polygonal numbers". Euler calls this "Bachet's problem", and gives his own solution of it in this note. Only 16 years later, in a letter dated April 3, 1753, does he communicate his solution to Gol′dbach [25, p. 369].

In notebook No. 132, l. 110, Euler investigates the problem of finding the pentagonal numbers that are perfect squares. The solution of this problem given in his *Universal arithmetic* (E388, Part 2, §89) differs from that in the notebook.

The first entry relating to the theorem on the representation of any natural number as a sum of four squares is contained in notebook No. 131, at the location l. 61 ob.–62 ob. Euler states that every number of the form $8n - 1$ can be decomposed as a sum of four squares, gives a rule for obtaining such a decomposition, and illustrates it with examples. However he states that there are counterexamples to this statement. It is clear that Euler had found exceptional cases and therefore refrained from including the statement in his published work[4]. In the same place Euler considers in succession the problem of representing a number given in one of the forms

$$(a^2 + b^2 + c^2 + d^2)(p^2 + q^2),$$
$$(a^2 + b^2 + c^2 + d^2)(p^2 + q^2 + r^2),$$
$$(a^2 + b^2 + c^2 + d^2)(p^2 + q^2 + r^2 + s^2),$$

as a sum of four squares. In establishing in each case that such a representation is possible, he points out the significance of the "roots" x of these four squares x^2, and in the last of these three cases arrives at his celebrated identity

$$(a^2 + b^2 + c^2 + d^2)(p^2 + q^2 + r^2 + s^2)$$
$$= (ap + bq + cr + ds)^2 + (bp - aq + dr + cs)^2$$
$$+ (cp - dq - ar + bs)^2 + (dp + cq - br - as)^2.$$

In this note the drift of Euler's reasoning is visible, whereas his published works on these questions are more opaque in this regard. Furthermore this entry in notebook No. 131 dates from well before May 4, 1748 when he communicated the above formula to Gol′dbach [25, pp. 288–291], since notebook No. 131 spans the period 1736–1740.

Euler often returns to the four-squares theorem. In notebook No. 132, at l. 142 ob., he formulates and proves the following theorem: If a is a number that is not decomposable into a sum of four squares but is a divisor of a number $P = A^2 + B^2 + C^2 + D^2$ [i.e., of a number that *is* decomposable as a sum of four squares], then there exists a number $b < a$ that is likewise not so decomposable, and which is a divisor of a number $Q < P$ that is so decomposable. Following the proof he states the following corollary: Among the numbers not representable as sums of four squares there are none that are divisors of sums of four

---

[4]Since every natural number *can* be represented as a sum of four squares of integers, presumably the counterexamples pertain to his rule for finding such representations.—*Trans.*

squares, i.e., every divisor of a sum of four squares must itself be [representable as] a sum of four squares. The proof of this result and its corollary appear as Theorem 4 in Euler's published paper of 1772 on the decomposition of numbers as sums of squares (E445), and it is on these that he bases his proof of the four-squares theorem given in that paper. In somewhat altered form this theorem appears also in [5] where it is reproved using certain of Euler's manuscript fragments.

Since the relevant entries in notebook No. 132 date from the period 1740–1744, it might be claimed that Euler had a proof of this important result (the four-squares theorem) as early as the 1750s, i.e., almost 30 years prior to publication. However his proof was incomplete. Following the publication by Lagrange in 1772 of a complete proof, Euler returned to the proof he had begun many years earlier and by means of the method described in notebook No. 132 succeeded in giving a complete proof of the four-squares theorem. His proof was in fact simpler than Lagrange's. There are other entries in the notebooks pertaining to the same problem.

Two notes have been discovered in the notebooks showing that Euler had formulated the so-called "Bertrand postulate" almost 100 years before Bertrand[5] [22]. Both of these entries are to be found in notebook No. 134, and so date from the period 1752–1755.

Thus at l. 120 of notebook No. 134 Euler states that between any number a whatever and its double $2a$ there exists at least one prime number (assuming, of course, that a is an integer and $a > 1$). In an attempt to prove this statement Euler seeks to eliminate all integers between a given integer a and $2a$ that are multiples of primes less than a, those remaining being just the primes in that interval, i.e, he uses in effect the "sieve of Eratosthenes". He considers the case $a = 24$. Between 24 and 48 there are 12 multiples[6] of 2, eight multiples of 3 (including four even numbers already excluded as multiples of 2), five multiples of 5 (including three already excluded as multiples of 2 or 3), three multiples of 7 (all already excluded), and, finally, three multiples of 11 (also already excluded). Hence the number of composite numbers between 24 and 48 is

$$12 + 8 - 4 + 5 - 3 = 18,$$

so that there must be six primes in the interval in question; they are 29, 31, 37, 41, 43, 47.

The second entry relating to this problem is to be found at l. 205 ob.–206 of notebook No. 134. Here Euler repeats the statement that between n and $2n$ there is at least one prime number and considers two further examples: he finds all primes between 30 and 60, and then between 50 and 100, performing the calculations in tabulated form. In the second and third columns of each of the tables he has entered opposite each prime in the interval $(0, n)$ the numbers in the intervals $(0, n)$ and $(n, 2n)$ that are multiples of that prime, omitting those numbers already encountered as multiples of smaller primes. Reasoning further on the basis of his tables, he attempts to arrive at a general method of calculating the set of primes in a given interval and to find some sort of regularity in the distribution of the primes, but without success. Euler goes on to consider the special case $n = 2p^2$, i.e., intervals of the form $(2p^2, 4p^2)$. He conjectures that the number of primes between $\sqrt{2n}$ and n is the same as the number between n and $2n$, or, equivalently (in this special case),

---

[5] Joseph Bertrand (1822–1900) conjectured that for every integer $n > 3$ there is at least one prime p satisfying $n < p < 2n - 2$. This "postulate" was first proved by P. L. Chebyshev in 1850.—*Trans.*

[6] One of the numbers 24, 48 must have been included in this count.—*Trans.*

that the number of primes between $2p$ and $2p^2$ is the same as the number between $2p^2$ and $4p^2$. However on checking this conjecture for $p = 2, 3, 4, 5, 6, 7, 8, 15$, he finds that it already fails at $p = 5$.

Thus examination of Euler's manuscript materials significantly broadens our conception of Euler's creative activity in the theory of numbers.

The unpublished manuscripts of Euler on number theory—which we have described here only in the most general terms—are being prepared by us for publication in Series IV of the *Opera omnia* [7] as a separate volume.

References

[1] Euler, L. *Opuscula analytica*. Petropoli, 1783–1785. Vols. 1–2.

[2] *Mém. Acad. Sci. Pétersb.* 1830. Vol. 11. pp. 1–94.

[3] *Correspondence mathématique et physique de quelques célèbres géomètres du XVIII siècle.* SPb., 1843. Vols. 1–2.

[4] Euler, L. *Commentationes arithmeticae collectae*. Petropoli, 1849. Vols. 1–2.

[5] ——. *Opera postuma mathematica et physica*. Petropoli, 1862. Vols. 1–2.

[6] Ozhigova, E. P. *Charles Hermite*. L.: Nauka, 1982. pp. 243–244.

[7] Euler, L. *Opera omnia*. Leipzig; Berlin, 1911. Vol. 1. (In four series.)

[8] Eneström, G. "Verzeichnis der Schriften Leonhard Eulers". Jahresber. Dtsch. Math. Ver. 1910–1913. Erganzungsb. 4, Lief. 1–2.

[9] ——. "Bericht an die Eulerkommission der Schweizerischen naturforschenden Gesellschaft über die Eulerschen Manuskripte der Petersburger Akademie". Jahresber. Dtsch. Math. Ver. 1913. Vol. 22, Issues 1–2. pp. 191–205.

[10] Mikhaïlov, G. K., Smirnov, V. I. "Unpublished materials of Leonhard Euler in the Archive Acad. Sci. USSR". In: *Leonhard Euler*. M.: Publ. Acad. Sci. USSR, 1958. pp. 47–79.

[11] Mikhaïlov, G. K. "The notebooks of Leonhard Euler in the Archive of Acad. Sci. USSR: General description and notes on mechanics". Research hist. math. (Ist.-mat. issled.). 1957. Issue 10. pp. 67–94.

[12] *Manuscript materials of L. Euler in the Archive of the Academy of Sciences of the USSR.* M.; L.: Publ. Acad. Sci. USSR, 1962–1965. Vol. 1: "Scientific description". Vol. 2: "Works in mechanics". Part 1.

[13] Euler, L. *Correspondence. Annotated index.* V. I. Smirnov, A. P. Yushkevich, eds. M.: Nauka, 1967. (More complete version: *Opera* IVA–1).

[14] Mikhaïlov, G. K. "Notizen über die unveröffentlichen Manuskripte von Leonhard Euler." In: *Leonhard Euler. Sammelband der zu Ehren des 250. Geburtstages Leonhard Eulers der Deutschen Akademie der Wissenschaften zu Berlin vorgelegten Abhandlungen.* Berlin: Akad.-Verl., 1959. pp. 256–280.

[15] Kiselev, A. A., Matvievskaya, G. P. "Euler's unpublished notes on partitio numerorum". Research hist. math. (Ist.-mat. issled.). 1965. Issue 16. pp. 145–180.

[16] ———. "Certain questions of number theory in Euler's correspondence with Gol'dbach". Hist. and methodology of the nat. sciences (Istoriya i metologiya estestv. nauk). 1966. Issue 5. pp. 31–34.

[17] Matvievskaya, G. P. "Unpublished manuscripts of Leonhard Euler on number theory". Lecture delivered at the third scientific conference of graduate students and junior scientific workers. IIEiT Akad. Nauk SSSR. M., 1957.

[18] ———. "Unpublished manuscripts of Leonhard Euler on number theory". Author's summary of dissertation submitted for degree of "Candidate" in physical and math. sci. L., 1958.

[19] ———. "Unpublished manuscripts of Leonhard Euler on Diophantine analysis". Proc. inst. hist. nat. sci. and technology Acad. Sci. USSR (Tr. In-ta istorii estestvozn. i tekhn. AN SSSR). 1959. Vol. 22. pp. 240–250.

[20] ———. "On the unpublished manuscripts of Leonhard Euler on Diophantine analysis". Research hist. math. (Ist.-mat. issled.). 1960. Issue 13. pp. 107–186.

[21] ———. "The notes on perfect numbers in Euler's notebooks". Proc. inst. hist. nat. sci. and tech. Acad. Sci. USSR (Tr. In-ta istorii estestvozn. i tekhn. AN SSSR). 1960. Vol. 34. pp. 415–427.

[22] ———. "Bertrand's postulate in Euler's notes". Research hist. math. (Ist.-mat. issled.). 1961. Issue 14. pp. 285–288.

[23] Mel'nikov, I. G. "On certain conjectures of Euler and Gol'dbach". Materials of the seventh conference on the history of science in the Baltic region. Vilnius, 1965. pp. 34–39.

[24] ———. "On certain number-theoretical questions in Euler's correspondence with Gol'dbach". Hist. and methodology of nat. sci. (Istoriya i metodologiya estestv. nauk). 1966. Issue 5. pp. 15–30.

[25] *Leonhard Euler und Christian Goldbach. Briefwechsel. 1729–1764.* Hrsg. A. P. Juškevič, E. Winter. Berlin: Akad.-Verl., 1965.

[26] Matvievskaja, G. P., Ožigova, E. P. "Leonhard Eulers handschriftlicher Nachlass zur Zahlentheorie". In: *Leonhard Euler. 1707–1783.* Basel, 1983. pp. 151–160.

[27] Mel'nikov, I. G. "Amicable numbers in Euler's manuscript legacy". Research hist. math. (Ist.-mat. issled.). 1983. Issue 27. pp. 10–24.

[28] Ozhigova, E. P. "Addendum to I. G. Mel'nikov's article 'Amicable numbers in Euler's manuscript legacy'". Research hist. math. (Ist.-mat. issled.). 1983. Issue 27. pp. 24–26.

[29] Matvievskaya, G. P. "Notes on polygonal numbers in Euler's notebooks". Research hist. math. (Ist.-mat. issled.). 1983. Issue 27. pp. 27–50.

[30] Ozhigova, E. P. "The Eulerian function in Euler's notebooks". Research hist. math. (Ist.-mat. issled.). Issue 27. pp. 50–63.

[31] Archive Acad. Sci. USSR, Leningrad Section (LO Arkhiva AN SSSR), f. 136, op. 1, Nos. 129–140.

[32] Dickson, L. E. *History of the theory of numbers. In 3 Vols.* Washington, 1919–1923; 2nd ed. New York, 1934.

[33] Ozhigova, E. P. *The development of number theory in Russia.* L.: Nauka, 1972.

Euler's Contribution to Algebra

I. G. Bashmakova

Introduction

It is usually thought that Euler's research has been of little significance for the development of algebra. In commemorative collections of articles dedicated to Euler, as a rule no mention is made of his work in algebra. This attitude is—as we shall attempt to show here—extremely unjust. As a matter of fact the problems that Euler solved in algebra influenced crucially the development in the 19th and 20th centuries of the theory of algebraic equations, the algebraic theory of numbers, and algebraic geometry.

However first of all one must frame the question clearly: just what did Euler understand by "algebra"? Which problems did he consider central to that discipline? From the 1930s on, mathematicians have been promoting the conception of algebra as the science of algebraic structures, i.e., of sets of elements of any kind whatever, on which there are defined one or more binary[1] operations, associating with each [ordered] pair of elements of the set in question a third element of that set. Although it might be debated as to whether in fact this definition encapsulates modern algebra in its entirety, it must be conceded that it characterizes its essence.

However this view of algebra is relatively recent. Over the course of its development that subject underwent frequent changes in content and methodology. Thus in ancient Babylon we find numerical algebra[2], and in ancient Greece geometrical algebra, so-called, using geometrical methods to establish algebraic propositions.

Then, beginning with Diophantus (ca. 2nd to 3rd centuries AD) algebra cultivates a new language: literal symbolism is introduced, at first to denote unknowns and later, in the 16th century, known quantities (parameters). From the time of Diophantus till that of Euler algebra becomes in essence the study of algebraic equations understood in a wide sense; i.e., including the investigation of determinate equations—in particular their solubility by radicals—as well as indeterminate equations[3].

---

[1] Actually nullary, unary, and in general n-ary operations are also very often considered—especially in universal algebra.—*Trans.*

[2] Presumably where general algebraic propositions concerning numbers are conveyed by means of particular numerical examples.—*Trans.*

[3] Such as Diophantine equations.—*Trans.*

Title page of Part I of Euler's *Algebra* of 1770.

A great many algebraic treatises dating from the long period beginning with Diophantus' *Arithmetic* and ending with Euler's *Complete introduction to algebra* (1770), were in
part—sometimes comprising as much as a half or more of the work—devoted to Diophantine equations, for example works of Abu Kamil, al-Karaji[4], Bombelli, and F. Viète. Euler
himself devotes the whole of the second section of Volume II of his *Complete introduction
to algebra* to "indeterminate analysis" (E388) [4].

Which problems were considered central to algebra thus widely conceived?

I. The Fundamental Theorem of Algebra, asserting that every algebraic equation

$$f_n(x) = x^n + a_1 x^{n-1} + \cdots + a_{n-1}x + a_n = 0$$

of arbitrary positive degree n , where $a_1, a_2, \ldots, a_n$ are real numbers, has at least one
real or complex root. In the 18th century the following equivalent formulation was more
often used: Any such polynomial $f_n(x)$ with real coefficients can be decomposed as a
product of factors of degrees one or two also having real coefficients. This theorem was

---

[4] Abu Kamil Shuja ibn Aslam ibn Muhammad ibn Shuja, b. ca. 850 AD in Egypt. Abu Bakr Muhammad ibn
al-Hasan al-Karaji, fl. ca. 1000 AD in Baghdad.—*Trans.*

first formulated in the 17th century by A. Girard (1629) and then in more rigorous form by R. Descartes (1637), but the first attempts at proving it date from the 18th century.

II. The solution of equations of degree five or more by radicals. Quadratic equations, it is now known, had been solved as early as four thousand years ago in ancient Babylon, and equations of degrees three and four in Italy in the 16th century. The solution by radicals of equations of degree five, the next case in line, defied all attempts made during the 17th and 18th centuries.

III. The investigation and solution in rational numbers of indeterminate[5] equations in two unknowns, primarily of the forms

$$y^2 = P_3(x), \quad y^2 = P_4(x), \quad \text{or } y^3 = P_3(x),$$

where $P_n(x)$ denotes a polynomial of degree n with rational coefficients. We shall relate the prehistory of this type of problem below.

IV. The solution of indeterminate equations in whole numbers[6], in particular equations of the forms[7]

$$x^2 - ay^2 = 1, \quad \text{and } x^2 - ay^2 = b,$$

where a and b are integers with a not a perfect square, and "Fermat's last theorem", i.e., the statement that the equation

$$x^n + y^n = z^n$$

has no non-trivial integer solutions if $n > 2$.

In connection with each of these four problems Euler obtained fundamental results in large part determining the further development of algebra, Diophantine analysis, and the algebraic theory of numbers.

1. The Fundamental Theorem of Algebra

The first "proof" of the fundamental theorem was published by J. R. d'Alembert in 1746; it used purely analytic methods and suffered from defects in rigor glaring even by 18th century standards.

In that same year Euler presented his proof of the fundamental theorem to the Berlin Academy of Sciences. This proof was subsequently included in his memoir "Investigations of imaginary roots of equations", which appeared in 1751 in the *Notes* of the Berlin Academy for 1749 [1].

Euler sought an algebraic proof of the theorem. Of course it is now known that a purely algebraic proof is impossible; use of results of an analytic character is unavoidable. It would seem that Euler came to this view also. In his proof he reduces the number of such results to two, namely:

(i) Every equation of odd degree with real coefficients has at least one real root;

---

[5]That is, Diophantine.—*Trans.*
[6]That is, Diophantine equations.—*Trans.*
[7]The first of these is "Pell's equation", so-called.–*Trans.*

(ii) Every equation of even degree with real coefficients and negative constant term has at least two real roots[8].

Euler's proof consists of a reduction procedure from the case of equations of degree $2^k m$, m odd, to those of degree $2^{k-1} m_1$, m_1 again odd [ultimately followed by an appeal to the result (i) above]. Thus he asserts that any polynomial $P_n(x)$ with real coefficients and even degree $n = 2\ell$ can be factored as follows:

$$P_{2\ell}(x) = f_\ell(x) g_\ell(x), \qquad (*)$$

where $f_\ell(x)$ and $g_\ell(x)$ are polynomials of degree ℓ, again with real coefficients.

He first observes that it suffices to consider the case $n = 2^k$, where lie the essential difficulties of the reduction. Indeed, if we have a polynomial $f_N(x)$ whose degree N is not a power of 2, then there is a positive integer k such that $2^{k-1} < N < 2^k$, and by multiplying $f_N(x)$ by $2^k - N$ degree-one factors, for example by $x^{2^k - N}$, we obtain a polynomial $P_n(x)$ of degree $n = 2^k$.

Euler establishes the factorization (*) first in the cases $n = 4, 8, 16$, and then proceeds to the general case $n = 2^k$. To convey Euler's method to the reader, we consider here the case $n = 4$ and then the general case $n = 2^k$.

Thus suppose we are given an equation of the form

$$x^4 + Bx^2 + Cx + D = 0. \qquad (1)$$

(It may be assumed that the equation is already reduced to this form, with the coefficient of x^3 equal to 0.) Euler assumes a factorization of the required sort:

$$x^4 + Bx^2 + Cx + D = (x^2 - ux + \lambda)(x^2 + ux + \mu), \qquad (2)$$

where the coefficients u, λ, μ are unknown. First, by comparing the coefficients on opposite sides of this identity he finds the following equation in the unknown u:

$$u^6 + 2Bu^4 + (B^2 - 4D)u^2 - C^2 = 0.$$

Since this equation has even degree in u and negative constant term, it follows from statement (ii) above that it has at least two real roots, either of which may be chosen as the value of u. Then, as Euler shows, the coefficients λ and μ can be expressed as rational functions of u and the coefficients of the original equation.

Euler next obtains the same result by general theoretical means, avoiding calculation. To this end he applies the following propositions, stated without proof, which were subsequently to figure centrally in Lagrange's work, and later in that of Galois:

(A) Every rational symmetric function of the roots of an equation is a rational function of the coefficients of that equation (the fundamental theorem of symmetric functions).

(B) A rational function $\varphi(x_1, \ldots, x_n)$ of the roots of an equation, taking on k distinct values under all possible permutations of those roots, satisfies an equation of degree

---

[8]Note that both of these results (as applied to a polynomial equation $f_n(x) = x^n + a_1 x^{n-1} + \cdots = 0$, say) are immediate consequences of the Intermediate Value Theorem: (i) since, n being odd, $f_n(x)$ is negative for large negative x and positive for large positive x; and (ii) since, n being even, $f_n(x)$ is positive for all sufficiently large positive and negative x ($f_n(x) \to \infty$ as $x \to \pm\infty$), and negative at $x = 0$, so that the graph of $y = f_n(x)$ must cut the x-axis at least two places.—*Trans.*

k whose coefficients are rational expressions in the coefficients of the original equation (subsequently proved by Lagrange).

These assumed, Euler argues as follows: To any equation $f_n(x) = 0$ of degree n with real coefficients, one may "ascribe" n "roots" $\alpha_1, \alpha_2, \ldots, \alpha_n$, and write

$$f_n(x) = x^n + a_1 x^{n-1} + \cdots + a_n = (x - \alpha_1)(x - \alpha_2) \cdots (x - \alpha_n),$$

where $\alpha_1, \alpha_2, \ldots, \alpha_n$ are merely symbols with which one may, however, operate as if they were ordinary numbers[9]; for instance one may write

$$\alpha_1 + \alpha_2 + \cdots + \alpha_n = a_1,$$
$$\alpha_1 \alpha_2 + \cdots + \alpha_{n-1} \alpha_n = a_2,$$
$$\vdots$$
$$\alpha_1 \alpha_2 \cdots \alpha_n = a_n.$$

The Fundamental Theorem of Algebra may then be understood as asserting that the "roots" $\alpha_1, \alpha_2, \ldots, \alpha_n$ are in fact all real or complex numbers.

Euler applies this reasoning in the case $n = 4$. Thus he supposes that the equation (1) has "roots" $\alpha, \beta, \gamma, \delta$, whence it follows that u is the sum of some two of these four roots, so that it takes on at most $\binom{4}{2} = 6$ distinct values under all permutations of the four roots. Hence by the proposition (B) above, u is a root of an equation of degree six. He goes on to observe that since

$$\alpha + \beta + \gamma + \delta = 0,$$

the values of u in question are

$$u_1 = \alpha + \beta =: p, \quad u_2 = \alpha + \gamma =: q, \quad u_3 = \alpha + \delta =: r,$$
$$u_4 = \gamma + \delta = -p, \quad u_5 = \beta + \delta = -q, \quad u_6 = \beta + \gamma = -r,$$

whence the desired equation for u has the form

$$(u^2 - p^2)(u^2 - q^2)(u^2 - r^2) = 0, \tag{3}$$

an equation of even degree with constant term $-p^2 q^2 r^2$. In order to apply statement (ii) above, one needs to show that $-p^2 q^2 r^2 < 0$, or, equivalently, that the product pqr is real. Euler proves this by showing that

$$pqr = (\alpha + \beta)(\alpha + \gamma)(\alpha + \delta)$$

is a symmetric function of the four roots, so that by the proposition (B) it can be expressed as a rational function of the coefficients of the original equation (1). This general argument shows that the coefficient u can indeed be chosen real.

---

[9]This can in fact be made rigorous using the later technique of constructing the "splitting field" of $f_n(x)$ over $\mathbb{R}$, involving the formation of successive finite field extensions of $\mathbb{R}$ as quotients of successive polynomial rings over intermediate fields, as described in textbooks on "modern algebra". However there still remains the (fundamental) question as to why the splitting field is contained in $\mathbb{C}$! (The author also makes this point below.)—*Trans.*

For the general case $n = 2^k$ (Theorem 7) Euler gives only an outline of the proof. He represents the given polynomial, assumed to have the form

$$f_n(x) = x^{2^k} + Bx^{2^k-2} + Cx^{2^k-3} + \cdots, \tag{4}$$

as a product of two factors of degree 2^{k-1} with coefficients to be determined:

$$f_n(x) = (x^{2^{k-1}} + ux^{2^{k-1}-1} + \lambda x^{2^{k-1}-2} + \cdots)(x^{2^{k-1}} - ux^{2^{k-1}-1} + \lambda' x^{2^{k-1}-2} + \cdots).$$

These unknown coefficients number $2^k - 1$, the same as the number of equations defining them.

Since u is a sum of 2^{k-1} of the 2^k "roots" of $f_n(x)$, the largest possible number of distinct values that u takes on under all permutations of the roots is $\binom{2^k}{2^{k-1}} = 2N$ where N is odd. Having shown this, Euler infers that u satisfies an equation of degree $2N$ with real coefficients. It is not difficult to see that this equation will have form similar to (3), namely

$$(x^2 - p_1^2)(x^2 - p_2^2) \cdots (x^2 - p_N^2) = 0.$$

Euler now asserts that the constant term $-p_1^2 p_2^2 p_3^2 \cdots p_N^2$ of the left-hand side is negative—which is in fact not difficult to show. He then infers from the result (ii) above that u may be chosen real. As far as the remaining coefficients $\lambda, \mu, \ldots, \lambda', \mu', \ldots$, are concerned, Euler merely claims that they can be expressed rationally in terms of u and the coefficients $B, C, D, \ldots$ of the original polynomial (4). It is difficult to say precisely what considerations led Euler to this conclusion. Lagrange subjected Euler's reduction argument to rigorous scrutiny in his memoir "On the form of imaginary roots of equations" (1772), making good all potential difficulties with the argument [16, 20]. In particular, in order to prove Euler's assertion concerning the possibility of expressing $\lambda, \mu, \ldots, \lambda', \mu', \ldots$ rationally in terms of u and the coefficients of the original polynomial (4), Lagrange invokes his theory of "similarity transformations", expounded in 1771 in his famous "Meditations on the algebraic solution of equations" [16][10].

Several 18th-century mathematicians tried to simplify Euler's reduction. The first such attempt was made by D. de Foncenet (1759), followed by Laplace and others. However in every such attempt—until the advent of Gauss—Euler's approach to the problem was imitated without question, i.e., it was always assumed that whatever the degree of an equation

$$x^n + Ax^{n-1} + Bx^{n-2} + \cdots + K = 0,$$

it could always be represented in the form

$$(x + \alpha_1)(x + \alpha_2) \cdots (x + \alpha_n) = 0,$$

where $\alpha_1, \alpha_2, \ldots, \alpha_n$, although merely symbols, can be operated with according to the same rules as hold for real numbers.

The first to reject this approach to the problem was the young C. F. Gauss. In his doctoral dissertation (1799), which was devoted to proving the Fundamental Theorem, he wrote: "Since over and above real and imaginary quantities $a + b\sqrt{-1}$ it is impossible to

---

[10]The article [16] seems to be a different one.—*Trans.*

conceive of further quantities, it is not completely clear how what we wish to prove differs from what is proposed as an initial assumption; however even if it were possible to invent other kinds of quantities F, F', F'', ..., then [the claim] that every equation is satisfied by some real value of x, or by some value $a + b\sqrt{-1}$, or by some value of the form F, or F', etc., would still be unacceptable without proof. Hence the initial assumption in question can have only the following meaning: that every equation is satisfied by a real value of the unknown, or an imaginary value of the form $a + b\sqrt{-1}$, or perhaps some value of another form, hitherto unknown, or else a value not accommodated to any form. How such quantities, of which we can form no conception whatever, these shadows of shadows, should be added or multiplied cannot be conceived with the clarity required in mathematics" [11, pp. 12–13].

In 1815 Gauss revisited the Fundamental Theorem, on this occasion giving a proof in large part algebraic, but not presupposing the existence of roots of any kind. In Gauss' opinion such a presupposition is, "at least when it is a question of giving a general proof of the decomposability of a polynomial into factors [of degrees one and two], nothing more than an instance of *petitio principii*" [11] [12, p. 40]

However, the charge that Euler's proof involves a vicious circle was unjust—and that this is so is most clearly brought out by an examination of Gauss' own second proof: in order to avoid presupposing the existence of symbolic "roots", he works with congruences modulo certain polynomials, i.e., in effect he constructs the splitting field of the given polynomial.

This method of Gauss was given pure form by Leopold Kronecker, who used it to give his celebrated construction of the splitting field of a real polynomial without even assuming the existence of the complex numbers (1870/1871). With this Euler's proof was fully "rehabilitated": although his assumption that every polynomial has a splitting field does constitute a gap in his proof, the argument contains no vicious circle.

It might be said that Euler and other 18th century mathematicians conceived the Fundamental Theorem of Algebra as asserting that the splitting field of any polynomial with real coefficients is isomorphic to some subfield of the field of complex numbers—however with the qualification that the existence of a "splitting field" was not proved but merely postulated.

It is remarkable that Euler's algebraic standpoint, rejected in the first half of the 19th century, was taken up again in the 1870s and 1880s. Thus his algebraic approach triumphed over the opposing analytic one, involving the construction of the field of complex numbers beforehand, and then the proof of the existence of a root in that field.

Note also that as formulated by Euler the Fundamental Theorem of Algebra coincides in essence with the Weierstrass–Frobenius theorem stating that the only finite-dimensional associative and commutative division algebras over the field of real numbers are 1) the field of real numbers itself, and 2) the field of complex numbers.

2. Solving equations by radicals

This problem occupied almost all 18th century mathematicians; for example Tschirnhausen, Bézout, Lagrange, Vandermonde—among many others—expended considerable effort on

---

[11] That is, begging the question.—*Trans.*

it. Euler considered the problem twice, first in a memoir presented in 1733 and published in 1738 (E30) [2] and then, 30 years later, in an article presented in 1759 and published in 1764 (E282) [3].

In the first of these works Euler begins by observing that the solution of an equation of degree $2, 3$, or 4 involves a reduction to an equation of degree $1, 2$, or 3 respectively, which he calls the *aequatio resolvens* (the *resolvent* of the original equation, i.e., that which solves it). He makes the suggestion that an equation

$$x^n = ax^{n-2} + bx^{n-3} + \cdots + g \tag{5}$$

of arbitrary degree, should have a resolvent of the form

$$z^{n-1} = \alpha z^{n-2} + \beta z^{n-3} + \cdots. \tag{6}$$

If the resolvent has roots $A_1, A_2, \ldots, A_{n-1}$, then the roots of the original equation should— so Euler maintains—have roots of the form

$$x = \sqrt[n]{A_1} + \sqrt[n]{A_2} + \cdots + \sqrt[n]{A_{n-1}}. \tag{7}$$

Of course we now know that he could never have found a general solution of equations of degree five by these means—in addition to which the proposed form (7) for the solutions has the crucial defect that since each term may take on n distinct values independently of the others, one obtains altogether as many as n^{n-1} distinct possible values for x.

In the same memoir Euler considers "reciprocal equations" and proposes for their solution his famous substitution

$$y = x + \frac{1}{x}.$$

In the second memoir Euler replaces the formula (7) by the following one:

$$x = w + A \sqrt[n]{v} + B \sqrt[n]{v^2} + \cdots + Q \sqrt[n]{v^{n-1}}, \tag{8}$$

where w is real and v satisfies an equation of degree $\leq n - 1$. Unlike the earlier formula, this one takes on at most n possibly distinct values. However since this also fails to yield a solution of the general equation of degree five, Euler considers instead special classes of degree-five equations that are so soluble. It turns out that the equations in all but one of these classes have cyclic Galois groups, while those in the exceptional class have metacyclic[12] Galois groups [24].

This work is of importance also because formula (8) is the most general in the sense that if the equation (5) *is* soluble by radicals, then its roots will indeed have the form (8). This was shown by N. H. Abel, who built on this formula in fashioning his celebrated proof of the insolubility by radicals of the general equation of degree five.

3. The problem of finding rational points on cubic curves of genus one

In this section we shall assume that the whole discussion takes place in the context of the field $\mathbb{Q}$ of rational numbers; thus, in particular, the coefficients of all equations considered are assumed rational.

---

[12] The original has "semi-metacyclic" here.—*Trans.*

D. Hilbert and A. Hurwitz proved [13] that any curve of genus zero and degree m is birationally equivalent to a straight line if m is odd, and to a conic section if m is even. In the first case there will always be infinitely many rational points on the curve, which may be expressed in terms of a rational function of a single parameter. In the second case the situation depends on whether there is at least one rational point on the given curve: if there is, then—as was already shown by Diophantus in his *Arithmetic*—again the curve will pass through infinitely many rational points with coordinates expressible in terms of a parameter:

$$x = \varphi(t), \ y = \psi(t).$$

Thus the set $\mathfrak{M}$ of rational points of a curve of genus zero has a simple structure. On the other hand, as has been shown relatively recently by G. Faltings using methods of A. N. Parshin, curves of genus ≥ 2 have only finitely many rational points (L. J. Mordell's conjecture) [10].

There remain, therefore, curves of genus one, of the very greatest of interest to researchers. Although the arithmetic of such curves was fully elucidated by Poincaré in the early 20th century [18], the roots of these investigations reach as far back as Diophantus, and the decisive step forward in this respect was taken by Leonhard Euler.

We confine our discussion to cubic curves (of genus one), i.e., to curves given by an irreducible equation

$$F_3(x, y) = 0 \tag{9}$$

of degree three without singular points. From the simple observation that every straight line intersects the curve (9)—which we denote by L—in exactly three points (if we count points according to their multiplicities, and also imaginary points and a possible point at infinity), we derive two methods of obtaining rational points of the curve L:

1) "The secant method": if we already have two rational points of the curve L, then the secant T passing through them will intersect L in a third rational point;

2) "The tangent method": if we know of just one rational point of L, then the tangent line T to L at that point will intersect L in exactly one other point, which will also be rational.

Both of these methods are contained in Diophantus' *Arithmetic*—not, it is true, in geometrical form, but formulated purely algebraically. Thus he speaks of "equations" instead of "curves", "solutions"—which he takes to be both rational and positive—rather than "rational points", an appropriate substitution instead of the tangent line T at a point (x_0, y_0), etc. Following Diophantus, this algebraic terminology continued to be used by the mathematicians of the Near East, and by Bombelli, Viète, Fermat, and Euler. Note also that Diophantus did not apply the "secant method" in its most general form but only in the situation where one of the known points is finite and the other a point at infinity. For example, for a curve of the form

$$y^3 = a^3x^3 + bx^2 + cx + d^3, \tag{10}$$

the line

$$y = ax + d \tag{11}$$

passes through the finite point $P(0, d)$ and the curve's point at infinity. In line with the mathematical tradition of antiquity, Diophantus did not provide "a general formulation" of these methods, but merely illustrated them with examples. Europeans became acquainted with his *Arithmetic* only during the second half of the 16th century. Its first 143 problems were included by Bombelli in his *Algebra* (1572) and its first translation into Latin was published in 1575.

Bombelli and Viète mastered Diophantus' methods to the extent that by means of the "tangent method" they were able to solve the equation

$$x^3 + y^3 = a^3 - b^3, \ a > b, \tag{12}$$

which Diophantus claims always has a solution. However Bombelli solves it only for the particular values $a = 4, b = 3$ of a and b, and Viète, in his more general solution, imposes additional conditions on a and b.

Pierre Fermat acquired a deeper understanding of these methods. The Abbot de Billy, who wrote out Fermat's investigations into the above problem on the basis of letters he received from Fermat, lauds him especially for his iteration of the "tangent method", whereby he was able to obtain infinitely many rational solutions starting with any particular known one. This enabled Fermat to solve the problem (12) of Diophantus quite generally, i.e., for any a and b with $a > b$.

In Volume II of his *Algebra* (E388) [4], Euler generalized and systematized everything known up to that time concerning the solution in rationals of indeterminate [i.e., Diophantine] equations

$$f_n(x, y) = 0 \ \text{for} \ n = 2, 3, 4.$$

He first clearly defines the principal distinction between the approaches to solving Diophantine equations of the second and third degrees. He applies the "secant method" in the same situation as Diophantus and iterates the "tangent method" just as Fermat did.

Towards the end of his life Euler revisited the problem of finding rational points on curves of genus one. In 1780 he presented to the Petersburg Academy a series of articles (E773, E774, E777, E778), which were only published more than 40 years after his death, in 1830 [5–8]. In these works he makes fundamental progress on the problem, consisting in the following:

1) He begins to apply the "secant method" to situations where both of the given points are finite, i.e., he accords the method its most general form;

2) He reformulates both the "secant" and the "tangent" methods so that by their means it becomes possible to define a group structure on the set $\mathfrak{M}$ of rational points of a given curve of genus one. This was achieved by using the operation of reflection of the rational points in an axis of symmetry of the curve.

In order to explain exactly what Euler achieved, we shall confine the discussion to the cubic curve Γ defined by the equation

$$y^2 = x^3 + ax + b. \tag{13}$$

(It was shown by K. Weierstrass that if a cubic curve has a rational point then it may always be reduced to this form.) Let $A(x_1, y_1)$ and $B(x_2, y_2)$ be two rational points on this curve. Let $C(x_3, y_3)$ be the third rational point of Γ obtained as the point of intersection of Γ with the straight line T through A and B. Of course this process does not lend itself to

iteration: if we connect C with either of A or B, we obtain the same straight line T—
this explains why, until the advent of Euler, only the "tangent method" was iterated. Euler
supplemented the "secant method" with the operation of reflection of a rational point in an
axis of symmetry of the curve in question. Thus in the case of our curve Γ, from the point
$C(x_3, y_3)$ we obtain by reflection in the x-axis the further rational point $\bar{C}(x_3, -y_3)$ of Γ,
and then by joining $\bar{C}$ to A or B we obtain a new straight line yielding yet another rational
point of Γ.

As a matter of fact Euler's innovation is of profound significance. If we take the point
C to be the "sum" of the points A and B: $C := A \oplus B$, we obtain an operation that is
not associative, so that it certainly fails to yield a group structure on the set $\mathfrak{M}$ of rational
points of a curve of genus one. However if instead one sets

$$A \oplus B := \bar{C},$$

then this sum does enjoy all the nice properties of the usual addition of numbers.

Euler exploited an analogous reflection in axes of symmetry of curves of genus one in
connection with the "tangent method", yielding the "sum" $A \oplus A := 2A$.

Thus it was precisely the ideas of these late works of Euler's on Diophantine analysis
that allowed later investigators to endow the set $\mathfrak{M}$ of rational points of a curve of genus
one with a group structure. This was done essentially by C. G. J. Jacobi in 1834 [15]—he
even showed that such groups may have torsion, i.e., they may in general have points of
finite order—, and in explicit form by Poincaré in the above-mentioned memoir of 1901. A
preliminary examination of these exceedingly important works of Euler was carried out by
J. E. Hofmann [14]; a detailed analysis of the whole series can be found in the paper [22]
of T. A. Lavrinenko.

It remains for us to mention that in his researches on curves of genus one Euler in
essence discovered a new approach to the study of elliptic integrals. As early as 1834 Jacobi, in the above-mentioned article of that year, observed that the methods developed by
Euler in those later articles were very closely analogous to the methods used to establish the
theorem on the sum of elliptic integrals. Thus in the case of the equation (13), if $A(x_1, y_1)$
and $B(x_2, y_2)$ are points of the curve Γ, then that theorem asserts the existence of a point
$P(x_3, y_3)$ of Γ such that

$$\int_0^A \frac{dx}{y} + \int_0^B \frac{dx}{y} = \int_0^P \frac{dx}{y},$$

where moreover x_3 and y_3 are rationally expressible in terms of x_1, y_1, x_2, y_2. It is not
difficult to show that in fact the point $P(x_3, y_3)$ coincides with the point $\bar{C}(x_3, -y_3)$. This
is historically the first time the connection between Diophantine analysis and the theory of
abelian functions is mentioned. Did Euler notice the connection? Jacobi was of the opinion
that the analogy between the formulae is so striking that it could surely not have escaped
the "great man's" notice (see also [21]). However no declaration by Euler on this issue is
known. Whatever the case may be, it remains true that in the 19th century Diophantine
analysis and algebraic geometry developed in the new directions determined by Euler. It is
noteworthy that in A. Weil's book [19], which appeared after the official commemoration
of the 200th anniversary of Euler's death, the works of the great scholar relating to the
solution of Diophantine equations are examined and analyzed with similar conclusions.

Fermat's last theorem and the first appearance of algebraic integers

Of the great many Diophantine equations considered by Euler, we single out here Fermat's last theorem.

Fermat formulated his theorem in the margin of his copy of Diophantus' *Arithmetic* as a note to Problem 8 of Book II: "To decompose a given square into two squares". Fermat wrote: "It is impossible to decompose a cube into two cubes, or a fourth power into two fourth powers, or in general any power greater than the second into two of the same powers. I have found a most marvelous proof of this. However the margin is too small to contain it"[13].

It is remarkable that in no other place did he state the theorem in this generality, while in his correspondence he often mentions the problem of proving it for cubes and fourth powers.

Fermat himself proved his "last theorem" in the case of fourth powers, using the method of infinite descent. This is in fact the only one of his number-theoretic proofs to have come down to us. It also was written in the margins of his copy of Diophantus' *Arithmetic* but in this case as a note to Problem 24 of Book VI.

Euler likewise began with a proof of the theorem in the case $n = 4$ in his memoir (E98), presented in 1738 and published in 1747 [9]. It was only 30 years later that he proposed a proof of the case $n = 3$ of the theorem. His letters to Ch. Goldbach show him to be intensely preoccupied with this proof. As early as August 4, 1753 Euler wrote to Goldbach that he had a proof of the "last theorem" in the case $n = 3$ [23, p. 374]. It seems likely that this proof was unsatisfying to him aesthetically. Note that Euler's and Lagrange's use of irrational and of imaginary expressions of the form $a + b\sqrt{\pm c}$, where $a, b, c \in \mathbb{Z}$, to solve problems in number theory, dates only from the late 1760s. On this topic Euler wrote to Lagrange: "I was delighted with your method whereby irrational and even imaginary numbers are applied in that area of analysis concerned only with rational numbers. Several years have passed already since similar ideas occurred to me, ..., in the course of publishing here the complete Russian version of my *Algebra*, I added a detailed exposition of this method, and showed that in order to solve the equation

$$x^2 + ny^2 = (p^2 + q^2 n)^\lambda,$$

it suffices to solve the following one:

$$x + y\sqrt{-n} = (p + q\sqrt{-n})^\lambda ."$$

(See [17, p. 215].) It appears that at that time Euler also conceived the idea of using expressions of the form $a + b\sqrt{-c}$ to solve Fermat's last theorem. For that purpose a preliminary decisive step needed to be taken, namely that of considering such expressions as if they behaved like integers.

What does this mean? In what does the difference between whole numbers and, say, rational numbers consist? Above all in the rich arithmetic of the whole numbers, which separate into primes and composite numbers, and have the property—the fundamental law

---

[13]"Cubum autem in duos cubos, aut quadratoquadratum in duos quadratoquadratos, et generaliter nullam in infinitum ultra quadratum potestatem in duos ejusdem nominis fas est dividere: cujus rei demonstrationem mirabilem sane detexi. Hanc marginis exiguitas non caperet".

of divisibility—that every nonzero integer is expressible as a product of primes in essentially only one way. For the natural numbers this fact was already known to Euclid—albeit in a slightly restricted form. From this result it follows in particular that if the product of two relatively prime integers is equal to some power of an integer, then each factor of that product must be a similar power: if $(a, b) = 1$ and $ab = \ell^k$, then $a = \ell_1^k$ and $b = \ell_2^k$. Euler carried this property over—without proof—to numbers of the form $m + n\sqrt{-3}$, $m, n \in \mathbb{Z}$, and used it to prove Fermat's theorem in the case $n = 3$.

Here is his argument. Suppose there exist positive integers x, y, z satisfying

$$x^3 + y^3 = z^3.$$

We may assume that two of x, y, and z are odd and one even (since the only other possibility is that they are all even, in which case we can divide out 2s from all three as long as this is possible). Suppose x, y odd and z even—otherwise consider the equation $x^3 = z^3 - y^3$ instead. Assuming $x > y$, we can always find relatively prime positive integers p and q of opposite parity such that $x = p + q, y = p - q$, whence

$$x^3 + y^3 = 2p(p^2 + 3q^2) = z^3.$$

Since z is even, z^3 is divisible by 8, whence p must be even—in fact divisible by 4—and q odd. We then have

$$\frac{p}{4}(p^2 + 3q^2) = \left(\frac{z}{2}\right)^3.$$

Euler now considers two cases: 1) x not divisible by 3; and 2) x divisible by 3. For our purposes—the appreciation of Euler's idea—it suffices to consider the first of these. Thus if 3 does not divide x, then the integers $\frac{p}{4}$ and $(p^2 + 3q^2)$ will be relatively prime, so that $\frac{p}{4}$ and $(p^2 + 3q^2)$ must both be perfect cubes. Next—and this is the crucial step in his proof—Euler factorizes $p^2 + 3q^2$:

$$p^2 + 3q^2 = (p + q\sqrt{-3})(p - q\sqrt{-3}) = r^3,$$

and from this he infers that each of these imaginary factors must be a cube:

$$p \pm q\sqrt{-3} = (u \pm v\sqrt{-3})^3,$$

whence

$$p = u(u - 3v)(u + 3v), \quad q = 3v(u^2 - v^2).$$

Since q is odd, v must be odd, whence u is even. From the fact that $\frac{p}{4}$ is a perfect cube it follows that $2p$ is also a perfect cube., i.e.,

$$2u(u - 3v)(u + 3v) = t^3.$$

Since p and q are relatively prime, so are u and v, and therefore $2u$, $(u - 3v)$, and $(u + 3v)$ are [pairwise] relatively prime, whence

$$u - 3v = f^3, \ u + 3v = g^3, \ 2u = f^3 + g^3 = t_1^3.$$

It is not difficult to see that f, g, t_1 are less than x, y, z, so that if the original equation has a solution in positive integers x, y, z, then it has a solution in smaller numbers f, g, t_1.[14] By means of an analogous argument we can find a further triple of smaller integers f_1, g_1, t_2 satisfying the original equation with f_1, g_1 odd and t_2 even, and so on[15]. This process must end in finitely many steps since there are only finitely many positive integers less than any given one.

What is new in this proof is Euler's transference of the usual divisibility properties of the ordinary integers to quantities of the form $p + q\sqrt{-3}$. It should be noted that Euler's proof is not rigorous[16]. Moreover it relies on an invalid assumption, since in fact in the ring of numbers of the form $m + n\sqrt{-3}$, $m, n \in \mathbb{Z}$, unique factorization as products of irreducible elements does not hold! This is shown for example by

$$4 = 2 \cdot 2 = (1 + \sqrt{-3})(1 - \sqrt{-3}).\text{[17]}$$

However it is well known that the ring of *all* algebraic integers[18] of the field of numbers of the form $r + s\sqrt{-3}$, $r, s \in \mathbb{Q}$, includes those numbers of the form $\frac{m+n\sqrt{-3}}{2}$, with $m \equiv n(\mathrm{mod}\ 2)$, and in this particular case this larger ring of algebraic integers does enjoy the unique factorization property. Euler managed to avoid error here only because his argument involves only numbers of the form $p + q\sqrt{-3}$ where $p \not\equiv q(\mathrm{mod}\ 2)$.

This proof of Euler's adumbrates two important new ideas which were taken up and developed by later mathematicians. The first of these is the idea that in order to prove Fermat's last theorem, the expression $x^\lambda + y^\lambda$ should be decomposed as the product

$$x^\lambda + y^\lambda = (x + y)(x + \zeta y) \cdots (x + \zeta^{\lambda-1} y),$$

where ζ is a primitive λth root of unity.

The second—and more important—idea is that for the investigation of certain properties of the ordinary integers, it is necessary to widen the concept of integer. The later development of this idea led in the 19th century to the creation of the theory of algebraic integers, where the arithmetic of the ring of algebraic integers of an algebraic number field is studied. Already in the period 1828–1830 Gauss, in his famous memoir "On biquadratic residues", considered the complex integers of the form $m + n\sqrt{-1}$, $m, n \in \mathbb{Z}$, showing that their arithmetic is completely analogous to that of $\mathbb{Z}$. This was followed in succession by the investigations of E. Kummer, P. Lejeune-Dirichlet, R. Dedekind, E. I. Zolotarev, L. Kronecker, and D. Hilbert. This line of development led—as just noted— to the notion of the ring of algebraic integers of an algebraic number field, and to the theory of ideals and their divisors, local methods, etc.—in short to the apparatus of commutative algebra.

---

[14] This is justified if $u > 0$. However if u is negative then it is not clear that $|f|, |g|, |t_1|$ are less than x, y, z in some order. However here t_1 is even and $|t_1| < z$, while f and g are odd, so perhaps an infinite descent relative to the even members of the triples solving the equation is possible. If this is indeed an error, presumably it is Euler's and not the author's.—*Trans.*

[15] Or rather, a triple of integers f_1, g_1, t_2 with $|t_2| < |t_1|$, and so on (see the previous footnote).–*Trans.*

[16] Does this refer to the possible error noted in the previous two footnotes?—*Trans.*

[17] It is not difficult to show that 2 and $1 \pm \sqrt{-3}$ are non-associated irreducibles of the ring in question.—*Trans.*

[18] That is, solutions of monic polynomials over $\mathbb{Z}$.—*Trans.*

Thus we see that in large part the work of Euler served as the source from which Galois theory and Diophantine analysis of curves of genus one developed, and also that it was he who took the first steps towards extending the domain of arithmetic.

References

[1] Euler, L. "Recherches sur les racines imaginaires des équations". Mém. Acad. Sci. Berlin (1749). 1751. Vol. 5, pp. 222–288; *Opera* I-6.

[2] ——. "De formis radicum aequationum cuiusque ordinis coniectatio". Comment. Acad. Sci. Petrop. (1732–1733). 1738. Vol. 6, pp. 216–231; *Opera* I-6.

[3] ——. "De resolutione aequationum cuiusvis gradus". Novi comment. Acad. Sci. Petrop. (1762–1763). 1764. Vol. 9, pp. 70–98; *Opera* I-6.

[4] ——. *Vollständige Anleitung zur Algebra*. SPb., 1770. Vol. 2; *Opera* I-1.

[5] ——. "Solutio problematis difficillimi, quo hae duae formulae $a^2x^2 + b^2y^2$ et $a^2y^2 + b^2x^2$ quadrata reddi debent". Mém. Acad. Sci. Pétersb. 1830. Vol. 11, pp. 12–30; *Opera* I-5.

[6] ——. "Investigatio binorum numerorum formae $xy(x^4 - y^4)$, quorum productum sive quotus sit quadratum". Mém. Acad. Sci. Pétersb. 1830. Vol. 11, pp. 31–45; *Opera* I-5.

[7] ——. "De resolutione huius aequationis $0 = a + bx + cy + dx^2 + exy + fy^2 + gx^2y + hxy^2 + tx^2y^2$ per numeros rationales". Mém. Acad. Sci. Pétersb. 1830. Vol. 11, pp. 58–68; *Opera* I-5.

[8] ——. "Methodus nova et facilis formulas cubicas et biquadraticas ad quadratum reducendi". Mém. Acad. Sci. Pétersb. 1830. Vol. 11, pp. 69–91; *Opera* I-5.

[9] ——. "Theorematum quorumdam arithmeticorum demonstrationes". Comment. Acad. Sci. Petrop. (1738). 1747. Vol. 10; *Comment. arith. collectae*. Vol. 1, pp. 24–34; *Opera* I-2.

[10] Faltings, G. "Endlichkeitssätze für Abelsche Variatäten über Zahlkörpern". J. Invent. Math. 1983. Vol. 73, pp. 349–366.

[11] Gauss, C. F. "Demonstratio nova theorematis omnem functionem algebraicam rationalem integram unius variabilis in factores primi vel secundi gradus resolvi potest" (1799). *Werke*. Göttingen, 1878. Vol. 3, pp. 1–30.

[12] ——. "Demonstratio nova altera theorematis omnem functionem algebraicam rationalem integram unius variabilis in factores reales primi vel secundi gradus resolvi posse" (1815). *Werke*. Göttingen, 1878. Vol. 3, pp. 31–56.

[13] Hilbert, D., Hurwitz, A. "Über die diophantischen Gleichungen vom Geschlecht Null". Acta. Math. 1890. Vol. 14, pp. 217–224.

[14] Hofmann, J. E. "Über zahlentheoretische Methoden Fermats und Eulers, ihre Zusammenhänge und ihre Bedeutung". Arch. Hist. Exact Sci. 1961. Vol. 1, No. 2, pp. 122–159.

[15] Jacobi, C. "De usu theoriae integralium ellipticorum et integralium abelianorum in analysi Diophantea". J. für reine und angew. Math. 1835. Bd. 13, pp. 353–355.

[16] Lagrange, J. L. "Sur la forme des racines imaginaires des équations". Nouv. Mém. Acad. Sci. Berlin (1772). 1774. pp. 479–516; *Oeuvres*. Paris, 1869. Vol. 3, pp. 479–516.

[17] ——. *Oeuvres*. Paris, 1892. Vol. 14.

[18] Poincaré, H. "Sur les propriétés arithmétiques des courbes algébriques". J. math. pures et appl. 1901. Vol. 7, pp. 161–233.

[19] Weil, A. *Number theory: An approach through history. From Hammurapi to Legendre.* Boston; Basel; Stuttgart: Birkhaüser, 1983.

[20] Bashmakova, I. G. "On the proof of the fundamental theorem of algebra". Research hist. math. (Ist.-mat. issled.). 1957. Issue 10, pp. 257–304.

[21] ——. "The arithmetic of algebraic curves (from Diophantus to Poincaré)". Research hist. math. (Ist.-mat. issled.). 1975. Issue 20, pp. 104–124.

[22] Lavrinenko, T. A. "The solution of indeterminate equations of degrees 3 and 4 in Euler's later works". Research hist. math. (Ist.-mat. issled.). 1983. Issue 27, pp. 67–78.

[23] *Leonhard Euler und Christian Goldbach. Briefwechsel. 1729–1764.* Hsrg. A. P. Juškevič, E. Winter. Berlin, Akad.-Verl., 1965.

[24] Maĭstrova, A. L. "The solution of algebraic equations in the works of Euler". Research hist. math. (Ist.-mat. issled.). 1985. Issue 29, pp. 189–198.

Diophantine Equations in Euler's Works

T. A. Lavrinenko

The part played by L. Euler in the development of number theory is well known: he laid the foundations of the theory of residues of powers, discovered the law of quadratic reciprocity, invented analytic methods to investigate number-theoretical questions, and so on (see, for example, [1–3]). Far less known are his achievements in the realm of Diophantine analysis, above all his results on the solution of Diophantine equations in rational numbers[1]. However it is precisely Euler's work in Diophantine analysis that represents the completion of the stage in the development of that science that might conditionally be termed "elementary algebraic". This stage was characterized by the use mainly of elementary algebraic means to investigate rational solutions of Diophantine equations; and within the limits of this approach important methods were developed for solving indeterminate equations of degrees two, three, and four, defining curves of genus zero or one (see [4, 5]). By "Diophantine analysis" we shall throughout mean only that portion of the subject concerned with finding rational solutions of indeterminate equations. Note also that occasionally we shall resort to using geometrical terminology—which is, of course, now standard for the subject—but this should not be taken as implying that Euler used such language.

The most important of Euler's results in Diophantine analysis are noted in the essay [6] of the present collection. Here we examine these results more closely, and also consider certain other little-known investigations of Euler's.

To the solution of Diophantine equations L. Euler devoted around 40 articles and several chapters of his textbook *Vollständige Anleitung zur Algebra* (which we shall henceforth refer to simply as his *Algebra*). There are also a great many notes on this topic among his manuscripts. The bulk of these articles are concerned with the investigation of particular Diophantine equations or systems of these. Faithful to the tradition of Diophantus and Fermat, Euler employs purely algebraic means—the apparatus of algebraic transformations, permutations, and substitutions—to obtain his solutions. In each particular case, in feeling his way towards the goal of eliciting rational solutions he shows extraordinary artistry in the application of the pertinent formulaic apparatus—in the present case algebraic, although this is characteristic of Euler's mathematical creativity quite generally.

---

[1]By "Diophantine equations" one usually means polynomial equations in two or more variables for which integer, or, more generally, rational solutions are sought.—*Trans.*

In addition to such particular solutions Euler obtained several fairly general results concerning Diophantine equations, chiefly of degree three or four in two variables. From the modern point of view the search for rational solutions of equations in two variables is classified under the arithmetic of algebraic curves. The basic facts concerning the structure of the set of rational points of an algebraic curve are given in the article [6] of the present collection, so that there is no need to repeat them here. We shall instead confine ourselves to the consideration of results on elliptic plane curves, i.e., of genus one, since the equations of degrees three and four investigated by Euler define elliptic curves and it is precisely Euler's methods of solution of such equations that are of chief interest for the present essay.

As is well known, there are two basic methods for finding new rational points of a cubic elliptic curve from known such points: the "tangent method", involving the tangent to the curve at a known rational point, and the "secant method", involving the secant determined by two known rational points of the curve. In each case the straight line in question— tangent or secant—intersects the cubic curve in a third point which is also rational—the *raison d'être* of each of the two methods. These methods may of course be formulated algebraically. They lie at the heart of the definition of the operation of addition on the set of rational points of a cubic elliptic curve, in terms of which that set becomes a finitely generated group. As early as Diophantus (3rd century AD) the algebraic version of the tangent method was known as it applies to equations of the forms

$$f_3(x) \equiv a + bx + cx^2 + dx^3 = y^2 \tag{1}$$

and

$$f_3(x) \equiv a + bx + cx^2 + dx^3 = y^3. \tag{2}$$

However Diophantus uses the secant method only to solve equations of the form (2), and then only in the special situation where one of the known points is finite and the other is an infinitely distant point of the curve—and of course his execution of the method is purely algebraic (see [4, 5] for greater detail). It was only with Euler that the secant method began to be applied to equations of the forms (1) and (2) using two finite points[2]. We shall show that Euler, striving to extract as much as possible from the elementary algebraic means for solving Diophantine equations at his disposal, arrived at the secant method via several algebraic avenues of approach.

Note that frequently Euler did not formulate his results in Diophantine analysis in general form—although he must surely have been aware that this was possible—but confined himself to examples illustrating them. This is the case, for example, for the equation

$$4 + x^2 = y^3, \tag{3}$$

considered in his *Algebra* [7, pp. 461–462]. This represents the first occasion when the idea (on which in fact the secant method is based) of using two known rational solutions of a cubic equation in order to obtain a third appears in print. Although Euler used two particular rational solutions of the equation (3), namely $x_1 = 2$, $y_1 = 2$ and $x_2 = 11$, $y_2 = 5$, his argument can without difficulty be carried over to the general case of equations of the

---

[2] Actually prior to Euler the secant method using two finite points was formulated—in geometrical form—by Newton. However Newton's notes on this topic came to light only relatively recently following on the publication of his mathematical manuscripts (see [4]).

form (2) for which any two rational solutions (x_1, y_1) and (x_2, y_2) are known (see [8] for details). Euler's idea consists in the transformation of the equation (2) by means of the substitution $x = (x_1 + x_2 t)/(1 + t)$ (or $x = (x_1 - x_2 t)/(1 - t)$—the difference is of no significance) to an equation of the form

$$a' + b't + c't^2 + d't^3 = y^3(1 + t)^3, \tag{4}$$

where

$$a' = a + bx_1 + cx_1^2 + dx_1^3 = y_1^3, \; d' = a + bx_2 + cx_2^2 + dx_2^3 = y_2^3.$$

Setting $y(1 + t) = u$ in (4), we obtain the equation

$$a' + b't + c't^2 + d't^3 = u^3, \tag{5}$$

where $a' = y_1^3, d' = y_2^3$. Since the coefficients a' and d' are cubes, rational solutions of this equation can now be found using known methods of Diophantus–Fermat (see [4, 5] for these methods). Thus the substitution

$$u = y_1 + y_2 t, \tag{6}$$

applied to equation (5) leads to a rational solution $t^*, u^* = y_1 + y_2 t^*$ of that equation, which yields in turn the rational solution

$$x^* = \frac{x_1 + x_2 t^*}{1 + t^*}, \; y^* = \frac{u^*}{1 + t^*} \tag{7}$$

of equation (2). Thus we have a method for constructing a new rational solution (x^*, y^*) of equation (2) from two previously known ones (x_1, y_1) and (x_2, y_2).

It is not difficult to see that geometrically speaking this method consists in the construction of the new rational point (x^*, y^*) of the given curve (2) as the third point of intersection with that curve of the straight line through its rational points (x_1, y_1) and (x_2, y_2), i.e., the secant method. Indeed, solving for t and u in terms of x and y, we obtain

$$t = \frac{x_1 - x}{x - x_2}, \; u = y \frac{x_1 - x_2}{x - x_2},$$

and replacement of t and u by these expressions in (6), yields

$$y \frac{x_1 - x_2}{x - x_2} = y_1 + y_2 \frac{x_1 - x}{x - x_2},$$

or

$$y = y_1 \frac{x - x_2}{x_1 - x_2} + y_2 \frac{x_1 - x}{x_1 - x_2}. \tag{8}$$

One can easily show that instead of making the substitution (6) in equation (5) in order to find (t^*, u^*) and then calculating (x^*, y^*) from the formulae in (7), one may find (x^*, y^*) directly by using (8) to substitute for y in the original equation (2). Since (8) is the equation of the straight line through the points (x_1, y_1) and (x_2, y_2), the latter procedure— substitution from (8) in (2) to find (x^*, y^*)—amounts to finding the third point of intersection of the secant (8) with the curve (2). Thus Euler's method is just the algebraic equivalent of the secant method using two finite points.

Perhaps because Euler did not expound his method in general form, confining the method to the above-mentioned example, it seems that later mathematicians failed to notice that in fact there is a general method lurking here. In any case we have been unable to find any further reference to it.

There is another interesting example from Euler's manuscripts [9, p. 112] where he solves the equation $mx^3 + k^2 = y^2$. We shall examine this example in detail since it has not hitherto figured in the historico–mathematical literature. It is assumed that we know one rational solution of this equation, say $x = a$, $y = b$, so that

$$ma^3 + k^2 = b^2. \tag{9}$$

The steps in Euler's reasoning are as follows: He first sets $x = at$ in the original equation, obtaining

$$ma^3t^3 + k^2 = y^2. \tag{10}$$

Since from (9) we have $ma^3 = b^2 - k^2$, substitution in (10) yields $(b^2 - k^2)t^3 + k^2 = y^2$, or

$$(b^2 - k^2)t^3 = y^2 - k^2. \tag{11}$$

Next Euler factorizes both sides of (11):

$$(b + k)t^2 \cdot (b - k)t = (y + k) \cdot (y - k),$$

and equates factors in pairs, obtaining

$$(b + k)t^2 = y + k, \quad (b - k)t = y - k, \tag{12}$$

or, alternatively,

$$(b + k)t = y + k, \quad (b - k)t^2 = y - k. \tag{13}$$

It is clear that any solution of either of these two systems (which are determinate!) will be a solution of the equation (11) (but not conversely). Consider the first system (12): subtracting the second equation from the first, Euler obtains

$$(b + k)t^2 - (b - k)t = 2k. \tag{14}$$

Since $t = 1$, $y = b$ is a solution of the system (12), $t = 1$ must be a root of equation (14), as indeed it clearly is. Then by Viète's theorem the second root of equation (14) is

$$t = \frac{-2k}{b + k},$$

which yields the rational solution $x = -2ak/(b+k)$ of the original equation[3]. Application of this procedure to the system (13) yields similarly $t = 2k/(b - k)$, and thence $x = 2ak/(b - k)$.

We now repeat this argument for the general equation (1), assuming $a = a_1^2$, and that a solution $x = \alpha$, $y = \beta$ is known, so that α and β satisfy

$$a_1^2 + b\alpha + c\alpha^2 + d\alpha^3 = \beta^2. \tag{15}$$

---

[3] And $y = (b - k)t + k = \frac{-2k(b-k)}{(b+k)} + k$.—Trans.

We first make the substitution $x = \alpha t$ in equation (1), obtaining

$$a_1^2 + b\alpha t + c\alpha^2 t^2 + d\alpha^3 t^3 = y^2. \tag{16}$$

Here in view of (15) the coefficient $d\alpha^3$ of t^3 may be replaced by $\beta^2 - a_1^2 - b\alpha - c\alpha^2$, yielding

$$(\beta^2 - a_1^2 - b\alpha - c\alpha^2)t^3 + c\alpha^2 t^2 + b\alpha t + a_1^2 = y^2,$$

or

$$(\beta^2 - a_1^2 - b\alpha - c\alpha^2)t^3 + c\alpha^2 t^2 + b\alpha t = y^2 - a_1^2. \tag{17}$$

By analogy with the special case considered above, we consider the following system implying, but not implied by, (17):

$$\left. \begin{array}{l} (\beta + a_1)t = y + a_1, \\[2mm] \left(\beta - a_1 - \dfrac{b\alpha + c\alpha^2}{\beta + a_1}\right)t^2 + \dfrac{c\alpha^2}{\beta + a_1}t + \dfrac{b\alpha}{\beta + a_1} = y - a_1, \end{array} \right\} \tag{18}$$

with the obvious solution $t = 1$, $y = \beta$. An alternative such system is:

$$\left. \begin{array}{l} (\beta - a_1)t = y - a_1, \\[2mm] \left(\beta + a_1 - \dfrac{b\alpha + c\alpha^2}{\beta - a_1}\right)t^2 + \dfrac{c\alpha^2}{\beta - a_1}t + \dfrac{b\alpha}{\beta - a_1} = y + a_1, \end{array} \right\} \tag{19}$$

also satisfied by $t = 1$, $y = \beta$. Eliminating y between the equations in (18), we obtain, as in the above special case, a quadratic equation in t:

$$\left(\beta - a_1 - \frac{b\alpha + c\alpha^2}{\beta + a_1}\right)t^2 + \left(-\beta - a_1 + \frac{c\alpha^2}{\beta + a_1}\right)t + \frac{b\alpha}{\beta + a_1} + 2a_1 = 0,$$

one of whose roots is known to be $t_1 = 1$. By Viète's theorem, the other root must be

$$t_2 = \frac{b\alpha + 2a_1(\beta + a_1)}{\beta^2 - a_1^2 - b\alpha - c\alpha^2} = \frac{2a_1^2 + 2a_1\beta + b\alpha}{-a_1^2 + \beta^2 - b\alpha - c\alpha^2},$$

and then the corresponding value of y can be obtained from the first equation in (18): $y = y_2 = (\beta + a_1)t_2 - a_1$. Clearly the pair $t = t_2$, $y = y_2$ is a rational solution of the equation (17) and therefore also of (16), whence $x = x_2 = \alpha t_2$, $y = y_2$ will be a rational solution of the original equation (1). By working similarly with the system (19) one again obtains a rational solution of equation (1).

It is not difficult to show that the point (x_2, y_2) of the curve (1) obtained in this way is in fact the third point of intersection of the curve with the straight line through the two known rational points (α, β) and $(0, -a_1)$ of the curve. To see this, observe first that Euler's method in effect consists in the simultaneous solution of the equations $(\beta + a_1)t = y + a_1$ (the first equation of the system (18)) and (16), or, in terms of x rather than t, the simultaneous solution of the equations (1) and

$$(\beta + a_1)\frac{x}{\alpha} = y + a_1. \tag{20}$$

However equation (20) is just the equation in rectangular Cartesian coordinates x, y for the straight line through the points $(0, -a_1)$ and (α, β), whence our claim. Thus we see that once again Euler has arrived by algebraic means at a method for finding a new rational solution of equation (1), obtainable geometrically as the third point of intersection of the curve (1) with the secant through two known rational points of it. Euler's reasoning in the example just considered is reminiscent of that used by Fermat–Diophantus in solving "double equations", i.e., systems of the form

$$f_2(x) = y^2, \quad g_2(x) = z^2.$$

Their method also involves factorization of the Diophantine equation of interest and setting equal respective factors on opposite sides of the equation (see [5]). However although they did not use this method to solve cubic Diophantine equations, Euler nonetheless refrained from publishing his method, with the result that it remained unknown to his contemporaries.

We encounter the secant method for a third time in one of Euler's notebooks, where he gives formulae for a new rational solution of the general equation (1) with square constant term $a = a_1^2$, in the case where just one rational solution $x = q$, $y = s$ is known[4]. It is shown in [10] that these formulae give the coordinates of the third point of intersection of the curve (1) with the straight line through the known rational points (q, s) and $(0, a_1)$ of that curve[5]. Thus again Euler's solution must have been found by a method representing an algebraic equivalent of the secant method. Note that if the constant term a of equation (1) is not a square (of a rational number) but any two rational solutions (x_1, y_1), (x_2, y_2) of that equation are known, then this case is readily reduced to the former by means of the substitution[6] $x = x_1 + t$—a procedure used widely by Euler. As a result we have[7] the secant method for curves of the form (1) in the general situation where any two rational points of it are known.

In [10] a reconstruction of Euler's method of obtaining his formulae is proposed. This reconstruction assumes that Euler's argument involved the use of degree-one substitutions of the form $y = a_1 + \alpha x$, and developed directly out of the ideas of Diophantus–Fermat for solving equations of the forms (1) and (2). We remind the reader that the latter methods consist in the use of degree-one and quadratic substitutions, with coefficients so chosen that once the substitution has been made in equation (1) (or (2)), a degree-one[8] equation in x results, yielding a rational value $x = x^*$. In our reconstruction we suggest that Euler chose a substitution converting equation (1) to a quadratic equation in x, with one obvious rational root, so that then the second root will also be rational, the finding of this second root constituting the aim of the method. Note that in this conversion to a quadratic equation in x, the fact that one already knows a rational solution $x = q$, $y = s$, is also exploited. (See [10] for full details of this reconstruction[9].)

---

[4] This is the problem just considered by the author as the general case of a special Diophantine equation solved by Euler. This presumably means that Euler also had the general version.—*Trans.*

[5] In other words, the point (x_2, y_2) obtained as above, but using $(0, a_1)$ as the second known rational point, instead of $(0, -a_1)$.—*Trans.*

[6] Since the constant term then becomes $a + bx_1 + cx_1^2 + dx_1^3 = y_1^2$.—*Trans.*

[7] That is, Euler had, in effect.—*Trans.*

[8] Or quadratic?—*Trans.*

[9] The author has earlier in this essay given us in considerable detail a solution of exactly this problem—solving the Diophantine equation (1) assuming $a = a_1^2$ and a given rational solution—and the procedure he uses there fits the present description. The question then arises: Is the solution he expounds earlier the same as his reconstructed one?—*Trans.*

The idea of using a degree-one or quadratic substitution to convert a given Diophantine equation into a quadratic equation in x with rational coefficients can also be found in Euler's published work in Diophantine analysis. For example in his *Algebra* [7, p. 444] he uses the substitution $y = a_1 + px$, where p is any integer, in his investigation of the same equation (1) with $a = a_1^2$. Having made this substitution in (1), he obtains a quadratic equation in x, and gives one of its roots as

$$x = \frac{p^2 - c + \sqrt{p^4 - 2cp^2 + 8da_1p + c^2 - 4bd}}{2d}.$$

He then observes that p must be chosen so that the quantity under the root sign is a square, i.e., such that

$$g_4(p) \equiv p^4 - 2cp^2 + 8da_1p + c^2 - 4bd = z^2 \tag{21}$$

for some rational number z. In this way Euler establishes a connection between the problem of solving the cubic Diophantine equation (1), and that of solving the fourth-degree Diophantine equation (21). In the article [11], in connection with the solution of an equation of the form

$$f_4(x) \equiv a + bx + cx^2 + dx^3 + ex^4 = y^2, \tag{22}$$

Euler gives explicit expression to the idea of making a degree-one or quadratic substitution to reduce to a quadratic equation with rational roots [11, p. 140] and then, using this idea, he derives in [11] two methods for obtaining rational solutions of a Diophantine equation of the form (22) under certain additional conditions on the coefficients.

To summarize: Euler was in possession of the algebraic equivalent of the secant method for equations of the forms (1) and (2), but nowhere did he expound that method in its general form, and the above particular cases remained unnoticed or at least unappreciated.

Valuable results in the arithmetic of elliptic curves were also obtained by Euler in his late works [12, 13], presented to the Petersburg Academy in 1780 but not published until 1830. In these two articles Euler considers from the outset equations of fairly general form, not arising from any particular case—as distinct from most of his work on Diophantine analysis where he investigates only special, concrete equations. In the article [12] he considers the equation

$$g_4(x, y) \equiv a + bx + cy + dx^2 + exy + fy^2 + gx^2y + hxy^2 + ix^2y^2 = 0, \tag{23}$$

with rational coefficients and in [13] equations of the forms (1) and (22). The results of the article [13] are discussed briefly in the essay [6] of the present collection; we shall now examine them in greater detail.

In essence the article [13] is concerned with the working out of an original algebraic procedure which, starting from one, two, or three given rational solutions of equation (1) (or (22)), yields a whole sequence of rational solutions, i.e., a new solution at each step in the procedure. As has been shown in [14], the geometrical version of this procedure consists in the iteration of the tangent, secant, and parabolic methods[10]. Euler's application of a

---

[10]By the "parabolic method" we mean the method whereby a new rational point of the elliptic curve (1) (or (22)) is found as the point of intersection of that curve with the parabola $y = \alpha x^2 + \beta x + \gamma$ through three given finite rational points of the curve in question. Diophantus, Fermat, and Euler all used this method—in algebraic guise—on an equal footing with the tangent and secant methods. It is now known that a solution of the Diophantine equation (1) found using the parabolic method may in fact always be found by means of successive applications of the tangent and secant methods.

variant of his procedure represents the historical first occasion when the successive points of a cyclic subgroup of the group of rational points of the curve (1) were constructed; thus one may say that the first algorithm for constructing cyclic subgroups of the group of rational points of the elliptic curve (1) was devised in the 18th century by Euler.

What were the means Euler employed to achieve these results? As in his other works on Diophantine equations, in the articles [12] and [13] he uses only elementary algebra. [Since [13] depends on [12], we first consider the latter article.] His goal in [12] is that of finding a succession of rational solutions of an equation of the form (23), starting from a single known such solution—an aim not envisaged by Fermat or Diophantus. Euler came to his procedure via his approach to solving a specific problem in [15], leading to the solution of a series of particular Diophantine equations of the form (23) [16–19], and thus ultimately to his procedure for solving (23) in general form in [12]. In essence the procedure is as follows: if (x_1, y_1) is a known rational solution of (23), then setting $x = x_1$ in that equation yields a quadratic equation in y with rational coefficients, and with the known root $y = y_1$. Since this root is rational, the second root y_2 say, will likewise be rational. The upshot is that we have found a new pair (x_1, y_2) satisfying (23), i.e., a new rational solution of (23). We now set $y = y_2$ in (23), obtaining a quadratic equation in x with the known rational root $x = x_1$ and on solving for the other root x_2 say, we arrive at yet another new rational solution (x_2, y_2) of (23). Iteration of this process yields an infinite sequence

$$(x_1, y_1), (x_1, y_2), (x_2, y_2), (x_2, y_3), \ldots, (x_k, y_k), (x_k, y_{k+1}), (x_{k+1}, y_{k+1}), \ldots, \quad (24)$$

of rational solutions of equation (23), the aim of Euler's procedure.

In the succeeding article [13]—whose presentation to the Petersburg Academy of Sciences followed only one week after that of [12]—Euler establishes a relationship between the solutions of equations of the form

$$f(x) \equiv Q^2(x) + P(x)R(x) = z^2, \quad (25)$$

where $Q(x)$, $P(x)$, and $R(x)$ are polynomials of degrees at most two with rational coefficients, with those of equation (23). To this end he makes the substitution

$$z = Q(x) + P(x)y \quad (26)$$

in (25), and after applying a few simple algebraic transformations and dividing the resulting equation by $P(x)$, arrives at a quadratic equation in x and y. Then, if one knows at least one rational solution (x_1, y_1) of the latter equation, one can apply the above-described algorithm to it to obtain a sequence of rational solutions (24), yielding in turn the following sequence of rational solutions of (25):

$$(x_1, z_1), (x_1, z_1'), (x_2, z_2), (x_2, z_2'), \ldots, (x_k, z_k), (x_k, z_k'), (x_{k+1}, z_{k+1}), \ldots, \quad (27)$$

where $z_k = Q(x_k) + P(x_k)y_k$, $z_k' = Q(x_k) + P(x_k)y_{k+1}$. In this way Euler extends his method for solving (23) to one for obtaining an infinite sequence of rational solutions also of equations of the form (25).

In the second half of this article Euler proves that if the equation

$$f_4(x) \equiv a + bx + cx^2 + dx^3 + ex^4 = z^2 \quad (28)$$

or

$$f_3(x) \equiv a + bx + cx^2 + dx^3 = z^2 \qquad (29)$$

(these being just equations (1) and (22) with z in place of y) has at least one rational solution, then the polynomial $f_4(x)$ (or $f_3(x)$ as the case may be) can be put in the form $Q^2(x) + P(x)R(x)$, so that the procedure just described can be applied. He goes on to present various methods for so representing $f_4(x)$ (or $f_3(x)$) in the cases when one, two, or three rational solutions of (28) (and one or two rational solutions of (29)) are known beforehand.

In the paper [14], the present author establishes that the algebraic procedure just described for obtaining a sequence (27) of rational solutions of an equation of the form (28) or (29) has the aforementioned geometrical interpretation. The geometrical version of the procedure consists in the iteration of the following two steps:

1) From the preceding (or initial) solution (x_k, z_k), infer the solution $(x_k, z'_k) = (x_k, -z_k)$; in other words, reflect the preceding rational solution (x_k, z_k) in the x-axis, an axis of symmetry of the curve;

2) Then find the solution (x_{k+1}, z_{k+1}) by applying the secant method or parabolic method, using the point (x_k, z'_k) together with other known points of the curve (28) or (29).

(Note that at the first application of the second step of the procedure, the new solution is to be found using the tangent method, if appropriate, rather than the secant or parabolic methods.)

From the second step we see that Euler had in [12] invented in effect yet another algebraic version of the secant method in the case where two finite points are known—figuring here not, however, as a method in its own right, but as merely one step in an iteration.

It should be emphasized that this procedure of Euler's represents the first occasion when the secant method was iterated. Such an iteration is made feasible by the introduction of the operation of reflection of known rational points in an axis of symmetry of the curve (see [6] for further discussion along these lines). (No such ploy is needed in order to iterate the tangent method and in fact Fermat knew how to carry out such iterations.)

Thus in his article [13] Euler, using an extreme paucity of means, obtained quite new results in the arithmetic of elliptic curves. However, the thorough-going algebraic nature of these means has long thwarted the appreciation of his procedure as comprised simply of the tangent and secant methods together with reflection in axes of symmetry, now considered basic tools for finding rational points of elliptic curves.

These investigations of Euler were important also in that they provided the impetus for the introduction into Diophantine analysis of radically new ideas relating to elliptic functions. In 1835 C. G. F. Jacobi [20] gave a fresh interpretation of the ideas developed by Euler in [13], in terms of the addition theorem for elliptic integrals. Jacobi's work showed that in investigating Diophantine equations defining elliptic curves, in addition to the elementary algebra used by Euler, results from the theory of elliptic functions might be fruitfully invoked (see [6] for further discussion of this theme.)

Euler's *Algebra* [7] was of the greatest importance for the development of Diophantine analysis. In that work, among much else he systematized the general results known up till that time concerning the solution in whole and rational numbers of indeterminate equations of degrees up to and including four. The Diophantus–Fermat methods for finding rational solutions of particular equations of the forms (1), (2), and (22) were given their first explication in this text in convenient symbolic form. It was in fact through their treatment in Euler's *Algebra* that the algebraic essence of the Diophantus–Fermat methodology was clarified, rendering it accessible to a much wider circle of mathematicians. Subsequently, certain mathematicians—and in their wake historians of mathematics—began to attribute these methods to Euler himself.

Soon after its appearance Euler's *Algebra* became generally acknowledged as the standard textbook on Diophantine equations. However apart from its systematic treatment of earlier results in this area, it contained many new results due to Euler: in addition to his innovative application of the secant method in the case of two given finite points, discussed earlier, in his *Algebra* Euler takes the first steps—always using purely algebraic means—in the study of such arithmetic properties of curves as birational equivalence of curves over $\mathbb{Q}$ and their uniformization in rational functions over that field. Of course he does not give strict definitions of these concepts but formulates the relevant results only in elementary algebraic language.

Thus for example he finds a sufficient condition for uniformizability of a curve of the form (1) or (2), stated as follows: if the polynomial $f_3(x)$ (the left-hand side of (1) or (2)) has multiple roots then one may obtain a general formula for the rational solutions of the equation in question in the form $x = P_m(t)/Q_n(t), y = R_k(t)/S_l(t)$, where $P_m(t), Q_n(t), R_k(t), S_l(t)$ are polynomials with rational coefficients. Another example: Euler shows that an equation of the form (1) with square constant term can be transformed into an equation of the form (21), i.e., of the form $g_4(x) = z^2$. (The procedure for achieving this was discussed earlier.) It is not difficult to verify that this transformation is reversible. What in essence is involved here is the birational equivalence of the curves (1) and (21).

Euler was the first to establish a connection between the solutions of equations of the forms (1) and (22), which had earlier been considered as representing quite separate problems. In retrospect, this result of Euler's is suggestive of the idea that Diophantine equations should not be classified by their degrees but according to some other feature taking more account of the relationships between the solutions of equations of possibly different degrees. An appropriate such property was discovered by Poincaré in 1901, namely birational equivalence of the corresponding curves over the field $\mathbb{Q}$.

Note also that Euler, although not in possession of the general notion of birational equivalence, was nonetheless the first to clearly formulate the idea of using one particular type of birational transformation to investigate Diophantine equations, namely linear-fractional transformations (see [21, p. 43]). Before Euler mathematicians had used only various special such transformations applied to particular equations (i.e., with specific numerical coefficients) without considering the question more generally. Thus we find in Euler's work the germs of those concepts and ideas of Diophantine analysis that were to come to fruition only at the beginning of the 20th century.

Euler also considered the question of the number of rational solutions a given Diophantine equation might have. He knew that equations of the forms (1), (2), and (22) may have infinitely many rational solutions, since the well known Diophantus–Fermat method for obtaining a new rational solution of an equation of any of these types might in principle be re-applied arbitrarily often. Nevertheless Euler gives examples of such equations having only a finite number of rational solutions. However we shall refrain from dwelling further on this topic here.

By way of a summation of our discussion of Euler's investigations in Diophantine analysis, we observe first that these belong to the stage in the development of that discipline that might be called "elementary", bearing in mind above all the tools available at that time for solving Diophantine equations. At that stage progress was of necessity limited mainly to the accumulation of methods of solution. Such was the situation also with respect to the corresponding stage in the development of the theory of determinate equations, where the main efforts of mathematicians of that era were directed towards the search for a method of solution of equations of arbitrary degree. At the beginning of the 18th century, Diophantine analysis had at its disposal certain methods for solving particular indeterminate equations in two unknowns of degrees up to and including the fourth. Of these, while the method for finding all rational solutions of quadratic Diophantine equations in two unknowns (assuming at least one such solution known beforehand), had been used already by Diophantus[11], so that it only remained for Euler to give it its general form [22], the situation was quite otherwise for equations of degrees three and four, which present considerably greater difficulties.

Euler continued the study of the basic types of Diophantine equations (1), (2), and (22) of degrees three and four, defining elliptic curves, that had been investigated earlier. To the known methods of Fermat–Diophantus for obtaining rational solutions of these equations, he added new ones, which—as we have indicated above—are closely related to modern methods of finding rational points of elliptic curves. In essence Euler exploited every possible procedure—namely the tangent and secant methods, and the reflection of a known rational point in an axis of symmetry of the curve in question, fundamental to the definition of the operation of addition on the set of rational points of an elliptic curve. And he was the first to iterate the secant method. We should however emphasize again that his results were all formulated algebraically and that not once did he provide any geometrical interpretation. We hope that the examples that we have provided in the present paper in order to illustrate the type of reasoning Euler used will also serve to convey some idea of the style characteristic of 18th-century Diophantine analysis and of the means available to Euler and his contemporaries for working in that discipline.

However not only did Euler invent new methods—in algebraic guise—for solving Diophantine equations, but his works also contain the germs of some of the concepts and ideas of Diophantine analysis now counted among the most basic—for instance the concept of birational equivalence of curves over $\mathbb{Q}$. Of course these concepts, just barely adumbrated in Euler's time, had to wait a rather long time for their fundamental role to be recognized.

In conclusion, we wish to remark that we have not attempted to provide an exhaustive characterization of the great mathematician's work in Diophantine analysis but rather to draw attention to some relatively little-known achievements of Euler in that discipline. Thus we have left certain aspects of his activity in Diophantine analysis out of account. In particular we have not considered Euler's correspondence with Gol′dbach, although in some of that correspondence problems concerning Diophantine equations are mentioned (see, for example, Euler's letters of September 2, 1747, June 25, 1748, April 15, 1749, and others). As a rule, in these letters he writes of particular Diophantine equations for which

---

[11] This claim seems rather exaggerated.—*Trans.*

he has found solutions, without going into his methods of solution. It seems that equations in general form were not considered in any of this correspondence. A closer examination of the correspondence would be of very great importance for an understanding of just how Euler's interest in Diophantine analysis developed, and for the history of individual problems of Diophantine analysis, among other things.

Of the historico-mathematical literature concerning Euler's work on Diophantine equations we mention, in addition to the articles already noted, the paper [23] by I. G. Bashmakova, which contains several important remarks on this theme, the papers [24, 25] of J. E. Hofmann, and the recent book [26] by A. Weil. Especially worthy of note is an interesting analysis contained in the articles [25, 26] of one of the most important of Euler's memoirs on Diophantine equations [13]. In particular Weil gives an interpretation of Euler's methods from [12] and [13] in terms of the theory of divisors. However in none of these works is the geometric meaning of Euler's procedure for solving equations of the form (29) elucidated in any detail.

References

[1] Venkov, B. A. "On the works of Leonhard Euler in number theory". In: *Leonhard Euler.* M.; L.: Publ. Acad Sci. USSR (Izd-vo AN SSSR), 1935. pp. 81–87.

[2] Gel'fond, A. O. "The role of Euler's works in the development of number theory". In: *Leonhard Euler.* M.: Publ. Acad. Sci. USSR (Izd-vo AN SSSR), 1958. pp. 80–95.

[3] Mel'nikov, I. G. "Euler and his arithmetical works". Research Hist. Math. (Ist.-mat. issled.). 1957. Issue 10. pp. 211–228.

[4] Bashmakova, I. G., Slavutin, E. I. *History of Diophantine analysis from Diophantus to Fermat.* M.: Nauka, 1984. 256 pp.

[5] Kauchikas, A. P. *Diophantus and indeterminate analysis in the works of European mathematicians from the 13th to the 16th centuries. Dissertation for degree of Kandidat of Physico-mathematical Sciences.* M.: Research Inst. Nat. Sci. and Technology Acad. Sci. USSR (IIEiT AN SSSR), 1979. 118 pp.

[6] Bashmakova, I. G. "Euler's contribution to algebra". In the present collection.

[7] Euler, L. *Vollständige Anleitung zur Algebra.* Stuttgart, 1959; *Opera* I-5.

[8] Lavrinenko, T. A. "Methods of solution of indeterminate equations in rational numbers in the 18th and 19th centuries". Research Hist. Math. (Ist.-mat. issled.). 1985. Issue 28. pp. 202–222.

[9] Euler, L. *Opera postuma mathematica et physica.* Petropoli, 1862. Vol. 1.

[10] Lavrinenko, T. A. *Indeterminate analysis in the works of L. Euler.* M., 1982. 37 pp. Filed with All-Union Inst. Sci. and Tech. (VINITI) December 22, 1983. No. 6988–83.

[11] Euler, L. "Dilucidationes circa binas summas duorum biquadratorum inter se sequales". *Opera* I-5. pp. 135–145.

[12] ——. "De resolutione huius aequationis $0 = a + bx + cy + dx^2 + exy + fy^2 + gx^2y + hxy^2 + ix^2y^2$ per numeros rationales". *Opera* I-5. pp. 146–156.

[13] ——. "Methodus nova et facilis formulas cubicas et biquadraticas ad quadratum reducendi". *Opera* I-5. pp. 157–181.

[14] Lavrinenko, T. A. "Solutions of indeterminate equations of the third and fourth degrees in Euler's later works". Research Hist. Math. (Ist.-mat. issled.). 1983. Issue 27. pp. 67–78.

[15] Euler, L. "Solutio problematis Fermatiani de duobus numeris, quorum summa sit quadratum, quadratorum vero summa biquadratum...". *Opera* I-5. pp. 77–81.

[16] ——. "Resolutio facilis quaestionis difficillimae, qua haec formula maxima generalis $v^2 z^2 (ax^2 + by^2)^2 + \Delta x^2 y^2 (av^2 + bz^2)^2$ ad quadratum reduci postulatur". *Opera* I-5. pp. 71–76.

[17] ——. "De insigni promotione analysis Diophanteae". *Opera* I-5. pp. 82–93.

[18] ——. "Solutio problematis difficillimi, quo hae duae formulae $a^2 x^2 + b^2 y^2$ et $a^2 y^2 + b^2 x^2$ quadrata reddi debent". *Opera* I-5. pp. 94–115.

[19] ——. "Investigatio binorum numerorum formae $xy(x^4 - y^4)$ quorum productum sive quotus sit quadratum". *Opera* I-5. pp. 116–130.

[20] Jacobi, C. G. J. "De usu theoriae integralium ellipticorum et integralium abelianorum in analysi Diophantea". *Gesammelte Werke*. Berlin, 1882. Vol. 2. pp. 53–55.

[21] Lavrinenko, T. A. *Indeterminate equations in the works of L. Euler and 19th-century mathematicians. Dissertation for degree of Kandidat of Physico-mathematical Sciences*. M.: Research Inst. Nat. Sci. and Technology Acad. Sci. USSR (IIEiT AN SSSR), 1984. 165 pp.

[22] Euler, L. "Resolutio aequationis $Ax^2 + 2Bxy + Cy^2 + 2Dx + 2Ey + F = 0$ per numeros tam rationales quam integros". *Opera* I-3. pp. 297–309.

[23] Bashmakova, I. G. "Arithmetic of algebraic curves (from Diophantus to Poincaré)". Research Hist. Math. (Ist.-mat. issled.). 1975. Issue 20. pp. 104–124.

[24] Hofmann, J. E. "Über eine zahlentheoretische Aufgabe Fermats". Centaurus. 1961. Vol. 16. pp. 169–202.

[25] ——. "Über zahlentheoretische Methoden Fermats und Eulers, ihre Zusammenhänge und ihre Bedeutung". Arch. Hist. Exact Sci. 1961. Vol. 1, No. 2. pp. 122–159.

[26] Weil, A. *Number theory: An approach through history. From Hammurapi to Legendre*. Boston; Basel; Stuttgart: Birkhäuser, 1983.

The Foundations of Mechanics and Hydrodynamics in Euler's Works

G. K. Mikhaĭlov and L. I. Sedov

Leonhard Euler was one of the greatest scientists of all time—as in mathematics so also in mechanics.

Euler's published output in mechanics is huge. The second series of his *Collected works* (*Opera omnia*), which is devoted mainly to his works in mechanics (and celestial mechanics), occupies more than 11,000 pages, distributed over 31 large quarto volumes (of which 27 have so far appeared)[1].

It should be emphasized that mechanics represents Euler's first serious scientific interest. The earliest of the extant notebooks, which he began to keep from the age of 18 or 19, already contain detailed plans for a series of general treatises on the dynamics of a point-mass and the motion of fluids. A comparison by subject of the relative numbers of Euler's publications over the first 10 years of his creative life (1726–1735) bears witness to the young Euler's lively interest in mechanics: of a total of around 1800 pages (i.e., pages of the *Opera omnia*) produced over that period, almost two-thirds are devoted to mechanics and only about a quarter to higher mathematics. Half of these are taken up by Euler's large treatise on the dynamics of a point-mass, which brought him to the attention of the international mathematical community for the first time.

The situation in the field of dynamics towards the end of the 1720s was, one may say, not static. Newton's *Philosophiae naturalis principia mathematica* (1687) represents a summation of the efforts of his precursors[2] to clarify the root concepts of mechanics and the fundamental laws of motion. The *Principia* provided a significant part of the foundations of the general edifice of dynamics and of the part of that structure concerned with the dynamics of a point-particle, and included several attempts at erecting other portions of it. However Newton lacked many elements crucial to the completion of the theoretical superstructure of mechanics—above all of the mechanics of systems, rigid bodies, and continuous media. Moreover the *Principia* was couched in the already essentially mori-

---

[1] That is, as of 1986. They have by now all appeared.—*Trans.*

[2] And himself!—*Trans.*

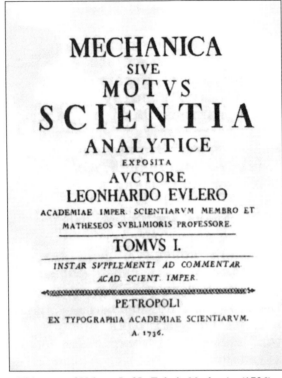

Title page of Volume I of L. Euler's *Mechanics* (1736).

bund geometrical terminology of the ancients, preventing anyone not privy to the secrets of analysis from advancing the subject further. Analytic expositions of the principles of mechanics made an appearance only at the beginning of the 18th century in works of P. Varignon[3] and in J. Hermann's *Phoronomia* (1716).

Such was the situation when Euler boldly took upon himself the task of reformulating the whole of the theory of the dynamics of a point particle in the language of mathematical analysis. This project reached its conclusion in the publication in 1736 of his two-volume *Mechanica sive motus scientia analytice exposita* as an appendix to an issue of the *Commentarii*, the journal of the Petersburg Academy of Sciences, which had been unable to cope with the flood of material submitted to it.

In the preface to Volume I of his *Mechanics*, concerning Newton's synthetic method Euler writes as follows:

"However if analysis is needed anywhere, then it is certainly in mechanics. Although the reader can convince himself of the truth of the exhibited propositions, he does not acquire a sufficiently clear and accurate understanding of them, so that if those questions be ever so slightly changed, he will not be able to answer them independently, unless he turn to analysis and solve the same propositions using analytic methods. This in fact happened to me when I began to familiarize myself with Newton's *Principia* and Hermann's *Phoronomia*; although it seemed to me that I clearly understood the solutions of many of

---

[3] A series of articles devoted to the dynamics of a point-particle were published by Pierre Varignon in the *Mémoires* of the Paris Academy of Sciences, beginning with the first issue in 1700. In his published work Euler never mentions these works of Varignon, which in fact are seldom referred to also in the later literature.

the problems, I was nevertheless unable to solve problems differing slightly from them. But then I tried, as far as I was able, to distinguish the analysis [hidden] in the synthetic method and to my own ends rework analytically those same propositions, as a result of which I understood the essence of each problem much better. Then in the same manner I investigated other works relating to this science, scattered here, there and everywhere, and for my own sake I expounded them [anew] using a systematic and unified method, and re-ordered them more conveniently. In the course of these endeavors not only did I encounter a whole series of problems hitherto never even contemplated—which I have very satisfactorily solved—, but I discovered many new methods thanks to which not only mechanics, but analysis itself, it would seem, has been significantly enriched. It was thus that this essay on motion arose, in which I have expounded using the analytic method and in convenient order both that which I have found in others' works on the motion of bodies and what I myself have discovered as a result of my ruminations" [1].

It is worth repeating Euler's statement—fully justified—to the effect that in his treatise on mechanics new methods of mathematical analysis arose incidentally; mathematical methods were for Euler primarily the means to the end of solving problems of mechanics.

With his *Mechanics* Euler was the first to systematically expound the dynamics of a freely moving point-mass and of a point-mass confined to a given surface, first in the absence of a resisting medium (i.e., in a vacuum), and then in a resisting medium. His exposition is formulated throughout in terms of natural coordinates afforded by the trajectory of the point-mass in question.

Having sorted out the dynamics of a point-mass, Euler next made an unsuccessful attempt to do the same for the motion of a rigid body. As he admits in the preface to his treatise, he found it necessary to postpone this project "in view of a insufficiency of [basic] principles"—which nonetheless did not prevent him from immediately conceiving a plan for producing an exposition of the whole of mechanics, as that subject presented itself to him in the mid-1730s. In his general remarks pertaining to the first chapter of Volume I of his *Mechanics* (§98) he wrote:

"To begin with we shall consider infinitesimally small bodies, i.e., such as may be treated as points. Then we shall proceed to bodies of finite size—those that are rigid, not admitting a change in their form. At the third stage [of our treatment] we shall discuss flexible bodies and fourthly, those bodies that can be expanded or contracted. At the fifth stage we shall subject to investigation the motion of several separated bodies, where some prevent others from pursuing their motion as they would tend to do by themselves. The sixth stage will be concerned with the motion of fluids. Concerning these [types of] bodies, we shall investigate not just how they continue to move by themselves but also how external causes, i.e., forces, act on them" [1].

How much of this grandiose program was Euler able to realize in the course of a long life devoted so unstintingly and with such singleness of purpose to science? Here we shall try to answer this question.

Of primary importance is Euler's role in laying the foundations of the dynamics of a rigid body and the hydrodynamics of an ideal fluid, which he gave their accepted present-day forms familiar to us from modern textbooks. He also contributed significantly to the development of the mechanics of flexible and elastic bodies, although the creation of a general theory of elasticity had to wait till the 19th century.

Before considering individual results of Euler's in mechanics it is essential to say a few words about the style of his research and to compare his work with that of his contemporary scholar-mechanicians. A leading figure among the latter was J. d'Alembert, and among the more youthful of them, J.-L. Lagrange. However with his *Mécanique analitique* Lagrange represents the next, more formalized, stage in the development of mechanics, following hard upon that of Euler, so that it is in this case easier to establish the succession of ideas—not, however, always correctly reflected in the subsequent scientific literature. On the other hand the work of d'Alembert—who was but 10 years younger than Euler and died in the same year as him—intersected with that of Euler over the whole range of subdisciplines of mechanics. Their investigations overlapped very extensively in the dynamics of a rigid body, celestial mechanics, the theory of vibrating strings, and in their approaches to the construction of a hydrodynamical theory of ideal fluids. It is certainly true that d'Alembert was a genius in the field of mechanics and a fitting rival for Euler. He conceived the most advanced ideas, sometimes before Euler, but failed to give them clear expression and practically never carried them through to a full realization. As early as the middle of the 19th century the mathematical language of d'Alembert, brilliant philosopher and *littérateur* though he was, had begun to seem ponderous, while on the other hand Euler heralded a new epoch with the style of his exposition of the exact sciences. The greater part of all that Euler wrote is phrased with extreme clarity, with his assumptions carefully noted, and with the methods employed transparently formulated. His language and style were greatly admired throughout the 19th century. He remains, perhaps, the only scientist of the mid-18th century whose works on mechanics can be read with ease today. Naturally this had the effect of entrenching Euler's direct influence on the subsequent development of that science for over 100 years. As a rule, his results were preserved in the later scientific literature practically in the form in which he originally cast them—even if his name was no longer attached to them. Moreover Euler possessed a rare talent for systematizing and generalizing scientific ideas, which allowed him to formulate in a relatively perfected form the theories pertaining to several of the main areas of mechanics.

But let us return to the crux of the matter. We begin with the mechanics of systems. In discussing the development of this topic in the mid-18th century, it is impossible to avoid mention of d'Alembert's principle[4]. How did that principle relate to Euler's methods? In the scientific literature this question is beset by disagreement as to the correct answer. This variety of opinion is connected with the ambiguity—preserved even till now—in the basic content of d'Alembert's principle. Already Lagrange had noted that the name "d'Alembert's principle" meant different things to different authors—such as the form d'Alembert himself had given it in his *Treatise on dynamics* (Paris, 1743) or the version taken less directly from d'Alembert but simpler to apply in methods of reduction of the laws of dynamics to those of statics, and "representing", in the words of Lagrange, "a return to the method of Hermann and Euler, which the latter applied to the solutions of many problems of mechanics" [2][5]. C. Truesdell writes in this connection: "Contrary to the usual claims, neither did

---

[4]"D'Alembert's principle permits the reduction of a problem in dynamics to one in statics, accomplished by introducing a fictitious force equal in magnitude to the product of the mass of the body and its acceleration and directed opposite to the acceleration, resulting in a condition of kinetic equilibrium. It was introduced by Jean le Rond d'Alembert in his *Traité de dynamique* (1743)". (Quoted with slight changes from the Hutchinson Encyclopedia.)—*Trans.*

[5]This remark was included in the second edition of Lagrange's *Mécanique analitique* (1811).

d'Alembert reduce dynamics to statics, nor did he, here [i.e., in his treatise] or anywhere, propose either of the two forms of the laws of dynamics now usually called 'd'Alembert's principle', these being due to Euler and Lagrange respectively, at a later period" [3].

Without going into detail, we note that as early as the 1730s Euler had considered pre-conditions for a method equivalent to d'Alembert's principle, and subsequently developed this method along Newtonian lines completely independently of d'Alembert. Perhaps this explains why Euler never referred to d'Alembert's principle and never treated his own approach to the solution of problems on the mechanics of systems as a separate method.

Throughout the first half of the 18th century the development of the theory of the dynamics of systems with a finite number of degrees of freedom was closely linked to the theory of vibrations. Not wishing to dwell on this theme here—important though it is for the general history of dynamics—, we note merely that Euler (simultaneously with D. and J. Bernoulli) contributed significantly to the development of that theory. But of course it is impossible to pass over in silence the tremendous significance for the theory of vibrations of Euler's discovery of a method of integrating linear differential equations, which figure so centrally in that theory. In particular, as early as 1739 Euler discovered the phenomenon of resonance for a sinusoidally excited harmonic oscillator. Of the later work of Euler in this area, fundamental to the theory of non-linear vibrations, we single out his second theory of lunar motion (SPb., 1772). His equations for the moon's motion, expressed in terms of rectangular Cartesian coordinates, turned out to be prototypical for the theory of non-linear vibrations. The continuation by G. Hill, at the end of the 19th century, of Euler's research on methods of integrating these equations, turned out in the sequel to be extremely significant for the development of the general theory of non-linear vibrations.

Note also that the first example of a differential equation for the motion of an n-body system, derived by means fully in the spirit of Newton's *Principia*, appears in Euler's work towards the end of the 1740s.

It is curious that in his understanding of the foundations of mechanics—as indeed in many other areas—Newton was not as strict a "Newtonian" as his successors. Thus while he correctly formulated the law of conservation of momentum—and, in essence, the law concerning kinetic energy[6]—in the very first edition of the *Principia* (1687), he in fact believed to the end of his days that the total momentum of an isolated system can change in the course of the system's movement. He even proposed as an instance where the law of conservation of momentum is violated a system consisting of two spheres connected by a rod, revolving uniformly about the center of mass of the system, which is at the same time moving uniformly in a straight line[7].

The dynamics of systems embraces, naturally, the classical problems of celestial mechanics, to which Euler devoted many works from the mid-1740s on. Of especial significance are those works concerned with one or another aspect of the three-body problem: theories of lunar motion, of perturbations of planetary orbits, and, finally, beginning in the 1760s, the three-body problem pure and simple. Euler's contribution to the establishment

---

[6]Truesdell [3, pp. 105–106] notes that Leibniz introduced the concepts of "live force" (kinetic energy) and "dead force" (potential energy), and "asserted that loss of dead force resulted in corresponding gain of live force". In Russian, kinetic energy is called "live force" (zhivaya sila).—*Trans.*

[7]An elucidation of the error in this example, which first appeared in the Latin edition of his *Optics* of 1706, has been published by V. A. Fabrikant [4].

and further development of celestial mechanics is of very great significance; it may even be maintained that the theory of perturbations of the planets' orbits began with Euler's memoir on the irregularities in the orbits of Jupiter and Saturn, entered in a competition of the Paris Academy of Sciences and published by that academy in 1749. This memoir of Euler's, together with d'Alembert's treatise on the precession of the equinoxes and the nutation of the earth's axis (published in Paris in 1749) and A.-C. Clairaut's *Théorie de la lune* (entered in an official competition of the Petersburg Academy of Sciences in 1750), may with full justification be considered as marking the birth of modern celestial mechanics[8].

Incidentally, in one of Euler's first general articles on celestial mechanics, presented in 1747 and published in 1749, for the first time there appear, in familiar and customary form, "Newton's equations" for the motion of a point-particle under the action of arbitrary forces [5, §18]:

$$\frac{2\,dd\,x}{dt^2} = \frac{X}{M}, \quad \frac{2\,dd\,y}{dt^2} = \frac{Y}{M}, \quad \frac{2\,dd\,z}{dt^2} = \frac{Z}{M}, \tag{1}$$

where X, Y, and Z are the components of the [resultant] force in the directions of the x-, y- and z-axes of a rectangular Cartesian coordinate system, and M is the mass of the particle. The presence in these equations of the apparently superfluous 2s should not perturb us; they arise from the notational system for mechanical quantities adopted by Euler at the time.

The fact is that in the mid-18th century there was as yet no clear notion of systems of units of measure and even less of dimension theory. Euler used essentially just two basic units of measure, namely a unit L of length or distance and a unit F of force. He took mass to be measured in units of weight (a force), i.e., the unit M of mass was defined by $[M] := F$. It follows from the basic law of dynamics[9] that acceleration is measured in multiples of the acceleration due to gravity near the earth's surface. One might infer directly from this that time should have dimension $[T] = L^{1/2}$.[10] However Euler came to this via the independence of his definitions of speed and time. His measure of speed was the square root of the height from which an object falling from rest attains the speed in question, and the time was then the ratio of the height to the speed, so defined. It follows from this that speed and time have the same dimension $L^{1/2}$. To translate from Euler's formalism into modern notation one must, while leaving force and distance as they are, replace the remaining quantities according to the following scheme: mass $\to mg$, speed $\to v/\sqrt{2g}$, and time $\to t\sqrt{2g}$, where g is the acceleration due to gravity near the earth's surface. Once this transformation is carried out in the equations (1) the offending 2s should disappear.

Euler's contribution to the creation of a general theory of the motion of a rigid body was outstanding. First, at the end of the 1730s, he worked on certain special problems of the dynamics of rigid bodies in connection with the preparation of his book *Scientia navalis*— which may be translated as *Nautical science*; in Russian it was usually called *Marine science*. (Difficulties relating to its publication in St. Petersburg delayed the appearance of this work till 1749.)

---

[8]Euler's works in celestial mechanics are discussed in greater detail in the essays of V. K. Abalakin and E. A. Grebenikov, and of N. I. Nevskaya and K. V. Kholshevnikov, included in the present collection.

[9]That is, Newton's second law: force=mass×acceleration.—*Trans.*

[10]In order that the second derivative of distance with respect to time—acceleration—come out dimensionless?—*Trans.*

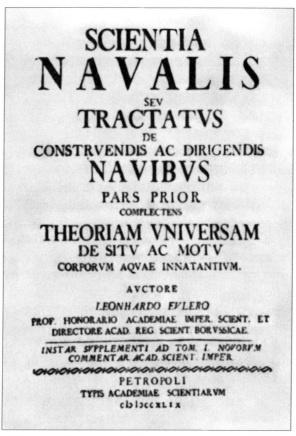

Title page of Volume I of Euler's *Nautical science* of 1749.

In this voluminous work we find the motion of a ship resolved into the component motions of pitching and rolling, an attempt to estimate the small vibrations of a ship on the water, an extensive investigation of the stability of the equilibria of floating bodies—or, as we now say, their "sea-stability"[11]—, and the elements of the theory of moments of inertia.

Euler returned to the general theory of the motion of a rigid body in 1749–1750. It is well known that his renewed interest in this theme was stimulated by d'Alembert's investigations into the precession of the equinoxes and the nutation of the earth's axis, which included clearly delineated avenues of approach to the theory of the rotation of a rigid body. The first, and decisive, step in the creation of a general dynamics of a rigid body was taken by Euler in his memoir "Discovery of a new principle of mechanics", presented to the Berlin Academy on September 3, 1750. In this memoir Euler expounds the principle he had discovered—"a general and fundamental principle of the whole of mechanics", as he describes it. The principle in question consists in the systematic mathematical application of the fundamental law of dynamics—i.e., Newton's second law, or the principle of an accelerating force—to the projections on each of the three axes of a rectangular Cartesian coordinate system at rest, of the motion of each infinitesimal particle (in particular an

---

[11] "Ustoĭchivost′" means simply "stability", while "ostoĭchivost′" means "sea-stability", according to Elsevier's Russian-English Dictionary.—*Trans.*

isolated point-particle) [6, §22], yielding the system

$$2Mddx = Pdt^2, \quad 2Mddy = Qdt^2, \quad 2Mddz = Rdt^2, \tag{2}$$

where M is the mass of the particle, and P, Q, and R the components of the [resultant of] the external forces. (Here the coefficient 2 arises for the same reason as before.) This principle may, according to Euler, be considered "as the unique basis of the whole of mechanics, as well as the other sciences dealing with the motion of arbitrary bodies". In his memoir Euler goes on to say that (§19):

"It is precisely on this principle that all other principles must be based, such as those already used to determine the motion of rigid and fluid bodies in mechanics and hydraulics and those as yet unknown that will be required in order to deal with both the aforementioned cases of rigid bodies, and the many other cases concerning fluid bodies".

Thus Euler's new principle consisted in the singling out of each elementary particle of a continuous medium and the application of Newton's second law to each of the projections of its motion on the axes of a rectangular Cartesian reference frame at rest. It is difficult for us to appreciate the leap forward that this idea represented for mechanics, since to us it seems self-evident. Be that as it may, it remains true that it was precisely this principle that opened the way to the systematic creation of a dynamical theory of rigid bodies and, most important, of continuous media generally.

In fairness it should be mentioned that the resolution of Newton's second law of motion component-wise in the directions of the axes of a coordinate system at rest as a means of investigating the motion of a point-mass, was proposed as a separate "principle" of mechanics by C. Maclaurin in his book *A treatise of fluxions*, published in 1742. (Maclaurin was still using the now forgotten[12] Newtonian term "fluxions".) Furthermore in the 1740s the resolution of equations of motion component-wise relative to a coordinate frame at rest was also used by several scientists on the continent, in particular J. Bernoulli, A.-C. Clairaut, J. d'Alembert, and Euler himself. However prior to Euler no-one had thought that the individual system of differential equations governing the motion of each and every infinitesimal particle of a medium or body might lead immediately to a mathematical formalism applicable to the general problems of mechanics. (It would seem that the need for an independent postulation of the law of conservation of angular momentum was realized by Euler only considerably later.)

With the aid of his new principle Euler was able to obtain at once, in the same memoir, general equations for a rotating rigid body, expressing them, however, in an asymmetric form inconvenient for further research, relative to a coordinate system at rest (§55):

$$
\begin{aligned}
\frac{Pa}{2M} &= \frac{ffd\lambda}{dt} - \frac{nnd\mu}{dt} - \frac{mmdv}{dt} + \lambda vnn - \lambda\mu mm \\
&\quad -(\mu\mu - vv)ll + \mu v(hh - gg), \\
\frac{Qa}{2M} &= \frac{ggd\mu}{dt} - \frac{lldv}{dt} - \frac{nnd\lambda}{dt} + \lambda\mu ll - \mu vnn \\
&\quad -(vv - \lambda\lambda)mm + \lambda v(ff - hh), \\
\frac{Ra}{2M} &= \frac{hhdv}{dt} - \frac{mmd\lambda}{dt} - \frac{lld\mu}{dt} + \mu vmm - \lambda vll \\
&\quad -(\lambda\lambda - \mu\mu)nn + \lambda\mu(gg - ff),
\end{aligned}
\tag{3}
$$

---

[12] In the English-speaking world not so much forgotten, as an obsolete but curious piece of terminology.—*Trans.*

where M is the mass of the body, Pa, Qa, and Ra are the moments (torques) of the external forces, λ, μ, and ν the angular velocities of rotation about the axes, and $f, g, \ldots, n$ are the radii of inertia of the body relative to the fixed axes, which therefore change during the rotation of the body and consequently depend on parameters not known beforehand.

A few years later (in 1755) J. A. Segner discovered that for every rigid body there are three principal axes of rotation[13]. Following on this discovery Euler returned to the general theory of the motion of a rigid body, and in articles written towards the end of the 1750s (but published in the *Mémoires* of the Berlin Academy of Sciences only in 1765) used as his underlying coordinate system the principal axes of inertia or "free axes of rotation". As a result he was able to recast his general dynamical equations for the rotation of a rigid body about its centre of mass in the following form, become classical (to within notational changes):

$$
\begin{aligned}
dx + \frac{cc - bb}{aa} \cdot yz\,dt &= \frac{2gP\,dt}{Maa}, \\
dy + \frac{aa - cc}{bb} \cdot xz\,dt &= \frac{2gQ\,dt}{Mbb}, \\
dz + \frac{bb - aa}{cc} \cdot xy\,dt &= \frac{2gR\,dt}{Mcc}.
\end{aligned}
\tag{4}
$$

Here M is the mass of the body, x, y, and z denote the projections of the instantaneous angular velocity on the fixed axes, a, b, and c the principal radii of inertia of the body, P, Q, and R the torques of the resultant of the external forces about the axes, and g is the displacement of the center of mass of the freely falling body in the first second of the motion, numerically equal to half the acceleration due to gravity. The presence of the factors $2g$ is explained by the fact that while Euler is here using time and velocity with their modern meanings, he is still taking mass to be measured in terms of the weight of the body. (The modern form of these classical equations dates from J.-L. Lagrange and S.-D. Poisson.)

Euler then turned immediately to the investigation of the celebrated first case of integrability of the equations describing a rigid body rotating about a fixed point, namely that where the net torque of the external forces about the fixed center of mass is zero.

Euler was also the first to work out the kinematics of a rigid body, including both forms of the kinematical equations of rotation (one of which is sometimes attributed to Poisson), and the first extensive investigation of moments of inertia (or "geometry of masses") is due to him.

This fundamental stage in Euler's research in dynamics culminated in the completion in 1760 of his treatise *Theory of motion of rigid bodies*, which he regarded as the third volume of his *Mechanics*. (Difficulties in finding a publisher—a perennial problem for Euler—meant that the book appeared only in 1765, in Rostock. However he continued to work on the dynamics of rigid bodies in later years. In particular, in his essay "A new method of determining the motion of rigid bodies", presented to the Petersburg Academy in 1775, for the first time one finds in one place the six equations of motion of an arbitrary body, representing the laws governing momentum and the moment of the momentum (i.e.,

---

[13]Truesdell writes [3, p. 118]: "In 1755 Segner proved that every rigid body has at least three axes of permanent rotation, and that these are mutually perpendicular; for special bodies there may be infinitely many axes".—*Trans.*

angular momentum) [8, §29][14]:

$$\int dM\left(\frac{ddx}{dt^2}\right) = iP, \qquad \int z\,dM\left(\frac{ddy}{dt^2}\right) - \int y\,dM\left(\frac{ddz}{dt^2}\right) = iS,$$

$$\int dM\left(\frac{ddy}{dt^2}\right) = iQ, \qquad \int x\,dM\left(\frac{ddz}{dt^2}\right) - \int z\,dM\left(\frac{ddx}{dt^2}\right) = iT, \qquad (5)$$

$$\int dM\left(\frac{ddz}{dt^2}\right) = iR, \qquad \int y\,dM\left(\frac{ddx}{dt^2}\right) - \int x\,dM\left(\frac{ddy}{dt^2}\right) = iU,$$

where P, Q, and R are the components of the net external force, S, T, and U are the corresponding torques, and i is twice the displacement of a freely falling body in the first second of motion, numerically equal to the acceleration due to gravity[15].

C. Truesdell calls the laws represented by these equations "Euler's general and final statement of the principles of linear momentum and moment of momentum", representing "*the fundamental, general, and independent laws of mechanics*, for all kinds of motions of all kinds of bodies" [3, p. 260], and an important first in the history of mechanics[16].

We shall refrain from examining the Eulerian principle of least action in mechanics since the wide spectrum of questions associated with the integral variational principle of mechanics is illumined in the essay by V. V. Rumyantsev included in the present collection. However for our survey of Euler's achievements in laying the foundations of general mechanics and contributing to our understanding of those foundations to pretend to completeness, it behooves us to at least draw attention to the fact that he contemporaneously laid the foundations of the calculus of variations. The creation by Euler of the analytical apparatus of Newtonian mechanics, his mathematical formulation of the first integral variational principle of mechanics, and his derivation of the basic equations of the calculus of variations, predetermined the course of the future development of general mechanics for at least a century and a half. Crucial to this further development was the idea of distinguishing the actual motions within the totality of all conceivable ones, which was widely applied after Euler by J.-L. Lagrange and many other prominent mechanicians of the 19th and 20th centuries.

We now turn to Euler's work in the mechanics of continuous media.

It was Euler who determined the appropriate mathematical basis for hydrodynamics. His interest in questions to do with the motion of fluids dates from his youth. Under the tutelage of Johann I Bernoulli he applied the law of kinetic energy[17] together with the "hypothesis concerning plane cross-sections" and the corresponding form of the principle of continuity[18]—which had been applied earlier by others—, to the investigation of the flow (or efflux) of liquids out of vessels. The 20-year-old Euler presented the results of these investigations to the Petersburg Academy in August 1727—two weeks after a similar announcement by D. Bernoulli. Their results coincided, and in this delicate situation

---

[14]These equations appear in the *Novi Commentarii Academiae Scientiarum Petropolitanae* for 1775.—*Trans.*

[15]Compare the remark above on Euler's units of measurement in connection with the equations (4).

[16]In [3, p. 260] Truesdell explains how Euler had to "abandon the attempt to regard the linear momentum principle, supplemented by restrictions upon the mutual forces, as sufficient for the whole science of mechanics".—*Trans.*

[17]Presumably that kinetic energy + potential energy is constant in a closed system.—*Trans.*

[18]Truesdell [3, p.76] writes: "According to the principle of continuity, as we know it today, the speed of steady flow varies inversely as the cross-sectional area of the channel".—*Trans.*

Euler conceded the right to publish them to his more senior colleague, thereafter ceasing completely to work in the field for a quarter-century. He returned to the investigation of general questions of hydrodynamics only at the beginning of the 1750s, by which time he had already fashioned the concepts essential to the creation of a general theory: a precise definition of the pressure in a flowing liquid, and of a continuous physical field (both vector and scalar), and a simple formulation of the fundamental law of dynamics (conservation of momentum) for the elementary particles of a medium. Euler's definition of pressure in a flowing liquid represented the pinnacle in the evolution of that concept, initiated by D. Bernoulli in 1730, and partially improved in Johann I Bernoulli's *Hydraulics* (1743). It was Johann Bernoulli the elder who also made the first attempt to apply the law of conservation of momentum to hydraulics. Certain of the preconditions for an approach to the study of the motion of fluids as continua may already be found in Euler's commentary to his translation of B. Robins' *New principles of gunnery* (1745), where in considering flow around bodies he makes use of a diagram of flow lines[19] [9]. However the first attempt of real significance to derive general equations for the motion of a fluid considered as a continuum was made by d'Alembert in work carried out in the late 1740s and published in 1752. All the same it was only somewhat later that for the first time a precise derivation of a full system of equations for the motion of an ideal[20] fluid was carried out—by Euler, using his above-mentioned "new principle" of mechanics. This is contained among other things in his fundamental essays on hydrostatics and hydrodynamics written in the period 1753–1755, and published in the *Mémoires* of the Berlin Academy in 1757.

In the first of these essays Euler generalized results of A.-C. Clairaut, expounding hydrostatics and aerostatics essentially in the form preserved in those disciplines till today. It is curious that here Euler obtained, in particular, the well known barometric formula for an isothermal atmosphere, and also suggested that it might be appropriate to define temperature as proportional to the pressure in a quantity of gas of fixed volume.

Euler presented the second essay, "General principles of the motion of fluids", to the Berlin Academy on September 4, 1755. He begins with a general statement of the problems needing to be solved in order to construct a theory of motion of an ideal fluid. Then he deduces via the elementary "fluid parallelepiped" with edges dx, dy, dz, familiar to us today, general hydrodynamical and continuity equations for a compressible fluid. These equations, in his notation and in their essentially still standard form, are as follows [10, § 21]:

$$P - \frac{1}{q}\left(\frac{dp}{dx}\right) = \left(\frac{du}{dt}\right) + u\left(\frac{du}{dx}\right) + v\left(\frac{du}{dy}\right) + w\left(\frac{du}{dz}\right),$$

$$Q - \frac{1}{q}\left(\frac{dp}{dy}\right) = \left(\frac{dv}{dt}\right) + u\left(\frac{dv}{dx}\right) + v\left(\frac{dv}{dy}\right) + w\left(\frac{dv}{dz}\right),$$

$$R - \frac{1}{q}\left(\frac{dp}{dz}\right) = \left(\frac{dw}{dt}\right) + u\left(\frac{dw}{dx}\right) + v\left(\frac{dw}{dy}\right) + w\left(\frac{dw}{dz}\right), \qquad (6)$$

$$0 = \left(\frac{dq}{dt}\right) + \left(\frac{d.qu}{dx}\right) + \left(\frac{d.qv}{dy}\right) + \left(\frac{d.qw}{dz}\right),$$

---

[19]Here also one finds essentially the first explanation of d'Alembert's paradox, so-called, concerning the absence of hydrodynamical resistance of a body to the continuous flow of an ideal fluid around it. However Euler never subsequently returned to this puzzle in later works—even to the extent of mentioning it.

[20]That is, inviscid with zero thermal conductivity.—*Trans.*

where p is the pressure, q the density, P, Q, and R are the components of the weight of an element of fluid in the directions of the axes, and u, v, and w are the components of the velocity at each point. (The only difference between the notation of these equations and the modern version consists in the modern use of the symbol ∂ in the partial derivatives in parentheses[21].)

In the same place Euler notes that these four equations should be supplemented by a fifth relating the pressure and density to an additional physical quantity that affects the pressure, and which of course is essentially just the temperature. The resulting five equations then, writes Euler, "contiennent toute la théorie tant de l'équilibre que du mouvement des fluides, dans la plus grande universalité qu'on puisse imaginer" [22] (§21).

After deriving the fundamental equations of hydrodynamics Euler introduces the gravitational potential S [23] and the speed of the flow W, and infers the formula (§§27, 28)

$$dp = q(dS - d\Pi - u\,du - v\,dv - w\,dw), \qquad (7)$$

where $\Pi := \partial W/\partial t$, and then integrates this equation in the case of an incompressible fluid ($q = $ const.), and more generally in the case of a barotropic flow[24], obtaining solutions which are today called the Cauchy–Lagrange integrals. The essay concludes with analyses of various special cases of fluid motion, and the remark that the equations he has obtained permit the reduction of the problems of fluid mechanics to problems of mathematical analysis.

In reading this article one is especially struck by the clarity and simplicity of exposition of the ideas—typical, indeed, of most of Euler's work. It is sometimes difficult to believe that more than two centuries separate him from us.

Following his first essays on the mechanics of liquids and gases Euler produced many further works devoted to hydrodynamics, and also to the theory of the propagation of sound. This work culminated in a compendious (516 page) summation and generalization of it, written towards the end of the 1760s and published in four parts over the period 1769–1772 in the *Novi Commentarii* of the Petersburg Academy of Sciences. The final chapter of the second part of this long essay is devoted, in particular, to the determination of the motion of a fluid given its initial state; here Euler derives the general equations of hydrodynamics in terms of so-called Lagrangian variables or "material" variables. (Note that Euler wrote to Lagrange about these variables in a letter dated January 1, 1760, reproduced by Lagrange in the published version of his own research on this theme in 1762 [11][25].) In the third part of the essay Euler considers flow in tubes of constant and of varying cross-sections,

---

[21] In modern vector notation these equations take the form

$$\frac{\partial \mathbf{v}}{\partial t} + (\mathbf{v} \cdot \vec{\nabla})\mathbf{v} = -\frac{\vec{\nabla} p}{\rho} + \mathbf{g}.$$

(Taken from *Fluid mechanics*, by L. D. Landau and E. M. Lifshitz, Pergamon Press, Oxford; etc., Second ed., 1987, p. 3)—*Trans.*

[22]"…contain the complete theory, both of the state of equilibrium and of motion of fluids, in the greatest generality imaginable".—*Trans.*

[23]The original has here "the potentials of the forces S".—*Trans.*

[24]That is, where the pressure is a function of density.—*Trans.*

[25]This letter of Euler's is published also in Lagrange's *Oeuvres* (Paris, 1892, Vol. 14) where the related articles of Lagrange may also be found (*Op. cit.*, 1867, Vol. 1). A historical enquiry into the origins of the term "Lagrangian variables" may be found in C. Truesdell's preface to Volume II-12 (p. CXX) of Euler's *Opera omnia*.

the raising of water by pumps, and flows under the condition of variable temperature. The fourth and final part contains generalized versions of a great many of his earlier results in acoustics and the theory of musical wind instruments.

Thus Euler laid the foundations for the whole of the hydrodynamics of an ideal fluid with the sole exception of supersonic aerodynamics, which was initiated only a century later and developed to a significant extent only in the 20th century. Lacking the concept of the stress tensor (introduced by A. L. Cauchy in 1823), Euler was unable to proceed to the investigation of more complex models of continuous media, such as the study of viscous fluids and elastic bodies requires. However every other mathematical tool necessary for the further development of mechanics had been prepared by him. (Here we exclude of course the later incorporation of thermodynamics into the mechanics of liquids and gases, a development also brought to completion only in the 20th century.)

Having examined one of the most splendid pages in Euler's *oeuvre*—his creation of mathematical hydrodynamics—, it behooves us to consider, however briefly, his work on the mechanics of flexible and elastic bodies[26]. He continued working in this field throughout his life.

Euler's interest in the mechanics of elastic bodies—more particularly, elastic rods—dates from his youth. It is curious that a brief note written by the young Euler when still in Basel (but published only posthumously, in 1862) contains the first-ever derivation of J. Bernoulli's bending law[27] for rods from Hooke's law—a result forgotten, it seems, by Euler himself and rederived by him much later. We shall pass over Euler's work on transverse vibrations of a rod, confining ourselves to a few words on his widely known investigations into the equilibrium shapes and buckling of elastic rods. His investigations on this theme were stimulated by D. Bernoulli's discovery in 1742 of the limiting elastic energy of a bent elastic rod. The resulting classical results obtained by Euler were published in 1744 in his famous treatise on the calculus of variations [12] where he carried out a complete analysis of the nine possible types of equilibrium that can be taken by an initially straight elastic rod of rectangular cross-section when stressed by forces and torques acting on its ends. It is here that one finds what is essentially the general formula for the critical force at which buckling of a rod commences. At that time Euler applied his formula only in the case of rods with ends secured by hinges but subsequently returned to the problem of the buckling of columns, and his last investigations on this topic, carried out in the 1770s, are devoted to the final resolution of certain difficulties that he had encountered in investigating the buckling of a column under its own weight. Euler's ideas and methods relating to bending have since been applied in a multitude of ways to the elucidation of a variety of mechanical effects and to the solution of important technological problems.

According to Euler, an equilibrium state of an elastic system is stable when the loads applied are subcritical, i.e., are such that when removed the system either returns to its original state or else undergoes acceptably small oscillations. Thus by definition loss of equilibrium takes place under hypercritical loading causing discrete elastic deviation of the system from its initial equilibrium state or from an oscillatory state involving small

---

[26] In the present collection the essay of N. V. Banichuk and A. Yu. Ishlinskiĭ is devoted to an examination of Euler's work on elastic systems.

[27] Might this be related to what is now sometimes called the "Euler-Bernoulli beam equation", where the Bernoulli in question is Daniel?—*Trans.*

amplitudes, or else observable inelastic deformations or, finally, rupture. As is well known, the consequences of instability of an elastic equilibrium depend to a great extent on the properties and means of application of the hypercritical loads.

In modern times investigations continue into the problems associated with elastic equilibria of rods and other systems with respect to variation of bending loads. Thus for example [recent] theoretical and experimental work of N. M. Belyaev and V. N. Chelomeĭ, has revealed that there exists a variety of forms of parametric instability and that it is possible to significantly increase the average external critical loads on elastic systems undergoing high-frequency vibrations. From these results it may be inferred, in particular, that a similar increase in the critical load is possible in the situation of shock loading, since in actual such situations vibrational elements are always present. (Note that of course for such loads[28] the forms of ensuing deformations are correspondingly altered.)

Euler actively participated in the debate over the vibrations of a string. The problem concerning small transverse vibrations of a string—as well as the resulting propagation of a sound—was essentially the first problem of dynamics to be considered with infinitely many degrees of freedom. It is remarkable that this problem began to be investigated long before the dynamics of systems with a finite number of degrees of freedom had been worked out. The classical (hyperbolic) wave equation for the vibration of a string was obtained by d'Alembert in 1746 and published in 1749. He obtained the solution of his equation in the very general form $f(ct + x) + g(ct - x)$, but somewhat arbitrarily restricted the class of functions f, g allowed by insisting they satisfy certain conditions of "continuity" and "smoothness". Euler immediately began to investigate d'Alembert's wave equation and concluded that the class of functions allowed in the solution of the vibrating string problem should properly include all piecewise smooth functions—a much larger class than that stipulated by d'Alembert. Euler's reasoning on this point was actually not completely rigorous—it could indeed hardly have been so given the level of mathematical analysis in the 18th century—but, as A. P. Yushkevich has stressed, his idea in this connection proved fruitful, and in fact the most up-to-date modern methods can be traced back to it [13, p. 166].

We mention also Euler's generalizing research in the 1770s into the mechanics of flexible filaments and elastic (one-dimensional) strips. Here he obtained general equilibrium equations and equations of motion of an elastic line (in the plane) in the absence of special assumptions about the nature of the material or the size of the deformations. In this connection he considered also transverse forces acting within cross-sections, thereby anticipating the later notion of tangential loading. Finally, we note that it was in these same years that Euler introduced the notion of a certain physical characteristic of a material fully equivalent to Young's modulus[29].

Thus as regards the program proposed by Euler in the 1730s for constructing a complete theory of mechanics, we may say that he successfully—and independently—managed over the course of his life to realize this aim for three of the six areas of mechanics he delineated, namely an analytic exposition of the mechanics of a point-particle (area 1), the mechanics of a rigid body (area 2), and hydrodynamics (area 6). To the study of flexible

---

[28] Exceeding the new critical load, presumably.—*Trans.*

[29] Essentially stress/strain. According to C. Truesdell this was anticipated to some extent by Johann Bernoulli in 1704; see Truesdell's *Essays in the history of mechanics.* 1968. Springer-Verlag, Berlin; etc. p. 103.—*Trans.*

bodies (area 3) and the mechanics of systems (area 5) he—contemporaneously with the other scientists mentioned above—made fundamental contributions. Although, as we have seen, Euler contributed substantially to the theory of elasticity (area 4), the development of this subject into an independent field had to wait till the following century and the efforts predominantly of the French school.

By way of a general response to the question we posed at the beginning of this essay as to the extent to which Euler had in the end realized his plan for constructing an all-embracing theory of mechanics, we may say that he succeeded brilliantly in fulfilling the grandiose program he set himself in his youth, given that he can hardly have appreciated then how improbably difficult a task it would be. It is to Euler more than to anyone else that we are indebted for our present understanding of mechanics. The period in the development of mechanics immediately following on this one had more to do with the formalization of its mathematical methods than with a deepening of its foundations.

References

[1] Euler, L. *Mechanica sive motus scientia analytice.* Vol. 1. *Opera omnia* II-1.

[2] Lagrange, J.-L. *Méchanique analitique.* Paris: Veuve Desaint. 1788.

[3] Truesdell, C. *Essays in the history of mechanics.* Berlin, etc.: Springer-Verlag, 1968.

[4] Fabrikant, V. A. "Isaac Newton, Johann Bernoulli, and the law of conservation of momentum". Uspekhi fiz. nauk (Progress in the physical sciences). 1960. Vol. 70, Issue 3. pp. 575–580.

[5] Euler, L. "Recherches sur le mouvement des corps célestes en général". Mém. Acad. sci. Berlin (1747). 1749. Vol. 3. pp. 93–143; *Opera omnia* II-25. pp. 1–44.

[6] ———. "Découverte d'un nouveau principe de méchanique". Mém. Acad. Sci. Berlin (1750). 1752. Vol. 6. pp. 185–217; *Opera omnia* II-5. pp. 81–108.

[7] ———. "Du mouvement de rotation des corps solides autour d'un axe invariable". Mém. Acad. Sci. Berlin (1758). 1765. Vol. 14. pp. 154–193; *Opera omnia* II-8. pp. 200–235.

[8] ———. "Nova methodus motum corporum rigidorum determinandi". Novi Comment. Acad. Sci. Petrop. (1775). 1776. Vol. 20. pp. 208–238; *Opera omnia* II-9. pp. 99–125.

[9] ———. *Neue Grundsätze der Artillerie. Opera omnia* II-12.

[10] ———. "Principes généraux du mouvement des fluides". Mém. Acad. Sci. Berlin (1755). 1757. Vol. 11. pp. 274–315; *Opera omnia* II-12. pp. 54–91.

[11] "Lettre de M. Euler à M. de la Grange contenant des recherches sur la propagation des ébranlements dans un milieu élastique". Mélanges philos. et math. Soc. roy. Turin (1760–1761). 1762. Vol. 2. pp. 1–10; *Opera omnia* II-10. pp. 255–263.

[12] Euler, L. *Methodus inveniendi lineas curvas maximi minimive proprietate gaudentes. (A method for finding curves enjoying maximum and minimum properties.) Opera omnia* I-24.

[13] Yushkevich, A. P. *History of mathematics in Russia before 1917.* M.: Nauka, 1968.

Leonhard Euler and the Variational Principles of Mechanics[1]

V. V. Rumyantsev

Euler's historical merit as far as the evolution of mechanics is concerned consists in his application of the methods of mathematical analysis to the problems of mechanics, so that he is quite appropriately counted among the founders of analytical mechanics—a science based on just a few fundamental postulates from which all the laws of equilibrium and motion follow. At the present time analytical mechanics rests on a few basic variational principles, i.e., fundamental initial postulates expressed mathematically in the form of variational relations, from which all of the differential equations of motion and laws of mechanics flow as logical inferences. Although these variational principles differ from one another in their form and the manner of variation, and also in their degree of generality, each of them represents within its domain of application a unique basis for the whole of the mechanics of that material domain, which is, as it were, synthesized in the principle. In other words, each of the fundamental variational principles of mechanics potentially encompasses within itself the whole content of the corresponding subdiscipline of science, uniting all of its basic postulates in a single formula.

Classical mechanics is founded, of course, on Newton's laws as they apply to freely moving material bodies, and on axioms of contiguity. The correctness of the variational principles has then to be inferred from these laws and axioms. Alternatively, each of the variational principles may be taken as axiomatic and the laws of mechanics deduced logically from them.

The variational principles subdivide formally into differential principles, characterizing the properties of motion at each instant of time, and integral principles, characterizing the properties of motion over arbitrary finite intervals of time. Euler was among the originators of one of the fundamental differential variational principles—that now known as the d'Alembert–Lagrange principle—and was the first to give a mathematically rigorous and useful formulation of the integral principle of least action, subsequently cast in a significantly more perfected and general form by J.-L. Lagrange and C. G. J. Jacobi.

[1]The translator would like to thank Prof. Ryu Sasaki of Kyoto University for his expert help with this essay.

The variational principles of mechanics led to the powerful mathematical formalism associated with them—that of Lagrange and Hamilton—and are closely bound up with transformational problems of mechanics, Lie groups, conservation laws, and other fundamental areas of research. Thanks to the efforts of many scientists, analytic mechanics has earned a reputation among the natural sciences as the standard of elegance, clarity, and orderliness in the development of its ideas.

1. A little prehistory

Vague feelings as to the likelihood of nature obeying certain general maximum or minimum laws must surely have arisen long ago. As Euler himself writes, certain philosophers of antiquity, pondering natural phenomena, reckoned that "nature does nothing in vain and in all its manifestations chooses the shortest or easiest path, and they took this as the chief finite principle towards which nature strives ... Aristotle often refers to this dogma; however it seems likely that he borrowed it from his precursors rather than inventing it himself. Subsequently this idea became so entrenched in schools of philosophy that it was made into the first canon of philosophy, until Descartes attempted to deny it" [1].

However for a long time no one gave definition to "that actual object, to whose extreme diminution nature constantly strives, not just in some but decidedly in all of its manifestations". It is, however, true that in certain particular situations there could be glimpsed "a certain shadow, as it were, of this general principle. Among these [particular situations] the most noteworthy is the reflection of light, in connection with which already Ptolemy, in explaining why the angle of reflection is always equal to the angle of incidence, showed that the path that a ray traces out in this manner is the shortest, so that if it were reflected otherwise it would trace out a longer path. However at the same time it was remarked that this explanation had no force as regards refraction of a ray of light [since] a bent line can have no relation to the shortest path" [1].

The first genuine variational principle to emerge in the history of science was the principle of shortest time in geometrical optics, proposed by the French scientist P. Fermat in 1662. Believing that "la nature agit toujours par les voies les plus courtes" [2] [2], Fermat took this principle as basic and derived from it the law of refraction of light discovered earlier experimentally by W. Snellius[3], by showing that the path traced out by a ray of light according to Snell's law is precisely that for which the time is least. A deeper explanation of Fermat's principle was later provided using the wave theory of light by Ch. Huygens (1629–1695), who showed that the fact that the time of transit of a ray of light from one point P to another Q is always least is a consequence of a more general principle[4].

Fermat's principle, which may be expressed as

$$\int_P^Q \frac{ds}{v} = \int_P^Q \mu\, ds = \min, \tag{1}$$

---

[2]"Nature always acts via the shortest ways".—*Trans.*

[3]The Dutch mathematician Willebrord Snellius (1580–1626).—*Trans.*

[4]Huygens' principle affirms that "the wavefront of a propagating wave of light at any instant conforms to the envelope of the spherical wavelets emanating from every point of the wavefront at the prior instant". Augustin-Jean Fresnel (1788–1827) later extended this to the so-called Huygens-Fresnel principle, according to which "among all possible waves along all possible paths, those with stationary paths contribute most owing to constructive interference". An analogous principle, involving probability amplitudes, is used in quantum electrodynamics; see for example R. P. Feynman's popularization *QED*.—*Trans.*

turned out to be the most general possible mathematical form of expression of the laws of geometrical optics; here $\mu(x, y, z)$ is the refractive index of an inhomogeneous but isotropic medium, through which light propagates from a point P to a point Q with speed v.

Fermat assumed that light travels faster the less dense the medium. Descartes was of the opposite opinion, and G. Leibniz also tried to repudiate Fermat's explanation. Starting from the assumption that nature chooses the easiest path, Leibniz introduced the concept of the "difficulty" to be overcome by a ray propagated through a medium, defining this as the length of the path multiplied by the "resistance"[5]. He concluded that a light ray will always follow the path for which the sum of all the difficulties is least. Like Descartes, Leibniz thought that the denser the medium the faster the speed of light. However he failed to apply his principle of the easiest path to any other situation.

All the same it is to Leibniz that we owe the first definition of mechanical "action" (1669): "The formal actions of motion are proportional to ... the product of the quantity of materials, the distances through which they move, and the speeds" [4, p. 354], i.e., $mv\Delta s$ or $mv^2\Delta t$.

In 1687 Newton published his famous *Philosophiae naturalis principia mathematica* [5], which laid the foundations of classical mechanics. In Lagrange's opinion, "In Newton's hands mechanics was transformed into a new science; his *Principia* began the epoch of this transformation" [6]. It is noteworthy that among the many problems Newton considers in his *Principia*, there is one concerning the shape a rotating body should have for its motion in a rarified medium in the direction of the axis of rotation to encounter least resistance. This would appear to represent the first instance of a variational problem in mechanics. More widely known is the problem of the brachistochrone, i.e., the curve of quickest descent, posed by Johann Bernoulli [7] in 1696, and solved by Johann Bernoulli himself and Leibniz, Newton, Jakob Bernoulli, and l'Hospital; they all found that the desired curve is an arc of a cycloid. Johann Bernoulli showed at the same time that this is likewise the shape of the trajectory of a ray of light [in a medium of uniformly changing density]. He wrote: "I have thus simultaneously solved two remarkable problems—one optical, the other mechanical ... I have shown that although these two problems are taken from completely different areas of science, they are nonetheless of the same nature". Further on he states that "Nature always acts in the simplest fashion, just as in the present situation by means of one and the same curve she fulfils two different functions" [8].

This was essentially the first instance of an optical-mechanical analogy—albeit a rather special case. The brachistochrone problem is considered as representing the source of the calculus of variations, one of the founding nurturers of which was the young Euler, Johann Bernoulli's former pupil.

2. On certain of Euler's works in the calculus of variations and analytical mechanics

The first substantial works of Euler were devoted to mechanics. In 1736, in two large volumes, his *Mechanica sive motus scientia analytice* (*Mechanics, or the science of motion, expounded analytically*) was published in St. Petersburg. In the preface to this work Euler

---

[5] Another Leibnizian property of matter.—*Trans.*

wrote: "However if analysis is needed anywhere, then it is certainly in mechanics. Although the reader can convince himself of the truth of the exhibited propositions [proved synthetically in Newton's *Principia*], he does not acquire a sufficiently clear and accurate understanding of them, so that if those questions be ever so slightly changed, he will hardly be able to answer them independently unless he turn to analysis and solve the same propositions using analytic methods. This in fact happened to me when I began to familiarize myself with Newton's *Principia* and Hermann's *Phoronomia*; although it seemed to me that I clearly understood the solutions of many of the problems, I was nevertheless unable to solve problems differing slightly from them. But then I tried, as far as I was able, to distinguish the analysis [hidden] in the synthetic method and to my own ends rework analytically those same propositions, as a result of which I understood the essence of each problem much better. Then in the same manner I investigated other works relating to this science scattered here, there, and everywhere, and for my own sake I expounded them [anew] using a systematic and unified method, and re-ordered them more conveniently. In the course of these endeavours, not only did I encounter a whole series of problems hitherto never even contemplated—which I have very satisfactorily solved—but I discovered many new methods, thanks to which not only mechanics, but analysis itself, it would seem, has been significantly enriched" [6][9].

In this work Euler derived the equations of motion of a point-mass in a natural form and applied them systematically to the solutions of many problems. In the course of characterizing Euler's *Mechanics* of 1736, J.-L. Lagrange was later to remark that it "must be acknowledged as the first substantial work in which analysis was applied to the study of motion"; the work is constructed around formulae describing motion in terms of the force acting, resolved into its tangential and normal components [6].

In order to solve problems in the theory of vibrations, Euler (1740) used a method that is essentially a generalization of a principle applied by Jakob Bernoulli and J. Hermann to the problem of a compound pendulum, affirming that "the forces acting on the weights comprising the pendulum so as to produce their movement in unison, are equivalent to those resulting from the action of gravity; hence the former, if re-directed oppositely, should just balance the latter" [6]. This principle anticipates in this special situation the general principle of d'Alembert in Lagrange's version.

At about the same time Euler was working on the isoperimetric problem, where one seeks to determine which among all simple closed plane curves of a given length encloses a region of greatest area, and was in the process of publishing a series of papers laying the foundations of a new mathematical discipline: the calculus of variations. In 1744 his celebrated treatise on the calculus of variations appeared [10], where, among other things, he gave a necessary condition for $y = f(x)$ to yield an extremum of a definite integral of the form

$$J = \int_a^b F(x, y, y')dx, \quad y(a) = \alpha, \quad y(b) = \beta. \tag{2}$$

Euler derives his condition by first subdividing the interval $[a, b]$ into small intervals of the same length, with abscissas

$$x_0 = a, x_1, x_2, \dots, x_n, x_{n+1} = b,$$

---

[6]Quoted at slightly greater length in the preceding essay by G. K. Mikhaĭlov and L. I. Sedov.—*Trans.*

Title page of Euler's monograph of 1744 on the calculus of variations

replaces the desired curve $y = f(x)$ by the continuous polygonal arc with vertex ordinates $y_s = f(x_s)$, the derivatives $f'(x_s)$ by the finite difference quotients

$$y'_s := \frac{(y_{s+1} - y_s)}{(x_{s+1} - x_s)},$$

and the integral (2) by the finite sum

$$S' = \sum_{s=0}^{n} F(x_s, y_{s+1}, y'_s)(x_{s+1} - x_s).$$

Since this is a function of the y_s only (which vary with changing f), at any extremum its partial derivatives with respect to the y_s must all vanish, whence the equations

$$\left[\frac{\partial F}{\partial y} - \frac{\Delta}{\Delta x} \left(\frac{\partial F}{\partial y'} \right) \right]_{x=x_s} = 0 \qquad (s = 0, 1, \ldots, n, n+1),$$

where $\Delta x = x_{s+1} - x_s$. Finally, letting $\Delta x \to 0$, we obtain Euler's equation

$$\frac{\partial F}{\partial y} - \frac{d}{dx} \frac{\partial F}{\partial y'} = 0 \qquad (a \leq x \leq b). \tag{3}$$

Euler's derivation of equation (3) is not completely rigorous since it involves unsubstantiated limit processes. A rigorous proof was later given by Lagrange.

3. The principle of least action

The principle of least action was first formulated by P.-L. Maupertuis in 1744. Pondering the behavior of direct and reflected rays of light, Maupertuis concluded that this "seems to depend on a metaphysical law to the effect that in performing her actions, nature always uses the simplest of means ... the path that light adheres to is that for which the quantity of action is least" [11]. He then goes on to show that all phenomena of refraction and reflection of light obey this principle.

Somewhat later in that same year 1744 Euler published as a second appendix to his treatise on the variational calculus [10] a work entitled "On the determination of the motion of projectiles in a resisting medium using the method of maxima and minima" [12], which begins as follows: "Since all natural phenomena obey some kind of maximum or minimum law, it is not to be doubted that the curves described by projectiles when acted upon by some force or other possess some maximum or minimum property ... It appears to be compatible with reality that ... the totality of all the motions manifested by a projectile should be a minimum. Although it may seem that this conclusion is insufficiently substantiated, however if I should show that it is in agreement with a reality already known *a priori*, then it will assume such weight that all doubts ... will be completely dispelled". Further on he writes: "I affirm that the curve described by a body will be such that among all curves with the same ends it will yield a minimum value of $\int M ds \sqrt{v}$, or, since M is constant, $\int ds \sqrt{v}$ ... Denoting the time during which the element of distance ds is traced out by dt, we shall then have, since $ds = dt \sqrt{v}$, that

$$\int ds \sqrt{v} = \int v dt,$$

so that for the curve described by a projected body the sum of all live forces[7] existing in the body at individual moments of time will be least" [12]. (Note that here Euler denotes the speed by $\sqrt{v}$.)

To establish his principle Euler begins by considering the motion of a body in the case of uniform gravity. He seeks the function $y = y(x)$ minimizing the integral $\int ds \sqrt{a + gx}$, or

$$\int dx \sqrt{(a + gx)(1 + p^2)},$$

where p is defined by $dy = pdx$, whence $ds = dx \sqrt{1 + p^2}$, and obtains the parabola

$$y = \frac{2}{g} \sqrt{c(a - c + gx)}.$$

He goes on to consider motion in various other situations, solving each problem by means of his principle of least action, thus establishing [or at least accumulating evidence for] the agreement of that principle with reality. In his next article [13] Euler considers the problem of the equilibrium shape of a liquid mass and several other problems.

---

[7]"Live force" is kinetic energy—reminiscent of Leibniz' "vis viva".—*Trans.*

In 1746 Maupertuis published a memoir[8] at the beginning of which he wrote in connection with the above-mentioned essay [12] of Euler: "The celebrated geometer proves that in the trajectories described by bodies under the action of central forces, the speed multiplied by an element of arc is always a minimum ... I shall now attempt to extract from the same source truths that are more important and of a higher order" [14]. Maupertuis begins with a critical summary of "the study of the evidence for the existence of God derived from the marvels of nature", and on this basis concludes that "the evidence for God's existence should be sought in the general laws of nature. The laws regulating how motion is preserved, distributed, and annihilated are founded on attributes of a higher reason". And then: "I ... have discovered a universal principle on which all laws are based, extending equally to rigid bodies and elastic bodies, on which the motion and rest of all material substances depend. This is the principle of least action, a principle so subtle as to be worthy of a supreme being. Nature obeys this principle constantly and persistently; she strives to observe it not only in all her variations but also when she is unchanging" [14]. Further on, Maupertuis at last formulates his principle: "When some change occurs in nature, the quantity of action essential to this change is the smallest possible. The quantity of action is the product of the mass of the bodies in question by their speed and the distance through which they move" [14]. He concludes the memoir with applications of this principle to collisions of rigid and elastic bodies.

Maupertuis' attempt to introduce teleology into mechanics was met with sharp rebuffs from several scientists, and his memoir generated a debate extending far beyond the confines of mechanics, involving mathematicians, mechanicians, philosophers, and publicists, including, in addition to Maupertuis, König, d'Arcy, Courtivron, Euler, d'Alembert, Voltaire, Friedrich II, and others. Mixed into the debate were issues of priority, questions of natural philosophy and physics to do with the measurement of motion, and fundamental questions concerning the ideological world view. At the center of the discussion was the question of the conditionality or causality of the phenomena of the material world, or their teleological predestination through the creator's wisdom. Euler entered the debate in support of Maupertuis. For example in his "Dissertation on the principle of least action" (1753) Euler criticized the attacks of S. König and certain other scientists on that principle and stressed Maupertuis' priority. He also wrote that he himself "had conceived this remarkable property, however not a priori, as they say, but a posteriori ... Only after a great many trials did I arrive at the formula that in motions of that type assumes its least value. For this reason I decided not to attribute greater efficacy to it than in those situations that I had investigated" [1].

Writing about Maupertuis' principle, however, Euler emphasized that "all of dynamics and hydrodynamics can with astonishing ease be developed by means of the single method of maxima and minima" [1]. He concludes this essay with the words: "Nature in all her manifestations strives to make something smallest, and this smallest thing ... is indisputably grasped by the concept of action" [1]. Voltaire also took part in the debate (1753), writing a missive [15] ridiculing Maupertuis and his teleology—and taking aim also at Euler for his support of Maupertuis. By order of the Prussian king Friedrich, Voltaire's pamphlet was publicly incinerated by the royal executioner.

---

[8]"Les loix du mouvement et du repos déduites d'un principe métaphysique".—*Trans.*

In his article on cosmogony (1754) in Diderot's *Encyclopedia*, J. d'Alembert wrote, by way of a summation of the debate, that Maupertuis' work "excited in 1752 a lively discussion ... Maupertuis was the first to show that in refraction the quantity of action is minimal ... Euler showed that this principle holds for the curves described by bodies attracted to or repelled from some fixed point. [His] beautiful proposition extends Maupertuis' principle also to the small curve described by a corpuscle of light in passing from one medium to another, so that from this point of view the principle turns out to be true quite generally and without restriction ... Euler has examined a large number of other cases to which the principle applies with ease and elegance" [16].

In 1751 König, disputing Maupertuis' priority in discovering the principle of least action, published an excerpt from a letter allegedly written by Leibniz, containing a statement of the principle. To this the Berlin Academy of Sciences reacted by declaring the fragment a forgery.

We give d'Alembert the last word: "In sum, all the theorems concerning action are nothing more than mathematical results of greater or lesser generality, and not philosophical principles ... Consequently, Maupertuis' principle, like the others of its kind, is just a mathematical principle ... This argument about action ... resembles to some extent certain religious controversies by the bitterness with which it is carried on, and by the number of people lacking all understanding of it who have taken part" [16].

Thus in summarizing Euler's role in connection with the principle of least action, it is appropriate to say that he was the first to formulate the principle rigorously in the situation of free bodies attracted to a fixed center, i.e., he was the originator of the mathematical formulation of the principle of least action in that particular case. The general formulation, applying in essence to all situations, was given only by J.-L. Lagrange, who played a decisive part in the establishment and development of the principle.

Over the period 1760–1761 Lagrange published the work [17], containing a proposed new method of determining maxima and minima of integrals together with applications of it to the solution of a variety of dynamical problems. With Lagrange there begins an essentially new epoch in the calculus of variations: not only did he simplify the method of solution of problems posed earlier through his invention of an efficient algorithm but he also applied this general method to the solution of a whole series of difficult unsolved problems of mechanics. In particular it was he who introduced the concept of the variation of a function, and rigorously derived equation (3)—now called the Euler–Lagrange equation—as a condition for stationarity both necessary and sufficient. Here, briefly, is his proof.

Let $y = f(x)$ be a differentiable extremal of the functional (2), and consider neighboring curves with the same endpoints, of the form

$$\bar{f}(x) = f(x) + \varepsilon\varphi(x),$$

where $\varphi(x)$ is an arbitrary differentiable function satisfying $\varphi(a) = \varphi(b) = 0$. Write $\delta y = \bar{f}(x) - f(x) = \varepsilon\varphi(x)$, the *variation* of the function $f(x)$. It follows that

$$\delta F(x, y, y') = \varepsilon\left(\frac{\partial F}{\partial y}\varphi + \frac{\partial F}{\partial y'}\varphi'\right),$$

and an integration by parts then yields

$$\delta J = \int_a^b \delta F dx = \int_a^b \left(\frac{\partial F}{\partial y} - \frac{d}{dx} \frac{\partial F}{\partial y'} \right) \varphi dx = 0,$$

whence, in view of the fact that $\varphi(x)$ is arbitrary, the Euler–Lagrange equation (3) follows.

In his next work, a continuation of [17], Lagrange generalized the principle of least action as formulated by Euler in [12], stating his "general principle" as follows: "Suppose we have arbitrarily many bodies $M, M', M'', \ldots$, acting on each other in some manner, and moving under the action of central forces proportional to arbitrary functions of distance; let $s, s', s'', \ldots$, denote the displacements undergone by these bodies in time t, and $u, u', u'', \ldots$, be their speeds at the end of this time; the expression

$$M \int u ds + M' \int u' ds' + M'' \int u'' ds'' + \cdots$$

will then always be a maximum or minimum" [18].

To establish this principle Lagrange assumes the law of conservation of energy

$$T - U = h = \text{const.} \tag{4}$$

for all motions considered, where T denotes the kinetic energy ("live force") of the system and U the force function of the forces acting[9]. In variational terms, then, $\delta h = 0$. Lagrange then solves several problems concerning the motion of a body attracted to fixed centers, of linked bodies, of an inflexible filament (also compressible and elastic), and problems relating to the laws of motion of inelastic and elastic fluids.

Thus by basing the theory on the law of conservation of energy, Lagrange was able to extend the principle formulated by Euler for a free point-mass to the situation of an arbitrary system of point-masses, possibly connected to one another, and acting on one another in any manner. It was on this that C. G. J. Jacobi based his remark that Lagrange's principle is the mother of analytical mechanics [19, p. 63].

It is noteworthy that Lagrange wrote, concerning his version of the principle of least action, that he regards it "not as a metaphysical principle but as a simple and general consequence of the laws of mechanics … This principle, when combined with the principle of live forces[10] and elaborated upon using the rules of the variational calculus, yields at once all the equations needed for the solution of each problem; from this we obtain a method as simple as it is general for solving problems concerning the motion of bodies" [6].

In 1788 Lagrange published his celebrated *Mécanique analitique*, which W. R. Hamilton was later to call "a kind of scientific poem" [11][20]. In this fundamental work Lagrange

---

[9]"The force function of the forces acting" is clearly the *negative* of the potential function $V (= -U)$ usually defined as the negative of a (variable) line integral of the (conservative) force field from a fixed initial point, in terms of which the total energy would then be $T + V$. The quantity $L := T + U (= T - V)$ is one form of the "Lagrangian" of the system, signifying that trajectories of the system are sought for which an appropriate integral of L is stationary (see below). See: L. D. Landau, E. M. Lifshitz. *Mechanics*. 1976. Pergamon Press. Oxford; New York, etc. Third ed. p. 2.—*Trans.*

[10]Presumably the law of conservation of energy.—*Trans.*

[11]C. Truesdell remarks that: "The historians delight in repeating Hamilton's praise of the *Mécanique Analitique* as 'a kind of scientific poem', but it is unlikely that many persons today would find Hamilton's recommendations in non-scientific poetry congenial". *Essays in the history of mechanics*. 1968. Springer-Verlag. Berlin; etc. p. 135. Perhaps Truesdell is referring to Wordsworth's reception of poems sent to him by Hamilton.—*Trans.*

took as his goal "the reduction of the theory of mechanics and the methods of solution of problems thereof to general formulae yielding by means of simple elaboration all the equations needed for the solution of each problem ... The expounded ... methods require neither constructions nor mechanical or geometrical reasoning; they require only algebraic operations subject to a systematic and uniform treatment. All lovers of analysis will delight in the conviction that mechanics has become a mere branch of analysis" [6]. Here Lagrange restates his version of the principle of least action as follows: "In connection with the motion of any system of bodies subject to the action of mutually attractive forces or forces directed to fixed centers and proportional to arbitrary functions of the distances, the curves described by the various bodies, as well as their speeds, must be such that the sum of the products of the individual masses by the integrals of their speeds multiplied by an element of arc is a maximum or minimum, under the condition that the initial and final point of each curve is taken as given, so that the variation of the coordinates of these points is equal to zero" [6]. In Lagrangian symbolism this may be expressed as follows[12]:

$$\delta \sum_{\nu} m_{\nu} \int v_{\nu} ds_{\nu} = 0,$$

or

$$\delta \int_{t_0}^{t_1} 2T dt = 0, \qquad \delta q_i = 0 \text{ for } t = t_0, t_1, \tag{5}$$

where in addition for each curve the law (4) above holds with one and the same value of h. Lagrange goes on to show that "the above theorem does not just express an interesting property of the motion of bodies but may in fact be used to determine that motion" [6]. He proves that the principle of least action has as a consequence the general Lagrangian dynamical formula, expressed in terms of the d'Alembert–Lagrange principle by

$$\sum_{\nu} (\mathbf{F}_{\nu} - m_{\nu}\mathbf{w}_{\nu}) \cdot \delta \mathbf{r}_{\nu} = 0, \tag{6}$$

where the $\delta \mathbf{r}_{\nu}$ are virtual displacements permitted by the configuration at an arbitrary instant[13]. It should be mentioned that the principle expounded by d'Alembert in 1743 in his treatise on dynamics [21], which he successfully applied to the solution of many problems—including that of the precession of the equinoxes—, is formulated by Lagrange as follows: "If motions are imparted to several bodies, which they are compelled to alter as a result of the presence of interactions between them, then ... these motions may be regarded as composed of those motions that the bodies undergo in fact, and others that are annihilated; from this it follows that the latter motions must be such that if the bodies were subject exclusively to them, they would exactly balance one another" [6].

He adds that this principle yields "a direct and general method by means of which it is possible ... to express in equational form all problems of mechanics ... However the difficulty in determining those forces that are supposed to be mutually annihilated, as well as the laws of equilibrium they obey, often makes the application of the principle inconvenient and wearisome" [loc. cit.]. Hence a different approach is preferable, based on the balance "between the forces and the motions they cause, which, however, are appropriately taken to

---

[12]Here the q_i are generalized coordinates of the system; see below.—Trans.

[13]And the $\mathbf{F}_{\nu}$ are the forces acting on the bodies of the system and the $\mathbf{w}_{\nu}$ their accelerations.—Trans.

be directed oppositely". This approach signifies, Lagrange notes, "a return to the method of Hermann and Euler, who applied it in the solutions of many problems of mechanics" [6]; the method in question is given symbolically by (6). In essence this principle represents a combination of d'Alembert's principle with the statical principle of possible displacements given by

$$\sum_\nu \mathbf{F}_\nu \cdot \delta \mathbf{r}_\nu = 0, \tag{7}$$

whose great generality and usefulness in solving problems of statics was first understood by Johann Bernoulli (1717), and was used by P. Varignon (1654–1722) in his work in statics. This principle—as it stood then—provided the impetus for the appearance of Maupertuis' "loi de repos", which Euler developed and generalized in 1751 [22]. Lagrange saw that this principle "possesses also the valuable and exclusive advantage over all other principles, that it can be expressed by means of a single formula embracing all problems that may be posed concerning the equilibrium of bodies" [6], having in view his general formula of statics (7).

From his general formula of dynamics (6) Lagrange derived two versions of the equations of motion of a mechanical system, called today Lagrange's equations of the first and second kind. The leading role in mechanics played by those of the second kind[14]

$$\frac{d}{dt} \frac{\partial L}{\partial \dot{q}_i} - \frac{\partial L}{\partial q_i} = 0 \quad (i = 1, \dots, n), \tag{8}$$

is a consequence of their invariance (covariance) under point transformations of the variables q_i[15]. (Here the q_i are the generalized independent coordinates of the system[16], in terms of which the rectangular Cartesian coordinates are expressed as explicit functions: $x_\nu = x_\nu(t, q_1, \dots, q_n)$.[17]) For appropriately chosen q_i, the equations (8) hold with respect to an arbitrarily moving frame of reference. Lagrange's equations provided the first instantiation of a "principle of invariance", an idea that has come to play such a fundamental role in modern physics.

The final stage in the elaboration of the principle of least action was taken by C. G. J. Jacobi. In his *Vorlesungen über Dynamik* (*Lectures on dynamics*) [23] he expressed the opinion that Lagrange's version of the principle of least action is insufficiently clear since it involves time as an independent variable, while in fact the time it takes for a system to arrive at a particular point [state] depends on the trajectory along which it moves [in the space of generalized coordinates] in accordance with the "law of live forces"[18], as a result of which the upper limit in the action integral[19] may turn out not to be constant. To resolve this difficulty, in 1837 Jacobi [24] proposed excluding the time from the integrand in (5) by using (4) to eliminate the total kinetic energy from that integrand, thereby obtaining the

---

[14]See below for the definition of L.—*Trans.*

[15]That is, transformations of the form $q_i \to q_i' = q_i'(q_1, \dots, q_n)$.—*Trans.*

[16]There are as many generalized coordinates as there are "degrees of freedom" of the system in question. For example a standard rigid pendulum has just one degree of freedom and as its single generalized coordinate one may take the angle it makes with the vertical.—*Trans.*

[17]Perhaps $\mathbf{r}_\nu = \mathbf{r}_\nu(t, q_1, \dots, q_n)$ would be better here.—*Trans.*

[18]Presumably the law of conservation of energy; perhaps the author is using Jacobi's terminology here.—*Trans.*

[19]See (5).—*Trans.*

following geometric version of the principle of least action:

$$\delta \int_P^Q \sqrt{2(U+h)} \sqrt{\sum_\nu m_\nu ds_\nu^2} = 0.$$

It should be noted that this exclusion of time from the integrand in (5) by means of (4), leaves the new Jacobian version of the principle equivalent to Lagrange's version, provided the supplementary condition (4) is assumed.

The kinetic energy

$$T = \frac{1}{2} \sum_\nu m_\nu \left(\frac{ds_\nu}{dt}\right)^2$$

of our system in terms of generalized coordinates, becomes

$$T = \frac{1}{2} \sum_{i,j=1}^n a_{ij}(q)\dot{q}_i\dot{q}_j \qquad \left(\dot{q}_i = \frac{dq_i}{dt}\right),$$

and in terms of the *metric of configuration space*, defined by

$$ds^2 = \sum_{i,j=1}^n a_{ij} dq_i dq_j, \tag{9}$$

Jacobi's principle takes the form

$$\delta \int_P^Q \sqrt{2(U+h)} ds = 0, \tag{10}$$

with $\delta h = 0$, and $\delta q_i = 0$ at the initial and final configurations P and Q of the system.

Jacobi's principle changes the general problem of investigating the motion of a holonomic[20] conservative system to the geometrical one of finding, in the Riemannian space defined by the metric (9), extremals of the variational problem (10) that represent actual trajectories of the system. Thus Jacobi's version of the principle of least action reveals a close connection between the motions of a holonomic conservative system and the geometry of a certain Riemannian space. In particular if the motion occurs in the absence of acting forces, so that $U = 0$, then the system moves along a geodesic of the Riemannian configuration space with constant speed—a result generalizing Galileo's law of inertia. When $U \neq 0$, the investigation of the motion of the system reduces again to the determination of the geodesics in the Riemannian space with metric

$$d\sigma^2 = 2(U+h)ds^2,$$

since then Jacobi's principle (10) has the form

$$\delta \int_P^Q d\sigma = 0.$$

---

[20] A *holonomic* dynamical system is one "for which the displacement represented by [any] arbitrarily small changes in the coordinates is in general a possible displacement." E. T. Whittaker. *A treatise on the analytic dynamics of particles and rigid bodies.* 1964. Cambridge Univ. Press. In *Mechanics*, by L. D. Landau and E. T. Lifshitz (Third ed. 1976. Pergamon. Oxford; New York, etc.) a holonomic system is defined as one for which the constraints impose relations among the coordinates q_i only.—*Trans.*

Note that in the case of a single point-mass, where the element of length ds becomes that of ordinary Euclidean three-dimensional space, Jacobi's principle takes on the aspect of a mechanical analogue of Fermat's principle (1) in optics. On comparing (1) and (10) we see that the trajectory of a point coincides with that of a ray of light passing through a medium with index of refraction $\mu = \sqrt{2(U + h)}$. This analogy applies, however, only to the actual trajectory of the point-mass in mechanics and the ray of light in optics: the evolutions of these two processes with respect to time are completely different [25].

We should also mention Jacobi's conclusion that a minimum according to the principle of least action "is achieved not between any two positions [in configuration space], but only when the initial and final positions are sufficiently close to each other" [23]. If on some trajectory of the system we choose a point A and consider another trajectory passing through A, making a small angle with the first at A and intersecting with it again at another point B, then the limiting position of B as the angle at A between the trajectories tends to zero is called a *kinetic focus* of the point A.[21] Jacobi showed that the action is least for a trajectory if the endpoint Q of the interval of integration is located closer to P than a kinetic focus of the point P.

4. Hamilton's principle and the optical-mechanical analogy

In the period 1834–1835 W. R. Hamilton [20, 26] established a new principle of least action, more general than that of Euler–Lagrange–Jacobi. Hamilton considered mechanical systems bound by stationary geometrical constraints and acted upon by forces with force function U,[22] which may explicitly depend on t, showing that in such situations a principle of least action of the form

$$\delta \int_{t_0}^{t_1} L\,dt = 0, \quad \delta q_i = 0 \text{ for } t = t_0, t_1, \quad \text{where } L = T + U, \tag{11}$$

is valid. Somewhat later (1848) M. V. Ostrogradskiĭ [27] extended the principle (11) to the situation of non-stationary constraints and non-conservative forces[23] $\mathbf{F}_\nu$; with this relaxation of Hamilton's conditions, the principle (11) becomes

$$\int_{t_0}^{t_1} \left(\delta T + \sum_\nu \mathbf{F}_\nu \cdot \delta \mathbf{r}_\nu \right) dt = 0. \tag{12}$$

Thus the principle (11) turned out to be valid also in cases where the Lagrangian $L(t, q, \dot{q})$ and the constraints imposed on the system depend explicitly on the time, and the "integral of motion" in (4) does not exist.

The equations for the extremals satisfying Hamilton's principle (11) turn out to have the form of Lagrange's equations (8) in the generalized coordinates q_i and the derivatives $\dot{q}_i$, together characterizing the state of the system. Hamilton proposed taking instead as basic variables the quantities q_i and the generalized momenta

$$p_i = \frac{\partial L}{\partial \dot{q}_i} \quad (i = 1, \dots, n). \tag{13}$$

---

[21] Relative to the first trajectory, assumed held fixed?—*Trans.*

[22] That is, the negative of the potential.—*Trans.*

[23] That is, for which the potential is undefined, since line integrals of the force field around closed paths need not be zero.—*Trans.*

It can be shown that the Jacobian $\left| \partial^2 L / \partial \dot{q}_i \partial \dot{q}_j \right|$ is non-zero, so that[24] the equations (13) define the $\dot{q}_i$ implicitly as functions of the q_i, p_i and t: $\dot{q}_i = \varphi_i(t, q, p)$. Then in terms of the function

$$H(t, p, q) = \sum_{i=1}^{n} p_i \dot{q}_i - L, \tag{14}$$

—now called the *Hamiltonian* of the system in question—Hamilton [20] showed that the equations (8) can be transformed into the following canonical form:

$$\frac{dq_i}{dt} = \frac{\partial H}{\partial p_i}, \quad \frac{dp_i}{dt} = -\frac{\partial H}{\partial q_i} \quad (i = 1, \ldots, n). \tag{15}$$

Note also that by replacing L in (11) using (14), the principle (11) may be rewritten as

$$\delta \int_{t_0}^{t_1} \left(\sum_{i=1}^{n} p_i \dot{q}_i - H \right) dt = 0, \quad \delta q_i = 0 \text{ for } t = t_0, t_1, \tag{16}$$

called Hamilton's second form of the principle. Thus the equations for the extremals of the problem (16) are Hamilton's equations (15). Although the equations (15) are fully equivalent to the equations (8), they have the great advantage over the latter that their right-hand sides involve the single function $H(t, p, q)$ not depending explicitly on the derivatives of p or q with respect to t, which appear only in the left-hand sides of the equations (15). Hamilton took these equations as basic in his remarkable investigations in dynamics.

It was in optics that Hamilton began his scientific research in 1824 [28], later extending the fundamental results he obtained to dynamics, thereby establishing an optical-mechanical analogy.

In the wave theory of light a mathematical description of the propagation of light can be founded on the idea of light rays, or on the method involving wave fronts, proposed by Huygens (1690) [3]. According to Huygens' principle, each point of a front Σ of a light wave at a time t serves as the source of a secondary wave, and the envelope of these secondary waves at time $t'(> t)$ is then the wave front Σ' an instant later. Taking this principle as basic, Hamilton introduced the *characteristic function* of the isotropic optical medium

$$V(x, y, z, x', y', z') = t' - t, \tag{17}$$

measuring the time taken for the wave to proceed from a point (x, y, z) to a point (x', y', z'). Let l, m, n denote the direction cosines of the normal to the surface Σ at the point (x, y, z), and l', m', n' those of the normal to the surface Σ' at the point (x', y', z'). Since both Σ and Σ' are envelopes of secondary waves, it follows that

$$\frac{1}{l} \frac{\partial V}{\partial x} = \frac{1}{m} \frac{\partial V}{\partial y} = \frac{1}{n} \frac{\partial V}{\partial z}, \tag{18}$$

and

$$\frac{1}{l'} \frac{\partial V}{\partial x'} = \frac{1}{m'} \frac{\partial V}{\partial y'} = \frac{1}{n'} \frac{\partial V}{\partial z'}. \tag{19}$$

---

[24] By the appropriate case of the Implicit Function Theorem.—*Trans.*

The equations in (17), (18), and (19), together with

$$l'^2 + m'^2 + n'^2 = 1, \tag{20}$$

form a system of six equations for determining the quantities x', y', z', l', m', n' as functions of x, y, z, l, m, n; they fully determine the behavior of light passing through a medium in terms of the function $V(x, y, z, x', y', z')$, so that every problem of optics reduces in the end to the determination of the characteristic function V of the medium [29].

From a mathematical point of view the function (17) defines a "tangent transformation" transforming each wave front Σ into the the wave front Σ' obtained from the former after excitation of the medium over the interval of time $t' - t$. Hamilton proceeded to show that an infinitesimal such transformation ("contact transformation"), i.e., a translation of the wave front from one position to another infinitesimally close to the first, is expressed analytically by equations of the form[25]

$$\frac{dx}{dt} = \frac{\partial H}{\partial \xi}, \dots, \frac{d\xi}{dt} = -\frac{\partial H}{\partial x}, \dots. \tag{21}$$

The equations (17), (18), (19), and (20) turn out to be integrals of an appropriate system (21), and the optical-mechanical analogy consists in the close formal resemblance between the latter system and Hamilton's dynamical equations (15).

On the basis of these ideas Hamilton [26] was able to show that the equations of dynamics could be expressed in an integrated form depending on the single unknown function of the action

$$S(t, q_i, q_i^0) = \int_{t_0}^{t_1} L \, dt, \tag{22}$$

where the q_i^0 denote the values of the q_i at time $t = t_0$, and the integral is taken over an arbitrary trajectory of the system (15). Hamilton called the function (22) the "principal function": it plays a role in mechanics analogous to that played by the characteristic function (17) in optics. To see this, observe first that in going from one motion of the system to another via a small change in the initial conditions, the change in S will by (22) and the definition of L be

$$\delta S = \sum_i p_i \delta q_i - \sum_i p_i^0 \delta q_i^0.$$

On the other hand we have immediately

$$\delta S = \sum_i \frac{\partial S}{\partial q_i} \delta q_i + \sum_i \frac{\partial S}{\partial q_i^0} \delta q_i^0.$$

Since these two expressions should be equal for all $\delta q_i, \delta q_i^0$, we infer that

$$p_i = \frac{\partial S}{\partial q_i}, \quad -p_i^0 = \frac{\partial S}{\partial q_i^0} \quad (i = 1, \dots, n). \tag{23}$$

Hence knowledge of the principal function S suffices for finding the motion of the system as given by $q_i = q_i(t, q_s^0, p_s^0)$ and the corresponding p_i, i.e., the general solution of the equations (15)[26].

---

[25]For certain variables ξ, η, ζ in addition to x, y, z and an appropriate function $H = H(x, y, z, \xi, \eta, \zeta)$.—*Trans.*

[26]Presumably one finds the p_i first from (23), and then the q_i using $p_i = \partial L / \partial \dot{q}_i$.—*Trans.*

In order to sidestep the difficulty with the definition of the principal function via (22) that it requires knowing the trajectories of the system in advance, Hamilton derived the first-order partial differential equation

$$\frac{\partial S}{\partial t} + H\left(t, q, \frac{\partial S}{\partial q}\right) = 0. \tag{24}$$

having S as a complete integral[27]. Jacobi [23] supplemented Hamilton's results, showing that given any solution $S(t, q_i, \alpha_i), |\partial^2 S/\partial q_i \partial \alpha_j| \neq 0$, the equations

$$\frac{\partial S}{\partial q_i} = p_i, \quad \frac{\partial S}{\partial \alpha_i} = \beta_i \qquad (i = 1, \ldots, n), \tag{25}$$

where α_i, β_i are parameters independent of t [28], represent a complete system of first integrals of Hamilton's canonical equations (15). From the second group of equations in (25) the functions $q_i = q_i(t, \alpha_j, \beta_j)$ can be calculated, and from the first the functions $p_i = p_i(t, \alpha_j, \beta_j)$, yielding a general solution of the equations (15). Hence the problem of integrating Hamilton's canonical equations (15) turned out to be equivalent to that of finding a complete integral of the equation (24). Although this second problem is, generally speaking, no easier than the original one, this approach has led to the solution of a large number of dynamical problems.

A highly effective method for investigating and solving the equations (15) arose in connection with "canonical (or contact) transformations" of the coordinates q_i, p_i of phase space. The essence of this procedure consists in the following: instead of attempting direct integration of the equations (15), one seeks a new system of coordinates Q_i, P_i in terms of which every system (15) is transformed again into a Hamiltonian system, but with simpler Hamiltonian.

Thus consider the replacement of the variables q_i, p_i by new variables Q_i, P_i satisfying the equations

$$\sum_i p_i \delta q_i - \sum_i P_i \delta Q_i = \delta W(t, q_i, Q_i). \tag{26}$$

Hamilton's principle of the second kind (16) then takes the form

$$\delta \int_{t_0}^{t_1} \left[\sum_i P_i dQ_i - \left(H + \frac{\partial W}{\partial t}\right) dt \right] + \delta \int_{t_0}^{t_1} dW = 0, \quad \delta Q_i = 0 \text{ for } t = t_0, t_1,$$

whence it follows that the differential equations in terms of the new variables Q_i, P_i have the same canonical form (15), with the new Hamiltonian

$$\tilde{H} = H + \frac{\partial W}{\partial t}. \tag{27}$$

In this context a transformation satisfying an equation of the form (26) is called a *canonical transformation*, and the function $W(t, q_i, Q_i)$ appearing there its *generating function*. From (26) we obtain as before

$$p_i = \frac{\partial W}{\partial q_i}, \quad -P_i = \frac{\partial W}{\partial Q_i} \qquad (i = 1, \ldots, n), \tag{28}$$

---

[27] Does this mean merely "solution"?—*Trans.*
[28] The original has simply "constants".—*Trans.*

which determine the canonical transformation, given its generating function $W(t, q_i, Q_i)$; and since the derivatives $\partial W / \partial q_i$ $(i = 1, \ldots, n)$, regarded as functions of the variables Q_i, are independent, it follows that $|\partial^2 W / \partial q_i \partial Q_i| \neq 0$.

Comparing (26) with the expression $\delta S = \sum_i p_i \delta q_i - \sum_i p_i^0 \delta q_i^0$ for the variation of Hamilton's principal function, we conclude that the values of the variables q_i, p_i at the initial time t_0 and the later time t are related by a canonical transformation, with the role of the generating function played by the principal function. It follows that the motion of the mechanical system described by the equations (15) may be looked upon as consisting of a chain of canonical transformations of the phase-space coordinates q_i, p_i. For an infinitesimally small canonical transformation $P_i = p_i + \Delta p_i$, $Q_i = q_i + \Delta q_i$, it is not difficult to derive equations of the form

$$\frac{\Delta q_i}{\Delta t} = \frac{\partial H}{\partial p_i}, \quad \frac{\Delta p_i}{\Delta t} = -\frac{\partial H}{\partial q_i} \qquad (i = 1, \ldots, n),$$

which become the equations (15) as $\Delta t \to 0$. Thus Hamilton's canonical equations lead us to the view of the motion of a system as composed of a continuous family of infinitesimal canonical transformations of the coordinates q_i, p_i.

Note that in the case of a generating function $W(t, q_i, Q_i)$ satisfying

$$\frac{\partial W}{\partial t} + H = 0, \tag{29}$$

i.e., $\tilde{H} = 0$, the canonical equations simplify to

$$\frac{dQ_i}{dt} = 0, \quad \frac{dP_i}{dt} = 0 \quad (i = 1, \ldots, n),$$

i.e., the Q_i and P_i are independent of t. In this situation, in view of the transformation formulae (28), the equation (29) takes on the form of the Hamilton-Jacobi partial differential equation (24) with the generating function $W(t, q_i, Q_i)$ as a complete solution, while the formulae (28) play the role of the equations (25) of Jacobi's theorem.

The foundations of the theory of canonical transformations were laid by Jacobi [24] in 1837, when he showed that under canonical transformations the Hamiltonian form of the equations of motion of any dynamical problem is preserved. As a consequence of these results, the motion of a dynamical system can be viewed as the continuous unfolding of canonical transformations, i.e., as a continuously changing mapping of the phase space onto itself. Thus on the basis of Hamilton's optical-mechanical analogy, in the mid-19th century there arose a new branch of analytical mechanics embracing the methods of the calculus of variations and the theory of transformation groups, and yielding a great many remarkable results—a survey of which, however, lies outside the confines of the present essay.

In the opinion of N. G. Chetaev [31] Hamilton's optical-mechanical analogy determined the course of analytical mechanics for a century. We conclude this section with a brief exposition of Chetaev's further investigations into that analogy. As stated above, the analogy in question was drawn between the dynamics of conservative systems and Huygens' wave theory of light. However that theory did not stay frozen in the form proposed by Huygens but was supplemented or replaced by those of Fresnel, Cauchy, and Maxwell.

Cauchy, having set himself the task of extending Hamilton's optical-mechanical analogy, developed a version of it in his theory of vibrations of an elastic medium. However with
this discovery Cauchy removed the question of the future development of that analogy—
now using post-Huygensian theories of light—from the purview of analytical dynamics.
Felix Klein was the first to bring this situation to the attention of scientists, and to re-state
the problem of the further evolution of the optical-mechanical analogy.

Following A. Fresnel light came to be understood as an oscillatory process, and it was
on these grounds that N. G. Chetaev [31] proposed seeking an extension of the optical-
mechanical analogy in the area of oscillatory motions, or stable repetitive (steady-state)
motions of conservative systems. Furthermore he formulated his new version of the analogy between the two kinds of phenomena in terms of the coincidence of the respective
transformation groups arising in the two theories.

Chetaev [32] showed that in connection with oscillatory motion of a conservative system Poincaré's variational equations admit a unimodular group of linear transformations
possessing a faithful representation in the full Lorentz group. Since the latter group is basic
to the theories of light of Cauchy and Maxwell, we may consider this result as constituting
a further chapter in the development of the optical-mechanical analogy.

Earlier, in 1958, Chetaev had published an article [30] in which he demonstrated an
analogy between Cauchy's mathematical theory of light and oscillatory motions of conservative systems. We now briefly consider his argument, with, as before, q_i, p_i $(i = 1, \ldots, n)$
denoting generalized coordinates and the corresponding generalized momenta of the given
system, and $U(q_1, \ldots, q_n)$ and $T = \frac{1}{2} \sum_{i,j}^{n} g_{ij} p_i p_j$ its force function[29] and kinetic energy.

In 1945 Chetaev had shown that if the unperturbed motion oscillates stably, then the
Poincaré variational equations can be reduced to a system of equations with constant coefficients, and moreover with all eigenvalues zero. Chetaev exploited this result by considering perturbations of the system's motion determined by changes in the constants β_i while
the α_i are held fixed in a complete integral $S = -ht + V(q, \alpha)$ of the Hamilton-Jacobi
equation (24), deriving the elliptic partial differential equation

$$\sum_{i,j}^{n} \frac{\partial}{\partial q_i} \left(g_{ij} \frac{\partial V}{\partial q_j} \right) = 0,$$

expressing the condition that the eigenvalues are all zero. In terms of an appropriate twice
differentiable function $\Phi(-ht + V)$ of the complete integral $-ht + V$ of Hamilton's equation, this condition takes the form of the wave equation

$$\frac{2(U + h)}{h^2} \frac{\partial^2 \Phi}{\partial t^2} = \sum_{i,j}^{n} \frac{\partial}{\partial q_i} \left(g_{ij} \frac{\partial \Phi}{\partial q_j} \right), \tag{30}$$

which by virtue of its form, establishes an analogy between Cauchy's theory of light and
oscillatory motions of conservative systems.

---

[29]That is, the negative of the potential.—*Trans.*

5. Nonholonomic systems

The terms "holonomic" and "nonholonomic" as applied to mechanical systems, were introduced by H. Hertz in 1894 [33]; the first kind of system is one which is subject only to finite constraints or those reducible to finite constraints, and the second kind where non-integrable differential constraints are present. All of the results expounded to this point pertain to the holonomic case.

The question of the applicability of the integral principles of mechanics to nonholonomic systems has a long history and a large bibliography, from which we single out just a very few works for mention. As is well known, the aforementioned principles were established in the first place for holonomic systems and attempts to extend them to nonholonomic ones encountered serious difficulties, which were, it would seem, first pointed out by Hertz [33]. He concluded that Hamilton's principle is not applicable to nonholonomic systems, observing that in such a system not every pair of points of its configuration space need be connected by a possible trajectory of the system.

In 1896 Otto Hölder [34] proposed a new integral principle in the form

$$\int_{t_0}^{t_1} \left(2T \frac{d\delta t}{dt} + \delta T + \sum_{\nu} \mathbf{F}_{\nu} \cdot \delta \mathbf{r}_{\nu} \right) dt = 0, \tag{31}$$

where now the symbol δ denotes a variation whereby each point of the initial trajectory at each corresponding time t undergoes a virtual displacement $\delta \mathbf{r}_{\nu}$ (with $\delta \mathbf{r}_{\nu} = \mathbf{0}$ for $t = t_0, t_1$), yielding a new trajectory with points in one-to-one correspondence with those of the original trajectory, but corresponding to times $t + \delta t$ for infinitesimal δt. By further specialization of the variation Hölder was able to derive from (31) the Hamilton-Ostrogradskiĭ principle (12) by taking $\delta T = 0$, and also an extended version of Lagrange's principle of least action in the form

$$\delta \int_{t_0}^{t_1} T dt = 0, \tag{32}$$

by taking

$$\delta T = \sum_{\nu} \mathbf{F}_{\nu} \cdot \delta \mathbf{r}_{\nu},$$

which determines how a continuous family of "variated" states of the system is to be traced out. In the case of the existence of an energy integral (4), the principle (32) reduces to (5). We note that the motions resulting from such variation turned out not to satisfy equations defining non-integrable constraints, whence it follows that even when the forces acting in nonholonomic systems are given by potential functions, Hamilton's principle for such systems will not have the form (11)—involving stationarity of the integral of the Lagrangian as in the case of holonomic systems—, but instead takes the form of the vanishing of the integral with respect to time of the variation of the Lagrangian:

$$\int_{t_0}^{t_1} \delta L dt = 0, \quad \delta q_i = 0 \text{ for } t = t_0, t_1. \tag{33}$$

In 1901 P. V. Voronets [35] and G. K. Suslov [36] simultaneously published works proposing, in the case of linear constraints, two new forms of integral principle apparently different from Hölder's, the first of which—i.e., Voronets'—was proffered unnamed and unsubstantiated, and the second characterized by its author Suslov as a variant of d'Alembert's principle but emphatically "not in the least representing Hamilton's principle".

The principle proposed by Voronets has the form

$$\int_{t_0}^{t_1} \left[\delta(\Theta + U) + \sum_{l=1}^{r} \frac{\partial T}{\partial \dot{q}_{k+l}} (\delta \dot{q}_{k+l} - \delta \varphi_l) \right] dt = 0, \tag{34}$$

where Θ denotes the kinetic energy T expressed using the constraint equations in terms of the independent generalized speeds $\dot{q}_l$, and where it is assumed that the equations

$$\frac{d}{dt} \delta q_i = \delta \dot{q}_i \qquad (i = 1, \ldots, n) \tag{35}$$

hold for all of the generalized speeds.

G. K. Suslov, on the other hand, proposed assuming condition (35) not for all generalized speeds, but only for a maximal set of independent ones q_s ($s = 1, \ldots, k = n - r$), while for the remaining dependent ones he assumed that

$$\frac{d}{dt} \delta q_{k+l} - \delta \dot{q}_{k+l} = \sum_{s=1}^{k} A_s^{k+l} \delta q_s \qquad (l = 1, \ldots, r), \tag{36}$$

where the quantities A_s^{k+l} are determined by the constraint equations. Suslov's principle has the form

$$\int_{t_0}^{t_1} \left(\delta L + \sum_{l=1}^{r} \frac{\partial T}{\partial \dot{q}_{k+l}} \sum_{s=1}^{k} A_s^{k+l} \delta q_s \right) dt = 0. \tag{37}$$

In 1911 S. A. Chaplygin [37] showed that the replacement of t by a new independent variable τ defined by an equation of the form $d\tau = N dt$, where the so-called "multiplier" $N = N(q, q_1)$ is a function of two free parameters q, q_1, leads to equations of motion of a nonholonomic system depending on the two independent parameters q, q_1, having canonical Hamiltonian form. If there exists such a multiplier N—determined by a certain pair of simultaneous equations—, then the variant of Hamilton's principle given by

$$\delta \int_0^t L N dt = 0 \tag{38}$$

is valid provided that the integral

$$\tau = \int_0^t N dt$$

is constant[30].

M. Kerner [38] compared the equations of motion of a system subject to differential constraints with the Euler-Lagrange equations for the stationarity of the Hamiltonian in the class of curves satisfying the constraint equations, and showed that these two systems of equations are equivalent if and only if the constraint equations are completely integrable.

---

[30]That is, independent of the path of integration?—*Trans.*

From this result he concluded that Hamilton's principle, considered as a principle of stationary action, is valid only for holonomic systems.

R. S. Capon [39] pointed out that the Hölder formalism differs from that of the calculus of variations since the curves used for comparison in the variational process do not satisfy the constraint equations, which restrict the permitted variations, and on these grounds declared Hölder's results to be unsubstantiated. H. Jeffreys [40] and L. A. Pars [41] came simultaneously to Hölder's defense. Jeffries, basing his argument on physical considerations, asserted that while Hamilton's principle does not represent a valid principle of stationary action in the case of nonholonomic systems or even holonomic systems with non-potential forces acting, on the other hand Hölder's principle (33) does. Pars subjected the question to a minute analysis, on the basis of which he concluded that while both Hamilton's principle (11) and Hölder's principle (33) are valid principles of stationary action in the case of holonomic systems, for nonholonomic systems only the latter is valid.

V. S. Novoselov [42] verified the validity of certain integral principles due to Chetaev [43] pertaining to the case of nonlinear constraints, showing that they do yield admissible trajectories for which the action is least provided that the interval of integration is sufficiently small, even though the comparison curves in the variation process need not satisfy the constraints. Yu. I. Neĭmark and N. A. Fufaev noted that the form of the principle of least action "... depends on one's attitude to the commutativity relations" [31][44, p. 180], and that for nonholonomic systems Hölder's form of the principle, in particular, is valid.

It is in fact not difficult to show that the principles (33), (34), and (37) are in fact equivalent to, and can be transformed into, one another [45]. Here is a sketch of the proof. Consider a nonholonomic system characterized by its Lagrangian $L = T + U$, and ideal non-integrable independent constraints

$$f_l(t, q, \dot{q}) = 0 \qquad (l = 1, \ldots, r), \tag{39}$$

and suppose that the independent speeds are the $\dot{q}_s$ with $s = 1, \ldots, k = n - r$, so that

$$\dot{q}_{k+l} = \varphi(t, q_i, \dot{q}_1, \ldots, \dot{q}_k) \qquad (l = 1, \ldots, r) \tag{40}.$$

The virtual displacements δq_i are assumed to satisfy Chetaev's condition [43]:

$$\sum_{i=1}^{n} \frac{\partial f_l}{\partial \dot{q}_i} \delta q_i = 0 \qquad (l = 1, \ldots, r).$$

Assuming also that the commutativity relations (35) hold for all of the generalized speeds, we infer that

$$\delta f_l = \sum_{i=1}^{n} \left(\frac{\partial f_l}{\partial q_i} - \frac{d}{dt} \frac{\partial f_l}{\partial \dot{q}_i} \right) \delta q_i \qquad (l = 1, \ldots, r). \tag{41}$$

If the constraints (39) were integrable, we should have $\delta f_l = 0$; however since we are assuming them non-integrable, the expressions (41) need not vanish, but it may all the same happen that they become zero as a consequence of the form of the equations of motion

$$\frac{d}{dt} \frac{\partial L}{\partial \dot{q}_i} - \frac{\partial L}{\partial q_i} = \sum_l \mu_l \frac{\partial f_l}{\partial \dot{q}_i} \qquad (i = 1, \ldots, n) \tag{42}$$

---

[31] Presumably those given by (35).—*Trans.*

of the system, if the constraints are non-linear [42]. Note that the non-vanishing of δf_l for at least one l signifies that the corresponding path of the variation does not satisfy all the constraints.

In Voronets' form of the principle (34) the quantity $\Theta(t, q_i, \dot{q}_1, \ldots, \dot{q}_k)$ denotes the kinetic energy $T(t, q_i, \dot{q}_i)$ from which the dependent speeds have been eliminated by means of (40). Since

$$\delta T = \delta \Theta + \sum_l \frac{\partial T}{\partial \dot{q}_{k+l}} (\delta \dot{q}_{k+l} - \delta \varphi_l), \tag{43}$$

by replacing δT by this expression for it in $\delta L = \delta T + \delta U$, we find (33) transformed into (34), i.e., into Voronets' form of Hamilton's principle.

We now turn to Suslov's form of the principle. Since in that version the variation is assumed to satisfy $\delta \dot{q}_{k+l} = \delta \varphi_l$, the equation (43) above simplifies to $\delta T = \delta \Theta$, whence Suslov's version (37) of the principle becomes

$$\int_{t_0}^{t_1} \left[\delta(\Theta + U) + \sum_{l=1}^{k} \frac{\partial T}{\partial \dot{q}_{k+l}} \sum_{s=1}^{k} A_s^{k+l} \delta q_s \right] dt = 0,$$

which in view of (36) is the same as Voronets' version (34).

Thus we have established the equivalence of the integral principles of Hölder (33), Voronets (34) and Suslov (37), purporting to be different forms of Hamilton's principle for nonholonomic systems.

Compare with the above the principle of Lagrangian type

$$\delta \int_{t_0}^{t_1} \left(L(t, q_i, \dot{q}_i) + \sum_l \kappa_l f_l(t, q_i, \dot{q}_i) \right) dt = 0 \tag{44}$$

for determining the extremals of the action integral (11) in the class of curves satisfying the constraint equations (39), where the κ_l are indeterminate Lagrange multipliers. The Euler-Lagrange equations (3) for the extremals of the problem (44) are then

$$\frac{d}{dt} \frac{\partial L}{\partial \dot{q}_i} - \frac{\partial L}{\partial q_i} = \sum_l \kappa_l \left(\frac{\partial f_l}{\partial q_i} - \frac{d}{dt} \frac{\partial f_l}{\partial \dot{q}_i} \right) - \sum_l \dot{\kappa}_l \frac{\partial f_l}{\partial \dot{q}_i} \qquad (i = 1, \ldots, n) \tag{45}$$

—partial differential equations of the second order in the q_i, and first order in the κ_l. The general solution of the system of equations comprised of (39) and (45) depends on $2n$ arbitrary parameters, whereas the general solution of the system comprised of (39) together with the equations of motion (42) depends only on $2n - r$ arbitrary parameters—from which we draw the obvious inference that the systems of equations (45) and (42) are not in general equivalent for nonholonomic systems.

In fact it is not difficult to prove [45] that the condition

$$\sum_{l,i} \kappa_l \left(\frac{\partial f_l}{\partial q_i} - \frac{d}{dt} \frac{\partial f_l}{\partial \dot{q}_i} \right) \delta q_i = 0 \tag{46}$$

is necessary and sufficient for the existence of a solution of the system comprised of (39) and (42), that is at the same time a solution of the system comprised of (39) and (45).

It follows that Hamilton's principle (11) constitutes a principle of stationary action for a *nonholonomic* system if and only if the motion of the nonholonomic system in question satisfies the condition (46). It is shown in [46] that this conclusion holds also for the Lagrangian and Jacobian forms of the principle of least action.

Observe that under condition (36) on the variation, the principle (37) has the form of a principal of stationary action if [47]

$$\sum_l \frac{\partial T}{\partial \dot{q}_{k+l}} A_s^{k+l} = 0 \qquad (s = 1, \ldots, k). \tag{47}$$

It should be pointed out that in fact for nonholonomic systems the conditions (46) and (47) are rarely satisfied.

The question of the stationarity of the Hamiltonian action for actual motions[32] is closely connected with the problem of extending to nonholonomic systems the Hamilton-Jacobi method of integration of the canonical equations of motion

$$\frac{dq_i}{dt} = \frac{\partial H}{\partial p_i}, \quad \frac{dp_i}{dt} = -\frac{\partial H}{\partial q_i} + \sum_l \mu_l \frac{\partial f_l}{\partial \dot{q}_i} \qquad (i = 1, \ldots, n), \tag{48}$$

which are equivalent to the equations (42). By means of the change of variables

$$\pi_i = p_i + \sum_l \kappa_l \frac{\partial f_l}{\partial \dot{q}_i} \qquad (i = 1, \ldots, n),$$

the Hamiltonian can be given the following form:

$$H(t, q_i, p_i) = H_1(t, q_i, \pi_i) - \sum_{l,i} \kappa_l \frac{\partial f_l}{\partial \dot{q}_i} \dot{q}_i.$$

The equations of the characteristics of the generalized Hamilton-Jacobi equation

$$\frac{\partial S}{\partial t} + H_1\left(t, q, \frac{\partial S}{\partial q}\right) = 0 \tag{49}$$

are then given by

$$\frac{dq_i}{dt} = \frac{\partial H_1}{\partial \pi_i}, \quad \frac{d\pi_i}{dt} = -\frac{\partial H_1}{\partial q_i} \qquad (i = 1, \ldots, n). \tag{50}$$

If $S(t, q, \alpha)$ is a complete integral of the equations (49), then by Jacobi's theorem the equations

$$\frac{\partial S}{\partial q_i} = \pi_i, \quad \frac{\partial S}{\partial \alpha_i} = \beta_i \qquad (i = 1, \ldots, n) \tag{51}$$

will represent integrals of the system of partial differential equations (50).

It can be shown [45, 48] that a solution of the equations (50) will constitute a solution of the equations of motion (48), (39) only if the condition (46) is satisfied. Thus the generalized Hamilton-Jacobi method combined with the method of Lagrange multipliers is applicable to the integration of nonholonomic systems if and only if Hamilton's principle preserves its character as a principle of stationary action.

---

[32] That is, of nonholonomic systems?—*Trans.*

6. The variational equation of L. I. Sedov

Hamilton's principle, like other integral principles, is applicable not only to systems with a finite number of degrees of freedom—our exclusive theme to this point—, but also to systems characterized by (continuous) families of parameters and to continuous media. In the case of reversible physical processes these versions of the principles take the form of variational principles involving stationarity of certain functionals, and in the case of irreversible processes, variational equations.

We consider briefly the variational equation of L. I. Sedov [49], which is fundamental to generalizations of the variational approach to irreversible processes. Sedov's equation has the form

$$\delta \int_{t_0}^{t_1} \int_V \Lambda \, dv \, dt + \delta W^* + \delta W = 0, \tag{52}$$

where V is the volume of an arbitrary region of the continuous medium in question, $[t_0, t_1]$ any time interval, $\Lambda = \rho \left(\frac{1}{2} v_i v^i - U \right)$ the Lagrangian (ρ denoting the mass density), δW^* and δW are functionals given by

$$\delta W^* = \int_{t_0}^{t_1} \left[\int_V \rho T \delta S \, dv - \delta G' + \delta A_{sol}^{(e)} \right] dt,$$

$$\delta W = - \left[\int_V \rho v_i \delta x^i \, dv \right]_{t_0}^{t_1} + \int_{t_0}^{t_1} \delta A_{sur}^{(e)} \, dt,$$

U and S are the energy and entropy per unit mass, the x^i are coordinates, the v^i the components of the velocity, T is the temperature, $\delta G'$ the uncompensated heat, and, finally, $\delta A_{sol}^{(e)}$ and $\delta A_{sur}^{(e)}$ are respectively the work done by the external bulk forces, and that done by the surface forces acting.

In equation (52) the Lagrangian Λ and the functional δW^* are given quantities, while δW is to be determined from the equation, its calculation amounting to establishing the state equations.

The variational equation (52) represents in essence a time-integrated general equation for the dynamics of a continuous medium, incorporating the first and second laws of thermodynamics. It was derived by L. I. Sedov in 1965 in connection with the construction of new models of continuous media with complex properties. Sedov's equation was proposed as a basic postulate for the mechanics of a continuous medium. The construction of a new model using the variational approach then involves prescribing a particular collection of characterizing functions together with the Lagrangian Λ and the functional δW^* [50].

7. Differential principles of mechanics

We look briefly also at differential principles of mechanics.

The d'Alembert-Lagrange principle (6) represents a variational condition not having the character of an extremal principle. In 1829 C. F. Gauss [51] converted the d'Alembert-Lagrange principle into an extremal principle. By introducing virtual motions differing from real ones only at the level of accelerations, Gauss was able to show that the actual motion of a system must satisfy the following minimal condition:

$$\sum_{\nu} \frac{m_{\nu}}{2} \left(\ddot{\mathbf{r}}_{\nu} - \frac{\mathbf{F}_{\nu}}{m_{\nu}} \right)^2 = \min .$$

In 1894 H. Hertz [33] based his "force-free" mechanics on the following "principle of the straightest path" (or "principle of least curvature):

$$\frac{1}{2} \sum_{\nu} m_{\nu} \ddot{\mathbf{r}}_{\nu}^2 = \min,$$

which is essentially just Gauss' principle in the special case $\mathbf{F}_{\nu} = 0$.

In 1941 N. G. Chetaev [52] introduced a significant modification of Gauss' principle, showing that for an actual motion the work done on an elementary cycle made up of a straight virtual motion in the force field and the retrograde (i.e., reverse) motion in the force field—which would suffice to determine the actual motion if the mechanical system in question were completely free—will be a maximum:

$$A = \max A_{\mu}, \qquad A_{\mu} = \sum_{nu} (\mathbf{F}_{\nu} - m_{\nu} \ddot{\mathbf{r}}_{\nu}) \left(\dot{\mathbf{r}}_{\nu} + \mathbf{w}_{\nu}^{\mu} \frac{dt}{2} \right) dt$$

where the $\mathbf{w}_{\nu}^{\mu}$ are the accelerations in the virtual motion.

This principle is interesting in that it represents a direct variant of the idea of Hermann and Euler, developed further by Lagrange in his exposition of d'Alembert's principle.

Chetaev's principle can also be extended to systems more general than those usually considered, by bringing to bear Carnot's thermodynamical principle. The extension of his principle to more general physical systems has the form

$$\delta \left\{ -\frac{dt^2}{2} \sum_{\nu} \frac{1}{2m_{\nu}} [(\mathbf{F}_{\nu} - m_{\nu} \mathbf{w}_{\nu})^2 + \Delta Q - \Delta U] \right\} = 0,$$

where U is the internal energy, ΔQ the influx of heat, and the first and second laws of thermodynamics are assumed [53, 54].

The principles of d'Alembert-Lagrange, Gauss, and Chetaev can also all be generalized to continuous media [55].

8. L. S. Pontryagin's maximum principle

We conclude this essay with a brief overview of the simplest problem of optimal control theory [56], namely that of finding necessary conditions on solutions of the problem

$$J = \int_{t_0}^{t_1} f^0(t, x^1, \ldots, x^n; u_1, \ldots, u_r) dt = \min$$

under conditions[33]

$$\frac{dx^i}{dt} = f^i(t, x^1, \ldots, x^n; u_1, \ldots, u_r), \qquad t_0 \leq t \leq t_1,$$

$$x(t_0) = x_0, \ x(t_1) = x_1, \ u(t) \in U.$$

---

[33] In the original the x-variables figuring as arguments in the f^i appear as $x_1, \ldots, x_n$, i.e., with subscripts rather than superscripts.—*Trans.*

Here the class of admissible control functions is that of all piecewise continuous, bounded vector-functions $u(t)$ having discontinuities only of the first order, with values in a certain closed region U. This is Pontryagin's version of variational problems of Lagrange type. Peculiar to this kind of problem is the need to prescribe the particular class of admissible control functions with values in the closed region U.

In terms of the "Pontryagin function"

$$\Pi(\psi(t), x(t), t, u(t)) = \sum_{\alpha=0}^{n} \psi_\alpha f^\alpha(t, x, u),$$

the equations for the phase variables x^i and the auxiliary variables ψ_i, which play the role of Lagrange multipliers, take the form

$$\frac{dx^i}{dt} = \frac{\partial \Pi}{\partial \psi_i}, \quad \frac{d\psi_i}{dt} = -\frac{\partial \Pi}{\partial x^i} \qquad (i = 1, \ldots, n).$$

We then have the following theorem [56]: *Let $u(t)$, $t_0 \leq t \leq t_1$, where t_0, t_1 are fixed instants of time, be an admissible control function carrying the phase point from position x_0 to x_1, and let $x(t)$ be the corresponding trajectory, so that $x(t_0) = x_0$, $x(t_1) = x_1$. The vector-function $u(t)$ constitutes a solution of the optimal problem for this fixed time interval, only if there exists a nonzero continuous vector-function $\psi(t) = (\psi_0(t), \psi_1(t), \ldots, \psi_n(t))$ corresponding to $u(t)$ and $x(t)$ such that:*

1) for all $t, t_0 \leq t \leq t_1$, the function $\Pi(\psi(t), x(t), t, u)$ attains a maximum over all $u \in U$ at this $u(t)$—in symbols

$$\Pi(\psi(t), x(t), t, u(t)) = M_{u \in U}(\psi(t), x(t), t);$$

and

2) the function $\psi_0(t)$ is nonpositive[34].

Clearly Pontryagin's maximum principle belongs to the category of differential variational principles. In the case where the the region of admissible values of the control function is open, the principle is equivalent to the well known Weierstrassian conditions. However in situations where the control function takes a value on the boundary of the region U, Weierstrass' conditions may fail, while Pontryagin's principle remains valid.

It can be shown [57] that the optimality condition may be reformulated as an integral variational principle analogous to Hamilton's principle:

$$\delta \int_{t_0}^{t_1} \left(\sum_{i=0}^{n} \psi_i \dot{x}^i - \Pi \right) dt \geq 0, \qquad \delta x^i = 0 \text{ for } t = t_0, t_1.$$

From this one can deduce the equations of motion, the boundary conditions, and the equations of transversality, as well as Pontryagin's maximum principle.

---

[34]The original has "nonpositive parameter" or "nonpositive constant"; it is not clear which of these is intended. (These final few pages of this essay manifest the symptoms of "arithmorrhoea", where theorems and formulae come indiscriminately thick and fast without the requisite definition of symbols, or provision of sufficient context for the non-expert to gain some kind of understanding.)—*Trans.*

In conclusion, we affirm yet again that the variational principles of mechanics give concise expression, in simple invariant form, to equations of motion and field equations, encapsulating both the discrete and continuous aspects of motion, and also constitute expressions of a general mechanical principle of causality. They possess vast heuristic potency, especially for the mechanics of continuous media, relativity theory, quantum mechanics and, one may hope, modern nuclear physics and elementary particle theory. Variational principles have been applied, generalized, and developed by a great many of the foremost scientists: H. Helmholtz, L. Boltzmann, J. Gibbs, H. Poincaré, A. Einstein, N. Bohr, A. Sommerfeld, L. de Broglie, E. Schrödinger, W. Heisenberg, P. Dirac, M. Born, and many others. The tremendous universality of the variational principles of mechanics, their aptness for generalization and extension, even to areas of science outside mechanics, their intimate connection with conservation laws and Lie groups, all tend to place them at the centre of much of physics. However further discussion of these themes lies beyond the confines of the present essay.

References

[1] Euler, L. "Dissertation sur le principe de la moindre action, avec l'examen des objections de M. le Professeur Koenig faites contre ce principe". ("Dissertation on the principle of least action, with an analysis of the proposed objections of Prof. König to that principle".) 1753. *Opera omnia*. Series II, Vol. 5. pp. 177–178.

[2] Fermat, P. Fermat's principle of refraction was first stated in attachments to a letter dated January 1, 1662, written to Marin Cureau de la Chambre. *Oeuvres de Fermat*. Paris. 1891–1912. Vol. I. pp. 132–179.

[3] Huygens, C. *Traité de la lumière*. 1690.

[4] Leibniz, G. W. "Dynamica, de potentia et legibus naturae corporeae". *Mathematische Schriften*. Hrsg. C. I. Gerhardt. Halle, 1860. Abt. 2. Vol. 2(6). pp. 281–314.

[5] Newton, I. *Philosophiae naturalis principia mathematica*. First ed. London, 1687, Second ed. 1713, Third ed. 1725/26. (English translation by F. Cajori: *Mathematical principles of natural philosophy and his system of the world*. Berkeley, Calif. 1934.)

[6] Lagrange, J.-L. *Mécanique analitique*. Courcier, Paris. Second augmented ed. 1811–1815. *Oeuvres*. Vols. XI, XII.

[7] Bernoulli, Johann. "Problema novum ad cujus solutionem mathematici invitantur". ("A new problem, which mathematicians are invited to solve".) Acta Erud. Leipzig. 1696, p. 269. *Opera omnia*. Vol. I, p. 161.

[8] ——. "Curvatura radii in diaphanis non uniformibus solutioque problematis a se in Actis 1696, p. 269, propositi, de invenienda linea brachystochrona, id est, in qua grave a dato puncto ad datumpunctum brevissimo tempore decurrit; & de curva synchrona, seu radiorum unda, construenda". ("The curvature of a ray in inhomogeneous transparent media and the solution of the problem proposed by me in the *Acta* for 1696, p. 269, concerning the finding of the brachistochrone curve, i.e., the curve along which a body passes from a given point to another in least time; and the construction of a synchronous curve, i.e., of a light wave".) Acta Erud. Leipzig. 1697. *Opera omnia*. Vol. I. pp. 187–193. (English translation in: D. J. Struik. *A source book in mathematics, 1200–1800*. pp. 391–396.)

[9] Euler, L. *Mechanica sive motus scientia analytice exposita*. 1736. *Opera omnia*. Series II, Vols. 1 and 2.

[10] ——. *Methodus inveniendi lineas curvas maximi minimive proprietate gaudentes, sive solutio problematis isoperimetrici lattisimo sensu accepti.* (*A method for finding curves with maximal or minimal properties, or the solution of the isoperimetric problem understood in its widest sense.*) Bousquet, Lausannae et Genevae 1744. *Opera omnia.* Ser. I. Vol. 24.

[11] Maupertuis, P. "Accord de différentes loix de la nature qui avaient jusqu'ici parues incompatibles". ("Agreement between different laws of nature that had hitherto seemed incompatible".) Mém. Acad. Sci. 1744. pp. 417–426.

[12] Euler, L. "On the determination of the motion of projected bodies in an unresisting medium by the method of maxima and minima".

[13] ——. "Réflexions sur quelques loix générales de la nature qui s'observent dans les effets des forces quelconques". ("Reflections on certain general laws of nature observed in the effects of arbitrary forces".) 1750. *Opera omnia.* Series II, Vol. 5. pp. 38–63.

[14] Maupertuis, P. "Les loix du mouvememt et du repos, déduites d'un principe metaphysique". ("The laws of motion and rest, deduced from a metaphysical principle".) 1746.

[15] Voltaire, F.-M. "Histoire du Docteur Akakia et du natif de St. Malo"[35]. 1753. Approx. 44 pages.

[16] d'Alembert, J. le Rond. "Cosmologie". In: Diderot's *Encyclopédie.* Paris. 1772.

[17] Lagrange, J.-L. "Essai d'une nouvelle méthode pour déterminer les maxima et les minima des formules intégrales indéfinies". ("Essay on a new method for determining the maxima and minima of indefinite integrals".) *Oeuvres.* Paris, 1867. Vol. 1. pp. 333–362.

[18] ——. "Application de la méthode exposée dans le mémoire précedent à la solution de différents problèmes de dynamique". ("Application of the method expounded in the preceding memoir to the solution of various problems of dynamics".) *Oeuvres.* Paris, 1867. Vol. 1. pp. 363–468.

[19] Polak, L.S. *The variational principles of mechanics, their development and application to physics.* M.: State publisher of physics and math. (Fizmatgiz). 1960. 599 pages.

[20] Hamilton, W. R. "On a general method in dynamics by which the study of the motions of all free systems of attracting or repelling points is reduced to the search and differentiation of one central relation, or characteristic function". Phil. Trans. Roy. Soc. 1834. Vol. 124. pp. 247–308. *The mathematical papers of Sir William Rowan Hamilton.* Cambridge, 1931–1940. A. W. Conway and J. L. Synge, eds. Vol. II. pp. 103–167.

[21] d'Alembert, Jean le Rond. *Traité de dynamique.* Paris. 1743.

[22] Euler, L. "Harmonie entre les principes généraux de repos et de mouvement de M. de Maupertuis". 1753. *Opera omnia.* Series II, Vol. 5. pp. 152–176.

[23] Jacobi, C. G. J. *Vorlesungen über Dynamik.* Supplementband der ges. Werke. G. Reimer. Berlin. 1884.

[24] ——. "Note sur l'intégration des équations différentielles de la dynamique". *Oeuvres complètes.* Vol. 4. pp. 129–136.

[25] Lanczos, C. *The variational principles of mechanics.* University of Toronto Press. Toronto. 1949. Reprinted by Dover Publ. 1970.

[26] Hamilton, W. R. "Second essay on a general method in dynamics". Phil. Trans. Roy. Soc. 1835. Vol. 125. pp. 95–144. *The mathematical papers of Sir William Rowan Hamilton.* Cambridge, 1931–1940. Vol. II. pp. 162–216.

---

[35] Maupertuis was a native of St. Malo.—*Trans.*

[27] Ostrogradskiĭ, M. V. "Differential equations of the isoperimetric problem". *Collected works.* Kiev. 1961.

[28] Hamilton, W. R. "The theory of systems of rays". Trans. Roy. Irish Acad. 1828. Vol. 15. pp. 68–174; 1830. Vol. 16. pp. 1–62; 1837. Vol. 17. pp. 1–144.

[29] Whittaker, E. T. *A treatise on the analytical dynamics of particles and rigid bodies.* Cambridge Univ. Press, Cambridge. 1964.

[30] Chetaev, N. G. "On an extension of the optical-mechanical analogy". In: *Stability of motion. Works in analytical mechanics.* M.: Publ. Acad. sci USSR (Izd-vo AN SSSR). 1962. pp. 404–406.

[31] ——. "On a problem of Cauchy". *Works in mechanics.* pp. 343–346. (See reference [30].)

[32] ——. "On the optical-mechanical analogy". *Works in mechanics.* pp. 393–403. (See reference [30].)

[33] Hertz, H. *Die Prinzipien der Mechanic. Gesammelte Werke.* Vol. 3. Leipzig. 1894.

[34] Hölder, O. "Über die Prinzipien von Hamilton und Maupertuis". In: *Nachrichten von der Königlichen Gesellschaft der Wissenschaften zu Göttingen. Mathematisch-Physikalische Klasse, aus dem Jahre 1896.* Göttingen. Dieterichsche Verlagsbuchhandlung. pp. 122–157.

[35] Voronets, P. V. "On the equations of motion for nonholonomic systems". Math. symposium (Mat. sb.). 1901. Vol. 22. No. 4. pp. 659–680.

[36] Suslov, G. K. "On a variant of d'Alembert's principle". Math. symposium (Mat. sb.). 1901. Vol. 22. No. 4. pp. 681–697.

[37] Chaplygin, S. A. "Towards a theory of motion of nonholonomic systems. The theorem on reducing multipliers". *Collected works.* M.; L.: Govt. Tech. Publ. (Gostekhizdat). 1948. Vol. 1. pp. 15–25.

[38] Kerner, M. "Le principe de Hamilton et l'holonomisme". Pràce mat.-fiz. (Warszawa). 1931. Vol. 38. pp. 1–21.

[39] Capon, R. S. "Hamilton's principle in relation to nonholonomic mechanical systems". Quart. J. Mech. and Appl. Math. 1952. Vol. 5. Part 4. pp. 472–480.

[40] Jeffreys, H. "What is Hamilton's principle?" Quart. J. Mech. and Appl. Math. 1954. Vol. 7. Part 3. pp. 335–337.

[41] Pars, L. A. "Variational principles in dynamics". Quart. J. Mech. and Appl. Math. 1954. Vol. 7. Part 3. pp. 338–351.

[42] Novoselov, V. S. *Variational methods in mechanics.* L.: Published by Leningrad State Univ. (Izd-vo LGU). 1966. 66 pages.

[43] Chetaev, N. G. "On Gauss' principle". In: *Works in mechanics.* pp. 323–326. (See reference [30].)

[44] Neĭmark, Yu. I., Fufaev, N. A. *The dynamics of nonholonomic systems.* M.: Nauka. 1967. 520 pages.

[45] Rumyantsev, V. V. "On Hamilton's principle for nonholonomic systems". Appl. math. and mechanics (Prikl. matematika i mekhanika). 1978. Vol. 42. No. 3. pp. 387–399.

[46] ——. "On the principles of Lagrange and Jacobi for nonholonomic systems". Appl. math. and mechanics (Prikl. matematika i mekhanika). 1979. Vol. 43. No. 4. pp. 583–590.

[47] Sumbatov, A. S. "On Hamilton's principle for nonholonomic systems". Bulletin Moscow State Univ (Vestnik MGU). Math. and Mechanics. 1970. No. 1. pp. 98–101.

[48] Rumjantsev, V. V., Sumbatov, A. S. "On the problem of a generalization of the Hamilton-Jacobi method for nonholonomic systems". Ztschr. angew. Math. und Mech. 1978. Vol. 58. Issue 11. pp. 477–481.

[49] Sedov, L. I. "Mathematical methods for constructing new models of continuous media". Results of math. sciences (Uspekhi mat. nauk). 1965. Vol. 20. No. 5. pp. 121–180.

[50] Berdichevskiĭ, V. L. *Variational methods of the mechanics of a continuous medium*. M.: Nauka. 1983. 447 pages.

[51] Gauss, C. F. "On a general principle of mechanics". *Werke*. Teubner. Leipzig. 1863–1929.

[52] Chetaev, N. G. "A variant of a principle of Gauss". *Works in mechanics*. pp. 327–328. (See reference [30].)

[53] ——. "From a notebook". *Works in mechanics*. p. 499. (See reference [30].)

[54] Rumyantsev, V. V. "On Chetaev's principle". Announcements acad. sci. USSR (Dokl. AN SSSR). 1973. Vol. 210. No. 4. pp. 787–790.

[55] ——. "On certain variational principles in the mechanics of continuous media". Appl. math. and mechanics (Prikl. matematika i mekhanika). 1973. Vol. 37. No. 6. pp. 963–973.

[56] Pontryagin, L. S., Boltyanskiĭ, V. G., Gamkrelidze, R. V., Mishchenko, E. F. *Mathematical theory of optimal processes*. M.: Physics and mathematics State Publ. (Fizmatgiz). 1961. 392 pages.

[57] Rumyantsev, V. V. "On certain questions of analytical mechanics". In: *Problems of analytical mechanics and control of motion*. M.: All-union Centre Acad. Sci. USSR (VC AN SSSR). 1985. pp. 20–36.

Leonhard Euler and the Mechanics of Elastic Systems[1]

N. V. Banichuk and A. Yu. Ishlinskiĭ

Although in histories of science the resistance of materials is usually mentioned in the first place in connection with its uses by the ancient Egyptians and Greeks, one can begin to trace the development of the basic ideas in this field only with the appearance of the works of Leonardo da Vinci—and in mathematical form only much later in the work of Hooke, Jakob and Daniel Bernoulli, and Leonhard Euler.

It is known that around the beginning of the 18th century the study of elasticity had available to it only Hooke's law and certain approaches to the measurement of the bending of beams. However with the advent of Leonhard Euler the field underwent a developmental surge. His contributions to the field were fundamental: he posed and solved the main problems of elasticity theory, resistance of materials, and structural mechanics. His work in the mechanics of elastic bodies is devoted to investigations of their flexion, stability, and vibrations. Below we devote a separate section to each of these three directions of research, indicating also the further developments of Euler's ideas in the works of succeeding researchers, with emphasis on the work of Russian and Soviet scientists.

1. Bending

Euler's contribution to the theory of the bending of elastic rods was fundamental. He expounded his results in the theory of bending, including the study of the shapes assumed by loaded elastic rods, in an article on the elastic curve formed by an elastic strip loaded with arbitrary forces at several of its points [1] and in his treatise *Methodus inveniendi lineas curvas maximi minimive proprietate gaudentes,...* (*A method for finding curves with maximal or minimal properties,...*) [2].

Prior to Euler's article [1] of 1732, Jakob Bernoulli had established the proportionality of the curvature of an [initially straight] elastic wire or filament at each of its points to the

[1]The translator would like to thank Dr. V. G. Hart of the University of Queensland for his expert help with the present essay.

bending moment at that point. In the paper [1] Euler considered the bending of a straight elastic wire (or "elastica") at one end of which a force P is applied[2], and over which a load F is distributed. Euler obtained the following general equation, synthesizing all of the contemporary theory on the topic[3]:

$$-P_y x + P_x y - \int_0^x Y\,dx + \int_0^y X\,dy = -\frac{Ek^2}{R},$$

where

$$Y = \int_0^s F_y\,ds, \quad X = \int_0^s F_x\,ds.$$

This equation equates the moments of the forces acting on the wire to the elastic moment. Here the system of coordinates is taken to have origin coinciding with the end of the wire to which the force P (with components P_x, P_y) is applied, Ek^2 is the "elastic rigidity" of the wire (in Euler's notation), R is the radius of curvature of the bent wire, and ds an element of its length.

In his treatise on the calculus of variations [2] Euler considers planar bending of thin elastic rods (i.e., elastic filaments) under forces applied to their ends. This problem was brought to his attention by Daniel Bernoulli, who suggested applying the apparatus of the variational calculus that he, Euler, had invented, to the investigation of the behavior of elastic rods, taking as basic to the analysis the [minimization of the] quantity

$$\int_A^B \frac{ds}{R^2},$$

proportional to the potential energy of the elastic deformation in question, and termed by Daniel Bernoulli the "potential force". Taking up Bernoulli's suggestion, Euler formulated the problem as follows: "Among all curves of prescribed length that not only pass through points A and B, but are also tangential at those points to prescribed straight lines [through A and B], determine that one for which the expression $\int \frac{ds}{R^2}$ is least" [2, p. 451].

A modern exposition of the solution of this variational problem would go as follows. Denoting by $y = y(x)$ the equation of the elastic curve joining A to B, and by L its (prescribed) length, we may formulate the problem as follows[4]:

$$\int_A^B \frac{ds}{R^2} = \int_{x_A}^{x_B} f\,dx \to \min, \quad \int_A^B ds = \int_{x_A}^{x_B} f_2\,dx = L,$$

$$f = f_1 + \alpha f_2, \quad f_1 = (y'')^2[1+(y')^2]^{-3}, \quad f_2 = [1+(y')^2]^{\frac{1}{2}},$$

where α is a Lagrange multiplier. In addition there are boundary conditions on the function $y(x)$ corresponding to the prescription of the points A and B and the slopes of the elastic curve at those points (Figure 1). The fact that the function f does not involve y explicitly

---

[2]Presumably this end can move and the other end is fixed.—*Trans.*

[3]In modern notation the expression Ek^2 in the right-hand side of the following equation would appear in the form EI, where E is Young's modulus (stress/strain) for the material of the wire or rod, and I is the "second moment of area of cross-section around the neutral axis" of the rod. Note that a comparison of dimensions of the two sides of the equation shows that the quantity Ek^2 must have dimension $[Ek^2] = ML^3T^{-2}$.—*Trans.*

[4]For a curve given in terms of rectangular Cartesian coordinates by $y = y(x)$, an element of Euclidean length is given by $ds^2 = dx^2 + dy^2 = (1+(y')^2)dx^2$, and the formula for the curvature ($= 1/R$) is $|y''|/(1+(y')^2)^{\frac{3}{2}}$.—*Trans.*

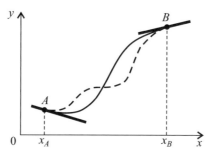

Figure 1.

allows one to infer the following first integral from Euler's equation[5]:

$$-\frac{\partial f}{\partial y'} + \frac{d}{dx}\frac{\partial f}{\partial y''} = C,$$

where C is a constant of integration. Next, taking into account that f involves only y' and y'' explicitly, so that $df = (\partial f/\partial y')dy' + (\partial f/\partial y'')dy''$, we may rewrite the above first integral in the form $Cdy' = -df + d(y''\partial f/\partial y'')$. A further integration then gives

$$Cy' + D = f_1 - \alpha f_2,$$

where D is another constant of integration. Solving this for the second derivative of the function sought, we obtain

$$y'' = (1 + (y')^2)^{\frac{3}{2}}[\alpha(1 + (y')^2)^{\frac{1}{2}} + Cy' + D]^{\frac{1}{2}}.$$

Seeking a parametric representation of the curve formed by the elastic wire, Euler reformulates this as follows ($p = y'$):

$$dx = \frac{dp}{(1 + p^2)^{\frac{3}{2}}[\alpha(1 + p^2)^{\frac{1}{2}} + Cp + D]^{\frac{1}{2}}},$$

$$dy = \frac{p\,dp}{(1 + p^2)^{\frac{3}{2}}[\alpha(1 + p^2)^{\frac{1}{2}} + Cp + D]^{\frac{1}{2}}}.$$

Not content with this result, Euler strives to push the integration further. Although he becomes convinced that neither of the preceding two expressions is integrable, he discovers an interesting fact, namely that the linear combination $Cdx - Ddy$ is the total differential of a certain expression, which allows him to carry out a further integration, yielding

$$Cx - Dy + E = \frac{2[\alpha(1 + (y')^2)^{\frac{1}{2}} + Cy' + D]^{\frac{1}{2}}}{[1 + (y')^2]^{\frac{1}{4}}},$$

---

[5]For a function $F(x, y, y', y'')$ Euler's (or the "Euler-Lagrange") equation for functions $y(x)$ for which $\delta \int_A^B F(x, y, y', y'')dx = 0$ is (compare (3) in the preceding essay)

$$\frac{\partial F}{\partial y} - \frac{d}{dx}\frac{\partial F}{\partial y'} + \frac{d^2}{dx^2}\frac{\partial F}{\partial y''} = 0,$$

the simplest case of a higher-order Euler equation. Since in the present situation $F(= f)$ does not involve y explicitly, the first term vanishes, and one can anti-differentiate.—*Trans.*

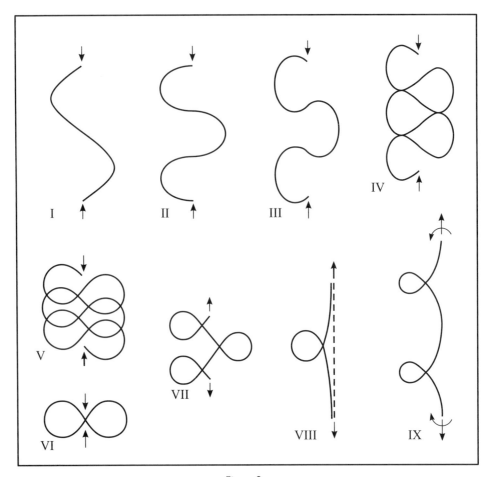

Figure 2.

where E is a constant of integration[6]. Noting that without loss of generality the constants D and E may be set equal to zero, and applying certain algebraic transformations, Euler arrives at the equation

$$y' = \frac{C^2 x^2 - 4\alpha}{\sqrt{16C^2 - (C^2 x^2 - 4\alpha)^2}}.$$

Finally, by changing x by an additive constant, and renaming the constants appearing, he obtains an expression for y' of the form

$$y' = \frac{a + bx + cx^2}{\sqrt{a^4 - (a + bx + cx^2)^2}},$$

where a, b, c are unknown constants[7].

In his treatise Euler also derives the fundamental expression giving the shape of the loaded elastic wire using a different method, based on a direct derivation of the equilibrium equation in terms of displacements.

---

[6]The original has δ rather than E.—*Trans.*

[7]Denoted by α, β, γ in the original.—*Trans.*

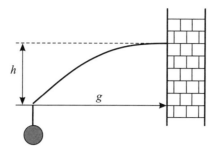

Figure 3.

Next Euler deduces a complete classification of the possible shapes of the loaded elastic wire, using, as he says, "the same method as is usually employed to enumerate the algebraic curves of a given degree" [2, p. 467]. It turns out that the bending of a thin elastic rod by means of loads applied at its ends has nine distinct equilibrium forms (Figure 2). Euler subjected each of these nine shapes to a detailed analytic examination, which, as we shall see below (§2), yielded an important finding.

The behavior of the elastic wire depends in an essential manner on the size of its bending rigidity, which Euler denoted by Ek^2. In order to determine this quantity Euler proposed the following experiment involving the bending of the wire by means of a load P applied to one end, as shown in Figure 3, allowing the calculation of its rigidity by means of the exact formula[8]

$$Ek^2 = \frac{Pg(2g - 3h)}{6h},$$

where h is the vertical displacement of the free end of the loaded wire, i.e., of the cantilever, and g its horizontal distance from the wall in which the other end is fixed.

It is not difficult to see that in the case of small deflections (vertical displacements) h this formula approaches the standard formula for the resistance of materials. In the case of a beam, in discussing the dependence of the bending rigidity on the properties of the material of the beam and the dimensions of a cross-section, Euler is inaccurate when he asserts that the quantity Ek^2 is proportional to the square of the thickness. This discussion, together with a detailed analysis of the above formula, is contained in the section of the treatise headed "The determination of absolute elasticity by means of experiments". In the treatise [2] Euler also derives results concerning the bending of inhomogeneous elastic rods and elastic rods whose unstressed form is not straight. For the latter he proposes his well known formula

$$M = Ek^2 \left(\frac{1}{R} - \frac{1}{R_0} \right),$$

where M is the moment and R_0 the radius of curvature of the rod in its unstressed state.

Note that Euler obtained this formula, which generalizes the basic formula for straight rods[9], during the early Basel period of his creativity (i.e., prior to 1727) in connection with his investigations of the vibrations of circular rings (see [3, 4]). Unfortunately Euler's youthful work [3] appeared in print only 60 years after his death.

---

[8] A comparison of dimensions of the two sides of the following equation shows that this formula must be incorrect as it stands; perhaps the g outside the parentheses should be g^2, but in any case the question remains as to the equation's correctness.—*Trans.*

[9] Due to Jakob Bernoulli; see earlier.—*Trans.*

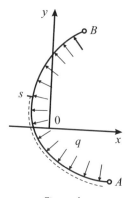

Figure 4.

In his treatise on the variational calculus [2] Euler also investigates the bending of initially straight elastic rods by loads applied at various points. Having derived the equilibrium equations, he goes on to consider specific problems and to discuss the case where the bending rigidity is vanishingly small and the elastic rod has become an essentially perfectly flexible filament. For a filament in equilibrium under the action of forces normal to its curve (Figure 4), Euler establishes the general property

$$\frac{R\,dq}{ds} = \text{const},$$

where dq is the force applied to an element of filament, for all "sail-shaped and scarf-shaped curves, and all those acted on in this way" [2, p. 518]. In speaking of Euler's work on the statics of elastic bending, one must mention also his solution of the problem of the bending of an elastic rod under its own weight. By means of ingenious transformations of this nonlinear problem, he was able to find an analytic solution giving the equilibrium form of a massive rod under transverse bending.

The above does not exhaust Euler's writings on elastic bending. There are others of his results that we might have described such as, for instance, those concerning the inverse problem of determining the load causing a given deformation of an elastic rod. The importance of Euler's contribution to the bending theory of elastic rods is generally acknowledged and is in fact usually called the Bernoulli-Euler theory.

Euler's work on the shapes assumed by loaded elastic rods was developed further in a memoir by Kirchhoff [5] published in 1859, in which he derived differential equations for the elastic curve[10] of a rod loaded at its ends. In the case where the axis of the rod in its unstressed state is straight, he also brought to light a mathematical analogy between the bending problem he was considering and the problem of the rotation of a massive rigid body about a fixed point. In this situation the integration of the [equilibrium] equations for the elastic rod is reduced to the integration of the equations describing the rotation of a [certain] massive rigid body about a fixed point, from which the shape of the elastic curve is calculated by integration.

Much later E. L. Nikolaï [6] carried out fundamental investigations of the elastic curve in the situation of twofold curvature, finding, in particular, the various shapes that may be

---

[10]"The curve assumed by the longitudinal axis of an originally straight elastic strip or bar bent within its elastic limits by any system of forces". *Webster's Third International Dictionary.* Springfield, 1964.—*Trans.*

assumed by the elastic curve of an isotropic rod having equal principal bending rigidities[11]. Note that in the article [6] one may also find references to other scientists' research on the shapes of elastic rods under spatial bending and on the analogy between the problem of the elastic deformation of a rod and that of a rigid body rotating about a fixed point.

More recent results on the problem of the statics of thin elastic rods may be found in the works of E. P. Popov [7]. Popov has found, in particular, exact solutions of the problem of substantial bending (i.e., for which the deflections are large) in the case where the forces acting depend on the shapes taken on by the rod during the bending process—for instance tracking forces. At the present time[12] Popov's results have found application in estimates relating to lamellate springs and other machine parts subject to considerable bending.

2. Stability

The science of the stability of elastic bodies was born in the works of Euler. Euler discovered the phenomenon theoretically, developed the concept of elastic stability, and solved the first problems of stability analysis—in both their linear and nonlinear formulations. The linearized formulation of the problem of elastic stability as a boundary problem for the eigenvalues remains relevant even today. In the majority of modern works concerned with the calculation of elastic stability, the bifurcational method proposed by Euler is still used. Without doubt, on the mathematical side the works of Euler on elastic stability were remarkable for their novelty and have influenced crucially the creation of the spectral theory of boundary problems, i.e., the theory of the eigenvalue spectra of differential equations. However it is not our aim here to dwell on the development of general mathematical ideas but rather to consider the investigations of elastic stability carried out by Euler and his successors.

The problem of elastic stability occupied Euler over a lengthy period of his scientific activity. The first work in which he considered questions concerning that concept appeared in 1744, when he was 37, and the last in 1778, when he was 71.

We begin with Euler's first theoretical investigation of elastic stability. In his treatise [2] on the calculus of variations, along with the derivation of equations for the shapes of elastic rods under bending, Euler carries out an analysis and classification of the possible such equilibrium shapes by means of an exhaustive procedure eliciting all of them. However the main outcome of this laborious process—which could hardly have been foreseen and which came as a surprise to Euler himself—lay in another direction. In isolating and investigating elastic curves of type I in Figure 2, he considered small deformations of a "strip" of length $2f$ and absolute rigidity Ek^2, and discovered "that the force required to procure in the strip this infinitely small deformation is finite, namely $(Ek^2/f^2)(\pi^2/4)$. This means that if the ends A and B are joined by a thread AB, then this thread will be tensed by a force of magnitude $\pi^2 Ek^2/4f^2$" [2, p. 476].

This effect brought to light by Euler—that of the finiteness of the compressive force needed to secure an infinitesimal further deformation—is correctly interpreted by him, and he goes on in the same treatise to apply his discovery to the problem of the stability of

---

[11] This is a literal translation. However modest changes in the original would yield instead: "... the elastic curves of isotropic rods of different principal bending rigidities".—*Trans.*

[12] Around 1987.

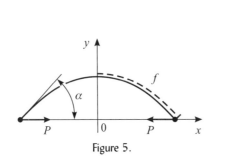

Figure 5.

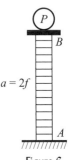

Figure 6.

columns. The next section of the treatise after that devoted to the classification of elastic curves is headed "On the strength of columns". There he considers a column of height a fixed at its base A and bearing a load P (Figure 6). The absolute rigidity of the column is denoted by Ek^2. Euler's main assertion is then as follows: "If the load P does not exceed $E\pi^2k^2/a^2$, then there is no need to fear a definite bending; but on the other hand if the load is greater, then the column will not withstand bending" [2, p. 492].

In the same section he makes the following inference of fundamental practical importance: "If the elasticity of a column, together with its thickness, remain the same, then the load that it is able to bear without danger is inversely proportional to the square of its height; hence a column twice as high can bear a load only a quarter as much" [*loc. cit.*].

The consummate ease with which Euler achieves his analysis and explanation of one of the most substantial concepts of mechanics is nothing less than astounding.

Note that Euler's exposition in the section in question is irreproachable, except for the minor caveat of the thrice-repeated inaccuracy in his calculation of the value of the critical load of the column; the value he gives is four times the actual value $\pi^2 Ek^2/4a^2$.[13]

In his succeeding works on elastic stability Euler encountered difficulties and his solutions were not always satisfactory. Thus he was only able to resolve the "paradoxes" associated with the stability of columns loaded by their own weight in later works on this topic appearing in 1778. However, as researchers into Euler's scientific development have often noted, the way in which he transcends each error to eventually arrive at the correct result is highly instructive.

Let us now analyze the linearized formulation of the problem of elastic stability as it appears in Euler's article "On the resistance of columns" [8] (see also [4, 9]) and certain solutions obtained by him within the framework of this formulation. Euler's approach is based on the consideration of equilibrium shapes of a slightly deformed compressed rod, in which case the curvature is approximated to sufficient accuracy by the absolute value of the second derivative of the deflection function $y(x)$.[14] (Euler had used this approximation earlier in his treatise [2] in connection with his investigation of small vibrations of a rod.) The linearized equation has the form

$$Ek^2 y'' + Py = 0.$$

---

[13]It might be surmised that Euler had in mind a column with top attached by a hinge; however his drawing shows no such attachment, and there is no mention of it in the text. Note that in the earlier problem represented by Figure 5, the boundary conditions are $y = 0$ at the two ends of the elastica, while in the present problem of a loaded column, they are $y(0) = y'(0) = 0$.

[14]Presumably the function with graph the elastic curve.—*Trans.*

The integration of this equation, taking into account the boundary conditions $y(0) = y(l) = 0$ at the supported ends of the rod, yields the following non-trivial distribution of deflections:

$$y = C \sin\left(\frac{\pi x}{l}\right),$$

for values of P that are multiples of the quantity $\pi^2 E k^2 / l^2$.[15] The least load for which slightly curved equilibrium forms appear is regarded by Euler as the critical force for loss of stability. The amplitude C of the distribution of deflections of the rod remains undetermined in this examination of the question of stability within the context of Euler's approach via linearization.

Euler carries through an interesting analysis—again within the framework of the linearized problem—in the situation of inhomogeneous columns with ends fixed by hinges. Here the bending rigidity varies along the column, i.e., $E k^2 = \varphi(x)$, where $\varphi(x)$ is a prescribed positive function. Euler applies the change of variables

$$y = e^{\int u\, dx},$$

to the equation for transverse bending, after which that equation, although remaining non-linear, becomes first-order:

$$u' + u^2 + \frac{P}{\varphi} = 0.$$

Euler considered functions for the bending rigidity of the form $\varphi = K_0(\alpha + \beta x/l)^\lambda$, and in the case $\lambda = 4$ was able to find the simple exact solution

$$y = A(\alpha l + \beta x) \sin\left(\frac{\sqrt{P_{cr}}\,lx}{\sqrt{K_0}\alpha(\alpha l + \beta x)}\right), \quad P_{cr} = \pi^2 \frac{\alpha^2(\alpha + \beta)^2 K_0}{l^2}.$$

Later in the same work Euler applies this result successfully to the investigation of the elastic stability of conical columns. It should be mentioned that in the article [8] Euler solves the more complicated problem of the stability of an inhomogeneous elastic rod with arbitrary degree of inhomogeneity λ.

Euler devoted considerable effort to investigating the stability of massive columns [8, 10–12], i.e., the transverse bending of such a column under its own weight. His starting point in these investigations is a certain third-order differential equation for equilibrium; however here he makes the initial mistake of omitting the transverse reactions at the points of support of the column, as a result of which he arrives at the incorrect conclusion that for a column acted on by distributed forces there is no loss of stability. This—at first glance simple—error was corrected by Euler in his next paper [12] on this topic. Euler's investigations into stability loss in columns under the action of their own weight are in fact highly instructive. Much later A. N. Dinnik [14] introduced greater precision into Euler's work on massive columns, giving in particular the correct value of the critical parameter measuring loss of stability. An extensive analysis of Euler's works on the stability of massive columns has been carried out by E. L. Nikolaï [9] and a popular exposition may be found in the book [13].

We note here a remarkable result discovered in the 1950s [15] in the course of using Eulerian methods to solve the problem of the elastic stability of a rectangular elastic plate

---

[15]Compare the situation of Figure 5, with $l = 2f$.—*Trans.*

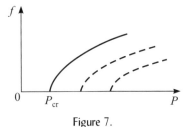

Figure 7.

undergoing compression on two secured (pinched) sides (of length l), the other pair of opposite sides being free to move. In [15] a precise expression is found for the critical compressive force P_{cr} at which stability loss occurs. It might have been supposed that as $l \to \infty$, i.e., for sufficiently long plates, the size of the critical load would approach arbitrarily closely the known limiting load P_{cr}^* at which the plate begins to buckle. However the result obtained in [15] disagrees with this conjecture: in fact $P_{cr} \to 0.9962 P_{cr}^*$.

Study of the hypercritical behavior of the elastic parameters of a loaded elastic structure leads to important though difficult problems. On the one hand such investigations provide a basis for the linearized versions of problems and on the other lead to a widening of the concept of stability. The very first results along these lines are due to Euler [2]; he derived a formula linking the maximal deflection f of a buckled rod of length l to the value of its absolute rigidity Ek^2, under the action of a compressive load P and appropriate conditions relating to the support of the ends of the rod:

$$
l = \pi \sqrt{\frac{Ek^2}{P}} \left\{ 1 + \left(\frac{1}{2}\right)^2 \frac{f^2 P}{4Ek^2} + \left(\frac{1 \cdot 3}{2 \cdot 4}\right)^2 \left(\frac{f^2 P}{4Ek^2}\right)^2 + \right.
$$
$$
\left. \left(\frac{1 \cdot 3 \cdot 5}{2 \cdot 4 \cdot 6}\right)^2 \left(\frac{f^2 P}{4Ek^2}\right)^3 + \cdots \right\}.
$$

Typical behavior of the maximal deflection for $P > P_{cr}$ is shown in Figure 7. For compressive loads P exceeding the critical load only by a negligible amount, it is more convenient to use the approximate formulae of R. Mises [16] (the formula of first approximation) and E. L. Nilolaï [9] (the formula of second approximation). As noted in [9], an analogous formula proposed by S. P. Timoshenko [17] turns out to be of insufficient accuracy.

The problem of transverse bending of an elastic rod in its nonlinear formulation is the subject of works by J.-L. Lagrange, F. S. Yasinskiĭ, A. E. Lyav, and A. N. Krylov [18–21]. A large proportion of this research is based on the representation of solutions by means of elliptic integrals.

Lagrange [18] investigates the behavior of a supported column of length a in the case when the value of the compressive load is not equal to any of the quantities $i^2 \pi^2 Ek^2/a^2$ $(i = 1, 2, \ldots)$. Basing his derivation on the precise nonlinear bending equation, he obtains an equation for the arclength of the elastic curve of the column in the form

$$
s = \int_0^\varphi \frac{d\varphi}{\sqrt{\frac{P}{Ek^2} - \left(\frac{P^2 f^2}{4Ek^2}\right) \cos^2 \varphi}},
$$

where $s = a$ for $\varphi = m\pi$, m the number of times the elastic curve crosses the axis

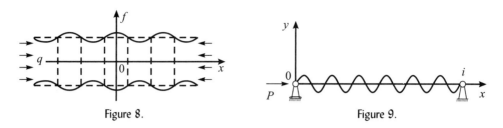

Figure 8.   Figure 9.

through the points of attachment of the column, i.e., the number of half-waves. Lagrange goes on to deduce a formula, which, after the introduction of corrections noted in the book of Todhunter and Pearson [22], has the form

$$a = i\pi\sqrt{\frac{Ek^2}{P}}\left\{1 + \frac{Pf^2}{4(4Ek^2)} + \frac{9P^2 f^4}{4\cdot 16(16(Ek^2)^2)} + \frac{9\cdot 25P^3 f^6}{4\cdot 16\cdot 36(64(Ek^2)^3)} + \cdots\right\},$$

from which one can calculate f, given a and P.

In this article Lagrange refers only to Euler's article of 1757 [8] on the stability of columns, leading one to suppose that he was unaware of Euler's earlier work on elastic stability in the nonlinear version, published in 1744 [2]. Acquaintance with another of Lagrange's memoirs [23] where, working with the exact formulation, he investigates the bending problem for an elastic rod acted on by transverse and compressive forces, lends weight to this surmise. In the case where only compressive forces are acting on the rod, Lagrange arrives at Euler's result that the achievement of an arbitrarily small deformation of the rod requires the application of a finite load.

As already noted, research into the longitudinal bending of columns in the nonlinear version of the theory, is for the most part carried out using elliptic integrals, in terms of which, in particular, the maximal deflection and the amount of settling of a column can be expressed [21]. Also of interest are certain simpler approximate expressions for the deflection function of a column and other characteristics of its behavior subsequent to loss of stability, found by A. Yu. Ishlinskiĭ:

$$y = 2\frac{m}{q}\sin\left[\left(1 - \frac{3}{4}m^2\right)qx + \frac{9}{4}m^2 \sin 2qx\right] + O(m^4),$$

$$f = \frac{2m}{q}, \quad m = \sin\left(\frac{\alpha}{2}\right), \quad q = \sqrt{\frac{P}{Ek^2}};$$

here α denotes the angle between the x-axis and the tangent line to the elastic curve of the column at an end-point, and f, as before, the maximal deflection.

In the article [5] G. Kirchhoff established a kinetic analogy with the theory—in its exact nonlinear form—of longitudinal bending, which has played a leading role in later investigations in the latter area. In particular, this analogy has allowed the application of results on the integration of the equations of motion of a massive rigid body rotating about a fixed point to problems of elastic stability.

The dissertation of L. Prandtl [24] is devoted to an interesting problem concerning the bending stability of a beam of small rectangular cross-section. He considers the bending of such a beam in the plane of maximal rigidity, and the occurrence of protrusion (buckling) outside this plane under sufficiently large loads. His theoretical analysis of stability in this context went hand-in-hand with experimentation, and his work has been recognized as of great practical importance, stimulating a great deal of research on transverse stability of beams and curved bars; we note in this regard the articles [25, 26] of S. P. Timoshenko.

Since the time of Euler problems concerning elastic stability had always been investigated in the context of models of rods, plates, or casings. The question of stability as it applies to other elastic models long remained unresearched, a situation eventually drawing attention to itself. In this connection a non-classical approach to the solution of problems concerning elastic stability was proposed independently by L. S. Leĭbenzon [27] and A. Yu. Ishlinskiĭ [28]. Their approach is based on the idea of using the linearized equilibrium equations in the interior of the region occupied by the material in question, and the nonlinear equations on its boundary. The resulting bifurcational problems are then investigated using classical methods. This approach has yielded results agreeing well with the known facts; thus, for example, the solution via this approach of the stability problem for a straight-edged elastic strip (Figure 8) as its width tends to infinity yielded the Eulerian expression for the compressive load [28]. The article [29] is devoted to a grounding of this approach.

We note in passing also the research of V. Z. Vlasov [30] and I. F. Obraztsov on stability in the case of thin-walled hollow rods and casings.

If a compressive load is applied suddenly (impulsively) to the ends of a rod, then the shape at which stability loss occurs differs essentially from that obtaining in situations of static loading. The main difference resides in the circumstance that under dynamic loading the form of the buckling develops a "high frequency" aspect (Figure 9), or, in other words, the rod loses stability in a harmonic range higher than in the statical Eulerian problem. Problems concerning the stability of elastic rods and casings under impulsive loading were first posed and investigated by M. A. Lavrent′ev and A. Yu. Ishlinskiĭ [32]. Their work was motivated by results of experiments examining the effects of explosions on rods and casings, where the form of stability loss could not be explained in terms of the statical analysis of stability. Instead, the equation of small deformations of the rod

$$\rho S \frac{\partial^2 y}{\partial t^2} + E k^2 \frac{\partial^4 y}{\partial x^4} + P \frac{\partial^2 y}{\partial x^2} = f(x)$$

is introduced (where S denotes the area of cross-section of the rod and f the function giving its initial deformation), and a dynamical method is used. One seeks a solution of the boundary problem with boundary conditions $y = \partial^2 y / \partial x^2 = 0$ for $x = 0$ and $x = l$ (corresponding to the situation where the ends of the rod are fixed), in the form of a Fourier sine series

$$y = \sum_{i=1}^{\infty} Y_i(t) \sin \frac{i \pi x}{l}.$$

If the function f giving the initial deformation is expanded in a sine series

$$f = \sum F_i \sin(i \pi x / l),$$

then one obtains the following equations for the time-dependent quantities $Y_i(t)$:

$$\rho S \frac{d^2 Y_i}{dt^2} + \frac{\pi^4 E k^2}{l^4} i (i^2 - n) Y_i = F_i,$$

where

$$n = \frac{P}{P_{cr}}, \quad P_{cr} = \frac{\pi^2 E k^2}{l^2} \quad (i = 1, 2, \ldots).$$

From these equations it follows that for $i^2 > n$, the dependence of the amplitude Y_i on the time t is sinusoidal. On the other hand for $i^2 < n$, the dependence is aperiodic:

$$Y_i(T) = A_i \sinh \lambda_i t + \frac{F_i}{\rho S \lambda_i^2}(1 - \cosh \lambda_i t), \quad \text{where } \lambda_i^2 = \frac{\pi^4 E k^2}{\rho S l^4} i(n - i^2).$$

From this we see that the rate of increase of the amplitude Y_i with respect to time depends crucially on the the quantity λ_i; the amplitudes Y_i of those harmonics with largest λ_i will increase fastest. From the definition of λ_i in the second of these two equations we see that in turn λ_i is greatest when $3i^2 = n$.[16] Hence the greatest instability will occur when the elastic curve of the rod is sinusoidal with the number of half-waves equal to the whole number closest to $\sqrt{n/3} = \sqrt{P/3P_{cr}}$.

Euler's work on the stability of elastic systems has been developed much further in modern times as a result of the need to solve new engineering problems. One direction of present research is that concerned with elastic systems acted on by non-conservative positional forces, i.e., forces not depending explicitly on the time or velocity. The most important kind of such non-conservative forces are the "tracking" forces, i.e., those whose direction changes as the system deforms. The flexible casing of a rocket under the action of the exhaust affords a typical example.

One of the pioneers in this area was E. L. Nikolaï, who was the first to consider, in the years 1928–1929 the stability of an elastic rod under the action of twisting moments. He discovered that Euler's method leads in such situations to a paradoxical result, namely that the straight shape of the rod is preserved for all values of the twisting moment. Nikolaï located the source of the paradox, and was the first to give expression to the idea that Euler's method is in fact of limited application and that for certain problems it must be replaced by a dynamical method, i.e., one involving the investigation of small vibrations of the elastic system in question in a neighborhood of the equilibrium position.

Nonetheless many subsequent attempts were made to apply Euler's method to problems of elastic stability in non-conservative systems, resulting in stormy discussions and the publication of erroneous results. B. L. Nikolaï, apparently influenced by his brother's work, took part in this debate, and in 1939 produced a correct solution of one of the problems of interest via a dynamical approach, thereby avoiding the apparent paradox. In spite of this the controversy continued right up to the 1950s, when V. V. Bolotin subjected the question to a many-sided investigation. Bolotin developed methods for solving such problems in terms of a theory of the stability of motion and investigated several newly arisen paradoxes, including the paradoxical effect of small frictional forces on the stability of an elastic system under non-conservative loading. Bolotin's book [33] on the subject, published in 1961, represents a summation of this stage in the development of the theory of elastic stability, bringing together within the framework of a general approach various non-conservative problems, including that of aeroelasticity (wing flutter and the elements of trimming) and problems of instability in high-speed rotors.

---

[16]The original has $2i^2 = n$ here, and 2s rather than 3s in the succeeding formula.—*Trans.*

In connection with stability problems it had been thought that at bifurcation points there always occurs either a replacement of one equilibrium shape by another or else the appearance of a new unstable shape. An unusual example, in this regard, of a mechanical system where the old shape and the branching shape are simultaneously stable, is examined in the article [34] by A. Yu. Ishlinskiĭ, S. V. Malashenko, and M. E. Temchenko.

Before turning to the next theme of Euler's investigations into elasticity, we note other directions that research in elastic stability has taken, together with some of the more fundamental works on this topic. Several investigations have been made into the stability of three-dimensional elastic bodies in the nonlinear formulation (V. V. Novozhilov [35], A. N. Guz′ [36]). Important work has been carried out in connection with the stability of casings (A. V. Pogorelov [37], V. I. Feodos′ev [38]; a detailed bibliography of works on the stability of casings and plates may be found in the book [39] of A. S. Vol′mir). The concept of stability has been generalized so as to take into account the inelastic behavior of certain structural elements. The need for such a generalization arose, in particular, in connection with the stability of short columns and also that of compressed plates and casings at high temperatures. A stability theory for rods and thin-walled elements of structures incorporating the plastic properties of materials has been created in works by Th. Kármán [40], F. Engesser [41], and A. A. Ilyushin [42, 43]. The stability of viscoplastic bodies has been investigated by A. A. Ishlinskiĭ [44, 45], and in the presence of creep by Yu. N. Rabotnov [46] and N. Hoff [47].

3. Vibrations[17]

Euler published a whole series of papers on the dynamics of elastic bodies devoted to investigations of vibrating strings, rods, membranes, and plates. He uses the situation of forced harmonic vibrations of an harmonic oscillator to study the phenomenon of resonance. On the whole Euler bases these investigations on the assumption that the vibrations are small and that the linearized versions of the various problems are viable. The assumption that the vibrations are small allows one to confine the investigation to "isochronous"[18] motions. In this case it turns out to be useful to correlate the vibrations of the elastic body being studied with those of a pendulum constructed so as to be synchronous. In addition, Euler makes wide use of d'Alembert's principle in his works on the dynamics of vibrating systems, but in a more modern formulation—so that one may, it would seem, consider Euler as one of the authors of that principle.

In describing Euler's research on the vibrations of elastic bodies, it is important to mention the fact that his activity in that direction was closely involved with that of another famous scientist, namely Daniel Bernoulli. For several years these two carried on a lively correspondence illuminating their current mutual interests in questions of the theory of elastic vibrations. For this reason it is not at all surprising that many of the problems of that theory were solved by them practically simultaneously and that some of the results of their research were complementary—notwithstanding the fact that they employed different methods. Taken together, their work on vibrating elastic bodies was foundational for

---

[17]Conspicuous by its absence, perhaps, from the discussion of this section is the controversy over d'Alembert's general solution of the vibrating string problem. However it is examined in other essays of the collection.—*Trans.*

[18]Meaning regular?—*Trans.*

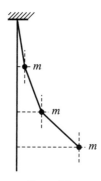

Figure 10.

that discipline. We shall refrain from considering here questions of priority, concentrating instead on Euler's published works on vibrations of elastic bodies (and, for the sake of brevity, mentioning Daniel Bernoulli's contributions only in passing, in connection with certain of Euler's works).

In his article [48], which was influenced by work of Bernoulli, Euler considers the problem of the vibration of a flexible string loaded with any number of weights (Figure 10). He notes that it is known from experiments that large oscillations of such systems are irregular and highly varied in form and expresses the opinion that the study of such problems is not just difficult but, in the absence of supplementary restrictions, beyond human capability. Thus in his articles [48–52] all vibrations are assumed small. As mentioned above, in tandem with the system under investigation Euler considers a mathematical pendulum whose oscillations are synchronous with those of the suspended weights. The problem to be investigated is that of finding the forms of those systems that vibrate in such a way that their separate parts arrive at the vertical position[19] simultaneously. It is also required to determine the length of the corresponding pendulum.

In his first work [48] on the problem, as also in those following it, the following situation turns out to be crucial. He notes that for the mathematical pendulum the ratio of the accelerating force to the weight is equal to that of the displacement [from the vertical] to the length of the pendulum. Using this equality of ratios, which, it is assumed, holds for every element of the system, Euler derives, in the case of simple oscillations of the system, equations for the balancing of the forces, and deduces the length of the pendulum. In the case of a string with just two suspended weights Euler obtains a quadratic algebraic equation giving the ratio of the displacements of the weights (the form of the oscillations), and a formula for calculating the equivalent length of the pendulum. In the same article he finds the analogous solution in the general case of a string with n weights attached.

In the article [49], devoted to dynamical problems and entitled "A new and simple method for small oscillations of rigid and flexible bodies", Euler proposes a new approach, possessing, he says "the greatest generality" and "well grounded in the principles of statics", so that with its help one may "solve with remarkable simplicity problems concerning the vibrations of rods and cables". Assuming again that the vibrations are small and that they are synchronous with the oscillations of a mathematical pendulum, Euler expresses the accelerating force in terms of the length of the pendulum and establishes an equation

---

[19]That is, of the corresponding mathematical pendulum.—*Trans.*

for the equality of the moments, again based on the above-mentioned equality of ratios. In this way he avoids the need to derive actual dynamical equations and the time is essentially excluded from consideration. For an elastic rod attached at its upper end and undergoing planar vibrations, the equation for the shape of the deflections is given in the form

$$Ek^2 y'' = \frac{g}{L} \int_0^x dx \int_0^x \sigma y\, dx, \quad \sigma = \rho A,$$

where L is the length of the pendulum, ρ the density, g the acceleration due to gravity, and A the cross-sectional area of the rod. On differentiating this equation twice Euler obtains the fourth-order equation

$$K^4 \left(\frac{d^4 y}{dx^4} \right) = y, \quad \text{where } K^4 = \frac{Ek^2 L}{\sigma g},$$

obtained earlier by Daniel Bernoulli. In the article we are considering Euler does not succeed in integrating this fourth-order equation but is able to establish a formula for the frequency of oscillation:

$$\nu = \frac{\zeta^2}{2\pi l^2} \sqrt{\frac{Ek^2}{\rho A}}, \quad \text{where } \zeta = \sqrt{\frac{5}{2}}.$$

From this he deduces that the period of oscillation of an elastic rod is proportional to the square of its length and inversely proportional to the square root of its absolute rigidity. In the opinion of C. Truesdell [4] the importance of this result of Euler's largely consists in its confirmation of G. W. Leibniz' notion of the relation between elastic and acoustic properties.

The concept of resonance plays an important role in the theory of forced oscillations. It seems that one of the very first works providing an explanation of the phenomenon of resonance is Euler's article "On a new type of oscillations" [50], where he considers forced linear vibrations of a simple oscillator under the action of a harmonic load and finds a solution in quadratures[20]. He shows that in a special case "the usual solution has to be replaced by another predicting unbounded growth in the vibrations". The special case in question is that where the frequency of the forces acting coincides with the characteristic frequency of the freely vibrating oscillator.

Euler carries out a detailed investigation of elastic vibrations in his treatise on the calculus of variations [2]. There he considers small vibrations of a homogeneous rod of length a and absolute rigidity Ek^2, and at the outset stipulates the assumptions to be made throughout. Thus gravity is neglected and the vibrational displacements of the rod are supposed "exceedingly small". The latter assumption allows the use of the approximation $R = |d^2 y/dx^2|$. For definiteness the rod is initially assumed to be attached by means of a bracket at $x = 0$.[21] Comparison with the oscillations of an isochronous pendulum of length f, yields the expression $My\,dx/af$ (M the mass of the rod) for the force causing the motion of an element of the rod of mass $M\,dx/a$. The expression for the elastic force arising in the bent rod is known from earlier investigations in statics. Euler asserts that "if

---

[20] That is, in integral form?—*Trans.*
[21] The original has $x = a$.—*Trans.*

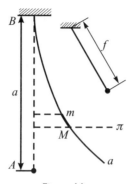

Figure 11.

to each individual element Mm[22] there were applied in the opposite direction $M\pi$ a force of the same magnitude $Mydx/af$, then the strip[23] would be in equilibrium in the position BMa [2, p. 526] (see Figure 11).

Euler concludes that "during its oscillation the strip undergoes the same amount of bending as it would if, in a state of rest, there acted on it at each of its points M a force of magnitude $Mydx/af$ in the direction $M\pi$" [*loc. cit.*]. With this Euler formulated a fundamental principle of mechanics whose significance for later work in dynamics is difficult to exaggerate[24].

Euler next sums the forces over all points of the elastic curve and differentiates twice, obtaining the fourth-order differential equation[25]

$$Ek^2\frac{d^4y}{dx^4} = \frac{My}{af}.$$

Writing $Ek^2af/M = c^4$, and applying results from his earlier work on the integration of differential equations of high order, Euler arrives at the following equation for the elastic curve:

$$y = Ae^{\frac{x}{c}} + Be^{-\frac{x}{c}} + C\sin\frac{x}{c} + D\cos\frac{x}{c}.$$

He now elaborates the process of determining the constants of integration A, B, C, D, and the constant c, using the conditions $d^2y/dx^2 = d^3y/dx^3 = 0$ at the origin $(x, y) = (0, 0)$, and $dy/dx = 0$ for $x = a$. In order to find the parameter f, in terms of which the period of oscillation is defined, Euler derives the equation

$$e^{\frac{a}{c}} = \frac{-1 \pm \sin\frac{a}{c}}{\cos\frac{a}{c}},$$

and shows that it has infinitely many solutions in c, leading him to the conclusion that there exist infinitely many different modes of vibration of one and the same strip—a fact unappreciated by Euler in his earlier works on dynamics. Another relevant remark concerns

---

[22]Here M and m denote points, not masses; see Figure 11.—*Trans.*

[23]The rod seems to have become a strip; actually in the original the word used here and in the other quotes below is "plate".—*Trans.*

[24]Is this Euler's version of the aforementioned "d'Alembert's principle"?—*Trans.*

[25]This equation also had been obtained earlier by Daniel Bernoulli—who had been unable to solve it, however.

the accuracy of the calculations he carries out: all to ten, and in some cases twelve, decimal places. Thus for the "frequency" of the oscillations he finds the value

$$\frac{a}{c} = 1.8751040813.$$

In addition to the first tone (major harmonic) of the vibrations, Euler finds the second tone and considers means for finding higher harmonics. According to his calculations the first tone is to the second as the square of the number 1.8751040813 is to the square of 4.6940910795, i.e., about 1 to 94/15, and on this basis he concludes that the second tone forms with the first an interval of a double octave and a fifth or almost a semitone. Euler carries out a similarly exhaustive investigation of the elastic oscillations of a freely vibrating rod and a rod securely attached at both ends.

Later, in his articles [53, 54], in connection with an investigation of transverse vibrations of an elastic rod, Euler finds an expression for the inertial force in terms of the second partial derivative of the deflection function y with respect to the time t. Taking this into account after equating moments, he obtains the following equation for small vibrations [53]:

$$-\int_0^x x\, dx \int_0^x \sigma \frac{\partial^2 y}{\partial t^2} dx = Ek^2 \frac{\partial^2 y}{\partial x^2}.$$

He finds a series solution of this integro-differential equation, satisfying appropriate boundary conditions. In the article [54] Euler extends these results to the case of an elastic rod vibrating under the action of normal and tensile forces.

The vibrating string problem dates from antiquity, when interest in it was stimulated in the main by the desire to perfect musical instruments. Euler's contribution to the creation of a mathematical theory of vibrating strings was crucial to that theory [55–59]. In his investigations on that topic he does not assume homogeneity. His first result concerning vibrations of an inhomogeneous string were published in the article [55], where the vibration equation is integrated in the special case of a string of variable thickness. In the paper [56] he develops an "inverse method" for computing the motions of inhomogeneous strings, and in [57] he discusses conditions that must be satisfied by the points where pieces of a string of differing thicknesses are joined. The articles [58] and [59] are likewise devoted to the investigation of vibrating strings of varying thickness, or else made up of pieces of different thicknesses. However in these works there are errors (see [4]) forestalling a correct formulation of the dependence of the frequency on the distribution of thickness along the string. In the article [60], completing the series and published posthumously, Euler discusses the results of Daniel Bernoulli on vibrating strings and establishes limits of applicability of his method.

All of the above-mentioned works of Euler and of other scientists are concerned with bending and vibrational problems exclusively of one-dimensional bodies—such as strings, columns, beams, and straight or curved rods. The passage to the study of two-dimensional bodies proved extremely difficult. It seems that Euler (see [52]) was the first to attempt a mathematical description of the deformation of an elastic surface, under the condition that one dimension is much smaller than the other. With the aim of establishing equations for small vibrations of a membrane Euler considered a pair of orthogonal families of elastic fibres respectively parallel to the x- and y-axes, and evenly spaced a distance δ apart. It is not difficult to see that the projection on the z–axis of the forces causing the fibres to flex are respectively $T_x \delta^2 \partial^2 z / \partial x^2$ and $T_y \delta^2 \partial^2 z / \partial y^2$. Equating the sum of these quantities to the product of the mass $\tau \delta^2$ of an element of the membrane with the acceleration $\partial^2 z / \partial t^2$, Euler arrives at an equation for the vibrations of the membrane, which in the case of equal elasticity of the fibres ($T_x = T_y$) has the form

$$\frac{1}{c^2} \frac{\partial^2 z}{\partial t^2} = \frac{\partial^2 z}{\partial x^2} + \frac{\partial^2 z}{\partial y^2}, \quad \text{where } c^2 = \frac{T}{\tau}.$$

The solution of this equation for a rectangular membrane under the assumptions that the initial velocities are zero and the edges of the membrane are secured, is given by

$$z = A \sin \omega t \sin \frac{\beta x}{a} \sin \frac{\gamma y}{b},$$

$$\frac{\omega^2}{c^2} = \frac{\beta^2}{a^2} + \frac{\gamma^2}{b^2}, \quad \nu = \frac{1}{2} c \left(\frac{m^2}{a^2} + \frac{n^2}{b^2} \right)^{\frac{1}{2}},$$

where some minor misprints, pointed out in [4], have been corrected. Euler also gives the vibration equation in polar coordinates r, φ:

$$\frac{1}{c^2} \frac{\partial^2 z}{\partial t^2} = \frac{\partial^2 z}{\partial r^2} + \frac{1}{r} \frac{\partial z}{\partial r} + \frac{1}{r^2} \frac{\partial^2 z}{\partial \varphi^2}.$$

In the case of a circular membrane Euler obtains z as a function of t, r, and φ, in the form $z = u(r) \sin(\omega t + C) \sin(\beta \varphi + D)$, where the function $u(r)$ is defined by him as a solution of a certain ordinary differential equation now known as Bessel's equation.

The theory of the vibrations of elastic bodies was subsequently very significantly developed further by Russian and Soviet scientists. The contribution of S. P. Timoshenko to this area of mechanics, both in its theoretical and practical aspects, is considerable, and the research of A. A. Andronov and his school has exerted a crucial influence on further developments in the theory of vibrations of elastic bodies. For the study of large-amplitude vibrations of elastic structural elements and nonlinear vibrations of elastic systems, the averaging methods devised by N. M. Krylov, N. N. Bogolyubov, Yu. A. Mitropol'skiĭ and their students have been of great importance.

There is yet one further direction of research originating in ideas of Euler, namely the extension of the analysis of vibrations to elastic systems acted on by periodic forces. Pioneering work in this direction was carried out by N. M. Belyaev [61] in 1924. Questions concerned with oscillations and stability of elastic systems subject to periodic action and the related topic of parametric resonance have been investigated mainly by Soviet scientists. Significant progress in this area has been made in particular by the following researchers: N. E. Kochin in his 1934 study of crankshafts; N. M. Krylov and N. N. Bogolyubov in their 1935 work generalizing Belyaev's results to a wide class of problems of elastic stability; and V. N. Chelomeĭ in his doctoral dissertation on the dynamical stability of elastic systems. Chelomeĭ discovered a curious phenomenon arising in connection with the stability of elastic systems acted on by periodic forces and brought to light new and interesting hydroelastic properties emerging in situations of parametric resonance [62; see also 63]. The theory of oscillations and stability of elastic systems under periodic actions was also studied by V. V. Bolotin, the first to tackle problems in this area in their nonlinear formulation.

In his work on elasticity, Leonhard Euler did not establish general equations for the theory as he did in hydrodynamics. Nevertheless the contributions he made to the mechanics of deformable rigid bodies entitles him to be counted among the founders of that science. For several generations of scientists Euler's achievements served as the standard of clarity in formulation of problems, of investigational depth, and of completeness of results. Many mechanicians have acquired their specialist education from reading Euler's works, taking them as the point of departure for their own research. This is especially true of Russian and Soviet scientists, successive generations of whom have devoted their professional lives to the further development of Euler's ideas and methods.

It should be noted, however, that it is far from being the case that Euler's creative heritage has been exhausted—in mechanics as in other areas of the mathematical sciences. Furthermore not all of his work has received due appreciation; several of his results have been "rediscovered" by later researchers. This is partly attributable to the difficulty of reading his works in the original arising from the circumstance of their being for the most part written in Latin. Only a very few—chiefly mathematical—works of Euler have been translated into Russian; it is a matter of urgent expediency that his collected works be so translated. In this regard it is appropriate to recall the words of the great scientist P.-S. Laplace: "Read Euler, read him: he is the teacher of us all".

The continuing publication by prominent scholars of analyses of Euler's works on the mechanics of elastic bodies is invaluable. It cannot be doubted that further examination of our Eulerian heritage will, through the persuasiveness of his work, accelerate scientific progress.

References

[1] Euler, L. "Solutio problematis de invenienda curva quam format lamina utcunque elastica singulis punctis a potentiis quibuscunque sollicitata". Comment. Acad. Sci. Petrop. (1728). 1732. Vol. 3. pp. 70–84; *Opera* II-10.

[2] ——. *Methodus inveniendi lineas curvas maximi minimive proprietate gaudentes (A method for finding curves possessing maximal and minimal properties)*. Lausanne; Genève, 1744; *Opera* I-24.

[3] ——. "De oscillationibus annulorum elasticorum". *Opera postuma*. SPb., 1862. Vol. 2. pp. 129–131; *Opera* II-11.

[4] Truesdell, C. *The rational mechanics of flexible or elastic bodies. 1638–1788. L. Euleri opera omnia* II-11$_2$. 1960. 428 pages.

[5] Kirchhoff, G. "Über das Gleichgewicht und die Bewegung eines unendlich dünnen elastichen Stabes". J. für reine und angew. Math. 1859. Vol. 56, Issue 4. pp. 285–313.

[6] Nikolaï, E. L. "On a problem concerning an elastic curve of twofold curvature". Pg., 1916; In: *Works in mechanics*. M.: Gostekhteorizdat (State technical-theoretical publisher), 1955. pp. 45–277.

[7] Popov, E. P. *Some problems concerning the statics of thin rods*. M.; L.: Gostekhizdat (State technological publisher), 1948. 170 pages.

[8] Euler, L. "Sur la force des colonnes". Hist. Acad. Sci. Berlin (1757). 1759. Vol. 13. pp. 252–282; *Opera* II-17.

[9] Nikolaï, E. L. "On Euler's works on the theory of longitudinal bending". Uch. zap. Leningr. un-ta (Scientific notes of Univ. of Leningrad). 1939. No. 44. pp. 5–19; *Works in mechanics*. M.: Gostekhteorizdat (State technical-theoretical publisher), 1955. pp. 436–454.

[10] Euler, L. "Determinatio onerum, quae columnae gestare valent". Acta Acad. Sci. Petrop. (1778:1). 1780. pp. 121–145; *Opera* II-17.

[11] ——. "Examen insignis paradoxi in theoria columnarum occurrentis". Acta Acad. Sci. Petrop. (1778:1). 1780. pp. 146–162; *Opera* II-17.

[12] ——. "De altitudine columnarum sub proprio pondere corruentium". Acta Acad. Sci. Petrop. (1778:1). 1780. pp. 163–193; *Opera* II-17.

[13] Panovko, Ya. G., Gubanova, I. I. *Stability and oscillations of elastic systems*. M.: Nauka, 1964. 336 pages.

[14] Dinnik, A. N. *Stability of elastic systems*. M.; L.: Izd-vo AN SSSR (Published by Acad Sci. USSR), 1950. 133 pages.

[15] Ishlinskiĭ A. Yu. "On a limiting transition in the theory of the stability of elastic rectangular plates". Dokl. AN SSSR (Announcements of Acad. Sci. USSR). 1954. Vol. 95, No. 3. pp. 477–479.

[16] Mises, R. "Ausbiegung eines auf Knickung beanspruchten Stabes". Ztschr. angew. Math. und Mech. 1924. Vol. 4. pp. 435–436.

[17] Timoshenko, S. P. *The stability of elastic systems*. M.; L.: Gostekhizdat (State technological publisher), 1946. 532 pages.

[18] Lagrange, J.-L. "Sur la figure des colonnes". Miscellanea Taurinensia (1770–1773). 1774. Vol. 5. pp. 123–166; *Oeuvres*. Paris, 1868. Vol. 2. pp. 125–170.

[19] Yasinskiĭ F. S. "On resistance to longitudinal bending". SPb., 1894; In: *Selected works on the stability of compressed rods*. M.: Gostekhizdat (State technological publishers), 1952. pp. 11–137.

[20] Love, A. E. H. *A treatise on the mathematical theory of elasticity*. Cambridge: Univ. Press, 1927. 642 pages; New York: Dover, 1944. 643 pages.

[21] Krylov, A. N. "On the equilibrium forms of compressed poles under longitudinal bending". Izv. AN SSSR. Otd. mat. i estestv. nauk (Results Acad. Sci. USSR. Section for math. and nat. sciences). 1931. No. 7. pp. 963–1012; *Collected works*. M.; L.: Izd-vo AN SSSR (Published by Acad. Sci. USSR), 1937. Vol. 5. pp. 181–226.

[22] Todhunter, I., Pearson, K. *History of the theory of elasticity*. Cambridge: Univ. Press, 1886–1893. Vols. 1, 2.

[23] Lagrange, J.-L. "Sur la force des ressorts pliés". Mém. Acad. Sci. Berlin (1769). 1771. Vol. 25. pp. 167–203; *Oeuvres*. Paris, 1869. Vol. 3. pp. 77–110.

[24] Prandtl, L. *Kipperscheinungen. Ein Fall von instabilem elastischen Gleichgewicht*. Diss. München, 1899. Nürnberg, 1900; *Gesammelte Abhandlungen*. Berlin etc., 1961. Vol. 1. pp. 10–74.

[25] Timoshenko, S. P. "On the stability of planar bending of a T-shaped beam under the influence of forces acting in the plane of greatest rigidity". Izv. SPb. politekh. in-ta (Results St. Petersburg polytech. inst.). 1905. Vol. 4. Issues 3, 4. pp. 151–219; 1906. Vol. 5. Issues 1, 2. pp. 3–34; Issues 3, 4. pp. 263–292; *Stability of rods, plates, and casings*. M.: Fizmatgiz (State publisher of math. and physics), 1971. pp. 9–105.

[26] ———. "On the stability of elastic systems. Application of a new method to the investigation of the stability of certain bridge structures". Izv. Kiev. politekh. in-ta (Results of Kiev polytech. inst.). 1910. Vol. 10. Book 4. pp. 375–560; *Stability of rods, plates, and casings*. M.: Fizmatgiz, 1971. pp. 208–383.

[27] Leĭbenzon, L. S. "On an application of harmonic functions to the question of the stability of spherical and cylindrical casings". Uch. zap. Yuryev. un-ta (Scientific notes of Yuryev univ.). 1917. No. 1; *Collected works*. M.: Izd-vo AN SSSR (Published by Acad. Sci. USSR), 1951. Vol. 1. pp. 50–85

[28] Ishlinskiĭ, A. Yu. "Examination of questions of the stability of the equilibrium of elastic bodies from the point of view of the mathematical theory of elasticity". Ukr. mat. zhurn. (Ukrainian math. journal). 1954. Vol. 6. No. 2. pp. 140–146.

[29] Balabukh, L. I., Yakovenko, M. G. "An equilibrium bifurcation equation for an isotropic body in terms of the rates of change of Lagrangian coordinates". Prikl. matematika i mekhanika (Applied math. and mechanics). 1974. Vol. 38. No. 4. pp. 693–702.

[30] Vlasov, V. Z. *Thin-walled elastic rods*. M.: Fizmatgiz (State publisher math. and physics), 1959. 568 pages.

[31] Obraztsov, I. F. "On the calculation of the stability of elastic rods under bending". Tr. Mosk. aviats. in-ta (Works Moscow aviation inst.). 1953. No. 26. pp. 5–85.

[32] Lavrient'ev, M. A., Ishlinskiĭ, A. Yu. "Dynamical forms of loss of stability of elastic systems". Dokl. AN SSSR (Announcements Acad. Sci. USSR). 1949. Vol. 64. No. 6. pp. 779–782.

[33] Bolotin, V. V. *Nonconservative problems in the theory of elastic stability*. M.: Fizmatgiz (State publisher math. and physics), 1961. 339 pages.

[34] Ishlinskiĭ, A. Yu., Malashenko, S. V., Temchenko, M. E. "On the branching of stable dynamical equilibrium positions in a certain mechanical system". Izv. AN SSSR. Otd. tekh. nauk (Results Acad. Sci. USSR. Section for technological science). 1958. No. 8. pp 53–61.

[35] Novozhilov, V. V. *Foundations of the nonlinear theory of elasticity*. M.: Gostekhizdat (State tech. publisher), 1948. 218 pages.

[36] Guz', A. N. *Stability of three-dimensional deformable bodies*. Kiev: Nauk. dumka, 1971. 276 pages.

[37] Pogorelov, A. V. *Geometrical methods in the nonlinear theory of elastic casings*. M.: Nauka, 1967. 280 pages.

[38] Feodos'ev, V. I. "On the stability of spherical casings under the action of external uniformly distributed pressure". Prikl. matematika i mekhanika (Applied math. and mech.). 1954. Vol. 18. No. 1. pp. 35–42.

[39] Vol'mir, A. S. *Stability of deformable systems*. M.: Nauka, 1967. 984 pages.

[40] Kármán, Th. *Untersuchungen über Knickfestigheit*. Diss. Göttingen, 1909. Berlin, 1909; *Collected works*. London, 1956. Vol. 1. pp. 90–140.

[41] Engesser, F. "Widerstandsmomente und Kernfiguren bei beliebigem Formänderungsgesetz (Spannungsgesetz)". Ztschr. VDI. 1898. Vol. 42. Issue 33. pp. 903–907; H. 34. pp. 927–931.

[42] Ilyushin, A. A. *Plasticity.* M.: Gostekhizdat (State tech. publisher), 1948. 376 pages.

[43] ———. "Elasto-plastic stability of plates". Prikl. matematika i mekhanika (Applied math. and mech.). 1946. Vol. 10. Nos. 5, 6. pp. 623–638.

[44] Ishlinskiĭ A. Yu. "On the stability of visco-plastic flow in a strip and a bar". Prikl. matematika i mekhanika (Applied math. and mech.). 1943. Vol. 7. No. 2. pp. 109–130.

[45] ———. "On the stability of visco-plastic flow in a circular plate". Prikl. matematika i mekhanika (Applied math. and mech.). 1943. Vol. 7. No. 6. pp. 405–412.

[46] Rabotnov, Yu. N. *Creeping elements of structures.* M.: Nauka, 1966. 752 pages.

[47] Hoff, N. *Longitudinal bending and stability.*

[48] Euler, L. "De oscillationibus fili flexilis quotcunque pondusculis onusti". Comment. Acad. Sci. Petrop. (1736). 1741. Vol. 8. pp. 30–47; *Opera* II-10.

[49] ———. "De minimis oscillationibus corporum tam rigidorum quam flexibilium. Methodus nova et facilis". Comment. Acad. Sci. Petrop. (1734–1735). 1740. Vol. 7. pp. 99–122; *Opera* II-10.

[50] ———. "De novo genere oscillationum". Comment. Acad. Sci. Petrop. (1739). 1750. Vol. 11. pp. 128–149; *Opera* II-10.

[51] ———. "De motu corporum flexibilium". Comment. Acad. Sci. Petrop. (1744–1746). 1751. Vol. 14. pp. 182–196; *Opera* II-10.

[52] ———. "De motu vibratorio tympanorum". Novi comment. Acad. Sci. Petrop. (1764). 1766. Vol. 10. pp. 243–260; *Opera* II-10.

[53] ———. "De motu vibratorio laminarum elasticarum, ubi plures novae vibrationum species hactenus non pertractatae evolvuntur". Novi comment. Acad. Sci. Petrop. (1772). 1773. Vol. 17. pp. 449–487.; *Opera* II-11.

[54] ———. "Investigatio motuum, quibus laminae et virgae elasticae contremiscunt". Acta Acad. Sci Petrop. (1779:1). 1782. pp. 103–161; *Opera* II-11.

[55] ———. "De motu vibratorio chordarum inaequaliter crassarum". Novi comment. Acad. Sci. Petrop. (1762–1763). 1764. Vol. 9. pp. 246–304; *Opera* II-10.

[56] ———. "Recherches sur le mouvement des cordes inégalement grosses". Miscellanea Taurinensia (1762–1765). 1766. Vol. 3. pp. 25–59; *Opera* II-10.

[57] ———. "Animadversiones in solutionen Bernoullianum de motu chordarum ex duabus partibus diversae crassitiei compositarum". Novi comment. Acad. Sci. Petrop. (1772). 1773. Vol. 17. pp. 410–421; *Opera* II-11.

[58] ———. "De motu vibratorio chordarum ex partibus quotcunque diversae crassitiei compositarum". Novi comment. Acad. Sci. Petrop. (1772). 1773. Vol. 17. pp. 422–431; *Opera* II-11.

[59] ———. "De motu vibratorio chordarum crassitiei utcunque variabili praeditarum". Novi comment. Acad. Sci. Petrop. (1772). 1773. Vol. 17. pp. 432–448; *Opera* II-11.

[60] ———. "Dilucidationes de motu chordarum inaequaliter crassarum". Acta Acad. Sci. Petrop. (1780:2). 1784. pp. 99–132; *Opera* II-11.

[61] Belyaev, N. M. "Stability of prismatic rods under the action of variable longitudinal forces". In: *Engineering equipment and structural mechanics.* M.: Izd-vo "Put'" (Published by "The Path"), 1924. pp. 149–167.

[62] Chelomeĭ V. N. "On the possibility of increasing the stability of elastic systems by means of vibrations". Dokl. AN SSSR (Announcements Acad. Sci. USSR). 1956. Vol. 110. No. 3. pp. 345–347.

[63] ——. "On the dynamical stability of elements of aviational structures". Grazhd. aviatsia (Civil aviation). 1940. No. 12. pp. 27–28.

[64] Bolotin, V. V. *Dynamical stability of elastic systems.* M.: Gostekhizdat (State tech. publisher), 1956. 600 pages.

Euler's Research in Mechanics during the First Petersburg Period

N. N. Polyakhov

In describing the work carried out by Euler during his first Petersburg period (1727–1741) it is essential to note that of the 28 papers he published under the auspices of the Petersburg Academy of Sciences between 1727 and 1734, nine were directly concerned with mechanics, and a further 11 related to the solution of problems of mechanics. This indicates Euler's preoccupation with mechanics from the moment of his arrival in St. Petersburg. However it is not the mere fact of his absorption in mechanics that is of such import but rather the appearance in 1736, as the culmination of his work over that period, of his two-volume treatise *Mechanica sive motus scientia analytice exposita* (*Mechanics or the science of motion, expounded analytically*), the first treatise on Newtonian mechanics written using the mathematical terminology and notation still fundamental today—the language of differential equations.

Although Newton founded classical mechanics, he failed to provide a general analytic method for solving particular problems of mechanics. In his solutions of such problems he used geometrical methods applied to infinitesimal figures with subsequent passage to limits; in other words, Newton refrained from explicit use of the apparatus and symbolism of the differential and integral calculus—which he himself had invented.

The first to fully exploit the power of that apparatus was in fact Euler, who may therefore be counted—on an equal footing with Newton—among the founders of classical mechanics. While Newton had founded mechanics as a branch of physics, it was Euler who created and developed the mathematical apparatus for adequately and efficiently expressing the physical essence of that mechanics.

The goal Euler set himself is most clearly described in his preface to the aforementioned treatise, where he writes:[1] "However if analysis is needed anywhere, then it is certainly in mechanics. Although the reader can convince himself of the truth of the exhibited propositions, he does not acquire a sufficiently clear and accurate understanding of them, so that

---

[1] This quotation appears also in the essay by G. K. Mikhaĭlov and L. I. Sedov, and, except for the last sentence, also in that by V. V. Rumyantsev.—*Trans.*

if those questions be ever so slightly changed he will not be able to answer them independently unless he turn to analysis and solve the same propositions using analytic methods.
This in fact happened to me when I began to familiarize myself with Newton's *Principia*
and Hermann's *Phoronomia*; although it seemed to me that I clearly understood the solutions of many of the problems, I was nevertheless unable to solve problems differing
slightly from them. But then I tried, as far as I was able, to distinguish the analysis [hidden]
in the synthetic method and to my own ends rework analytically those same propositions,
as a result of which I understood the essence of each problem much better. Then in the
same manner I investigated other works relating to this science scattered here, there, and
everywhere, and for my own sake I expounded them [anew] using a systematic and unified
method and reordered them more conveniently. In the course of these endeavors not only
did I encounter a whole series of problems hitherto never even contemplated—which I have
very satisfactorily solved—but I discovered many new methods, thanks to which not only
mechanics, but analysis itself, it would seem, has been significantly enriched. It was thus
that this essay on motion arose, in which I have expounded using the analytic method and
in convenient order both that which I have found in others' works on the motion of bodies
and what I myself have discovered as a result of my ruminations".

Euler goes on to state that in his treatise he will "consider infinitely small bodies, those
that may be regarded as points".

The first volume of that work is devoted to the motion of a free point-mass and the
second to that of a point-mass not moving freely. As far as the physical presuppositions
or axioms are concerned, in general Euler takes the same position as Newton, although in
some particular situations his view diverges from Newton's. He begins by defining what
the expression "uniform motion of a particle" is to mean, and then examines non-uniform
motion under the assumption that infinitesimal elements of the path of the particle represent
uniform motion. Considering the law of [conservation of] momentum[2], he notes that it is
most unfortunate that momentum is also called "force" since it is not the same sort of
entity as other so-called forces—such as the gravitational force—and is not comparable
with them. By the term "force" Euler understands the same thing as Newton, according
to whom "an applied force is an action effected on a body in order to change its state of
rest or uniform rectilinear motion". In Euler's words, "force (potentia) is that effort (vis)
which changes a body's state of rest to a state of motion, or changes the form of its state
of motion". Euler defines the direction of a force to be that straight line in the direction of
which it tends to move the body. He does not explicitly formulate the axiom represented by
the parallelogram of forces, indicating instead that if several forces are acting on a particle
then the motion (motus) they cause in it is the same as it would be if a certain uniquely
defined force were acting, equivalent to them all. It is evident from such formulations that,
for Euler, in dynamics it is not so much forces that are combined but the various motions
they cause.

Concerning mass, Euler holds the same point of view as Newton, namely that the mass
of a body is to be inferred from its weight, which is proportional to the amount of matter
contained in the body.

Following on his exposition of the basic concepts of mechanics, occupying the first
chapter of the treatise, Euler establishes in the second a formula expressing Newton's sec-

---

[2]The original has "the law of inertia."—*Trans.*

ond law of motion in the form of a differential equation:

$$dv = \frac{F}{m}dt,$$

which is to be integrated.

The third chapter contains a large collection of problems concerning the rectilinear motion of a particle under the action of forces varying with distance. The point of these problems is to demonstrate methods of integration of the above equation of motion, the fundamental equation of dynamics. Many of these problems serve somewhat as examples of integration of differential equations, i.e., of that mathematical discipline created to a large extent by Euler himself. An example of a purely mechanical such problem is that where Euler considers harmonic oscillation of a particle, and its motion under an attractive force inversely proportional to the square of the distance.

In the fourth chapter Euler examines rectilinear motion of a particle in a resisting medium under the assumption that the force of resistance is proportional to the first, second, or, finally, nth, power of the speed.

The two final chapters of the first volume of the *Mechanics* contain a treatment of circular motion of a particle in a vacuum and in a resisting medium. The differential equations for such motions are given by Euler in the form now termed "natural", consisting of an equation for each of the projections of the motion on the tangent and the normal to the trajectory at a general point. It is important to note the appearance here of the problem of the motion of a particle under the action of a central force given by Newton's universal law of gravity; in essence Euler's treatment of this problem constitutes an introduction to celestial mechanics in analytical form. Subsequently Euler investigated far-reaching elaborations of this problem, to the extent that it is appropriate to count him as one of the founders of analytical celestial mechanics. Here one also finds another fundamental problem, namely that of the curvilinear motion of a particle in a resisting medium under the action of gravity, i.e., the problem of exterior ballistics. Euler returned to a more detailed investigation of this problem in 1753, when he was in Berlin, publishing his results in an article entitled "Investigation of the actual curve described by a body projected into the air or some other medium". Since this paper was not composed while Euler was in St. Petersburg, we shall not discuss it here beyond noting the undoubted influence on it of the ballistics problem Euler investigated in the *Mechanics*.

We conclude our brief summary of the content of the first volume of Euler's *Mechanics* by noting the interesting circumstance that nowhere in that work does he formulate the equations of mechanics in terms of projections on "fixed" rectangular Cartesian axes, with the result that in several cases—as he himself remarks later in his treatise *The theory of motion of rigid bodies* (1765)—he "became involved in excessively tangled calculations". The approach whereby mechanical equations are formulated in terms of projections of the motion on fixed Cartesian axes was initiated by Newton's successor Maclaurin in his book *A treatise of fluxions* (1742).

The second volume of Euler's *Mechanics* is devoted to an investigation of the constrained motion of a point-mass along curves and on surfaces (prescribed), where, as it turns out, the solution of problems in terms of the natural equations of motion is indeed natural and convenient; it is precisely in these equations that the normal reaction of the curve or surface figures naturally. In the problem of the pendulum, in connection with the reaction Euler follows Huygens in proposing that the centrifugal force acts on the string. It is in this context that Euler derives the formula for the curvature of a curve lying in a surface and establishes the differential equation for the geodesics of a surface.

From the above it is clear that Euler reduced the solution of problems of mechanics to problems of integration of the corresponding differential equations, thereby reaching the goal he set himself in the preface to his treatise. It is essentially in this achievement that his enormous scientific merit consists: only subsequent to the appearance of the *Mechanics* was mechanics transformed into a science equipped with a mathematical apparatus in terms of which its physical essence, brought to light by Newton, could be satisfactorily expressed.

The fundamental idea introduced in Euler's *Mechanics*—that of deriving equations expressing Newton's second law as it applies to the situation at hand and then solving them— was to have extensive repercussions, both in succeeding works of Euler himself and in the work of his successors in the mechanics of rigid bodies, in fluid mechanics, and in celestial mechanics.

It is interesting that already in his *Mechanics* Euler formulates his plan for future research in the various branches of mechanics, comprising the writing of treatises on the mechanics of a point, of rigid bodies, and of fluid bodies[3]. The second and third of these projects were realized only when he had left St. Petersburg for Berlin; *The theory of the motion of rigid bodies* was published in 1765, and his works on the motion of fluids appeared in the interval 1755–1771 in four memoirs of the Berlin and Petersburg Academies of Sciences.

Thus we conclude that Euler should be considered an equal of Newton's among the founders of classical mechanics insofar as he furnished that discipline with an analytical framework transforming it into a powerful instrument of modern science—and the fundamental pioneering work in this regard was his *Mechanics*, composed entirely in St. Petersburg.

---

[3] According to an earlier essay this plan was formulated when Euler was still in Switzerland.—*Trans.*

The Significance of Euler's Research in Ballistics

A. P. Mandryka

Out of the enormous number of Euler's works—exceeding 800—only two are devoted to ballistics. His first investigations on this topic are contained in the extensive commentary appended to his translation of Benjamin Robins' treatise "New principles of gunnery" [1] from English into German. The second are published in his memoir of 1753 [2]. Although uncharacteristically few in number, these two works turned out to be the most important of the plethora of books and articles on ballistics published in the 18th century by artillerists, mathematicians, and mechanicians—including some of the most eminent European scientists. Such was the quality of the 1753 article that it retained its importance for ballistics for over a century. It is chiefly with the results of this article that the present essay is concerned.

The basic problem of exterior ballistics consists in the determination of the trajectory of the center of mass of a shell under the action of gravity and the resistive force of the air, which is directed opposite to the velocity vector. So formulated, i.e., with the air resistance taken into account, the problem was first solved for all practical purposes by Euler.

Of course he had predecessors. Beginning in antiquity, and later in the middle ages and Renaissance, there had been thinkers concerned with the question of the shape of the trajectory of the center of mass of a projectile launched—by catapult or more modern artillery—at an angle to the horizontal. At first the motion of an idealized projectile was considered, i.e., of a point-mass, and the resistance of the air was left out of account on the grounds that it was negligibly small. The most substantial of the earliest results appeared in works by Tartaglia in 1537 and 1546 [3, pp. 27–34]. His solution of the problem as just formulated, yielded a trajectory in the form of two straight-line segments, one coinciding with the line of projection, making the same angle with the horizontal, and the other vertical, i.e., in the direction of the force of gravity, these two straight portions being joined by a circular arc. He argued that during the first straight portion of the trajectory (the "forced" part of the motion) the projectile was impelled by an "impetus", a concept close to what we now call "momentum"[1]. The technical characteristics of the projectile—its mass, weight,

---

[1] A literal translation of one of the Russian terms for "momentum" is "quantity of motion".—*Trans.*

and diameter—were ignored. Tartaglia compiled a firing table based on his theory for use in practical problems of gunnery; however his solution rested on assumptions ignoring real factors and in practice for a long time artillerists continued using tables compiled on the basis of experimental firings.

In 1632 Galileo introduced the notion of an abstract spherical shell or cannonball into the theory of the motion of a point-mass [3, pp. 74–86]. His solution of the problem relied on developments in dynamical theory—stimulated, however, by those same artilleristic interests. Galileo's chief merit consisted in his rejection of the concept of "impetus" as a factor causing the shell to continue to move, and his establishment of the notion of the acceleration due to gravity, i.e., the kinematical manifestation of gravity. As a result he was able to take into account on solid grounds the action of gravity on the projectile as determining the curvature of the trajectory, and showed that in the case where a point-mass is projected horizontally, its trajectory is a descending branch of a parabola. The fact that this was the only situation he considered would seem to indicate in particular that Galileo did not have applications to problems of artillery directly in view.

Toricelli went further, compiling in 1644 on the basis of Galileo's results tables of sines of double angles of projection, which were actually used in practical artillery [3, pp. 86–95]. By the 17th century mortars were used extensively for projecting bombs. However the low level of the technology of their production resulted in considerable eccentricity of the barrels, significantly increasing the dispersion of the points of impact, so that artillerists, in the hope of achieving better results, rejected the empirical firing tables in favor of those compiled using Galileo's theory, which was given the name "parabolic" [4].

The parabolic theory, like Tartaglia's, ignored the geometric characteristics and weight of the projectile. In situations where it was possible to neglect air resistance, the trajectory was determined by the initial speed and the angle of projection alone, and this theory applied in practice only in a few isolated limiting situations. One such situation arose in the 17th century: at that time if, in firing bombs from a mortar, the initial speed was small and the weight of a bomb relatively large, then the air resistance could indeed be neglected. Another such situation arose in the 20th century in connection with very long-range projectiles. Since the upper portion of the trajectory of such a projectile passes through rarefied layers of the atmosphere, where the air resistance is essentially negligible, that portion of the trajectory may be assumed parabolic.

The end of the 17th century witnessed a new stage in the development of exterior ballistics, when many scientists began to pay attention to air resistance as a factor affecting the trajectory of a point-mass projected at an angle to the horizontal. It is of great importance in this regard to pay due attention to the formula for the force due to the resistance of the air proposed by Newton in 1687 [5]. Newton introduced into the problem characteristics determining the force of air resistance, i.e., various parameters associated with the projectile such as its cross-sectional area, physical properties of the atmosphere such as its density, and the speed of the projectile. He concluded that, all other factors being held fixed, the resistive force of the air is proportional to the square of the projectile's speed. Newton's formula contains a coefficient which at that time was reckoned constant. Half a century later, in 1745, Euler showed that that coefficient does in fact depend on the speed—as indeed Robins' experiments had foreshadowed. Somewhat later Euler called this coefficient the "resistance function".

Many mathematicians and mechanicians of the 17th and early 18th centuries tried to solve the basic problem of ballistics with the force due to air resistance, expressed by a quadratic formula, taken into account. However their results were either too specialized, like for instance Newton's, or of a too general mechanical-mathematical character, such as those of J. Hermann (1716) and J. Bernoulli (1719).

Euler's first investigations along these lines were included in his *Mechanics* of 1736, and concerned the solution of special cases of the problem of determining the trajectory of a point-mass projected at an angle to the horizontal in a resisting medium [6]. He obtains a solution in terms of quadratures, which he does not calculate explicitly—which explains why this result was not used to solve the practical problems arising at that time in ballistics, when it had become clear that the parabolic theory was not of use in the compilation of firing tables for mortars in view of the significant effect of air resistance on the shells.

In 1753 Euler wrote, and in 1755 published, the above-mentioned memoir [2], devoted to the solution of the problem, which turned out not to be reducible to one of merely calculating quadratures. However he did propose a method for calculating solutions explicitly and this was used by F. Graewenitz at the end of the 18th century to compile ballistics tables for spherical shells, and in the middle of the 19th century by J. C. F. Otto and A. F. Siacci[2] to do the same for elongated shells. These tables could be used to compute features of trajectories provided the initial speed did not exceed 240 meters per second [3, pp. 186–193].

Of all the memoirs on ballistics published in the course of the 18th century, this one of Euler's alone found practical application—albeit only in the situation of spherical shells[3], and for a restricted range of speeds.

Following on the introduction of air resistance into the basic problem, researchers in exterior ballistics began to consider more realistic technical objects—actual shells—than just abstract point-masses. Until the middle of the 19th century only spherical projectiles were considered, with the result that until then the shape of the front portion of the shell was preserved in order to hold fixed the coefficient of resistance for the shell figuring in contemporary versions of the differential equations for the the motion of its center of mass. Since Euler's equations involved other characteristics of the shell, such as its weight and cross-sectional area, it is clear that he was considering not just the simple case of a point-mass, but that of a more realistic technical object, and that he had therefore effectively solved the basic problem of exterior ballistics for a spherical shell. Newton's formula for the force of air resistance, which was used by Euler in his solution, thus prepared the way for the use of that solution in analyzing more general cases, such as that of elongated shells with the density gradient of the air taken into account, projected at supersonic speeds. Thus Euler introduced into exterior ballistics an essentially new element which was to find application in the second half of the 19th century and later, when smooth-walled barrels gave way to rifled or grooved ones.

It behooves us to say in conclusion that Euler, in his solution of the fundamental problem of exterior ballistics, was the first to bring together three objects for consideration:

---

[2]Francesco Siacci (1839–1907) taught mechanics at the University of Turin. From 1872 till 1892 he concurrently held the post of Professor of Ballistics at the Military Academy in Turin, and in 1893 was made a Senator in Rome.—*Trans.*

[3]And elongated ones?—*Trans.*

the shell, the air, and the gravitational field. It is this—together with the various factors enumerated above—that in essence justifies us in regarding him as the founding father of exterior ballistics as a modern technical science.

References

[1] Robins, B. *Neue Grundsätze der Artillerie enthaltend die Bestimmung der Gewalt des Pulvers nebst einer Untersuchung über den Unterschied des Widerstands der Luft in schnellen und langsamen Bewegungen, aus dem Englischen des Hrn. Benjamin Robins übersetzt und mit den nötigen Erläuterungen und vielen Anmerkungen versehen von Leonhard Euler.* Berlin, 1745; *Opera* II-14.

[2] Euler, L. "Recherches sur la véritable courbe que décrivent les corps jettés dans l'air ou dans un autre fluide quelconque". Mém. Acad. Sci. Berlin. (1753). 1755. Vol. 9. pp. 321–352; *Opera* II-14.

[3] Mandryka, A. P. *History of ballistics (to the mid-19th century).* M.; L.: Nauka, 1964.

[4] ——. "Teoria paraboliczna a strzelanie bombami z moździerzy". Kwart. hist. nauki i techn. 1960. Vol. 5. Nos. 3/4. pp. 361–368.

[5] Newton, I. *Philosophiae naturalis principia mathematica.* First ed. London, 1687. (English translation by F. Cajori: *Mathematical principles of natural philosophy and his system of the world.* Berkeley, Calif. 1934.)

[6] Euler, L. *Mechanica sive motus scientia analytice exposita.* Petropoli, 1736. Vols. 1, 2; *Opera* II-1,2.

[7] Mandryka, A. P. *Sketches of the development of the technological sciences. Mechanics series.* M.; L.: Nauka, 1984.

Euler and the Development of Astronomy in Russia

V. K. Abalakin and E. A. Grebenikov

In all historical epochs the creative work of the great scientists has been primarily motivated by the desire to understand the world about them. Among the geniuses who dedicated their spiritual strength and learning to the attainment of the bliss of scientific discovery, Leonhard Euler occupies an honored place.

Euler played a prominent role in the development of the whole of the natural science of his time. Many of his ideas and results became the starting points of independent scientific disciplines, and have not faded in importance over the centuries. This holds true in particular of his work in classical astronomy, which in the 18th century, thanks mainly to Euler and his colleagues at the Petersburg and Berlin Academies, flourished as never before.

Euler's first Petersburg period lasted from 1727 till 1741 and the second from 1766 to 1783. Between 1741 and 1766 he lived in Berlin, whither he had been invited by the Prussian king Friedrich II to work at the Berlin Academy. However throughout the Berlin period of his life Euler maintained constant scientific contact with the Petersburg Academy.

It is hardly possible in a short lecture to examine in detail Euler's outstanding contribution to astronomy, if only because his astronomical works number over a hundred. An inventory of these works may be found in the extensive survey prepared by M. F. Subbotin in 1957 on the occasion of the jubilee meeting of the Academy of Sciences of the USSR dedicated to the 250th anniversary of Euler's birth [1].

To begin with, we shall attempt to single out just those trails of astronomical discovery blazed by Euler that have by today developed into highways of that science. This goal is all the more realistic in view of the continuing lack of a detailed history of the development of astronomical theory in the 18th century, rendering an exhaustive and accurate representation of Euler's astronomical activity impossible. In any case his astronomical interests were so varied that it would be extremely difficult to "cover" them all satisfactorily within the scope of a single lecture.

We considered it expedient also to describe the content of certain of Euler's astronomical works which stand apart from the rest by virtue of their methods and results, yet are

245

of independent methodological or scientific value. And there is the further non-trivial circumstance that many of Euler's astronomical works were published in the official journals of the Petersburg and Berlin Academies which are not now[1] readily accessible and have hitherto not been reprinted[2] or translated, with the result that not only were they little read by succeeding generations of astronomers but many were even forgotten. Most 19th and 20th century astronomers became acquainted with Euler's ideas chiefly through the works of P.-S. Laplace and C. F. Gauss, where they appeared in altered form, with the result that his ideas and discoveries have often not been linked to his name, his priority has in many cases been forgotten, and his influence on astronomical progress, as recorded in the historico-scientific literature, has been less than it might have been.

The study of the original astronomical works of Euler can for many reasons be strongly recommended to the modern astronomer. In the first place, many of his neglected ideas have ramifications not apparent in their rediscoveries by later scientists. And then his works in astronomy—as indeed in other fields—are distinguished by the unsurpassed mastery of a great pedagogue and thinker, in addition to which the reader always senses Euler's own interest in the solution of the problem to hand, and the spiritual uplift it evokes in him. As one reads, a deep emotional unconscious "accompaniment" of Euler's train of thought helps to grasp his ideas, always clearly and crisply minted.

Of the projected 72 volumes of Euler's *Opera omnia*, which are to contain all of his published works, 10 are to be devoted to his astronomical investigations. (Three are yet to appear[3].) In terms of sheer quantity of astronomical output Euler is rivalled only by Laplace, whose collected works number 14 volumes all but three of which are devoted to astronomical investigations.

The majority of Euler's astronomical works are concerned with problems of theoretical astronomy, i.e., that branch of astronomy dedicated to the study of the motion of celestial bodies and systems of bodies under the influence of forces obeying the universal law of gravitation, or, more precisely, the problems of celestial mechanics concerned with the motion of point-masses—*not* extended bodies—attracted to one another in accordance with the universal law of gravity.

It is noteworthy that certain of Euler's ideas were so far in advance of his time that science has been able to utilize them only in modern times with the aid of computer technology. Many of his achievements are at such a high theoretical level that they transcend the boundaries of astronomy proper to form an integral part of theoretical science.

1. A general survey of Euler's astronomical works

Leonhard Euler began to investigate astronomical problems in his first Petersburg period. During that period he published several first-rank works on the theory of both unperturbed and perturbed motion of celestial bodies, on calculating the attraction of an "elliptical" spheroid, on various questions of spherical astronomy, and on astronomical problems related to optics and other branches of physics. Such was the breadth of his interests that

---

[1] As of 1986.

[2] Presumably these works are now available in the *Opera omnia.—Trans.*

[3] This was in 1986. The project is now complete.—*Trans.*

practically no current problem of astronomy, optics, or physics generally escaped his attention.

In his investigations into unperturbed motion Euler went from the study of the heliocentric motion of planets and comets to the much more difficult problem of determining the orbit of a celestial body relative to an observer on the earth's surface, which of course is simultaneously orbiting the sun. In his own lifetime his method of determining the orbits of comets was applied in calculating the actual orbits of many comets, chiefly by A. J. Lexell[4]. The main progress in this area depended on a particular equation of Euler's characterizing parabolic motion.

Euler devoted an even larger number of works to the elaboration of a theory of perturbed motion of celestial bodies. Of these fundamental achievements, the following are of greatest significance: his method of variation of the parameters determining the features of orbits, basic to the study of the larger planets' motion, the fundamentals of the method of expansion of the perturbation function, illustrated by practical examples, and the introduction of the highly effective notion of intermediate orbits differing from Keplerian ellipses. Euler was essentially the first to point out the connection between the efficiency of iterative methods as applied to the solution of the nonlinear equations arising in celestial mechanics and the choice of the initial approximation (an intermediate orbit). We shall examine these works in more detail below.

Of those of Euler's works in areas neighboring on astronomy we should mention his monograph *An investigation of the physical causes of incoming and outgoing sea-tides* [2], concerned with a theory of tides incorporating dynamical aspects, inertia, and oscillations of bodies of water. For this work Euler shared the prize of the Paris Academy for 1740 with Daniel Bernoulli and Colin Maclaurin, who had submitted essays on the same topic.

In 1747 Euler won the prize again for his essay "Thoughts on the question of how most conveniently and accurately to determine the time by means of observations made at sea by day, by night, or by twilight" [3], which once again he shared with Daniel Bernoulli. In particular, Euler proposes in this essay that in order to obtain an accurate measurement at sea of the elevations of the stars at night or, more generally, when the horizon is obscured, one should attach to a freely suspended quadrant a pendulum affording a reference line for the zenith, thus providing a fix on the horizontal independently of the ship's roll. In the same essay Euler considers problems of spherical astronomy to do with the calculation of angles of elevation[5] of celestial bodies.

Somewhat later, between 1748 and 1751, Euler published the two memoirs [4, 5] in which he gives a complete exposition of aberration of light and parallax. These memoirs played a leading role in astronomy over the following decade. In 1754 Euler derived, and in 1756 published, the following differential equation for an element of astronomical refraction of a light ray reaching an observer on earth [6]:

$$d\varphi = \frac{a\alpha^{(q-k)/c}\sin\zeta dx}{x(x^2 - a^2\alpha^{(2q-2k)/c}\sin^2\zeta)^{1/2}};$$

here a denotes the radius of the earth, x the distance of the front of the light ray from the

---

[4] Anders Johann Lexell (1740–1784) was a Finnish-Swedish-born Russian astronomer and mathematician. He immigrated to Russia in 1768, where he was known as Andreĭ Ivanovich Leksel'.—*Trans.*

[5] The original has "horary angles" or "hour angles".—*Trans.*

earth's center, ζ and φ the zenith distances[6] of points on tangent lines to the light ray, at respective distances a and x from the earth's center, q and k the respective densities of the atmosphere at those two points, and α the index of refraction for a light ray passing through the interface between a vacuum and a region of air of standard density c, which Euler gives as $\alpha = 3324/3325$. He remarks that this formula does not yield a general universal expression for angles of refraction, and goes on to investigate the geometric properties of the path of a light ray passing through the earth's atmosphere, calculating, among other things, its initial and final radius of curvature. He also establishes formulae for the angle of refraction of a light ray passing through the atmosphere under a variety of assumptions about the structure of the atmosphere, i.e., the pressure and temperature gradients and the location of the observer.

Euler's article "On finding the longitude of a location by means of an observation of the angular distance between the moon and a known star" [7], published in St. Petersburg in 1784, was of great practical importance for navigation in that era. Here Euler first solves the problem of calculating the actual geocentric angular distance between the moon and a known star from the elevations of the moon and the star and the angular distance as measured by an observer on the earth's surface, and then shows how, by comparing the geocentric angular distance between moon and star at the location in question with the angular distances calculated at two given times at locations on two standard meridians (Greenwich and the Faroes), one may calculate the time at which the observation at the location of interest was made, and thence the longitude of that location.

Euler was also interested in questions concerning the physical nature of celestial bodies, devoting a great many articles and memoirs to the topic. After examining hypotheses current at the time he came to the conclusion—remarkable for the time—that the tails of comets, the polar auroras, and the zodiacal light[7] are attributable to a single cause, namely the pressure of light on particles in the atmospheres of celestial bodies. He thus predicted the existence of the pressure of light, which was later detected and measured experimentally by the the noted Russian physicist P. N. Lebedev (1886–1912).

Euler's second Petersburg period is likewise characterized by intensive work on astronomical problems, even though from the very beginning of that period he was practically blind.

He continued to perfect the known methods of determining the orbits of comets as part of the theory of perturbed celestial motion. In collaboration with A. Lexell, then an adjunct of the Petersburg Academy, he completed a substantial work on the determination of the elliptical orbit of the 1769 comet. Around the same time Euler brought to completion his extensive investigations, spanning many years, of lunar motion, and began looking into the possibility of applying ideas fundamental in the theory of lunar motion to the study of planetary motion. This possibility was thought of anew, and successfully exploited, in the 1950s and 1960s.

Euler actively participated in the processing of the data from observations of the transit of Venus across the face of the sun in 1769, which had been organized under the auspices of

---

[6]The *zenith distance* of a celestial object is the angular distance along a great circle on the celestial sphere between the zenith (i.e., point of the celestial sphere directly above the observer on earth) and that celestial object.—*Trans.*

[7]"The zodiacal light is a faint, roughly triangular, whitish glow seen in the night sky, which appears to extend up from the vicinity of the sun along the ecliptic or zodiac".—*Trans.*

the Petersburg Academy with the aim of accurately measuring the size of the sun's parallax, a fundamental astronomical constant. Since precise measurement of solar parallax requires that observations be made at widely spaced locations on the earth's surface, the Academy dispatched expeditions to the Kolsk peninsula, Yakutsk, Orenburg, Orsk, and Guryev. The observations of the transit of Venus, and also of the eclipse of the sun that took place on the day, were carried out successfully. The solar eclipse made possible the solution of a problem which in those days presented enormous difficulties, namely that of determining the precise longitudes of the observation posts.

To this period in Euler's life belong also articles on the shape of Saturn's ring. Assuming that Saturn's motion occurs in the plane of the ecliptic[8], Euler calculated the position and size of the ellipse representing the apparent shape of Saturn's ring. In connection with the simplicity of the ultimate formulae Euler made the following comment, wonderfully apt for researchers of any era (and quoted by M. F. Subbotin in his lecture on Euler's works in 1957): "The solution of this problem is highly significant", he writes, "in that via a path involving complicated calculations we finally arrived at very simple formulae; there can be no doubt, therefore, that there is another, much simpler, path to the solution which in fact might have been anticipated; however there is no shame in expounding the present solution since it involves noteworthy devices that may be of use in other researches". And immediately following this pronouncement Euler sets out a very much simpler four-page solution.

2. Unperturbed motion and the calculation of orbits

We now turn to Euler's work in theoretical astronomy.

The laws of unperturbed celestial motion were established by Kepler (1609) and Newton (1687). By the time Euler began his scientific activity, the properties of unperturbed celestial motion emerging from those laws had been investigated sufficiently thoroughly for most practical purposes, but by means of methods belonging in essence to the science of antiquity. In order to solve the immeasurably more complex problems concerning perturbed motion, it was first necessary to reformulate the theory of unperturbed celestial motion in terms of analytic methods using the language of infinitesimals. This program was carried out to completion by Euler in his works on the perturbed motions of the moon and planets. In his earliest works on theoretical astronomy, presented to the Petersburg Academy in the years 1734–1735, he applied the method of infinitesimal analysis to the solution of a few special problems—for example, he deduced Kepler's celebrated equation[9] from the latter's law of areas. As an illustration of the analytic approach, so brilliantly elaborated by Euler in the theory of equations, the following solution by him of a transcendental equation considered by Kepler will serve. Kepler's equation is

$$x = E + \kappa \sin E,$$

where the unknown is the "eccentric anomaly" E. Euler considers that the best approach

---

[8]That is, the plane of the earth's orbit.—*Trans.*

[9]"Kepler's equation gives the relation between the polar coordinates of a celestial body and the time elapsed from a given initial point".—*Trans.*

involves the iterated formula

$$E = x - \kappa \sin(x - \kappa \sin(x - \kappa \sin(x - \cdots)),$$

and to calculate the real anomaly z, given by the standard formulae

$$\cos z = \frac{\cos E + \kappa}{1 + \kappa \cos E}, \qquad \sin z = \frac{(1 - \kappa^2)^{1/2} \sin E}{1 + \kappa \cos E},$$

makes extensive use of the series

$$z = E - \frac{\kappa}{1 \cdot 1} \sin E + \frac{\kappa^2}{2 \cdot 2} \sin 2E - \frac{\kappa^3}{3 \cdot 4}(\sin 3E + 3 \sin E)$$

$$+ \frac{\kappa^4}{4 \cdot 8}(\sin 4E + 4 \sin 2E) - \cdots .$$

Among Euler's works on theoretical astronomy there are several papers containing discussions of "the greatest equalization of the centers of the planets"[10], questions to do with the parabolic paths of comets, and planetary motions.

It is well known that Newton obtained an approximate solution of the problem of the parabolic motion of a comet. Although his graphical solution was found unsatisfactory by many astronomers, nevertheless E. Halley used it to determine the elements[11] of 26 comets. The discovery of comets with elliptical orbits gave rise to the problem of determining orbits of comets without assuming *a priori* that they must be parabolical. Here Euler's ideas provided the basis for a method of determining the orbits of comets in use today; as extended by J.-H. Lambert and J.-L. Lagrange, those ideas were basic to the method devised by C. F. Gauss which served astronomers for over a century.

In connection with the problem of inferring orbits from observational data, Euler's work [9] is noteworthy; here he examines in detail the calculation of the coefficients of the "conditional equations", i.e., the partial derivatives of the geocentric coordinates with respect to tentative elements of the orbit, showing how to overcome difficulties of a technical character related to the complexity of the functions being differentiated. In order to appreciate fully the true significance of this work one need only recall the methodology in use prior to its implementation: Earlier not all the available observational data was used, but only those observations corresponding precisely to the moment of opposition[12], thus providing the geocentric longitude of the planet, or else those giving the planet's longitude at an instant when its latitude becomes zero and its longitude coincides with that of a node[13] of its orbit. Euler's method, on the other hand, opened the way to the use of all of the available observational data.

In this connection Euler's monograph *The theory of the motion of comets and planets* [10] is especially valuable. It is in this book that we find the solution of the fundamental problem of theoretical astronomy: given two radius-vectors, the angle between them, and

---

[10]The original reads literally "the greatest equality of [the] center of [the] planets".—*Trans.*

[11]That is, characteristics.—*Trans.*

[12]A planet is in opposition when the earth lies directly between it and the sun.—*Trans.*

[13]On the celestial sphere, either of the points at which the great circle representing the orbital plane intersects the great circle corresponding to the reference plane (usually the ecliptic). *Encyclopedia of astronomy and astrophysics.* Nature Publishing Group. London, etc., 2001.—*Trans.*

the interval of time taken by the planet or comet to describe this angle, determine a parameter of the orbit from which the whole orbit can be determined. Euler derived an approximate formula of the form (in modern notation)

$$\sqrt{p} = \left(\frac{r_1 r_2}{\tau} + \frac{\tau}{6\sqrt{r_1 r_2}} \right) \sin 2f,$$

where $\tau = k(t_1 - t_2)$.[14] An exact solution of the problem was found by C. F. Gauss in 1809.

The year 1742 marks the appearance for the first time (in the memoir [11]) of the celebrated equation

$$6k(t_2 - t_1) = (r_1 + r_2 + s)^{3/2} - (r_1 + r_2 - s)^{3/2},$$

which provided a basis for the calculation of parabolic orbits, and which subsequently became known as "Euler's equation"[15]. Around the same time Euler published several works on the determination of the orbits of comets, including the popularization *Answer to various questions concerning the structure, motion, and action of comets* [12].

So far we have been considering some of the more concrete, relatively simple problems of celestial mechanics investigated by Euler. In the next section we turn our attention primarily to certain general scientific ideas foreshadowed in Euler's works and playing a key role in the later development of astronomy.

3. The theory of perturbed motion of celestial bodies

In Newton's monumental work *Philosophiae naturalis principia mathematica* there are, among many other things, certain groundbreaking ideas which gave rise to the most universal and effective means of solving nonlinear equations, now known as "perturbation theory". Nevertheless one is fully justified in regarding Euler as the founder of that theory, which thanks to the efforts of leading mathematicians and mechanicians of succeeding eras—especially J.-L. Lagrange, P.-S. Laplace, C. F. Gauss, A. M. Lyapunov, and H. Poincaré—became the basic method of solution of problems of theoretical mechanics, nonlinear oscillations, and theoretical astronomy—including that subject's new offshoot astrodynamics. In Euler's memoirs on celestial mechanics one finds the ideas of perturbation theory formulated in general terms, equally applicable to other branches of science employing dynamical mathematical models, as well as solutions of particular problems of astronomy, remarkable for being "reduced to numbers" for comparison with available observational data.

By the middle of the 18th century the mathematical theory of the unperturbed motion of the celestial bodies of the solar system was to all intents and purposes complete. However, the practical requirements of government—above all with respect to navigation, greater accuracy in time-keeping, and various sorts of calendars—demanded more, and more precise, information about the surrounding world, in particular about the motion of celestial bodies.

---

[14]It would have been considerate of the authors to supply the meanings of all symbols. Such omission is one symptom of the disease known as "arithmorrhoea". Compare the earlier essay on the variational calculus.—*Trans.*

[15]One of many!—*Trans.*

This situation, in conjunction with the intrinsic laws of development of science, provided the impetus for the development of a more substantial branch of theoretical astronomy—the theory of perturbed motion. Without wishing to detract from the importance of the pioneering work of J. d'Alembert and A.-C. Clairaut in this field, one still has to say that it is in the fundamental works of Euler on celestial mechanics that one finds a treatment of the mathematical methods of the theory of perturbed motion unsurpassed in breadth and depth.

To begin, we discuss briefly his theories of lunar motion. Following Newton's discovery of the universal law of gravitation scientists produced around 20 dynamical lunar theories, of which two were Euler's. This overabundance of theories is explained by the moon's exceptional dynamical complexity, and of all these theories only Euler's satisfied the requirements of practice. On the positive side, by virtue of the mathematical complexities of the problem of lunar motion every prominent mathematician was given a rare opportunity to demonstrate the power and scope of his mathematical gift. Furthermore, a substantial moral and mathematical stimulus to tackle the problem of lunar motion was supplied by the thematic prize competitions regularly announced by the science academies of various European states, enhanced by the high professional level of the scientist-academicians who judged the essays and treatises entered in these competitions. In fact one can say that because of the significance of the force of attraction exerted by the sun on the moon, the problem of lunar motion is of a very rare type in celestial mechanics. There is in this problem neither an actual "small parameter" nor a "large parameter" which would allow the formulation of a theory of perturbed motion for the moon using classical expansions in series of positive powers of the small parameter or negative powers of the large one. In modern mathematical terminology, one may say that the problem of lunar motion is amenable to treatment neither as a regular perturbation problem nor as a singular one, requiring instead the creation of some special mathematical method different from classical ones for its solution—a circumstance rendering the problem one of the most complex in celestial mechanics. Many of the eminent mathematicians among Euler's contemporaries—already effectively immortalized by their ingenious solutions of other problems—were defeated by the problem of lunar motion precisely because they sought to solve it as a problem involving small perturbations, i.e., as a regular perturbational problem, when in fact it is simply not possible to obtain a solution in this way since the series arising in this approach diverge, thus yielding no successively more accurate approximations.

Just what did Euler propose in connection with this problem? The short answer is that he made the only possible reasonable suggestion for a problem of this type: if the classical expansion theory of perturbations is not effective, then one needs must create a variant of that theory in which the initial approximation already incorporates a part of the perturbation—preferably a part having greatest impact—so that the further, higher, perturbations applied over and above such an initial approximation are appropriately small. In other words one should, without altering the model in its essentials, reformulate it in such a way that either the regular or singular perturbational approach becomes viable, i.e., reduce the problem to one that has already been solved or to which known procedures apply. This general idea first appears in connection with Euler's lunar theories; it remains as fruitful now as in the 18th century.

A hundred years after the publication of these works of Euler's on lunar theory, G. W. Hill perfected the approach. He proposed taking as the initial approximation of the lunar

Title page of Euler's first *Theory of lunar motion* of 1753.

orbit a so-called variational curve, rather than a Keplerian ellipse, i.e., an initial approximating curve in which part of the perturbing action of the sun on the moon has already been taken into account. Using this approach E. W. Brown (1866–1938) was able to create a sufficiently accurate and long-standing—enduring almost 100 years—description of the moon's motion, now called the Hill–Brown theory of lunar motion.

Euler's first theory of lunar motion, created in the early 1750s, represents the first such theory incorporating as far as possible the extensive observational data available. It was precisely by using these observations in a semi-empirical fashion and disregarding the elegance of the apparatus of mathematical analysis that Euler achieved a result that was exceptional in answering the demands of practice. His first approximation to the solution of the differential equations for lunar motion was used by the Prussian astronomer Tobias Mayer to compute lunar tables of exceptional accuracy for the time. No other contemporary lunar theory was capable of yielding comparable accuracy.

In his first lunar theory Euler substantially developed his method of variation of orbital parameters, subsequently applied not only to the theory of lunar motion—we have in mind here the remarkable works of C.-E. Delaunay (1816–1872) from the mid-19th century—but also to planetary theory. In the latter connection it suffices to recall the fundamental works of J.-L. Lagrange, P.-S. Laplace, Urbain Le Verrier, and Simon Newcombe, who, starting from Euler's ideas, together created a highly accurate theory of motion for the large planets, still forming today the basis of astronomical almanacs. The most complete exposition of Euler's first lunar theory is to be found in his fundamental monograph *Theory of lunar motion* [13], written in Latin, and published in Berlin in 1753 at the expense of the Petersburg Academy.

It is interesting that following the publication of his first lunar theory Euler continued for almost three decades to perfect it, publishing on that theme over ten original works of great scientific importance.

When Euler's "second lunar theory" is mentioned, it is usually with reference to the two essays [14, 15], awarded prizes by the Paris Academy in 1770 and 1772, and published in Paris in 1777. In these works his idea of including in the initial approximation a main perturbational element, i.e., a part of the perturbing action of the sun, is even more prominent. He achieves this by means of an ingenious choice of coordinate system and the analytic form of representation of the functions sought. After determining the initial approximation, he finds the solution for the remaining correction—"the main problem" of the theory of lunar motion and representing a particular case of the three-body problem—in the form of trigonometric series, subdividing the terms of these series into five classes embracing all perturbations caused by the forces of mutual attraction of the system of point-masses sun-earth-moon. In terms of the standards of mathematical exposition Euler's treatment of his second lunar theory serves as a model. And he was the first to demonstrate that in principle it is possible to create a perfected theory even for the most complex of dynamical problems. In 1772 the Petersburg Academy of Sciences published a voluminous treatise with the lengthy, but fully descriptive, title *The theory of lunar motion, treated by means of a new method, including astronomical tables from which the moon's position at any time may easily be calculated. An essay composed under the supervision of Leonhard Euler with incredible zeal and untiring labor by the three academicians J.-A. Euler, W. L. Kraft, and A. J. Lexell* [16]. With this essay, astronomers and mathematicians had at their disposal a clear and accurate algorithm for elaborating and formulating a theory of motion of celestial bodies. However the lunar tables compiled on the basis of this second Eulerian lunar theory were in fact not as accurate as the above-mentioned semi-empirical tables of Mayer, which circumstance may be taken as confirming the invariable law associated with the solution of complex nonlinear problems: an effective solution satisfying the demands of practice can be obtained only by combining in a rational manner abstract theory—i.e., the choice of mathematical model—with a correct interpretation of experimental results— in the present case observational lunar data. Euler more than anyone else understood that the most elegant mathematical model, the most ingenious mathematical fabrication, cannot long survive divorced from practical considerations. This creative credo of Euler's shows through not only in his astronomical works but even in his works in areas so distant from practice, seemingly, as the theory of numbers, of conformal mappings, and of analytic functions of a complex variable.

In assessing Euler's merit in astronomy, one may assert with complete confidence that if he had made no other contribution to astronomical theory than his theories of lunar motion, these would more than suffice for him to be counted among the founders of post-Newtonian classical astronomy. However, his prodigious genius did not confine itself to this single— albeit extremely complex—astronomical problem. In strictly chronological terms, Euler's work on the theory of perturbed motion began with his solution of other, then very topical, problems of planetary motion.

A. J. Lexell
Silhouette from a portrait by F. Anthing, from the 1780s

Two such works were written on a theme prescribed in connection with prize competitions of the Paris Academy, and indeed Euler's researches on the irregularities of the orbits of Saturn and Jupiter [17, 18] won him the prize in 1748 and 1752. In deriving the basic differential equations for these planets' motions Euler assumes that their orbits are contained in the plane of the ecliptic, so that the position of each is determined by its radius-vector and longitude[16]. The projections of the force of attraction in each case on the radius-vector and on a normal to it, yields four second-order differential equations whose right-hand sides are simple expressions in the coordinates of the planets and their distance apart. After analyzing the conditions under which the mean, eccentric, or actual anomaly of one of the planets may be taken as the independent variable, Euler decides on the angle ω between the radius-vectors of the two planets as independent variable, calling it "a new form of anomaly", and then with consummate mastery transforms the basic differential equations accordingly. Turning then to the solution of the equations so obtained by means of a "new method" for which he has prepared his reader at the outset, and which amounts to a sequence of successive approximations, Euler introduces the extremely fruitful yet simple idea of dividing the perturbations into classes, and for each class devises a method for determining the appropriate perturbations. The method he thus develops for finding the perturbed values of the eccentricities and the longitudes of the perihelia formed the starting point for the theory of secular perturbations in trigonometric form, brilliantly worked out subsequently by J.-L. Lagrange in 1788. Nevertheless, no explanation emerged of the large irregularities observed in the motion of Saturn and Jupiter, since Euler confined himself to second-order accuracy in the eccentricities of the orbits of the two planets. That this was the source of the conundrum became clear only in 1784, when P.-S. Laplace found among the third-order perturbations of the longitudes terms of the form

$$
\left.\begin{array}{l} A_1 \\ A_2 \end{array}\right\} \left\{ \frac{1}{5n_\varsigma - 2n_\upsilon} \right\} \begin{array}{l} \sin \\ \cos \end{array} (5n_\varsigma - 2n_\upsilon)t,
$$

---

[16]That is, as observed from earth—on the celestial sphere.—*Trans.*

containing so-called "small denominators"[17]; here $n_\varsigma = 120.4547''$ and $n_\upsilon = 299.1283''$[18] are the mean motions[19] of Saturn and Jupiter respectively.

Euler's third long memoir on planetary perturbations [19] was awarded a double prize by the Paris Academy in 1756. This work is made up of a preface, where the indispensability of creating a gravitational theory of the Earth's motion is explained, a first part, where a general method for calculating planetary perturbations is expounded, and a second part "containing an application of the theory to the motion of the earth and the perturbations of it arising from the action of other planets".

It should be noted that Euler was the first to pay attention to the displacement through perturbations of the plane of the ecliptic, i.e., of the coordinate plane basic to the theory of planetary motion, a problem which he pursued in the article "On the alteration of the latitude of the fixed stars and the inclination of the ecliptic" (1756) [20], and in the article "On establishing a fixed great circle on the celestial sphere for ... the orbits of planets and comets" (1776) [21], written 20 years later!

Euler notes the large discrepancies between the various astronomical tables then in existence; for example, for the secular motion of a node of Saturn's orbit, Cassini's[20] tables give $1°35'11''$, while Halley's tables give $0°30'0''$. He was also the first to propose taking the plane of the ecliptic as it was at some fixed time (such as the beginning of the 1700s) as the basic coordinate plane, to which all observations and tables would be referred. This proposal was taken up and remains in effect today.

In the 1770s Euler devoted several articles [22–24] to the theory of the perturbations of the earth's motion under the influence of Venus. His earlier paper "A more accurate investigation of the perturbations of the earth's motion caused by the moon" [25] has the same general theme.

In the last years of his life Euler published several more articles on planetary motion, amongst which the following two generalize earlier work: "A new method of producing astronomical tables for the motion of the planets" [26] and "A new method of determining the motion of the planets" [27]. In the latter work he applies to the study of unperturbed planetary motion his own "lunar" method, whereby the position of a planet in the plane of its orbit is given by heliocentric coordinates X, Y of a synodic[21] reference frame, where the axis of abscissas passes through the mean position of the planet. Then, setting $X = a(1 + x), Y = ay$, where a denotes the major semiaxis of the orbit, he obtains

$$x = eP + e^2Q + e^3R + \cdots, \qquad y = ep + e^2q + e^3r + \cdots,$$

where e is the eccentricity of the orbit, and the coefficients $P, Q, R, \ldots, p, q, r, \ldots$ are periodic functions of the mean anomaly, which is determined from the solution of a certain system of differential equations with constant coefficients.

Euler also devoted a series of outstanding memoirs to general mathematical questions

---

[17] "So-called" perhaps because $5n_\varsigma - 2n_\upsilon$ is indeed small.—*Trans.*

[18] In the original, instead of the subscripts ς and υ, standard stylized symbols for the two planets are used.—*Trans.*

[19] The *mean motion* of a planet is "the average number of degrees the planet moves in one solar day".—*Trans.*

[20] G. D. Cassini (1625–1712).—*Trans.*

[21] The term "synodic" is also used in the phrase "synodic period", which is "the time interval between two successive alignments of two celestial bodies. The moon's synodic period is the time between successive recurrences of the same phase".—*Trans.*

concerning translational and rotational motion of celestial bodies, the settling of which was crowned with the laying of the foundations of two important subdisciplines of mechanical science: the mechanics of a rigid body and the theory of the Newtonian potential. In particular, he applied with panache the equations of motion of a rigid body that he himself had established to an analysis of the earth's rotational motion about its axis.

4. The problem of the integrability of the equations of celestial mechanics

Around a dozen of Euler's works concern the very important and interesting problem of the integrability of the equations arising in the three-body problem, considered as an abstract mathematical model independently of any detailed consideration of possible applications to astronomy. Note that the problem of the integrability of differential equations generally, and of the differential equations arising in the natural sciences in particular, is of course as topical in our day as it was in Euler's. If crowned with success, investigations into integrability in the first place very significantly enhance the accuracy of calculations, and secondly economize on the labor of researchers and the use of computational technology.

In one of his memoirs Euler states that "the celebrated three-body problem, which arises from the study of lunar motion, is still too far beyond the power of analysis for one to be able to hope to find a complete solution", and in another [28] he writes: "From this I draw the undeniable conclusion that one cannot hope to solve the general case of the three-body problem while no means is known for solving it even in the case where the three bodies move along one and the same line". Pursuing this idea, Euler did manage to find solutions in special cases of the one-dimensional three-body problem, today called "collinear libration[22] points". In the lecture mentioned at the beginning of the present essay, delivered at the meeting dedicated to the 250th anniversary of Euler's birth, M. F. Subbotin noted Euler's priority in the discovery of these particular solutions. In 1772 J.-L. Lagrange found "triangular libration points", and subsequently the adjective "Lagrangian" became a widespread, though not completely accurate, term in celestial mechanics embracing all known particular solutions of the three-body problem. It would be more appropriate to use the terms "Eulerian collinear solutions" and "Lagrangian triangular solutions" in specifying the particular solutions of the three-body problem known at the time.

Solutions of special cases of the Newtonian three-body problem played a key role in the development of stability theory in Hamiltonian mechanics, the theory of dynamical systems. In order to appreciate the difficulty of stability theory as it applies to the three-body problem, it suffices to point out that the complete solution of the problem of the stability of the Lagrangian triangular solutions in one special model of the three-body problem— now known as the "bounded circle variant"[23] of that problem—was found only 200 years after the discovery of those solutions by Lagrange. Over this long time span many emi-

---

[22]*Libration points* (or *Lagrangian points*) are the five positions in interplanetary space (or in the three-body problem) where a small object affected only by gravitation can theoretically be stationary relative to two larger bodies".—*Trans.*

[23]"In the 'restricted' three-body problem one of the masses is assumed to be small enough not to influence the motions of the other two, which are assumed to move in circular orbits about their centres of mass". Is this what is intended here?—*Trans.*

nent scientists, such as Lagrange himself, P.-S. Laplace, C. F. Gauss, C. G. J. Jacobi, H. Poincaré, and A. M. Lyapunov, worked on the problem. To Lyapunov belong the most general results on the stability to a first approximation of the Lagrangian triangular libration points. The passage from "stability to a first approximation" to "stability in the sense of Lyapunov" in this remarkable astronomical problem required the application of a new mathematical apparatus known as "the metrical theory of differential equations", created by the German mathematician C. L. Siegel and independently by our brilliant compatriot A. N. Kolmogorov. This "metrical theory" has since evolved into one of the most powerful tools of the modern qualitative and analytic theory of differential equations and finds application in modern celestial mechanics and the mechanics of rigid bodies.

The subsequent development of celestial mechanics showed how great a debt astrodynamics, the newest branch of astronomy, owes to a memoir of Euler's published in 1766 under the title "On the motion of a body attracted to two fixed centers of force" [29]. Here Euler found for the first time a general integral of the planar problem of two fixed centers, expressed in terms of elliptic integrals and elliptic Jacobi functions. A few years later J.-L. Lagrange found the general solution of the spatial problem of two fixed centers, again in terms of elliptic integrals and elliptic Jacobi functions. The "problem of two fixed centers" is of course the special case of the Newtonian three-body problem where one studies a passively gravitating body in the field produced by two fixed attractive centers. An analogous problem from physics is that of the motion of a charged particle in the field of an electric dipole.

Notwithstanding the elegant solution obtained by Euler and then in the general situation by Lagrange, for two centuries the problem of two fixed centers failed to find a significant application. Part of the reason for this is as follows: Two fixed centers with real masses situated in real Euclidean space produce a potential field close to that in the exterior of a single stretched body, i.e., a body shaped somewhat like a cucumber or the solid of revolution of an ellipse about its major axis. However all of the larger planets and satellites of the solar system are somewhat oblate in shape rather than elongated solids of revolution—resembling more closely solids of revolution of ellipses about their minor axes—so that the use of the Euler-Lagrange classical solution of the problem of two fixed centers in order to obtain an approximation to the external potential field of a planet or a planet's satellite, was not appropriate. But on the other hand if instead one considers the so-called "generalized problem of two fixed centers", i.e., the mathematical model where two points with *complex* masses produce a potential field in *complex* Euclidean space, through which a third passive gravitating point-mass moves, it turns out that the resulting field does approximate to sufficient accuracy the potential field of oblate solids of revolution flattened at their poles, like, roughly speaking, the larger planets of the solar system and their natural satellites.

Thus thanks to the generalized problem of two fixed centers, a procedure became available for obtaining sufficiently good approximate mathematical descriptions of the external gravitational fields of the planets, in particular the earth. And furthermore the solutions in elliptic quadratures of the problem of two fixed centers, obtained by Euler and Lagrange, furnished investigators with a splendid mathematical tool—an integrable model—for adaptation to research into the dynamics of the satellites of the nearby planets.

Of primary interest in this regard are the artificial satellites of the earth (ASEs), which undergo large perturbations of their orbits caused by the earth—considered now as a sphere flattened at the poles and no longer as a point-mass. And the earth's atmosphere represents a second very significant cause of perturbations of ASEs.

Because these two sources of perturbations of ASEs are so significant, the problem of modelling the dynamics of ASEs constitutes a large perturbation problem, and a practical solution can be found only by means of the methodology briefly described above in connection with the Eulerian lunar theories. In the problem of the motion of ASEs the classical expansions are no longer effective and a "reformulation" of the problem is called for similar to that effected in connection with the calculations of lunar motion. The difference in the present situation consists in the difference in origin of the force causing the predominating perturbations: in the case of the ASEs it is the oblate shape of the earth, while for the moon it is the gravitational attraction of the sun. The realization of these ideas in practical astrodynamics—more specifically in the high-altitude theory of artificial earth satellites created chiefly by Soviet scientists—serves as a monument to the scientific achievements of Euler, and to a definite degree demonstrates the unbreakable organic connection between the science of the mid-18th century and that of our own epoch.

5. Numerical integration. Works in applied astronomy

In 1763 Euler published an article in the *Memoirs* of the Berlin Academy with the purely astronomical title "Remarks on the three-body problem" [30], which was destined to play, however, a leading role in the development of new methods of numerical integration of differential equations. In this paper Euler invented one of the most efficient difference methods of integration of the differential equations of celestial mechanics, formulated in rectangular Cartesian coordinates, and rediscovered a century and a half later, at the beginning of the 20th century, by J. D. Cole. The method is now called Cole's method, although it would be more just to call it "the Euler-Cole method". The importance of this work of Euler's extends well beyond the confines of celestial mechanics—in fact it contains an exposition of the basic preconditions which later led to the well-known method of representing solutions in the form of series in powers of the time, i.e., the series method of Sophus Lie. It is worthwhile dwelling for a moment on these two aspects of this work since his analytic method, regarded as the Euler-Cole method on the one hand and on the other as the forerunner of the method of time series, is ideally suited for computer implementation. This in essence explains the popularity of the method among specialists in celestial mechanics.

We have above briefly illumined certain of the more interesting or more important of the achievements of Leonhard Euler in theoretical astronomy and celestial mechanics. These have proved invaluable for astronomy, mathematics, and mechanics, determining many of the paths of development of these fields a century ahead. However in addition to those main works, Euler wrote several dozen scientific articles devoted to a wide variety of questions of applied astronomy, geodesy, and geography. The breadth of his scientific interests and his extraordinary professionalism in fields of knowledge quite different from one another is simply flabbergasting. Here are a few examples illustrating this.

Returning to Euler's essay [2], we note that it contains the fundamentals of the dynamical theory of the tides, thanks to which that science took a large step beyond the Newtonian statical theory of a half-century earlier. As mentioned above, in 1740 Euler was awarded a prize by the Paris Academy of Sciences for this work, the second of a dozen so honored by that Academy. At roughly the same time, in 1739, he published an article with the interesting title "The determination of the degree of warmth and cold at various places on the earth and at various times" [31], the first-ever work, it would seem, on meteorology.

By appointment of the Petersburg Academy of Sciences, Euler headed the Academy's Geography Department for a number of years, and his work in this post—as indeed in every position he held—was in the highest degree useful to Russia. Apart from the series of outstanding treatises on mathematical cartography and geodesy that he published, Euler collaborated with G. Heinsius on a project to prepare a map of Russia, and then participated directly in the production of the resulting *Geographic Atlas of the Russian Empire*, published in 1745. That rare figure, a universal scientist of genius, Euler set great store not only by abstract theoretical research but also—and to the same degree—by the results of experiment and observation, intimately comprehending the importance of the methods used to analyse and interpret such data. In fact he was the first of the great scientists of his age to formulate the fundamental problems arising in connection with the processing of experimental data, i.e., those relating to the theory of errors, subsequently solved so brilliantly by P.-S. Laplace and especially C. F. Gauss. Here the astronomical observations made during the various eclipses that took place over his lifetime played their part, as did the data brought back from the geographical expeditions of the mid-18th century to far-flung reaches of the Russian empire.

Conclusion

Euler's correspondence with his contemporaries must be included as part of his priceless creative heritage. In those letters one finds a wealth of fresh scientific questions, solutions of new problems, conclusions concerning a great variety of topics, ranging from the philosophical to the everyday, and, finally, intelligent thoughts of a general nature. Reading these letters, one cannot but be impressed by his philanthropic nature, and by his sincere respect for every correspondent irrespective of title, authority, or social standing. On the other hand, the manner in which he defends his scientific views and results in the face of criticism not always just or well-founded, is instructive and worthy of emulation. Although he always showed respect for the point of view of an appropriate opponent, he did not permit his own take on a subject to be undervalued or derided and refused to compromise his principles. The single and unwavering motive behind his abundant correspondence with contemporary scientists was the attainment and defense of scientific truth. In a large number of the letters there are discussions of astronomical topics. Many of Euler's letters to the mathematician and mechanician P.-L. Maupertuis, and the astronomers N. L. de la Caille[24], T. Mayer, G. Heinsius, J. N. Delisle, and others, constitute by themselves remarkable discourses on astronomy, cogently and elegantly argued and dealing with the most topical scientific problems of the time, for instance the problem of the earth's shape—then a burning question in all academies—and the related question of the interpretation of the measurements of longitude made on several continents, as well as problems at the juncture of astronomy and mechanics and a great many other exceedingly interesting problems of natural science.

---

[24]Nicholas Louis de la Caille (1713–1762).—*Trans.*

We would like to say in conclusion that the works of Leonhard Euler continue in our day to serve as an almost inexhaustible source of fresh creative ideas, so that studying them today is just as appropriate and useful as it was during the lifetime of the great scientist.

References

[1] Subbotin, M. F. "The astronomical works of Leonhard Euler". In: *Leonhard Euler*. M.: Izd-vo AN SSSR (Publ. Acad. Sci. USSR), 1958. pp. 268–376.

[2] Euler, Leonhard. "Inquisito physica in causam fluxus et refluxus maris". Rec. pièces remp. prix Acad. Sci. Paris (1738–1740). 1741. Vol. 4. pp. 235–350; *Opera* II-31.

[3] ———. "Meditationes in quaestionem, quibusnam observationibus mari, tam interdiu quam nocti, itemque durante crepusculo verum temporis momentum commodissimo et certissimo determinari queat?" Rec. pièces remp. prix Acad. Sci. Paris (1745–1748). 1750. Vol. 6. pp. 111–167; *Opera* II-20.

[4] ———. "Mémoire sur l'effet de la propagation successive de la lumière dans l'apparition tant des planètes que des comètes". Mém. Acad. Sci. Berlin (1746). 1748. Vol. 2. pp. 141–181; *Opera* III-5.

[5] ———. "De la parallaxe de la Lune, tant par rapport à sa hauteur qu'à son azimut, dans l'hypothèse de la Terre sphéroidique". Mém. Acad. Sci. Berlin (1750). 1751. Vol. 5. pp. 326–328; *Opera* II-30.

[6] ———. "De la réfraction de la lumière en passant par l'atmosphère selon les divers degrées tant de la chaleur que de l'élasticité de l'air". Mém. Acad. Sci. Berlin (1754). 1756. Vol. 10. pp. 131–172; *Opera* III-5.

[7] ———. "De inventione longitudinis locorum ex observata Lunae distantia a quadam stella cognita". Acta Acad. Sci. Petrop. (1780:2). 1784. pp. 301–307. *Opera* II-30.

[8] ———. "De apparitione et disparitione annuli Saturni". Acta Acad. Sci. Petrop. (1777:1). 1778. pp. 288–316; *Opera* II-30.

[9] ———. "Emendatio tabularum astronomicarum per loca planetarum geocentrica". Comment. Acad. Sci. Petrop. (1740). 1750. Vol. 12. pp. 109–221; *Opera* II-29.

[10] ———. *Theoria motuum planetarum et cometarum, continens methodum facilem ex aliquot observationibus orbitas cum planetarum tum cometarum determinandi.* Berolini, 1744; *Opera* II-28.

[11] ———. "Determinatio orbitae cometae qui mense Martio hujus anni 1742 potissimum fuit observatus". Miscellanea Berolinensia. 1743. Vol. 7. pp. 1–90; *Opera* II-28

[12] ———. *Beantwortung verschiedener Fragen über die Beschaffenheit, Bewegung und Wirkung der Cometen, Forsetzung dieser Beantwortung.* Berlin, 1744; *Opera* II-31.

[13] ———. *Theoria motus Lunae exhibens omnes eius inaequalitates etc.* Berolini, 1753; *Opera* II-23.

[14] ———. "Théorie de la Lune". Rec. pièces remp. prix Acad. Sci. Paris (1764–1772). 1777. Vol. 9; *Opera* II-24.

[15] ———. "Nouvelles recherches sur le vrai mouvement de la Lune, où l'on détermine toutes les inégalités auxquelles il est assujetti". Rec. pièces remp. prix Acad. Sci. Paris (1764–1772). 1777. Vol. 9; *Opera* II-24.

[16] ——. *Theoria motuum Lunae, nova methodo pertractata, una cum tabulis astronomicis, unde ad quodvis tempus loca Lunae expedite computari possunt, incredibili studio atque indefesso labore trium academicorum: J. A. Euler, W. L. Krafft, J. A. Lexell, opus dirigente Leonhardo Eulero*. Petropoli, 1772; *Opera* II-22.

[17] ——. "Recherches sur la question des inégalités du mouvement de Saturne et de Jupiter". Rec. pièces remp. prix Acad. Sci. Paris (1748). 1749. Vol. 6. (The memoir appears only in certain copies of this collection.) *Opera* II-25.

[18] ——. "Recherches sur les irrégularités du mouvement de Jupiter et de Saturne". Rec. pièces remp. prix Acad. Sci. Paris (1751–1761). 1769. Vol. 7; *Opera* II-26.

[19] ——. "Investigatio perturbationum, quibus planetarum motus ob actionem eorum mutuam afficiuntur". Rec. pièces remp. prix Acad. Sci. Paris (1753–1760). 1771. Vol. 8; *Opera* II-26.

[20] ——. "De la variation de la latitude des étoiles fixes et de l'obliquité de l'écliptique". Mém. Acad. Sci. Berlin (1754). 1756. Vol. 10. pp. 296–336; *Opera* II-29.

[21] ——. "De circulo maximo fixo in coelo constituendo, ad quem orbitae planetarum et cometarum referantur". Novi comment. Acad. Sci. Petrop. (1775). 1776. Vol. 20. pp. 503–508; *Opera* II-30.

[22] ——. "De perturbatione motus Terrae ab actione Veneris oriunda". Novi comment. Acad. Sci. Petrop. (1771). 1772. Vol. 16. pp. 426–467; *Opera* II-26.

[23] ——. "Réflexions sur les inégalités dans le mouvement de la Terre, causées par l'action de Vénus". Acta Acad. Sci. Petrop. (1778:1). 1780. pp. 297–307; *Opera* II-27.

[24] ——. "Investigatio perturbationum, quae in motu Terrae ab actione Veneris producuntur". Acta Acad. Sci. Petrop. (1778:2). 1781. pp. 308–316; *Opera* II-27.

[25] ——. "Quantum motus Terrae a Luna perturbatur accuratius inquiritur". Novi comment. Acad. Sci. Petrop. (1747–1748). 1750. Vol. 1. pp. 428–443; *Opera* II-23.

[26] ——. "Nova methodus motus planetarum principalium ad tabulas astronomicas reducendi". Novi comment. Acad. Sci. Petrop. (1773). 1774. Vol. 18. pp. 354–376; *Opera* II-29.

[27] ——. "Nova methodus motum planetarum determinandi". Acta Acad. Sci. Petrop. (1778:2). 1781. pp. 227–301; *Opera* II-29.

[28] ——. "Tria capita ex opera quodam majori inedito de theoria Lunae". *Opera postuma*. Petropoli, 1862. Vol. 2. pp. 365–390; *Opera* II-24.

[29] ——. "De motu corporis ad duo centra virium fixa attracti". Novi comment. Acad. Sci. Petrop. (1764). 1767. Vol. 11. pp. 152–184; *Opera* II-6.

[30] ——. "Considérations sur le problème de trois corps". Mém. Acad. Sci. Berlin (1763). 1770. Vol. 19. pp. 194–200; *Opera* II-26.

[31] ——. "Determinatio caloris et frigoris graduum pro singulis Terrae locis ac temporibus". Comment. Acad. Sci. Petrop. (1739). 1750. Vol. 11. pp. 82–99; *Opera* III-10.

Euler and the Evolution of Celestial Mechanics[1]

N. I. Nevskaya and K. V. Kholshevnikov

Introduction

The works of Leonhard Euler on astronomy — especially those relating to celestial mechanics — have by now been examined in sufficiently great detail; see, for example, the excellent sketch by M. F. Subbotin [1], which contains an almost exhaustive bibliography. Here we consider only the most important of those works Euler produced as a St. Petersburg academician, i.e., those recognized as fundamental to modern celestial mechanics. According to S. I. Vavilov [2, p. 144] Euler's mathematical genius obviously "lacked physical intuition", so that the mathematician in him suppressed the physicist. However it is impossible to maintain this of Euler the astronomer. In his astronomical work one senses a profound astronomical intuition and professionalism at work. Although the sheer quantity of his astronomical output — over ten volumes' worth — would defeat any modern astronomer, the main consideration in assessing that work is of course acuity of astronomical intuition, so often lacking in mathematicians who move into our field. Here are some examples.

In resorting to series for calculating the ephemerides[2] of celestial bodies, Euler never wrote down unnecessary terms, always keeping a close check on the accuracy. In this connection he would look for variables in terms of which the series expansions converged more quickly. Euler the mathematician did not hesitate to use divergent series. Euler the astronomer insisted on their rapid convergence.

The theory of lunar motion involves six parameters. Euler left the number at eight, having failed to discover two relations connecting them. (These were found much later by G. W. Hill.) However, appreciating the extreme importance of quickly producing a lunar theory in order to meet the urgent requirements of the imperial fleet, the practical Euler

---

[1]Definitions of some of the astronomical terms appearing in this essay may be found in footnotes to the preceding one.—*Trans.*

[2]"An *ephemeris* (pl. *ephemerides*) was a table providing the positions of the sun, moon, planets, asteroids, or comets in the sky at a given moment of time. A *modern* planetary ephemeris comprises software generating the positions of such bodies at any desired time".—*Trans.*

decided to simply assign values to two of those quantities "out of the blue", as he put it—a step no pure mathematician would dare to take. Here "out of the blue" meant that the mean motions of the lunar perigee[3] and nodes were taken straight from observational data.

After long-lasting and unsuccessful attempts to find integrals of the equations arising in the three-body problem not reducing to classical ones, Euler ultimately came to the conclusion that such integrals, even if there were any, would not be of any practical use—in the first place because they would have to be of such complexity that solving them would likely be just as difficult as direct integration of the original equations. Now, almost a quarter of a millenium later, there remains no doubt whatever as to the truth of this contention. In the second place, the advantage, often emphasized, of analytic methods over numerical ones, consisting in the possibility of calculating the state x of a system at any time t, in practice turns out to be—with rare exceptions—illusory, essentially because the most minute error $\Delta x|_{t=0}$ in the initial conditions can easily lead to a large one Δx for sufficiently large t. This idea of Euler's was in fact far in advance of his time, when Laplacian determinism was the slogan, and became generally accepted only at the end of the 19th century or even later, when scientists ceased perceiving the difference between classical and quantum mechanics as an impassable abyss.

We now turn to specific problems considered by Euler in order to convince the reader that the basis of almost every branch of modern astronomy was established by him.

1. The two-body problem

Euler produced a simple analytic solution of this problem, accessible to every student, in contrast with Newton's geometrical solution.

He also found most of the presently known exact series expansions of the basic functions arising in the two-body problem, for example the following Fourier series in the real anomaly for the equalizing of the center[4]:

$$\Theta - M = \sum_{k=1}^{\infty} \frac{2(-1)^{k+1} e^k (1 + k\sqrt{1 - e^2})}{k(1 + \sqrt{1 - e^2})^k} \sin k\Theta. \tag{1}$$

(Incidentally, J. Fourier had not yet been born at that time, and his article on "Fourier series" appeared only in 1822.)

Euler's contribution to the problem of determining elliptic, and especially parabolic, orbits is invaluable. He posed and solved a great many special problems, such as that of determining an orbit from two positions and a single parameter, and from three geocentric directions. In the parabolic case he discovered a relation later termed the "Euler-Lambert formula". He also solved a quite large number of practical problems, such as that of determining a parabolic orbit from five and from four observations, the orbit of a comet intersecting the plane of the ecliptic twice, and so on.

Euler provided a mathematical formulation of the problem of improving approximations of orbits both purely within the framework of the two-body problem, and taking perturbations into account.

---

[3] The position of closest approach to the earth.—*Trans.*

[4] As in the preceding essay, the phrase "equalizing of the center" (of the planetary orbits?) demands explanation for the uninitiated; and it would be mere courtesy to tell us (the uninitiated) exactly what Θ and M stand for precisely.—*Trans.*

There is perhaps only one problem pertaining to Keplerian motion on which research began after Euler, namely that of determining the domains of convergence of the basic series expansions of celestial mechanics.

2. The problem of two fixed centers

To Euler belongs the merit of having reduced this problem to quadratures. However he desisted from a detailed investigation of these integrals when the irksome circumstance came to light that the motion [of a point-mass] in the gravitational field of two fixed point-masses is inadequate as a model of such three-body systems as Sun-Earth-Moon and Sun-Jupiter-Saturn. It seems likely that here the astronomer in him took precedence over the mathematician, for whom any practical application of a problem's solution is a matter of indifference.

Very much later E. P. Aksenov, E. A. Grebenikov, and V. G. Demin published articles revealing an unexpected application of the solution of this problem when complex values of the parameters are admitted, namely to the motion of a satellite of an aspherical planet. The Department of Celestial Mechanics of Leningrad University resolved to call motion in a field produced by two fixed centers "Eulerian", a term now standard.

3. The abstract three-body problem

Having convinced himself of the non-integrability of the general three-body problem, Euler turned to the investigation of special cases—to the collinear and restricted[5] versions of the problem.

The collinear case yielded for the first time exact particular solutions of the three-body problem. By rotating the line on which the three bodies are assumed to lie with appropriate angular velocity, one discovers states of relative equilibrium, the "collinear libration centers". Euler was also the first to find special solutions of the restricted problem, including certain triangular libration points[6].

By means of a close examination of his special solutions Euler was led to the fruitful idea of limiting the motion of satellite or planetary type to a certain surface containing the points of libration L_1 and L_2. Later developments of this idea ultimately led to the concept of gravitational spheres, or spheres of attraction or action, due to Hill and others, which now play a significant role in celestial mechanics and astrodynamics.

4. Osculating elements[7]

This, one of the central concepts of modern celestial mechanics, was first given a strict definition by Euler. He was also the first to derive analytically the relations defining the change in the osculating elements with time—the so-called "Eulerian differential equations", giving rise to the analytic theory of perturbed motion in terms of osculating elements, so successfully applied in his investigations of the orbits of Jupiter, Saturn, Earth, Venus, and other celestial bodies.

---

[5] "In the 'restricted' three-body problem one of the masses is taken to be small enough not to influence the motions of the other two, which are assumed to move in circular orbits about their centers of mass".—*Trans.*

[6] "The *libration points* are the five positions in interplanetary space where a small object affected only by gravitation can theoretically be stationary relative to two larger bodies".—*Trans.*

[7] *Osculating elements* are parameters specifying the instantaneous position and velocity of a celestial body in its perturbed orbit. Thus they determine the unperturbed (two-body) orbit that the body would follow in the absence of perturbations.—*Trans.*

5. The theory of lunar motion

Euler's contributions to the theory of lunar motion are also of inestimable value, along with resulting applications to practical problems such as the determination of a ship's longitude by means of lunar distances and the occultation of stars by the moon.

Euler's lunar theory in its final form defined the subject for a century after his death. One can find in it the ultimate sources of the modern theory of nonlinear oscillations and the method of averaging. We limit ourselves to these brief comments since an exhaustive commentary can be found in the essay [3] by A. N. Krylov.

6. The problem of two rigid bodies of finite dimensions

Having become interested in the motion of the Earth-Moon pair and the Galilean satellites of Jupiter, Euler began simultaneously to elaborate a theory of gravitational potential for bodies nearly spherical in shape, and a theory of the motion of a free rigid body. What, in modern terminology, Euler found in the first of these theories was the expansion of the gravitational potential in solid spherical harmonics up to and including the second-order term in the general case, and specifically for an ellipsoid. On the other hand on the basis of his dynamical and kinematical equations for the motion of a rigid body as applied to the earth, he was able to explain the precession and nutation of the earth's axes. In particular, he predicted free oscillations, i.e., independent of the action of the moon, of the earth's rotational axis, detected much later by Chandler[8].

Although in the case of the earth the flattening at the poles has little effect on the moon's motion, this is not at all the case with Io. In the first place, Io is much closer to Jupiter than the moon to the earth, relative to the radii of the respective planets, and in the second, Jupiter is much more oblate in shape than the earth. With the Galilean moons of Jupiter specifically in mind Euler created a theory of motion for satellites of oblate planets, finding in particular the secular motion of the line of apsides[9] and the line of nodes.

One can point to more than a dozen papers published just before and just after the launching of the first artificial satellites of the earth (ASEs), in which these results of Euler were rediscovered—initially in less accurate form. It follows that we are fully justified in considering the great Petersburg academician the founder of the first analytic theories of motion of ASEs.

7. Integration by numerical methods and by power series in the time

It is well known that with his method involving continuous broken line segments, Euler laid the foundations of the numerical approach to the integration of the differential equations of celestial mechanics. In general use for this purpose at the present time is the Taylor-Stevenson method, involving the representation of the solution of an equation of the form

$$\frac{dx}{dt} = f(x, t), \tag{2}$$

---

[8]"Seth Carlo Chandler, an American astronomer, discovered in 1891 a small variation in the earth's axis of rotation, known as the 'Chandler wobble'".—*Trans.*

[9]The major axis of an elliptical orbit.—*Trans.*

where x and f are vector-functions and t is a scalar (the time), as a power series in the time t:

$$x = \sum_{k=0}^{\infty} c_k (t - T)^k. \tag{3}$$

The coefficients c_k are calculated directly from the given right-hand side f and the initial conditions $X = x|_{t=T}$: in terms of the differential operator

$$D = f(X, T)\frac{\partial}{\partial x} + \frac{\partial}{\partial t}, \tag{4}$$

the coefficient c_k is given by

$$c_k = \frac{1}{k!} D^k X. \tag{5}.$$

The series (3) with coefficients (5) is call a "Lie series" after the famous Norwegian mathematician who so minutely studied its properties. However the formula (5) is essentially the same as one appearing in Euler's article [4], so that the Lie series should actually be called the "Euler-Lie series".

8. The determination of the mass of Halley's comet

The determination of the masses of comets was a popular problem of 18th century astronomy. Early in the century Buffon[10], on the basis of the comets' gigantic sizes, proposed that their masses must be comparable with that of the sun. In order to estimate the mass of Halley's comet Euler calculated the perturbation of the earth's orbit caused by the passage of that comet in April-May 1759. Here analytical methods are ineffective. Euler first considered the situation under the assumption that the mass m_{com} of the comet is equal to the earth's $m_{\oplus}$. He also assumed that the comet's orbit was elliptical, as the observational data suggested, from which it followed that there should exist first-order perturbations of the earth's orbit. Since these perturbations would be linear in m_{com}, this would allow the mass of the comet to be calculated from the computed and observed changes in the elements of the earth's orbit[11], which Euler was able to find. We quote the most arresting passage from [5, p. 315], since its translation from Latin in [1, pp. 333–334] is inaccurate: "...If the comet is equal to the earth [in mass], the year will lengthen by 27 minutes... But then all the more if the comet exceed the earth 100 times, the year will lengthen by 45 hours ...". Since observations revealed no measurable changes in the earth's motion, Euler had demonstrated that in fact the masses of the comets are less than planetary masses by several orders.

9. The anthropic principle

Although properly speaking this topic does not relate immediately to celestial mechanics, we found ourselves unable to refrain from mentioning it here. This principle states that the fact of the existence of observable objects and the observer himself imposes severe restrictions on the possible kinds of laws of nature and their realization. Usually this principle

---

[10]Georges-Louis Leclerc, Comte de Buffon (1707–1788).—*Trans.*

[11]Or, conversely, a change in an element estimated from a given hypothesized value of m_{com}.—*Trans.*

is considered as emerging recently. However, the anthropic principle—albeit less precisely stated, tinged with the spirit of the time, and given a religious slant—can be found in Euler. He asks what would happen if the moon "was created" in an orbit significantly closer or more distant from the earth. In the first case it would cause very strong tides mortally dangerous to humanity. In the second case, under the action of solar perturbations the moon would ultimately leave the earth's vicinity. If Euler had asked himself what would happen if the inclination of the moon's orbit were changed to close to 90°, he would certainly have concluded—just as M. L. Lidov did in 1961 [6]—that in this case the moon would have fallen to earth. Thus it would not be possible for us to observe the moon in any position substantially different from its present one—and is this not essentially the anthropic principle?

We would like to summarize our little essay by saying that the seeds of analytic, numerical, qualitative, and applied celestial mechanics are all to be found in the works of Leonhard Euler.

References

[1] Subbotin, M. F. "The astronomical works of Leonhard Euler". In: *Leonhard Euler.* M.: Izd-vo AN SSSR (Published by Acad. Sci. USSR), 1958. pp. 268–376.

[2] Vavilov, S. I. "The physical optics of Leonhard Euler". In: *Collected essays.* M.: Izd-vo AN SSSR (Publ. Acad. Sci. USSR), 1956. Vol. 3. pp. 138–147.

[3] Euler, L. *A new theory of the motion of the moon.* Translation from Latin into Russian, with comments and explanations, by Academician A. N. Krylov. L.: Izd-vo AN SSSR (Publ. Acad. Sci. USSR), 1934. 208 pages.

[4] ——. "Methodus universalis serierum convergentium summas quam proxime inveniendi". Comment. Acad. Sci. Petrop. (1736). 1741. Vol. 8. pp. 3–9; "Inventio summae cuiusque seriei ex dato termino generali". Ibid. pp. 9–22; "Methodus universalis series summandi ulterius promota". Ibid. pp. 147–158; *Opera* I-14.

[5] ——. "Astronomia mechanica: Digressio, qua effectus cometae A.1759 expectati in motu Terrae perturbando investigatur". *Opera postuma.* Petropoli, 1862. Vol. 2. pp. 294–316; *Opera* II-27.

[6] Lidov, M. L. "On an approximate analysis of the evolution of the orbits of artificial satellites". In: *Problems of the motion of artificial celestial bodies. Lectures delivered at a conference on general and applied questions of theoretical astronomy, Moscow, 1961.* M.: Izd-vo AN SSSR (Publ. Acad. Sci. USSR), 1963. pp. 119–134.

New Evidence Concerning Euler's Development as an Astronomer and Historian of Science

N. I. Nevskaya

Leonhard Euler's position in science is mainly that of a pre-eminent mathematician and mechanician. However he also worked on problems of physics and astronomy—especially celestial mechanics, which he called "astronomical mechanics" [1, p. 84]. In this field his contribution was exceptionally large; along with A.-C. Clairaut, J. d'Alembert, and P.-S. Laplace, he must be counted one of the founding-fathers of modern celestial mechanics. His works in this field have been examined in detail in M. F. Subbotin's survey [2], which is all the more valuable in that its author was one of the most outstanding specialists in the field. Euler's works in geodesy and cartography [3, 4], in physical and astronomical optics [5–10], and in astrophysics [11, 12], have also been studied. However his astronomical output is far from having been fully examined. The aim of the present essay is to fill this gap in the scientific biography of the great scientist.

The fact of the matter is that while in the very first years of the Petersburg Academy's existence an astronomical school peculiar to the time and place was formed with which Euler became closely associated, up till now that association has been inadequately examined. An analysis of the peculiarities of that scientific school and a history of its emergence can be found in the monograph [13]. Without going into detail, we note only that it was typical of the representatives of the 18th century Petersburg astronomical school to be deeply interested in a large range of astronomical subjects over and above celestial mechanics, such as astrometry, astrophysics, geodesy, cartography—and even the history of the calendar.

New evidence of Euler's astronomical activities has recently been found in the logbooks recording the astronomical observations of the St. Petersburg observatory, believed lost until they came to light by chance in 1977. On examination, these logbooks, comprising around 1000 pages, turned out to include comments on data from observations made by Euler between March 11, 1733 (new style) and the end of his first Petersburg period in 1741 [14]. The observations made by Euler and his colleagues at the St. Petersburg astronomical

Joseph Nicholas Delisle
Engraving by K. Westermeyer

observatory had been till then completely forgotten, since the logbooks containing notes on
these observations, dating from March 11, 1726 to May 21, 1747, had never subsequently
been examined in any detail.

Study of these materials reveals that from June 12, 1727 to May 21, 1747 (new style),
those collaborating in the work of the St. Petersburg observatory included—apart from
one or two permanent staff—many volunteers from among the teachers and students of the
Academic gymnasium and the university, the professors and adjuncts of the Academy itself
and the Naval Academy, and also the geodesists and navigators completing their training
at the Academy. Euler was himself one of these volunteer helpers, and in this capacity
carried out many interesting and important observations—only a negligible proportion of
which, however, appeared in the Notes of the Petersburg Academy of Sciences, i.e., the
Commentarii, published in Latin. This fact was noticed long ago; from it P. P. Pekarskiĭ
and then others concluded that the observations of the Petersburg astronomers of the first
half of the 18th century had remained unknown to their contemporaries and had had no
influence on the development of science [15, p. 129].

A close examination of long-forgotten articles in the newspaper *The St. Petersburg
News* [16] and its supplement the *Notes on the news* [17] appearing between 1727 and
1742, when these were published by the Petersburg Academy of Sciences, showed that
in fact a very much broader group of people visited and worked in the observatory than
is suggested by the notes accompanying the record of observations in the logbooks. And
comparison of these notes with the newspaper articles helped also to confirm the natural
circumstance that Euler's association with the observatory preceded his active participation
in astronomical observations.

There is nothing surprising in the fact that a young Swiss mathematician, having just
arrived in Russia in the spring of 1727, should take an interest in the work of the Academic

astronomers. After all, as a rule all "members of the mathematical class" of the Academy were expected to be present when observations were being made [16, 1728, No. 64, p. 260]. This tradition was introduced by the academician Joseph Nicholas Delisle, the first director of the Petersburg Observatory, who himself came to St. Petersburg only in the winter of 1726. Delisle understood that for the proper development of astronomy it is essential to use mathematical methods, and to maximize effectiveness in this respect the people involved should be mathematically highly competent—all that would then be needed would be to acquaint them with the methodology of observational astronomy. It was for this reason that he strove to interest such personnel in the activities of the observatory and secure their participation. Thus the mathematicians F.-Ch. Meyer and G. W. Kraft were involved in the work of the observatory from the time of their arrival in Russia, and later became members of the observatory's permanent staff [13, pp. 36, 68, etc.]. Euler's compatriots J. Hermann and Daniel Bernoulli also participated in the astronomical investigations; Daniel Bernoulli, we now know, even made observations: commencing on March 2, 1728, he took measurements of the noon elevations of the sun [13, p. 68]. Very soon after Euler's arrival, Delisle enlisted him also as a collaborator in the work of the observatory.

At that time the observatory was accommodated on three floors of the tower above the building housing the art gallery, where the Academy of Sciences had been relocated in September 1726. From September 9 of that year astronomical observations were regularly carried out in the observatory even while still under construction. Euler was a member of the astronomical group continuously from February 1727 on. Initially he functioned merely as spectator of the astronomers' observations—which, judging by the notes in the logbooks, consisted of measurements of the sun's noon elevation, observations of eclipses of Jupiter's moons and occultations of Saturn and various stars by the moon, and, on September 15, 1727, of an eclipse of the sun [18]. In addition the observatory staff carried out regular meteorological observations and observations of the polar auroras, which continued without interruption from February 26 (March 9), 1726, and in which many of the Academic personnel participated.

The very first—extremely brief—piece of information in printed form we have concerning the observations made in the observatory while still under construction, and when Euler was involved, appeared in the issue of the *St. Petersburg News* for Saturday, August 10, 1728. (Note that the dates and the days of the week attached to them were then "old style".) That issue contains the following announcement: "Two days ago [i.e., on August 8/19, 1728] astronomers and other members of the mathematical class of the Imperial Academy of Sciences gathered in a certain large chamber (since the Academic observatory is still under construction), from which they witnessed, and carried out observations of, the occurrence of a complete eclipse of the moon, all phases of which were fully observed in accordance with their wishes, although at times clouds were a hindrance" [16, 1728, No. 64, p. 260].

At that time observations of solar and lunar eclipses were of particular importance since they were used to determine and check the accuracy of measurements of the geographical longitude of the Academic Observatory, and hence of the new capital of Russia. In St. Petersburg on August 8/19, 1728 observations of the lunar eclipse were made by Delisle and the permanent staff of the observatory—Kraft, Meyer, and P. Vignon—in the presence of all "members of the mathematical class", and in Moscow by A. D. Kantemir, a former student of the Academic university. A detailed description of the phenomenon, accompanied by sketches and assembled by the Moscow and Petersburg observers, has been preserved [13, pp. 35–37].

The published material concerning observations of two other lunar eclipses is much more detailed. The first of these occurred in the evening of February 2/13, 1729 (reported in the issue for February 8, 1729 of the newspaper [16, No. 11, p. 44]) and the second during the night from July 28/August 8 to July 29/August 9, 1729 (reported in the issue for August 2, 1729 [16, No. 61, p. 244]). Both eclipses, as reported by the *St. Petersburg News*, "were observed at the local Imperial Observatory very truly by astronomers and mathematicians of the Academy of Sciences" [*loc. cit.*].

On September 8/19, 1729 Euler was present when observations were made of the long-awaited lunar occultation of Venus. The issue of September 9, 1729 of the *St. Petersburg News*—a peculiar type of "express information" organ providing in particular the latest information on the work of the observatory—contained a detailed note on these rare observations, beginning as follows: "Yesterday mathematicians of the local Academy of Sciences gathered at the Imperial Observatory to assist ... at the observation on a fine day of the occultation of the planet Venus caused by the moon... This occurred in the middle of the day at 3 o'clock, under a very bright sky, so that that planet could be seen with the unaided eye until it was covered by the moon". Concerning the aim of the Petersburg observers, the newspaper states that "... They hoped to discover by means of the telescopes if it be impossible to see such colors on this planet as it approaches the moon as certain astronomers in Paris saw on it 14 years ago when carrying out the same observation, which [colors] seemed to establish [the existence of] an atmosphere on the moon" [16, No. 72, p.288].

An anonymous article concerning these same observations, appearing in the *Notes on the news* under the heading "The conjunction of Venus with the moon" [17, 1729, Part 78, pp. 313–316], contained a detailed explanation of the physical nature of the phenomenon and of the methods and aim of the observations. Noting that usually such phenomena are exploited for investigating planetary motion and orbits, the author of the article (who according to P. P. Pekarskiĭ must have been G. W. Kraft [15, p. 465]) stresses that in 1729 the Petersburg astronomers had "an altogether different intention", namely that of detecting a lunar atmosphere. In connection with the history of this question, Kraft mentions Kepler, who considered that the moon does have an atmosphere, Hevelia, who "... ascribes a vaporous surround [i.e., atmosphere] not only to the moon but to all other planets" [17, 1729, Part 78, p. 315], and also opponents of this view. He pays particular attention to the Paris observations of 1715, of especial importance for the settling of the question, during which "... certain astronomers noticed ... at the instant when Venus went behind the moon red and blue colors on the body of Venus ..." [17, 1729, Part 78, p. 315].

Having explained in detail the refraction of light "according to the extraordinary discovery of Mr. Newton" who explained how white light decomposes into its component colors, Kraft writes: "Certain among those who observed these colors hoped that they were caused by the moon's vaporous surround, while others [thought] they had some other cause. But on both sides the matter remained very doubtful". Noting that in 1729 "... no colors whatever could be discerned ...", Kraft ends his article with the following words: "... It is likewise not possible to truly infer the non-existence of a vaporous surround about the moon from the lack of colors. Time will tell better than anything else ..." [17, 1729, Part 78, p. 316].

Although the Parisian observers are identified neither in the newspaper nor in the log-book notes, the Petersburg scientists knew very well that they were none other than J. N. Delisle and his Parisian friend J. E. Louville[1]. It was they who, with the aid of a seven-foot-long telescope, observed curious chromatic effects at the instant of second contact[2] of the moon's occultation of Venus on June 28, 1715. Here is how Delisle described them in his paper published in the *Mémoires* of the Paris Academy of Sciences for 1715: "...The strangest thing of all that I observed was that Venus, when close to the illuminated edge of the moon, for a short time before it went [behind the moon] seemed to me to be red on the side towards the moon and blue on the opposite side ... M. Chevalier de Louville, who also saw these colors, attributes them to the atmosphere that he believes the moon to have, but it seems to me that they can be very well explained without such an atmosphere ..." [19, p. 136].

The chromatic effects observed at the instant of contact of Venus with the moon may have arisen from diffraction or refraction of light in either of the atmospheres of the two celestial bodies involved. Louville, believing in the existence of a lunar atmosphere, took them as proof of his opinion. On the other hand Delisle, convinced that the moon has no atmosphere since none is observed during lunar occultations of stars, proposed explaining the effects as caused by the diffraction of light at the edge of the lunar disc, or by the atmosphere of Venus. However to decide the issue a more detailed knowledge of the diffraction of light than was then available would have been necessary—in fact at that time many scientists disputed the reality of the phenomenon of diffraction. In the course of the ensuing debate the director of the Paris Observatory J. Cassini[3] conjectured that the chromatic effects observed by Delisle and Louville were caused by optical defects in the telescope. After all, at that time achromatic optical instruments did not exist and in order to avoid chromatic aberration of an image it had to be kept constantly at the centre of the field of vision. Cassini was suggesting that Delisle and his friend Louville had broken this rule [20, pp. 139–140].

Thus the Petersburg astronomers were attempting in 1729 to settle an old argument that had flared up in Paris 14 years earlier. The matter was made that much more difficult by the fact that Louville's opinion was at that time shared by many scientists, including E. Halley, who had been a very close friend of the recently deceased Newton. The Petersburg observations were also lent exceptional importance by the accidental circumstance that in 1729 the lunar occultation of Venus was not visible anywhere in Europe. It is therefore not surprising that in St. Petersburg in 1729 preparations for these observations were made well in advance.

In the course of these preparations the Petersburg scientists examined the literature on optics and the diffraction of light, and also carried out laboratory experiments on diffraction in a specially equipped *camera obscura* attached to the observatory. There the image of the sun obtained using a telescope was projected onto a screen and the diffractional image was examined in turn by means of a multi-lens microscope. For such delicate experiments it was essential that the images obtained using the telescope and microscope be free of

---

[1] Jacques Eugène d'Alonville, Chevalier de Louville par Fontenelle (1671–1732).—*Trans.*

[2] The first, second, and third contacts refer to stages in Venus' occultation (see below).—*Trans.*

[3] Jacques Cassini (1677–1756), son of the more famous astronomer Giovanni Domenico Cassini (1625–1712).—*Trans.*

chromatic aberration. It was hoped that by studying the eyes of animals and human beings preserved in the famous anatomical collection of F. Ruish, and by using multi-lens systems, chromatic aberration might be significantly reduced. It was then that Euler began for the first time to turn his thoughts to achromatic microscopes and telescopes, leading eventually to his achromatic theory of these instruments [13, pp. 137–143].

As part of the preparations for the 1729 observations, trial firings of cannon were also performed, at which, in the Summer of 1727, Euler, Daniel Bernoulli and a few of their Academic colleagues attempted to measure the speeds at which the flash and sound of a cannon-shot were transmitted through the air. Subsequently these Petersburg trial firings, which repeated in somewhat varied form experiments performed by Delisle in Paris in 1712 and were considered as confirming the wave theory of light—supported also by the existence of the phenomenon of diffraction—, formed a basis for the formulation of a mathematical model of the atmosphere of the earth, and then of other celestial bodies. Euler proposed such a model, involving so-called "atmosphere-clouds", in September 1727. He assumed that the atmospheric particles are enclosed in extremely thin liquid "shells" closely packed together. In terms of this model, expounded in his 1729 paper "An attempt at an explanation of atmospheric phenomena", he was able to explain the air's elasticity and several other properties of the atmosphere. However from a distance such an atmosphere would have been completely opaque and no chromatic effects would have been observed.

According to the notes in the observatory logbook, the lunar occultation of Venus of September 8/19, 1729 was observed using in turn telescopes of lengths 13, 15, and 23 feet, in order to allow for optical defects. The greatest pains were taken at the instants of second and third contact, when the image was kept at the center of the telescope's field of vision. The observers saw no trace of any chromatic effects such as had been reported in the aforementioned earlier publications. On the other hand, against the entry in the observatory logbook for that day there is an interesting note to the effect that at the instant of third contact, i.e., a few seconds before Venus completely disappeared behind the moon, Delisle, using the largest of the telescopes, registered a slight trembling of the image [18, l. 104].

These results threw doubt on the existence of an atmosphere on the moon, but significantly increased the likelihood of one existing on Venus. The importance of these conclusions was enhanced by the favorable conditions prevailing when the observations were made: the weather was clear, and the moon and Venus were visible in daylight at a satisfactorily high elevation ($h = 46°7'30''$)—and in addition, as already noted, the occultation was not accessible to the astronomers of other countries. The Petersburg observations of September 8/19, 1729 served to bring into stark relief the problem of detecting planetary atmospheres. As described in detail in the *Notes on the news*, an observational method had been developed, and the required instruments and conditions of their use carefully chosen. From that time Euler would on more than one occasion turn his attention to this problem.

Judging by remarks made in those years in his notebooks [22], Euler became engrossed by the observations of the sun and sunspots. A detailed note concerning the first Petersburg observation of sunspots is included in the entry in the logbook of the Petersburg observatory for July 15, 1730 [18, l. 143–144 ob.]. Early that morning, in accordance with established custom, "all members of the mathematical class" gathered in the observatory in anticipation of the imminent solar eclipse. Having set their clocks beforehand, aimed the 14-foot and 15-foot telescopes at the sun, and produced its image on the screen using the 14-foot telescope equipped with a filament micrometer, Delisle and Kraft, together with other observers not named in the logbook, were amazed to see so many spots on it[4]. With painstaking accuracy they measured the diameter of the sun, timed the emergence from behind the lunar disc of each individual spot, and made a careful sketch of the phenomenon, so rare at that time.

This unprecedented and majestic spectacle made a powerful impression on Euler and his friends, all of whom were seeing sunspots for the first time. From that day on they kept careful track of the distances of the spots from the edge of the solar disc and their movement, using these data to estimate the angular speed of rotation of the sun and to check the basic tenets of the theory of motion of the spots that Delisle had proposed in Paris in 1713. That theory was published in St. Petersburg only in 1738 [23] although, according to G. F. Müller [24, p. 497], it was ready for publication as early as 1735. Euler actively participated in all of this work. Among the entries in his notebooks for those years one finds versions of formulae for computing the distances of the sunspots from the edge of the solar disc, as well as other material having to do with the elaboration of the details of the above-mentioned theory [22]. These notes were eventually to form the basis of a paper by Johann-Albrecht Euler, Leonhard Euler's eldest son, entitled "On the rotation of the sun about the axis determined by the observed movement of the sunspots", published in 1768 [25]. Judging by the observatory's logbook, observations of the sunspots, begun in 1730, continued to be made until the end of 1735, reaching a maximum in 1734. It was during this period that Euler began to use the Petersburg observatory to carry out his own independent observations.

As noted in the logbook, on March 11, 1733 (new style) he arrived at the observatory with a group of geodesists left unnamed. They synchronized four clocks, observed the setting of Jupiter's main satellite, and then, sometime between 8 a.m. and 9 a.m. and then again between 2 p.m. and 3 p.m., measured the elevations of the sun's upper edge. From these observations Euler calculated the time when the sun passed the meridian of longitude at that location using a small quadrant, and gave an estimate of the error involved in his result, which was as follows: true noon occurred at $11^h 30^m 4.5^s \pm 2.5^s$ [18, 1. 272 ob.-273]. From that time right up to his departure from St. Petersburg in 1741 Euler regularly carried out such observations, becoming "his own person" in the observatory.

The determination of the noon solar elevations was important in connection with the Time Service provided by the Petersburg observatory throughout most of the 18th century, and especially so following the launching of the second expedition to Kamchatka (1731–1744). Since at that time astronomers had no tables or statistical annuals, the usefulness of any observations carried out in Siberia depended on their being compared somehow with similar observations made simultaneously at some other location—whence the great care taken by the astronomers of the Petersburg observatory between 1731 and 1744 over the accuracy of all observations made by them. Euler participated very actively in these observations. His help was all the more valuable in that most of the astronomers and geodesists were away with the expedition, while F.-Ch. Meyer had died, Daniel Bernoulli had quit Russia, and G. W. Kraft had been appointed professor of physics in 1731 and was thus unable to give much of his time to astronomical observations.

Euler was directly associated with the organization at the Petersburg observatory of

---

[4]Although Delisle, at least, had observed the sunspots long before—see below.—*Trans.*

the first ever Time Service in our country. Very soon after his arrival in Russia he began to frequent the observatory, and noticed how much trouble clocks going at different rates were causing the astronomers [18]. This prompted him to work on their regulation, and then on the invention of new astronomical means of measuring time. Taking as an initial model of an accurate chronometer Christiaan Huygens' design of a pendulum clock he, together with Daniel Bernoulli and J. Hermann, worked out a theoretical basis for the construction of new clocks for the Petersburg observatory, which they published in the *Commentarii* of the Petersburg Academy of Sciences over the period 1727–1736. In 1735 J. G. Leutmann, professor of mechanics, actually constructed such clocks, which were thenceforth used in connection with the observational work of the observatory [13, p. 57].

Euler's proposed method for compiling tables relating to the sun's position at true noon was also of considerable importance. With his help tables of solar elevations at St. Petersburg for each hour were compiled, and later a table to be used for calculating true noon from two solar elevations measured before and after midday. From 1733 on, these tables were used by all those working in the observatory, and in 1735 his work on this topic was presented to the Academic "Conference". In the minutes of the meeting of January 27/February 6, 1735 it is stated that "Mr. Prof. Euler presented his calculation of the time, by means of which at any given latitude one can determine what the time is for that person, provided only that he has at his disposal some fixed collection of elevations of the sun. Since this paper is to be published in the *Notes [on the News]*, the author has taken it back home with him in order to make a number of accurate copies" [26, Vol. 1, p. 142].

Euler's work was of great interest to Delisle. The minutes of the meeting of the Academic conference of January 31/February 11, 1735 state that "Mr. Prof. Delisle took away with him, in order to examine it, the 'Method for quickly and simply calculating tables of times of true noon' worked out by Mr. Prof. Euler and presented here ... The same Mr. Prof. Delisle also took home ... a table of [real] noontimes compiled by that same Mr. Prof. Euler, calculated to within a few minutes' accuracy from any two elevations of the sun observed before and after midday taken from observations made at intervals during each of which the sun's elevation changes by one degree over any hour of the 18 available at the location of the Petersburg observatory"[5] [26, Vol. 1, p. 146]. This work of Euler's was later published in Latin [27]; however it is likely that in part it provided the content of the series of articles published in 1731 in Russian and German under the running head "On time and its subdivision" in the *Notes on the news* [17, 1731, Parts 2–10]. Although it has generally been thought that the author of these anonymous articles was G. W. Kraft [15, p. 465], it seems evident that Euler was a co-author, and even that he wrote two of the nine papers, namely those printed in Parts 7 and 8 [17, 1731, Parts 7, 8, pp. 25–32] where questions are discussed concerning the calculation of corrections to pendulum clocks by means of observations of the sun and stars—the same as are addressed in the work [27]. The fact that Euler had earlier collaborated successfully with Kraft both in the observatory and in the writing of articles for the *Notes on the news* adds weight to this conjecture.

Thus by 1735 fundamental research had been completed on the construction of new clocks for the Petersburg observatory, and also on the regulation of all clocks in the observatory and on methods for making corrections of the time they indicated. This meant that

---

[5]The latter part of this is just one interpretation of a somewhat obscure passage.—*Trans.*

Delisle could propose to the Senate the founding in Russia of its first Time Service. In the minutes of the meeting of the Academic conference of January 20/31, 1735 it was noted that henceforth the firing of the signal cannon should every day inform the citizens of St. Petersburg of the precise moment of the solar zenith [26, Vol. 1, p. 139]. These traditional midday firings continue in Leningrad[6] to this day, reminding the public of the founders of our first Time Service.

Euler was also actively involved in the organization of a Solar Service in our country. The discoverers of sunspots in the 17th century had noticed that the appearance of sunspots coincided with an increase in the frequency of the manifestations of the *aurora borealis*. This correlation was confirmed visually by Delisle in Paris in 1713, in the course of studying the sunspots that appeared at that time. Before leaving Paris for Russia in 1725 he arranged for his friend J. J. Dortoux de Merand to carry out systematic observations of weather patterns, polar auroras, and sunspots in Paris while he, Delisle, did the same in Russia, in order to settle the question of the interdependence of these phenomena. A. Celsius was co-opted to carry out similar observations in Sweden and other Scandinavian countries.

In St. Petersburg these observations began on the day following Delisle's arrival, namely February 26/March 8, 1726, with the participation of all permanent staff of the observatory and a number of volunteers. The majority of the observations were carried out by F.-Ch. Meyer and G. W. Kraft, and the data they obtained formed the basis of the theory of polar auroras proposed by Meyer in October 1726, and published in 1729 [28]. As noted earlier, sunspots were observed in St. Petersburg for the first time only in 1730. While it remains unknown if Euler carried out independent observations of the *aurora borealis*, it has been discovered recently that he did publish the results of his theoretical investigations of the phenomenon. Up till then G. W. Kraft had been considered the author; according to P. P. Pekarskiĭ, at that time Kraft published five anonymous articles on the polar aurora in the *Notes on the news* [15, p. 465]. And indeed in the *Notes* for 1730 one finds ten articles under the general heading "On the recent large northern polar aurora" [17, 1730, Parts 14–17, 21, 25, 32, 35, 77, 78]. If in fact Kraft wrote five of these ten articles, then which ones? And who wrote the others? The answers to these questions have recently been determined: Yu. Kh. Kopelevich[7], in the course of examining letters from G. F. Müller to J. D. Schumacher, found in them a reference to a paper of Müller's shedding light on the history of the question [17, 1730, Parts 14–17, 21]. It turned out that Euler was in fact the author of the articles on the cause of the polar auroras [17, 1730, Parts 25, 32, 35] and that only the two final articles of the series—containing an analysis of the Petersburg observations of the aurora—had been written by Kraft [17, 1730, Parts 77, 78].

The articles by Euler contain a critical examination of the explanations of the physical nature of the polar auroras that had been proposed by various scientists from Aristotle to F.-Ch. Meyer. It is the latter's theory, accompanied by the observational data on which it was based, that forms the content of the final two articles written by Kraft. Although Euler made no essentially new proposals, he was able to demonstrate in the light of the observational data obtained by the Petersburg scientists the inadequacy of various conjectures on

---

[6] And in present-day St. Petersburg.—*Trans.*

[7] See Kopelevich's essay "Leonhard Euler, active and honored member of the Petersburg Academy of Sciences", in the present collection.

the nature of the polar auroras, including that of Aristotle, which "…fills man's memory more with empty words than with reason and real concerns" [17, 1730, Part 25, p. 97]. The later explanation offered by Descartes, who, it is stressed, never actually witnessed a polar aurora, "…is derived …from the ancient Aristotelian opinion, which is no longer considered of any significance" [17, 1730. Part 35, p. 138].

In one of the articles Euler pays particular attention to the most popular explanation then current, namely that the polar lights arise from the reflection of the light of erupting volcanoes from ice in the process of breaking up. However "…here in St. Petersburg this opinion is not greatly supported …since the ice of Lake Ladoga[8] breaks up every Spring, yet no such lights have ever been observed there. Moreover it is unclear just how Hekla[9] and other volcanoes could illuminate the ice of Lake Ladoga powerfully enough for its gleam to be visible throughout Europe" [17, 1730, Part. 25, pp. 98–99].

Having examined all proposed explanations of the polar lights, Euler concludes that each proponent judges the universe "by means of a science personal to him, in which he is skilled. The same consideration applies to the fact that a general sees a field of battle where a merchant sees a fair, a poet Parnassus, a doctor a hospital, and an astronomer a planet! …However those reason better"—thus he concludes his section—"who respect the universe for what it truly is …—a real chemical laboratory …" [17, 1730, Part 35, p. 140].

It should be noted that the Petersburg scientists were very fortunate in their choice of a time to draw the attention of the readers of the *Notes* to the "chemical laboratory of the universe". While the observation of the anomalous polar aurora of February 5, 1730, served as the formal reason for the publication of articles by G. F. Müller, Euler, and G. W. Kraft in the *Notes on the news* over the period from February 16 to September 28, 1730, the real stimulus of these publications was the broader one of the heightened activity of various processes in the earth's vicinity observed by the Petersburg astronomers. This manifested itself in particular through an increase in the frequency of polar auroras and very changeable weather [29], leading to an expectation of enhanced activity of processes taking place in more distant regions of the "chemical laboratory of the universe". This prognosis was spectacularly fulfilled: starting on July 15, 1730 sunspots were observed from St. Petersburg, and thereby the interconnection between solar activity and processes taking place in the earth's vicinity convincingly demonstrated.

The article entitled "On comets", published anonymously in the *Notes on the news* for 1733 [17, 1733, Part 86], represents the first publication asserting the identity of the physical processes involved in the polar auroras and the tails of comets—a claim based on the observations of the anomalous polar aurora of 1733, sunspots, and other phenomena. From then on, authors of all papers published in the *Notes on the news* touching on polar auroras [17, 1738, Parts 70–75], the tails of comets [17, 1742, Parts 33–40], sunspots [17, 1735, Parts 23–27], the zodiacal light [17, 1739, Parts 45–48], and other similar phenomena, were always careful to emphasize their identical physical nature. The most detailed analysis of all of these phenomena is given in Euler's article "Physical investigations of the cause of the tails of comets, the northern lights, and the zodiacal light" [30], published in 1748, when Euler was in Berlin.

Thus, once having begun to work on a theory of the polar auroras in 1730, Euler went

---

[8] An historically famous lake not far from St. Petersburg.—*Trans.*
[9] A volcano in Iceland.—*Trans.*

on to investigate sunspots and other manifestations of solar activity, pondering the interconnections of these phenomena over many years. This provides the justification for believing that it was indeed Euler who authored the articles in which for the first time the similarity of the processes involved in the tails of comets and the polar lights was clearly broached [17, 1733, p. 86]. Till now all three of the papers under the general heading "On the unusual northern lights of 1733" [17, 1733, Parts 85–87] had been attributed to G. W. Kraft [15, p. 465]. However, on examining these articles one discovers that only the first and last parts are devoted to a description of the anomalous polar aurora of 1733, while the middle one has the independent heading "On comets", and deals with a different topic—chiefly the physical nature of the tails of comets. Among the various views as to the physical nature of the tails of comets explicit preference is given to Newton's theory of efflux. The stripe-like formations observed in the tails of comets are considered as arising in the atmospheres of the comets in the same way as the pillar-like formations of the polar lights arise in the earth's atmosphere. Also noted is the importance of studying comets for the purpose of future elaboration of methods of determining the orbits of comets and planets, and to shed light on the question of the density of interplanetary space [17, 1733, Part 86, pp. 343–346]. The point is that it was precisely these questions that occupied Euler above all in St. Petersburg in the early 1730s. One may surmise that, given his authorship of articles of the polar auroras, he was also entrusted with writing one on the cause of comets' tails.

The likelihood of this conjecture was confirmed much later by G. F. Müller. Relating events of 1729 in his *History of the Academy of Sciences*, Müller wrote about the publication of the *Notes on the news* as follows: "...Now [i.e., in 1729] Messrs. Euler, Gmelin, Kraft, and Weitbrecht also decided to write for these pages, as a result of which, by virtue of the variety of the material collected in them, they became more and more popular. A society was formed, meeting every Saturday evening at Mr. Schumacher's home, where the themes of discussion were those on which articles were to be published" [24, p. 181].

And what precisely did Euler publish in the *Notes on the news* in 1729? Taking into account that at that time he was working on questions of navigational astronomy, lunar motion, the ebb and flow of tides, and on methods of determining longitude [1, pp. 76–78, 85–96], it is natural to suppose that he authored the following articles in that journal: "On finding the longitude of locations at sea" [17, 1729, Part 16, pp. 61–64]; "On navigation" [17, 1729, Parts 2, 29]; and the section entitled "On the ebb and flow of the sea" [17, 1729, pp. 90, 91] of the series of articles appearing under the general heading "On the rising and falling of water in the river Neva" [17, 1729, Parts 86, 88–91], apparently written by G. W. Kraft, who was responsible for keeping track of the changing level of the waters of the Neva.

It is noteworthy that the general arrangement of material and the basic line of argument in the article "On the ebb and flow of the sea" coincide with those of the corresponding portions of Euler's "Letters to a German princess" [31, Vol. 1, letters 62–67]; there is even textual coincidence of the passages where he expounds Descartes' opinion that the tides are caused by action of "lunar whirlwinds". Characteristically, this article in the *Notes* [17, 1729, Part 90], which appeared on November 11, 1729, contains the first published criticism opposing Descartes' views.

By comparing the Cartesian "pressure of the lunar whirlwind" on the earth's atmosphere and oceans with the Newtonian theory of their attraction by the moon, the Petersburg scientists were able, on the basis of their astronomical and meteorological observations, to make a choice between the two theories. Here is how it is put in the above-mentioned article: "...The cause of the action proposed by Descartes cannot be correct. For if the air under the moon were pressed so severely, then there would have to be a corresponding pressure of it on the mercury of barometers causing it to rise significantly higher, but this is not observed. Then also it would be appropriate for the sea to fall when the moon passes across the zenith, while in fact quite the opposite occurs ...It was for this reason that Kepler ascribed the cause [of the tides] to the attractive force of the moon rather than pressure, and Mr. Newton proved this very rigorously with the aid of higher mathematics..." [17, 1729, Part 90, p. 363].

It is likely that Euler participated in the preparation of two further articles published in the *Notes on the news* for 1729. The first of these, entitled "On Johann Leo" [17, 1729, Parts 32, 33], is devoted to the adventures of the well known Scottish adventurer John Law, an unusually lucky gambler—this serving as a motive for including a passage on the theory of probability in the article. Law was of good education, including mathematical. In 1695 he was forced to flee England after yet another duel. In the course of travels through the countries of continental Europe he studied finance and banking, and in 1705 published a book expounding his theory on the connection between trade and the circulation of money[10]. Law's system caught the attention of Philip of Orleans, regent for the child Louis XV, and he invited Law to France, where for five years (1716–1721) he pursued a dazzling career, first as director of a private bank, then as founder in 1717 of the Company of the West for the exploitation of a trade monopoly with Louisiana, soon extended to the French colonies in the West Indies and Canada; finally, in 1720, he was appointed Controller-General of Finances for France. He was also made a Count and an honored member of the Paris Academy of Sciences. However Law's bank and his company, now merged and in charge of the royal debt, collapsed into bankruptcy, and Law fled to Venice, whither he had transferred his money in good time[11].

The second article, "On the general congress of the Swiss cantons" [17, 1729, Part 66], had no connection with mathematics. It contains a description of the political structure of the Swiss republic, accompanied by a brief but very warmly expressed panegyric on that type of government. Although there were several Swiss on the staff of the Petersburg Academy, it is thought that Euler is likely to have been the author—or at least co-author—of this passage, since in his correspondence he often expresses himself in this spirit.

Of the remaining articles in the *Notes on the news* a further four series may be attributed to Euler: "On the earth" [17, 1732, Parts 6–12, 49, 50]; "On finding the longitude of any location on the earth" [17, 1734, Parts 53–55, 57–59]; "On the external shape of the earth" [17, 1738, Parts 27–32, 103, 104]; and "How the ebb and flow of sea-tides should be observed" [17, 1740, Parts 9, 10]. The third of these is actually signed by Euler and the last, although unsigned, coincides textually, as Yu. Kh. Kopelevich has noticed[12], with part of the manuscript of Euler's work on the tides (E57) [32; 1, p. 72].

---

[10]"In the financial capitals of Europe [Law] had learnt that money was only a means of exchange, that real national wealth depended on trade and that trade depended on money. The shortage of currency, he truly saw, was one of the chief handicaps to the French economy". Alfred Cobban. *A history of modern France*. Vol. 1. Penguin Books Ltd. 1963. p. 23.

[11]According to Alfred Cobban in his *A history of modern France* (see the previous footnote), in December 1720 Law fled to London, not Venice—or at least not immediately.—*Trans.*

[12]See the essay by Yu. Kh. Kopelevich in the present collection.

However the first three of these series of papers are of greater interest. In them, on the basis of the discoveries of Copernicus-Kepler-Newton, the view of the earth as just one planet among others in the solar system is promoted, and methods of determining latitude and longitude on its surface, as well as its shape, are discussed. The similarity of the series entitled "On the earth" with that entitled "On the external shape of the earth" is striking both in respect of the headings of the individual articles, and their content. The problems considered are essentially the same in both although separated by a six-year interval. It is as if in the later version the author wished to acquaint his readership with the new scientific discoveries of the intervening six years. The two series are, however, very different in the manner of exposition of their subject-matter. Thus the earlier series, of 1732, is written in a somewhat argumentative tone: passionately defending the discoveries of Copernicus and Newton, the author often resorts in the heat of the argument to anti-clerical attacks. One of these involves an ironic retelling of the story of St. Augustine's denial of the existence of the antipodes—which story, incidentally, figures also in the 1729 article "On navigation" [17, 1729, Part 2]. It is interesting to observe that in the earlier of the two articles we are comparing, namely "On the earth", one finds the clearest extant quotations from speeches made by Delisle and D. Bernoulli in defense of Copernicus at the public meeting of the Academy of Sciences held in March 1728 [33]. This is the more curious in that it had hitherto been thought that the Russian translations of these speeches had been withheld from publication by interdiction of the Synod [34, p. 202].

However the Synod certainly did not forbid the publication in 1732 of the harshly polemical outbursts directed at the church contained in the article "On the earth". For example, having described the structure of the universe and its gigantic dimensions, the author remarks caustically: "Now I would like to know, should anyone judge all of this according to reason, if he could continue to be of that opinion which many hold, that our earth alone represents the Creator's blessedness, and that all else that exists has been created by God for the sake of our earth and its inhabitants?" [17, 1732, Part 10, p. 39]. The author goes on to remark that such an opinion is equivalent to the conviction of that well-known Athenian who thought that every ship entering the harbor at Piraeus had arrived especially for him, and that on similar grounds a spider, having spun its web in a corner of the huge building housing the opera, would be justified in considering it expressly built just for him to conveniently catch flies there!

On the other hand the cycle of articles "On the external shape of the earth", published in 1738 after the death of Theophan Prokopovich, the constant defender of the Petersburg scientists in the Synod, is written in an altogether different tone. Although, as in the earlier paper, the author argues from the position of Copernicus and Newton, he now refrains from all disparaging or anti-clerical asides. The Copernican doctrine is taken as a self-evident fact which no one doubts, so that it is not necessary to defend it. The exposition proceeds in scientifically rigorous fashion, sometimes rather too drily, but on the other hand very succinctly and clearly.

Euler's work at the Academy, like that of all the other Petersburg astronomers, was closely involved not only with the observatory but also with the Geography Department where, from August 25, 1735 he was appointed a member of the permanent faculty. The main sources of information concerning Euler's activities there consist of various surviving documents of the Academy of Sciences published by M. I. Sukhomlinov [35], the published minutes of the meetings of the Academic Conference [26], the collection [4] of his articles, as well as others of his papers, and his correspondence [36–39]. Of great value also in this respect is the description of Euler's surviving manuscripts with annotations relating to his documents and the aforementioned minutes, kept in the Archive of the Academy of Sciences of the USSR[13] [1].

However the minutes of the meetings of the Geography Department represent the most significant source. These have been preserved almost entirely, and contain abundant and valuable lodes of information on Euler's work in Russian cartography and geodesy in both his first and second Petersburg periods. A survey of Euler's activities in the Geography Department for the period 1735–1737 based on the departmental minutes for that interval, which were written in French, has been compiled by V. F. Gnucheva [3, pp. 37–38, 52]. More detailed information, derived from the departmental minutes for 1735–1747, can be found in the book [13] by the present author (see [13, Chapter 3]). However this likewise far from exhausts the available material.

The recorded minutes are also useful for the purposes of commentary on Euler's correspondence and fixing the dates of a number of the undated letters. One of these deals with a proposed visit by Euler and Delisle "on Thursday" to the office of engineers, with the aim of looking at maps there. The publishers of the correspondence conjectured—correctly—that this letter (No. 7) was written towards the end of 1735 [39, p. 131]. However examination of the minutes of the Geography Department allows us to date the letter much more precisely and moreover to determine the circumstances surrounding the visit. Thus it turns out that on November 3, 1735 (old style) Delisle, Euler, and their colleagues were invited to visit the Chancellery of the Main Administration of Artillery and Fortification [40, l. 27] and on December 4, 1735—which was indeed a Thursday—they made the visit. The minutes contain a detailed description of this visit. From them we infer that letter No. 7 must have been written between November 27 and December 3, 1735 (old style).

Another letter, No. 4 of the correspondence between Euler and Delisle, tentatively assigned to October 1735 [39, p. 139], has been even more accurately dated using notes contained in the departmental minutes. This letter informs Euler that a map begun by him is being sent back to him with a request that it be completed as soon as possible, by order of the president of the Academy J. A. Korff [41, p. 129]. The minutes of the meeting of the Geography Department held on January 14, 1736 contain the following note: "Since it is precisely Mr. Professor Euler who, after agreeing to this with Mr. Professor Delisle, has taken on the preparation of the map of Russia's boundaries requested by the Supreme Senate, Mr. Delisle sent him the map (which he, Mr. Euler, had earlier begun to draw up) yesterday [i.e., 13.01.1736!] with a note begging him ... to complete said map ... " [41, l. 4-4 ob.]. Thus letter No. 4, from Delisle to Euler, was sent on January 13, 1736. Incidentally, the order from Korff mentioned in the letter was received by the Geography Department on January 10, 1736 [41, l. 3 ob.]. Such examples can be multiplied.

Research into the history of astronomy was also popular in St. Petersburg; according to Delisle such investigations were capable of "significantly stimulating the inspiration of a genius, this being essential for the progress of astronomy by means of new discoveries" [13, p. 161]. While carrying out observations of the sun and compiling tables on the solar

---

[13]Presumably these materials are still there in what is now the Archive of the Russian Academy of Sciences.—*Trans.*

zenith Euler became interested in the problem of the length of the tropical year. Assuming that interplanetary space was filled with a material resistant to the motion of celestial bodies through it, he hoped to detect that material by means of an examination of calendars; for if the earth's speed round the sun does actually decrease, then the average length of the tropical year should also gradually decrease[14]—a circumstance that could scarcely have escaped the notice of compilers of calendars.

In the early 1730s Euler, together with Delisle, Kraft, Müller, V. K. Trediakovskiĭ, T. S. Bayer, and others made a comparative analysis of the calendrical systems of ancient and contemporary China, Japan, and other countries of the Far, Middle, and Near East, including Northern and Southern India. The results of these collective investigations were brought together in a book published by Bayer, lacking exact indications as to the author of each section [13, pp. 184–190]. Only in his last book, *History of the Bactrian kingdom of the Greeks, in which simultaneously the ancient chronicle of the Greek colonies in India is expounded...* [42], which appeared in 1738, did Bayer break his pet rule by including as an appendix an article signed by Euler, entitled "On the astronomical solar year of the Hindus" (E18) [43].

The research for this article involved careful examination of manuscripts sent to St. Petersburg by the Danish missionary C. T. Walther, who lived for 20 years on the Malabar Coast[15]. Subsequently Walther's manuscripts were published as another appendix to Bayer's book [42]. The date at which Euler began working on the South Indian calendar may be inferred from a surviving letter written by Bayer to Euler, dated January 17/28, 1736, enclosed with two manuscript books of Walther's. He writes: "In the two enclosed books there is new information on the Malabar art of counting. On examining how they operate with fractions, Mr. La Croze[16] noted that the same method was used by the ancient Greeks and Romans, and proves this by means of references to Petronius[17] and Horatius[18].

"But I suggest that just as for example in the case of their own addition and division, so also for the other [operations] the Hindus may have arrived [independently] at the same [procedures] as the Greeks and Romans. I beg you to show me the kindness of pondering this further... I would be very grateful to you for these labors ... " [44].

Euler and his colleagues also examined the folk calendars of Russia, the Ukraine, and Georgia. They always strove to obtain their material from original sources. Thus information on Chinese calendars was sent to them from Peking and on Tamil calendars from Malabar, while they learned about Japanese and Georgian calendars from Japanese and Georgians living in St. Petersburg. They were informed about North Indian calendars by a Hindu merchant come to St. Petersburg from Astrakhan[19], whom Bayer called "Zongbara", while Trediakovskiĭ called him "Sungir Pritomovich". He taught Sanskrit and the elements of Hindu science to all comers in return for learning Russian from them [15, p. 183]. On his departure from St. Petersburg he was presented with a Russian grammar compiled by Trediakovskiĭ as a parting gift.

---

[14]Presumably because the earth would at the same time gradually move closer to the sun.—*Trans.*

[15]The southwest coast of India.—*Trans.*

[16]Mathurin Veyssière de La Croze (1661–1739), French orientalist, Huguenot, worked in Berlin.—*Trans.*

[17]Gaius Petronius Arbiter (27(?)–66 A. D.), Roman writer, aristocratic epicurean, confidant of Nero.—*Trans.*

[18]Quintus Horatius Flaccus (65–8 B.C.), Roman poet.—*Trans*

[19]The present city of this name is situated in the delta of the Volga.—*Trans.*

After moving to Berlin in 1741 Euler continued with all of the research projects he had begun in St. Petersburg. In 1746 a collection of various of his papers was published (E80) [45], concerning the preparation of which he wrote to Delisle on February 2, 1746 as follows: "I hope that this collection will bring you pleasure, for, apart from a few papers on analysis and mechanics, there are some on astronomy and physics that are for the most part rather interesting, for example a new theory of light and color quite different from Newton's, which explains all phenomena perfectly. Concerning the earth's motion . . . I apply a new lunar inequality, but it seems to me that I have made an altogether more important discovery, namely that the period of revolution of the earth is not constant but decreases little by little . . . This shortening of the year is a consequence of the resistance of the ether . . . if the ether gives resistance, then the periods of revolution of the planets must decrease, along with their eccentricities . . . for comets this effect will be very significant . . ." [39, pp. 240–241].

The articles appearing in the second and third volumes of the *Notes* of the Berlin Academy of Sciences in 1748 and 1749 represent a notable culmination of the astronomical investigations begun in St. Petersburg. In this regard mention must be made above all of the work on the physical nature of the tails of comets, the polar auroras, and the zodiacal light (E103) [30]. The papers on celestial mechanics are also of great interest (E112–E115): "Investigations into the motion of celestial bodies in general"; "A method for determining the true times at which new moons and half-moons appear"; "A method for determining the true geocentric location of the moon by observing lunar eclipses of a fixed star"; and also "A method for determining the longitude of locations by observing lunar eclipses of fixed stars" [46]. And, finally, there appeared the article expounding the basics of the Eulerian theory of achromatics, entitled "On the perfection of the objective lenses of telescopes" [47].

Nor did Euler forget the astronomical observations he had carried out in St. Petersburg: after all it was precisely this observational experience that brought out in him that rare and amazing—for a mathematician—astronomical-observational intuition, demonstrated many times over in his research in celestial mechanics[20]. However the Berlin observatory of the 1740s was not to be compared with the Petersburg astronomical observatory, which, after the publication in 1741 of a description of its instruments and equipment [48] and a survey of the work carried out there appearing in I. Weidler's *History of astronomy* [49], came to be regarded as one of the world's best. The spread of this opinion was largely effected through the reports of foreign scientists working for a time in the Petersburg observatory and then returning to their countries of origin; prominent in this regard were D. Bernoulli (Basel), G. W. Kraft (Tübingen), G. Heinsius (Wittenberg), and J. Ch. Liebert (Berlin).

For the astronomers of Prussia and other medium-sized European states the Petersburg observatory long remained an unattainable model by virtue of its ample equipment and the variety of scientific research carried out there. This view was held also by Euler, who on March 18/29, 1746 wrote to J. D. Schumacher as follows: "One may take real pride in the observatory in St. Petersburg as a result of the unstinting provision over so many years of the means for acquiring all essential instrumentation. Furthermore the building in which it is housed is so well adapted to astronomical aims that we are unable to propose a better model in that respect" [37, Vol. 2, p. 86].

---

[20] See the essay by N. I. Nevskaya and K. V. Kholshevnikov in the present collection.

Making extensive use of his Petersburg experience, Euler invested considerable effort in resuscitating the Berlin observatory, which had fallen into desuetude following the death in 1740 of its last director Ch. Kirch. On April 4, 1743 he wrote to Delisle as follows: "From the death of Mr. Kirch till now the Observatory of the Society has remained in a sorry state, so that it has been almost impossible to make observations of anything. It was believed that the king had in mind another use for that locale, and for this reason there was no desire to spend anything on astronomical requirements. However now they are beginning to restore greater order to the observatory and equip it with the necessary instruments. We have once again located the meridian of longitude you traced here on your way through[21], and if not for that we would have been completely unable to locate it" [39, p. 163].

As Euler informed Delisle on April 4, 1748, the restored Berlin observatory was equipped with a *camera obscura* resembling the one in St. Petersburg. There Euler and the Berlin astronomer J. Kies carried out experiments with "artificial eclipses" using Delisle's methodology [39, p. 260]. There also, on July 25, 1748 Euler, together with Kirch's sisters, made observations of a ring-shaped solar eclipse in the course of which the question of the existence of a lunar atmosphere was settled once and for all—the same question that had exercised the young Euler and his Petersburg colleagues as long ago as 1729 [50, 51]. In continuing with investigations begun in Russia, Euler always felt himself still a representative of the Petersburg Academy of Sciences. This explains why, when in 1749 Friedrich II asked him where he had acquired his knowledge, he responded with: "...I am grateful for everything to my time at the Petersburg Academy" [37, Vol. 2, p. 182].

References

[1] *L. Euler's manuscript materials in the Archive of the Academy of Sciences of the USSR.* M.; L.: Izd-vo AN SSSR (Publ. Acad. Sci. USSR), 1962. Vol. 1. 427 pages.

[2] Subbotin, M. F. "The astronomical works of Leonhard Euler". In: *Leonhard Euler.* M.: Izd-vo AN SSSR (Publ. Acad. Sci. USSR), 1958. pp. 368–375.

[3] Gnucheva, V. F. *The Geography Department of the Academy of Sciences in the 18th century.* M.; L.: Izv-vo AN SSSR (Publ. Acad. Sci. USSR), 1946. 446 pages.

[4] Euler, L. *Selected cartographical articles: Three articles on mathematical cartography.* M.: Izd-vo geodez. lit. (Publ. geodesic lit.), 1959. 80 pages.

[5] Vavilov, S. I. "The physical optics of Leonhard Euler". In: *Collected works.* M.: Izd-vo AN SSSR (Publ. Acad. Sci. USSR), 1956. Vol. 3. pp. 138–147.

[6] Dorfman, Ya. G. "The physical views of Leonhard Euler". In: *Leonhard Euler.* M.: Izd-vo AN SSSR (Publ. Acad. Sci. USSR), 1958. pp. 377–411.

[7] Minchenko, L. S. "Euler's physics". Trudy in-ta istorii estestvozn. i tekhn. AN SSSR (Works of the inst. for the hist. nat. sci. and technology Acad. Sci. USSR). 1957. Vol. 19. pp. 221–270.

[8] Slyusarev, G. G. "Euler's *Dioptrics*". In: *Leonhard Euler.* M.: Izd-vo AN SSSR (Publ. Acad. Sci. USSR), 1958. pp. 414–420.

[9] Pogrebysskaya, E. I. *The dispersion of light: an historical sketch.* M.: Nauka, 1980. pp. 51–66.

---

[21] Delisle had visited Berlin in 1725 on the way from Paris to St. Petersburg.

[10] Nevskaya, N. I. "On the history of achromatic instruments: the works of L. Euler". *Proceedings of the Xth and XIth conf. of grad. students and junior scientific workers.* IIEiT AN SSSR (Inst. hist. nat. sci. and tech. Acad. Sci. USSR). M., 1968. pp. 3–9.

[11] ——. "The first works on astrophysics of the Petersburg Academy of Sciences (18th century)". Ist.-astron. issled. (Historical-astronomical research). 1969. Issue 10. pp. 121–157.

[12] ——. "The diffraction of light in the works of 18th century astrophysicists". Ist.-astron. issled. (Hist.-astronomical research). 1977. Issue 13. pp. 339–376. (On Euler: pp. 363–376.)

[13] ——. *The 18th century Petersburg astronomical school.* L.: Nauka, 1984. 238 pages.

[14] LO Arkhiva AN SSSR (Leningrad Section of Archive of Acad. Sci. USSR), r. 1, op. 50, No. 1, l. 1–44; r. 1, op. 44, No. 1, l. 1–662; r. 1, op. 44, No. 2, l. 1–368.

[15] Pekarskiĭ, P. P. *History of the Imperial Academy of Sciences in St. Petersburg.* SPb., 1870. Vol. 1. 774 pages.

[16] *The St. Petersburg news.* SPb., 1727–1742.

[17] *Monthly historical, genealogical, and geographical notes on the St. Petersburg news.* SPb., 1728–1742.

[18] LO Arkhiva AN SSSR (Leningrad Section of Archive of Acad. Sci. USSR), r. 1, op. 44, No. 1, l. 9 ob., 22 ob., 35 ob.–36, 40–43 ob., 12–13 ob., 21 ob., 43 ob., etc.

[19] Delisle, J. N., le cadet. "Observation de l'éclipse de Vénus par la Lune, faite en plein jour au Luxembourg le 28 Juin 1715". Mém. Acad. sci. Paris (1715). 1718. pp. 135–137.

[20] Cassini, J. "Extrait de l'observation de l'éclipse de Vénus du 28 Juin 1715, faite à Montpellier par Mrs. De Plantade et De Clapier. Avec quelques réflexions sur les apparences qui ont pu donner lieu de juger qu'il y avoit une atmosphère autour de la Lune". Mém. Acad. Sci. Paris (1715). 1718. pp. 139–140.

[21] Euler, L. "Tentamen explicationis phaenomenorum aeris". Comment. Acad. Sci. Petrop. (1727). 1729. Vol. 2. pp. 347–368; *Opera* II-31.

[22] LO Arkhiva AN SSSR (Leningrad Section Archive Acad. Sci. USSR), f. 136. op. 1, No. 130, 131.

[23] Delisle, J. N. "Théorie du mouvement des taches du Soleil". In; *Mémoires pour servir à l'histoire et au progrès de l'astronomie, de la géographie, et de la physique.* SPb., 1738. pp. 143–179.

[24] Müller, G. F. "History of the Academy of Sciences with additions by I. G. Shtritter 1725–1743". In: *Materials relating to the hist. of the Imperial Academy of Sciences.* SPb., 1890. Vol. 6.

[25] Euler, J.-A. "De rotatione Solis circa axem ex motu macularum apparente determinanda". Novi comment. Acad. Sci. Petrop. (1766–1767). 1768. Vol. 12. pp. 273–286; *Opera* II-30.

[26] *Minutes of the meetings of the Conference of the Imperial Academy of Sciences from 1725 to 1803.* SPb. 1897–1899. Vols. 1, 2.

[27] Euler, L. "Methodus computandi aequationem meridiei". Comment. Acad. Sci. Petrop. (1736). 1741. Vol. 8. pp. 48–65; *Opera* II-30.

[28] Meyer F.-Ch. "De luce Boreali". Comment. Acad. Sci. Petrop. (1726). 1729. Vol. 1. pp. 351–367.

[29] Kraft, G. W. "A brief description of the most noteworthy observations of the weather and changes in the air here in St. Petersburg from the beginning of 1726 till the end of 1736". In: *Monthly historical, genealogical, and geographical notes on the St. Petersburg news*. SPb., 1738. pp. 70–75.

[30] Euler, L. "Recherches physiques sur la cause de la queue des comètes, de la lumière boréale, et de la lumière zodiacale". Mém. Acad. Sci. Berlin (1746). 1748. Vol. 2. pp. 117–140; *Opera* II-31.

[31] *Lettres à une princesse d'Allemagne sur divers sujets de physique et de philosophie*. SPb., 1768–1772. Vols. 1–3; *Opera* III-11, 12.

[32] Euler, L. "Inquisitio physica in causam fluxus et refluxus maris". In: *Rec. pièces remp. prix Acad. Sci. Paris (1740)*. 1741. Vol. 4 pp. 235–350; *Opera* II-31.

[33] *Discours lû dans l'Assemblée publique de l'Académie imp. des sciences, le 2. Mars 1728, pars Mr. De L'Isle, avec la Réponse de Mr. Bernoulli*. SPb., 1728.

[34] Raĭkov, B. E. *Sketches on the history of the heliocentric world-view in Russia: From the past of Russian natural science*. M.; L.: Izd-vo AN SSSR (Publ. Acad. Sci. USSR), 1947. 391 pages.

[35] *Materials relating to the history of the Imperial Academy of Sciences: in 10 volumes*. SPb., 1885–1900.

[36] Euler, L. *Correspondence. Annotated index*. L.: Nauka, 1967. 391 pages.

[37] *Die Berliner und die Petersburger Akademie der Wissenschaften im Briefwechsel Leonhard Eulers*. Hrsg. A. P. Juškevič, E. Winter. Berlin, 1959–1976. Vols. 1–3.

[38] Euler, L. *Letters to scientists*. M.; L.: Izd-vo AN SSSR (Publ. Acad. Sci. USSR), 1963. 397 pages.

[39] Yushkevich, A. P., Klado, T. N., Kopelevich, Yu. Kh. "L. Euler and J. N. Delisle in their correspondence 1735–1765". In: *Russian-French scientific connections*. L.: Nauka, 1968. pp. 119–279.

[40] LO Arkhiva AN SSSR (Leningrad Section Archive Acad. Sci. USSR), f. 3, op. 10, No. 2/1.

[41] LO Arkhiva AN SSSR (Leningrad Section Archive Acad. Sci. USSR), f. 3, op. 10, No. 2/2.

[42] Bayer, Th. S. *Historia regni graecorum Bactriani.* . . . Petropoli, 1738. 214 pages.

[43] Euler, L. "De indorum anno solari astronomico". In: Bayer, Th. S. *Historia regni graecorum Bactriani.* . . . Petropoli, 1738. pp. 201–213; *Opera* II-30.

[44] LO Arkhiva AN SSSR (Leningrad Section Archive Acad. Sci. USSR), f. 136, op. 2, No. 5, l. 32–33.

[45] Euler, L. *Opuscula varii argumenti*. Berolini, 1746. ("De motu corporum in superficiebus mobilibus". pp. 1–136; *Opera* II-6. "Tabulae astronomicae Solis et Lunae". pp. 137–168; *Opera* II-23. "Nova theoria lucis et colorum". pp. 169–245; *Opera* III-5. "De relaxione motus planetarum". pp. 245–276; *Opera* II-31. "Enodatio questionis. Utrum materiae facultas cogitandi tribu possit nec ne? ex principiis mechanicis petita". pp. 277–286; *Opera* III-2. "Recherches physiques sur la nature des moindres parties de la matière". pp. 287–300; *Opera* III-1.)

[46] Mém. Acad. sci. Berlin (1747). 1749. Vol. 3. ("Recherches sur le mouvement des corps célestes en général". pp. 93–143; *Opera* II-25. "Méthode pour trouver les vrais moments tant des nouvelles que des pleines lunes". pp. 144–173; *Opera* II-30. "Méthode de trouver le vrai lieu géocentrique de la Lune par l'observation de l'occultation d'une étoile fixe". pp. 174–177; *Opera* II-30. "Méthode de déterminer la longitude des lieux par l'observation d'occultations des étoiles fixes par la Lune". pp. 178–179; *Opera* II-30.)

[47] Euler, L. "Sur la perfection des verres objectifs des lunettes". Mém. Acad. Sci. Berlin (1747). 1749. Vol. 3. pp. 274–296; *Opera* III-6.

[48] *Musei Imperialis Petropolitanae*. Petropoli, 1741. Vol. 2. 796 pages.

[49] Weidler, I. F. *Historia astronomiae*. Vitembergae, 1741. 664 pages.

[50] Euler, L. "Réflexions sur la dernière éclipse du Soleil du 25 Juillet a. 1748". Mém. Acad. Sci. Berlin (1747). 1749. Vol. 3. pp. 250–273; *Opera* II-30.

[51] Euler, L. "Sur l'atmosphère de la Lune prouvée par la dernière éclipse annulaire du Soleil". Mém. Acad. Sci. Berlin (1748). 1750. Vol. 4. pp. 103–121; *Opera* II-31.

Leonhard Euler in Correspondence with Clairaut, d'Alembert, and Lagrange[1]

A. P. Yushkevich and R. Taton

Introduction

As is well known, L. Euler's scientific activity from the age of 20 falls into three periods: first the years 1727–1741 spent at the Petersburg Academy of Sciences, then the interval 1741–1766 at the Berlin Academy of Sciences, where he continued to maintain close contact with the Petersburg Academy, and finally the last stage of his life back at the Petersburg Academy (1766–1783). And concurrently with his work at the Petersburg and Berlin Academies, there was the collaboration with the Paris Academy, which began in his early years and was regularized at the end of the 1730s. In 1727 Euler entered an essay (E4) in the competition of the Paris Academy which, although it did not win the prize, was favorably received and published a year later in Paris[2]. Between 1738 and 1772 Euler won prizes in competitions organized by the Paris Academy twelve times—some for essays on navigation, technology, and physics, and others for fundamental research in celestial mechanics[3]. Euler's lively correspondence with French scientists also played an important role in his scientific activity—as indeed in theirs. He personally knew only two Parisian academicians well: first and foremost the astronomer and geographer J. N. Delisle, who worked in the Pe-

---

[1] The present essay is a significantly abridged version of the authors' introduction to Euler's correspondence with the three named French scientists, published in French in Volume 5 of Series IV of Euler's *Opera omnia* [1]. The translation from French into Russian was made by N. S. Ermolaeva.

[2] In the original all quotations, including those of Euler, were given only in Russian translation, and it is these, rather than their originals, that have been used for the present English version. The works of Euler and his son Johann-Albrecht are labelled by the letter E followed by a number, indicating their place in the well known list compiled by G. Eneström, giving the year and place of first publication of each item, as well as the volume of the *Opera omnia* [1] where it can be found (Eneström, G. *Verzeichnis der Schriften Leonhard Eulers*. Jahresber. Dtsch. Math. Ver. 1910–1913. Erganzungsb. 4, Lief. 1–2). Each of Lagrange's works mentioned is indicated by the letter L followed by the numeral assigned to it in the list [22] compiled by R. Taton.—*Authors' note, modified by trans.*

[3] A list of the prizes awarded to Euler in competitions of the Paris Academy has been published by Yu. Kh. Kopelevich; see [6, pp. 60–61].

289

tersburg Academy from 1725 to 1747, and later P.-L. Moreau de Maupertuis, president of
the Berlin Academy from 1746 to 1759. He had also met J. d'Alembert when the latter was
a guest of the Prussian king Friedrich II at Potsdam in 1763. Euler's contact with all other
French scientists was purely epistolary. Although the number of his French correspondents
was not large—only 20 out of a total of nearly 300 correspondents of all nationalities—
in terms of significance his correspondence with them represents a very weighty portion.
This assessment is determined primarily with reference to his correspondence with such
first-rank "geometers", as they put it in the 18th century—meaning scientists active in the
physico-mathematical sciences—as Clairaut, d'Alembert, and Lagrange, but to some extent also with those not of the first rank, such as Delisle and Maupertuis. We exclude J.-H.
Lambert from the French group of Euler's correspondents, since, as a native of Alsace who
travelled extensively throughout Europe, he was less connected to France than to Germany
and the German part of Switzerland and spent the last years of his life working in Germany,
from 1765 as a member of the Berlin Academy (see the essay by K.-R. Biermann in the
present collection).

Euler's correspondence with Clairaut, d'Alembert, and Lagrange is of exceptional interest for the history of science, and of course for the biographies of all four scientists. In
richness of ideas it does not yield to the three other most important sets of Euler's correspondence: with his teacher Johann Bernoulli, with Daniel Bernoulli—from 1727 to 1733
his colleague and closest friend in St. Petersburg—, and with Ch. Gol'dbach, who was also
a Petersburg academician and, though not as gifted as the others, was nonetheless a perceptive mathematician and able like no one else to stimulate Euler in his investigations.
Note that Euler's correspondence with Johann Bernoulli is published in its entirety, and
with Daniel Bernoulli in part; together these comprise Volumes 2 and 3 of Series IV of
Euler's *Opera omnia* [1]; the correspondence with Gol'dbach has been published twice, in
1843 and 1965, and will soon appear a third time as Volume 4 of Series IV of the *Opera
omnia*[4]. Also, Euler's extensive correspondence with Delisle has been published in French
with a Russian translation [8]. His correspondence with Maupertuis comprises the first part
of Volume 6 of Series IV (1986); however it is in large part devoted only to administrative issues since Maupertuis was president of the Berlin Academy and Euler his closest
aide in its management. It is natural to associate this correspondence with that conducted
by Euler from 1741 to 1766 between Berlin and St. Petersburg, published as the three-
volume collection [3]. Euler's correspondence with Clairaut, d'Alembert, and Lagrange
had earlier been published only partially; a complete edition with commentary and in the
original languages—for the most part French (all letters written in Latin being included
with French translations)—appeared in 1980 [2]; unfortunately some letters, albeit only a
few, have been lost. Of the letters published in [2], 138 have not yet been translated into
Russian. Below we give a general characterization of these letters, going into detail when
appropriate. As in the publication [2] we consider each of the three sets of correspondence
separately; they differ greatly from one another in content and in how Euler relates to each
of the three scientists, at that time all members of the Paris Academy; it is noteworthy also
that Clairaut was from 1754 simultaneously a foreign member of the Petersburg Academy,
d'Alembert from 1764, and Lagrange from 1776—all three at the initiative, or at least with

---

[4]This was as of 1986; by now it has appeared.—*Trans.*

the complete approval, of Euler. Euler, on the other hand, was elected foreign member of the Paris Academy in 1755 with the support of Clairaut and d'Alembert, Lagrange living then in Turin.

1. The correspondence with A.-C. Clairaut

Euler's correspondence with Clairaut consists of 61 letters. Clairaut had very early on been elected to the Petersburg Academy on the strength of his remarkable early publications, especially his fundamental *Investigations into curves of double curvature*, completed in 1729 and published two years later (Paris, 1731), and was elected to the Paris Academy in that same year, at the age of 18. In the first of his letters to Euler, dated September 17, 1741 (all dates are new style), Clairaut discusses the question of the integration of differential expressions of the form $Adx + Bdy$, where A and B are functions of x and y, informs Euler that he has dispatched to him manuscripts of his work on that question, and asks him to inform him, Clairaut, of his own publications on the topic, which he had heard of through D. Bernoulli. In addition Clairaut considers the problem of differentiation of an integral with respect to a parameter, which his colleagues A. Fontaine and P. Bouguer were working on, and also the theorem on the differentiation of homogeneous functions. Clairaut's memoir "On the integral calculus", published with a somewhat different title in the *Notes* of the Paris Academy for 1740 (1742), contains an approach to integrating the above expression, together with a partial differential equation for its integrating factor (see [9, p. 246]). Although none of this was especially new to Euler, all the same the area of enquiry was of the greatest interest to him. In his reply of October 30, 1741 he first expresses his satisfaction at the commencement of correspondence with Clairaut, describes his work on the questions raised by Clairaut, noting in this connection certain earlier results of J. Hermann and N. Bernoulli, and refers in terms of praise to the published works of Clairaut that he is familiar with, in particular two articles of 1739 on the shape of the earth [9, p. 245]—at that time a problem of the liveliest interest to the scientific community since the question of the correctness of the Newtonian theory of gravitation depended on one or another of its practical or theoretical solutions. In the same letter Euler gives a very positive appraisal of Clairaut's approach to the problem of integrating differential expressions of the form $Adx + Bdy$—which then Clairaut promptly generalizes to its many-variable analogues (see his letter of December 26, 1740 and Euler's of March 6, 1741), publishing a paper on this theme in the Paris *Notes* for 1740 (1742), in which he acknowledges the works of Euler that have now become known to him, as well as work of Fontaine [9, p. 247].

We thus see that from the very beginning of their correspondence both Euler and Clairaut discussed topical problems of mathematical analysis—in the present case certain aspects of the theory of functions of several variables, subsequently extensively elaborated by Euler in his *Differential calculus* (1755, E212) [10] and in his work in theoretical mechanics. Later Euler would develop much further the method using an integrating factor, especially in Volumes 1 and 2 of his *Integral calculus* (1768, 1769) [11], and in 1743 Clairaut would publish his classical *Theory of the earth's shape* [12]. The latter work represents the culmination of a series of investigations into hydrodynamics, concerned with the equilibrium (and stable equilibrium) shapes of rotating liquid masses, work ultimately completed by Henri Poincaré, and above all by A. P. Lyapunov.

Alexis-Claude Clairaut

It is worth emphasizing the mutually friendly and respectful manner of address observed by both correspondents throughout their correspondence: there are neither disagreements nor claims of one against the other; all ideas are described openly, often significantly prior to their publication. They have very many mutual mathematical interests—for example, the theory of ordinary differential equations. When Euler, in a letter dated October 31, 1741 and others, notes tersely certain pivotal points in his approach to the solution of linear homogeneous differential equations, appearing in print only two years later (E62), Clairaut fills in the gaps in Euler's description. He expounds the method at the meetings of the Paris Academy of August 1743 and February 1744, in this way collaborating in its dissemination, Euler's published version not having yet become widely known.

There is, however, one point on which Euler and Clairaut diverge, namely in their attitudes to number theory. For Euler this is an extremely important discipline on at least an equal footing with other parts of mathematics, which he unwaveringly regards as a united whole. On the other hand in Clairaut's eyes number theory is more an amusing pastime; he is completely oriented towards applications of mathematical analysis and those of its branches that may prove useful, and considers time and energy spent on the rest wasted. In letters to Clairaut, as in his other correspondence, Euler tries to stimulate Clairaut's interest in his published—and even more in his unpublished—results of the purer sort, but often in vain. Thus Clairaut refrains from taking up Euler's ideas for a method of solving problems of the calculus later known as variational (see the section devoted to Lagrange below), adumbrated in the aforementioned letter of March 6, 1741. (Here Euler writes about matters that he would later expound in detail in his celebrated monograph on the calculus of variations (E65) [13].) Very interesting letters of January and February 1742 and March 23, 1742 show Clairaut to have been indifferent to the summation of series of reciprocals of powers of integers. Euler is aware of this difference in their approaches to mathematics and in a letter of April 1742 concerning his current preoccupation with problems posed by Fermat, in which he asks indirectly whether perhaps any relevant papers of Fermat have been preserved, he apologizes for bothering Clairaut with "such dry matter" which, though perhaps boring to Clairaut, from his own point of view merits attention. (It is well known how much time and effort Euler devoted to his arithmetic investigations.) In his reply of May 29, 1742 Clairaut writes that such problems are "not fashionable", adding patronizingly that they may serve incidentally for "exercising the mind"; in response to Euler's query about Fermat he says that he has never heard what became of Fermat's papers and expresses astonishment at Euler's wishing to engage in extensive investigations of such sorts of questions. It is noteworthy that the cast of mind demonstrated by Clairaut was likewise that of D. Bernoulli and many other mathematicians of that time.

However the divergence of their views as to the relative value of different mathematical problems and theories did not in the least diminish the amity of the relations between Euler and Clairaut. Clairaut kept Euler informed of the various activities of the Paris Academy, and participated in the awarding of some of that Academy's prizes to Euler over the period 1746–1752 (including those awarded for his investigations into the irregularities in the orbits of Jupiter and Saturn; see E120 and E384); it may be surmised that he played no small role in Euler's election to foreign membership of the Paris Academy. On the other hand we have Euler's proposal of Clairaut's candidature for foreign membership of the Berlin Academy. The election was approved on February 2, 1744 and Clairaut expressed his gratitude to Euler in particular and to the Academy as a whole on March 14 of that same year.

Beginning with Clairaut's letter of August 23, 1744, the correspondence between him and Euler deals almost exclusively—i.e., apart from various special problems of mechanics —with celestial mechanics. Since this portion of the correspondence is discussed directly or indirectly in other essays of the present collection, we give but a brief summary. The disagreement between the observed lunar motion and that calculated on the basis of Newton's gravitational theory initially caused Clairaut, Euler, and d'Alembert to doubt the correctness of the universal law of gravitation, and they began to consider that perhaps a correction is needed, namely the replacement of the square of the distance between the gravitating masses by the square plus a supplementary term depending on the distance. Then in December 1748 Clairaut discovered the source of the disagreement—the invalid neglecting of certain terms in the process of integrating the relevant second-order differential equations—announcing his discovery forthwith to the Paris Academy and informing Euler in a letter dated June 19, 1749. However Euler was not wholly convinced that Clairaut was correct, and, since the Petersburg Academy had opportunely just at that time asked him to name themes suitable for its first international competition, he suggested as the topic for the 1751 prize the question of the agreement of the moon's motion with that predicted by Newtonian theory, advising Clairaut to enter his essay—which he did. As a member of the jury appointed by the Petersburg Academy, Euler recommended awarding the prize to Clairaut. Naturally, Euler's opinion was decisive and Clairaut's entry "Lunar theory deduced from a single principle of attraction" [9, p. 249] won the prize and was published in French in St. Petersburg in 1752. Yet Euler was still not entirely satisfied. With astonishing speed he worked out his own lunar theory (E187), which soon (1753) appeared in Berlin. Many years later, once again living in St. Petersburg, Euler revisited this problem and published—now with the necessary aid of his colleagues in view of the almost total blindness that had overtaken him some years earlier—a new and more complete theory of the moon (1772, E418), destined to become the starting point for later research on that theme. The crucial sections of that work have been translated into Russian and supplemented with commentaries by academician A. N. Krylov [14].

Following the letter of June 24, 1752 relating to the conclusion of the Petersburg competition and the publication of the above-mentioned essay by Clairaut, there is a hiatus in the correspondence lasting many years. It is renewed only a little over ten years later in connection with Clairaut's work, preceded by that of Euler, on problems to do with the construction of optical instruments. Two of Clairaut's letters on this subject, of 1763 and 1764, are preserved, but nothing is known of Euler's replies. Clairaut died a year later.

2. The correspondence with J. d'Alembert

Euler's relations with Jean d'Alembert, member of the Paris Academy since 1741, were much more complicated and uneven than with Clairaut. Forty letters of their correspondence have survived, the first of which, dated August 3, 1746, related to the forwarding of two of d'Alembert's essays to Euler through Maupertuis. One of these, entitled "Thoughts on the general cause of winds", published a year later in both Berlin and Paris, had just been awarded the prize in a competition of the Berlin Academy of Sciences by a jury headed by Euler. The decision to award d'Alembert's essay the prize was taken on June 2, 1746, and on that very day d'Alembert was elected to foreign membership of the Berlin Academy. Incidentally, Euler's old friend Daniel Bernoulli had also entered an essay in that competition, and the fact that it was not awarded the prize led to a cooling of their relations, notwithstanding Euler's undoubted support of Bernoulli's election to foreign membership of the Berlin Academy on June 30, 1746. Euler's reply to d'Alembert's letter, dated December 29, 1746, is concerned with disagreements between d'Alembert and D. Bernoulli with respect to a particular question, which Euler attempts to smooth over by appropriately re-interpreting their different treatments of the question—whose precise nature we need not go into here[5]. In that same letter Euler praises d'Alembert's "Investigations in the integral calculus", just received in Berlin and published in 1748 in the issue of the *Notes* of the Berlin Academy for 1746—which journal was published in French at the behest of Friedrich II, admirer of French culture. Euler goes on to praise in particular the proof of the Fundamental Theorem of Algebra proposed in that memoir—that result being required in connection with the integration of rational functions considered in the memoir—and notes that he himself has just given a different proof of that theorem in a memoir of his own (E170) (eventually published in the *Notes* of the Berlin Academy for 1750 (1751)). These two proofs are both incomplete but capable of being made rigorous, and are essentially different from one another (see [15, pp. 70–76]). Euler also lavishes praise on certain transformations of elliptic integrals that d'Alembert applies but brings objections against d'Alembert's treatment of logarithms of negative numbers—thereby setting the scene for their first bout of polemics, fated to last many years. To put the matter briefly, d'Alembert, like Leibniz and Johann Bernoulli before him, did not have a precise, unambiguous notion of the logarithm function and its connection with the exponential function and consequently, following Bernoulli, considered that the logarithm of an arbitrary negative number

---

[5] As is clear from his letters to Euler, D. Bernoulli did not rate d'Alembert's research in mechanics very highly, in fact considering it completely unsatisfactory—at least on the physical plane. Compare the description of his correspondence with Euler in Volume 1 of Series IV of the *Opera omnia* [1], and in [4].

should be just the logarithm of the absolute value of that number. Leibniz had long before announced that the logarithm of a negative number should be imaginary, but did not have at his disposal the means for discovering the exact nature of this "imaginariness". Euler, on the other hand, had framed as early as the beginning of the 1740s an essentially modern theory of the logarithm function in the complex plane, as is clear from certain arguments he makes in the second volume of his *Introduction to the analysis of infinitesimals* (E102), which was delivered to the printer in 1744 and appeared in 1748 [16]. The dispute went on in their correspondence till 1748, from which time it was continued by d'Alembert in print, and until almost the end of the 18th century certain mathematicians remained in doubt as to which of the participants in the argument was correct [15, pp. 324–328]. As a matter of fact Euler's reasoning as set out in the memoir E168 published in the Berlin *Notes* for 1749 (1751) is unobjectionable, while d'Alembert's notion is unsound.

Gradually the debate between Euler and d'Alembert became more heated as various mutual suspicions and recriminations multiplied. Thus in 1750 by decision of a jury of the Berlin Academy headed by Euler, d'Alembert's entry in the Academy's competition—on the theme of the resistance to the motion of bodies in a fluid—was rejected. There can be no doubt that this decision was unjust since d'Alembert's essay was to become one of the foundational works in the development of mathematical physics, in particular because in it the theory of functions of a complex variable is applied successfully and for the first time there appeared in print what we now call the Cauchy-Riemann equations—soon to be applied also by Euler. A deeply disappointed d'Alembert published his essay with the title "Attempt at a new theory of resistance of fluids" in Paris (1752). The competition was postponed to the next year, when the prize went to the vacuous essay of a certain amateur mathematician by the name of J. Adami, thereby giving d'Alembert serious offense. D'Alembert writes of these matters in his letters of January 4 and September 10, 1751.

In these same years many other questions arose for debate, often involving mutual criticism and eventual clarification, such as for example that concerning second-order cusp points (discussed in d'Alembert's letter of September 7, 1748 and Euler's reply of September 28, 1748), or those concerning the precession of the equinoxes (in d'Alembert's letters of February 22 and March 30, 1750; the corresponding letters of Euler have been lost). Celestial mechanics is much discussed in their correspondence. And then there is a long gap in the correspondence—filled, however, with printed polemics instead of the epistolary sort, mainly controversy concerning problems of the very first importance of mathematical physics and its analytic apparatus; incidentally, these polemics were destined to continue even after the renewal of the two opponents' correspondence in 1763. In a letter dated September 10, 1751 d'Alembert expresses a profound sense of injury at the decision of the Berlin Academy to award the prize to Adami, and from then till 1763 ceases corresponding with Euler.

We now turn to the most famous disagreement between Euler and d'Alembert, soon involving other scientists, namely the debate over the solution of the particular problem of the theory of elasticity concerning a string whose transverse vibrations are expressed through a second-order partial differential equation of hyperbolic type later called the wave equation. This problem had long been of interest to mathematicians. The first approach to a solution worthy of note was proposed by B. Taylor, and in the interval 1729–1732 by Johann Bernoulli. A decisive step forward was made by d'Alembert in two articles

Jean le Rond d'Alembert
From a pastel by De la Tour

appearing one after the other in the Berlin *Notes* for 1747 (1749) containing the differential equation for the vibrations, its general solution in the form of a sum of two "arbitrary functions" arrived at by means original with d'Alembert, and a method of determining these functions from any prescribed initial and boundary conditions. These two papers are mentioned with title omitted in the first of d'Alembert's letters to Euler, and then again in a letter dated January 6, 1747 together with a title almost exactly the same as that under which they were to be published: "Investigations into the curve whose shape is assumed by a stretched string made to vibrate". D'Alembert's method, later called the "method of characteristics", was undoubtedly highly regarded by Euler, yet in his surviving letters from the period 1747–1751 there is absolutely no mention of this problem; they are instead completely taken up with the questions noted above: logarithms, cusp points, problems of celestial mechanics—including the theory of lunar motion—, academic competitions, the simplest form of the water turbine invented by J. A. Segner, and so on. All of these letters are of interest on both the scientific and personal levels. Especially noteworthy is d'Alembert's high estimation expressed in a letter of June 17, 1748 of the expansion of the function $(1 - g \cos \omega)^{-\mu}$ in a trigonometric series used by Euler in his prize-winning essay on the irregularities in the motions of Saturn and Jupiter (E120), this being the first example of its type. (Clairaut had written of the expansion in equally glowing terms in his letter to Euler of September 11, 1747.) However in not a single letter written by Euler during the aforementioned period is there any mention of d'Alembert's method of characteristics; it is of course possible that some such letter or letters have been lost: for example of their correspondence during the year 1747, only five of d'Alembert's and three of Euler's letters have survived.

However that may be, Euler in the main certainly approved—with one crucial caveat— of the method of characteristics, using it in a note of moderate length presented to the Berlin

Academy on May 16, 1748 and published first in Latin in the issue of a certain Leipzig journal for 1749 (E119) and then in French in the Berlin *Notes* for 1748 (1750) (E140). The most crucial issue dividing d'Alembert and Euler in connection with the vibrating string problem was that of the compass of the class of functions admissible as solutions of the wave equation, and of the boundary problems of mathematical physics generally. D'Alembert regarded it as essential that the admissible initial conditions obey stringent restrictions or, more explicitly, that any function giving the initial shape and speed of the string should over the whole length of the string be representable by a single analytic expression (i.e., should be "continuous", in Euler's terminology) and furthermore be twice continuously differentiable (in our terminology). He considered his method invalid otherwise.

However Euler was of a different opinion. In his article published in the Berlin *Notes* for 1748 (E140) he rejects d'Alembert's restrictions on the initial conditions, maintaining that for the purposes of physics it is essential to relax these restrictions: the class of admissible functions or, equivalently, curves should include any curve that one might imagine traced out by a "free motion of the hand", such as for instance one broken into straight-line segments (he calls such piecewise smooth functions variously "discontinuous", "mechanical", and also "mixed"). Although in such cases the analytic method is inapplicable, Euler proposed a geometrical construction for obtaining the shape of the string at any instant. This disagreement was given expression in print; in the correspondence between the two, which is the topic of the present essay, it is mentioned only when that correspondence is resumed, for instance in Euler's letter to d'Alembert of December 20, 1763. This date indicates the long duration of the debate on the question, made the more complicated when shortly thereafter Daniel Bernoulli entered the fray: in an article published in the Berlin *Notes* for 1753 (1755) Bernoulli proposed finding a solution by the method of superposition of simple trigonometric functions, i.e., using trigonometric series, or, as we would now say, Fourier series. Although Daniel Bernoulli's idea was to turn out extremely fruitful—in others' hands—, he proved unable to develop it further. Both d'Alembert and Euler immediately began to dispute the generality of such a solution, the chief practical difficulty with which was, incidentally, their inability at that time to determine for an arbitrary function the coefficients of its expansion in a trigonometric series in sines or cosines or both. The general integral formulae for the "Fourier coefficients" were derived much later by Euler in papers written in 1777 (E707 and E704) and published posthumously in the Petersburg *Notes* for 1793 (1798). These formulae were re-derived 25 years later, using a different method, by J. B. J. Fourier, evidently in ignorance of Euler's priority. Following D. Bernoulli, Lagrange also became involved in the "argument over a string" (see below), as did other scientists (see [15, pp. 312–318]).

Returning to Euler's correspondence with d'Alembert, we note that their polemical discussions resumed their initial epistolary form for a short period between 1763 and 1764, when they touched also on Lagrange's work on the question. As far as that whole debate is concerned we will confine ourselves here to just a few remarks of a summarizing character. Thus without going into the details of the "argument over a string" (more of which can be found in [17] and the brief treatment [15, pp. 412–419]), which lasted for a half-century and was in some sense resolved satisfactorily—though not completely—only in the 19th century (through the work of Fourier and others), we mention only that ultimately d'Alembert moved away from his initial position by demanding only that for an nth order differential equation the class of admissible "arbitrary" functions [representing initial conditions] be only n times continuously differentiable [18]. Euler, on the other hand, continuing to develop his ideas on this theme, anticipated in a certain well known sense the introduction of generalized functions by S. L. Sobolev (called "distributions" in the Western literature at the initiative of L. Schwartz)—this circumstance has been noted earlier by A. P. Yushkevich [19], and a detailed exposé on it written by S. S. Demidov [20]. By way of an historical analogue, one might take divergent series, which Euler applied sometimes with great success albeit unrigorously, but which now form part of modern summability theory, where all the generally accepted modern demands of rigor apply. And, finally, the "argument over a string" has had a tremendously stimulating effect on the subsequent development of many mathematical concepts and fields, such as: the methodology of the theory of partial differential equations; the theory of trigonometric series and expansions in terms of other orthogonal families of functions; the concept of a function, and so on—right up to the modern theory of sets and functions between them.

Apart from the disagreements we have so far briefly described, sometimes concerning questions of priority, the relations between Euler and d'Alembert were influenced by other circumstances. Euler was quite negative in his attitude towards the ideas of the Enlightenment, these being foreign to his rather standard religious and philosophical views, and the francomania of Friedrich II was repugnant to him. Moreover he saw in d'Alembert a possible successor to Maupertuis as president of the Berlin Academy of Sciences—and in fact on the death of Maupertuis in 1759 the king did achieve his goal of appointing d'Alembert to that post. However Euler was completely ignorant of d'Alembert's true intentions. D'Alembert did not at all find the king's offer attractive; on the contrary, when he heard that purportedly Euler intended to quit Berlin and return to St. Petersburg he wrote a series of letters between July 29, 1763 and April 28, 1766 in which he attempted to dissuade Euler from leaving, offering more than once to act as mediator between Euler and the king. However d'Alembert allowed for the eventuality of Euler's quitting Berlin by simultaneously carrying on negotiations with Lagrange as a suitable successor to Euler in Berlin.

Correspondence between Euler and d'Alembert ceased after April 28, 1766, and on June 9, 1766 Euler bade farewell to the Prussian capital to return to St. Petersburg forever. There is however one more letter from d'Alembert to Euler, dated February 27, 1773, relating to the dispatch of a certain one of d'Alembert's books to St. Petersburg.

3. The correspondence with J.-L. Lagrange

In the persons of Clairaut and d'Alembert Euler had correspondents who, although extraordinarily gifted, were nevertheless not his equals and not congenial to him. It was only in Joseph-Louis Lagrange that Euler found an associate of the same quality and breadth of interests—although Lagrange yields to Euler in terms of quantity, i.e., of the number though not the quality of their works. (Euler's collected works, excluding his correspondence, fill 75 volumes, Lagrange's 14.)

A native of Turin, then the capital of the kingdom of Sardinia, but French on his father's side (in Italy his surname was for a long time written as Lagrangia and pronounced accordingly), Lagrange had at a youthful age already mastered the basic mathematical literature and even forged far ahead in certain directions, largely using Euler's works as a guide. The correspondence of the two scientists is not extensive, comprising only 37 letters written between 1754 and 1775, but is extremely rich in ideas. (The correspondence was interrupted for two and a half years during the seven years' war, when the mail service between Turin and Berlin was suspended.) We confine ourselves here to examining just a few of the questions considered in that correspondence.

We observe first of all that in reading the correspondence one senses clearly the generation gap between the correspondents, an evolving style of thought, and a resulting gradual alteration in manner of exposition—none of which was so evident in the two sets of correspondence examined above. The reasons for this are perhaps to be sought in the generational difference, the rapid progress in mathematics, and in changes in social attitudes: Lagrange was younger than Euler by 30 years, while Clairaut was only six years younger and d'Alembert ten. Euler characteristically explained his ideas in a detailed and verbose style, illustrating them with many special examples, while Lagrange tended towards broad and where possible tersely formulated generalizations, appropriately systematized. Although the latter aspect of Lagrange's mathematical style was not foreign to Euler—indeed to the very end of his life he was the greatest of mathematical systematizers—, nevertheless Lagrange emerged as the first to frame great monolithic mathematical systems based on a minimal set of postulates—occasionally a mere singleton. Here we have in mind above all his *Analytical mechanics* of 1788 and his *Theory of analytic functions* of 1797 (L102). In this respect A. N. Kolmogorov drew a parallel between Lagrange and the originators of the broad and innovative ideas pertaining to all spheres of knowledge propagated in France at the height of the 18th-century enlightenment, with the qualification that perhaps ("undoubtedly" would be more apt) Lagrange yields to Euler in terms of sheer number and variety of problems solved [21, pp. 473–474]. Lagrange's criteria of rigor were also more stringent than Euler's, though in many ways still a long way from meeting the stricter standards of the early 19th century towards which they were evolving. Although acknowledging Euler's tremendous merit, Lagrange regarded his manner of exposition as completely out of date; in a letter to d'Alembert of January 1, 1766 he describes the style of Euler's *Theory of motion of rigid bodies* (1765, E289) as "grandiloquent". For the future author of the *Analytical mechanics* (1788; second ed. 1811; L97) [23] this reaction is entirely natural; however Lagrange's contemporaries often found Euler easier to understand than Lagrange, the reason being again not far to seek: familiarity with the new mathematical language required an elapse of time. Euler and Lagrange were representatives of different epochs, different styles.

In our examination of the topics discussed in the correspondence we shall confine ourselves mainly to three questions, touching on others only in passing[6].

The first letter from Lagrange to Euler, dated June 28, 1754, deals with the analogy between the structure of the binomial expansion of $(x + y)^n$ and the formula for $d^n(xy)$ for any rational exponent n. (Lagrange first heard of Leibniz' result in this direction, to be found in his correspondence, from Euler's reply.) Later reflection along these lines was to

---

[6]There is in existence an inventory of Lagrange's works, namely that prepared by R. Taton [22], analogous to Eneström's index of Euler's works. We refer to an item in that inventory by means of the letter L followed by the appropriate numeral.

lead Lagrange to results important in operational calculus and his later theory of analytic functions, expounded in the memoir "On a new form of calculus relating to differentiation and integration of variable quantities", published in the Berlin *Notes* for 1772 (1774, L33). At the end of the letter, in the course of enumerating particular remarkable works of Euler's that he had studied, including E65[7], Lagrange notes by the way that he will shortly communicate to Euler certain of his ideas on maximal and minimal values and on surfaces. This means that the 18-year-old Lagrange must have already embarked—and perhaps even made significant progress—on his reformulation of Euler's variational calculus [13] and certain of its applications.

Without waiting for an answer, on August 12, 1755 Lagrange sends Euler an extremely condensed sketch, comprehensible to Euler alone, of his new algorithm for solving isoperimetric problems, i.e., problems of the calculus of variations (a name soon to be introduced by Euler), in which he introduces, alongside the usual sign for differentiation, the new symbol δ for the variation, just one of his departures from Euler's treatment. The few rules of his algorithm allow Lagrange to derive purely analytically and very economically the necessary condition established by Euler for functions $y = y(x)$ to be extremals of a functional of the form $\int_a^b Z(x, y, y', \ldots, y^{(n)})dx$, where Z is a given expression in $y(x)$ and its first n derivatives. The problem of carrying out such a derivation was stated by Euler himself in his memoir [13], his own derivation, which reduced the problem to the solution of what is now called the Euler equation[8], depending to a significant extent on geometrical considerations[9].

Euler responded with an ecstatic reply (dated September 6, 1755) to the effect that his youthful correspondent had brought the theory to the highest degree of perfection. And in fact Lagrange's algorithm allows the solution of problems not solvable by Euler's method—such as the determination of extremals of integrals whose value depends on two functions $y(x)$ and $z(x)$, of double integrals, and other such variants. Thanking Euler for his encouraging reply, in his letter of November 20, 1755 Lagrange gives by way of an example his solution of the brachistochrone problem with one end-point free—the problem in this formulation having been solved earlier only in a very special case. He also tells Euler that he has obtained the position of adjunct professor of mathematics at the Turin School of Artillery—where as it turned out he was to remain for the next 10 years. In another letter—unfortunately lost—he informed Euler that he had used his method to obtain a generalization of the principle of least action as formulated by Euler in an appendix to the work [13]. As is well known, Maupertuis considered himself the discoverer of this principle, to which he imputed the greatest importance.

We know of this lost letter of Lagrange's from a letter of Euler's dated April 24, 1756 in which he informs Lagrange that he has shown Maupertuis that letter and that Maupertuis expressed his great satisfaction at this "justification" of his principle, and intended to propose first the election of Lagrange to foreign membership of the Berlin Academy (which shortly thereafter came to pass, on September 2, 1756) and secondly that Lagrange

---

[7]This is the Eneström number attached to Euler's memoir "Methodus inveniendi lineas curvas maximi minimive proprietate gaudentes, sive solutio problematis isoperimetrici lattisimo sensu accepti", introducing the calculus of variations (see [13]).—*Trans.*

[8]Or Euler-Lagrange equation.—*Trans.*

[9]Euler's derivation is given in §2 of the essay "Euler and the variational principles of mechanics" by V. V. Rumyantsev, included in the present collection.—*Trans.*

be invited to work at that academy—which, so Euler wrote, reflected also his own wishes. From this letter of Euler's and the minutes of the meeting of the Berlin Academy of May 6, 1756 it emerges that together with the lost letter, Lagrange had sent Euler a work in two parts, one mathematical, the other on mechanics, containing an exposition of his discoveries. This has also been lost; it is possible that Maupertuis took both this work and the letter with him when around that time he left Berlin for France, fated never to return through the ill-health that dogged him till his death. It is known that Maupertuis wrote a personal letter to Lagrange in which he promised to have his work published in Berlin, since Lagrange told a Milanese friend, the mathematician P. Paolo Frisi, of this on May 4, 1756, and so before receiving Euler's letter of April 24. Incidentally, in his letter to Frisi Lagrange also mentioned that he had not been able to complete the preparation of two memoirs for publication since he was busy preparing lectures on mechanics. On May 19 of that same year Lagrange wrote a letter to Euler in which he called his generalized principle of least action the "universal key" to all problems of mechanics. However, as is clear from the text of his *Analytical mechanics* unlike Maupertuis, he did not conceive of it as a general metaphysical world-law.

Among his other correspondence, Lagrange's letter of October 5, 1756 merits particular attention. Here the double-integral formulae

$$\iint dy dx \sqrt{1 + p^2 + q^2},$$

for surface area, and $\iint z dy dx$, for the volume of a right cylinder with lid given by a surface $z = f(x, y)$, appear for the first time. Although Euler is usually credited with introducing the double integral, this would seem to contradict that attribution. On the other hand the fact that Lagrange presents both formulae without any explanation perhaps indicates that they were known to the leading mathematicians of the time. Whatever the case may be, the fact remains that the formulae written down in Lagrange's letter represent the very first known appearance of double integrals in manuscript or printed form—although calculations essentially equivalent to the computation of repeated integrals had earlier been carried out by Newton in his classical work of 1687 on celestial mechanics[10], and the concept of the double integral had been mentioned, albeit only in passing, in correspondence between Leibniz and Johann Bernoulli in 1697. In this letter also, Lagrange formulates for the first time the problem of finding minimal surfaces and derives the partial differential equation for such surfaces, the solution of which would defeat both him and Euler; at that time the only known examples of such surfaces were regions of spheres. It is true, however, that the wide dissemination of the concept of the double integral was brought about by means of a paper (E391) expressly written by Euler on the topic, presented to the Petersburg Academy in 1768 and published in the *Novi Commentarii* for 1769 (1770); here one finds also the formula for changing coordinates in a double integral[11].

There now followed a hiatus in the correspondence between Turin and Berlin caused by the seven years' war—an hiatus which was to have distressing consequences for Lagrange. To begin with, the question of his move to Berlin, so attractive to him, was for obvious reasons deferred. In the meantime, in 1757 he had organized, together with a few other sci-

---

[10]That is, the *Principia.—Trans.*

[11]Presumably the one involving the Jacobian.—*Trans.*

entists, a private scientific society, later to become the Turin Academy of Sciences, and had begun to arrange the publication of the society's *Notes*, the first issue of which appeared in 1759. Incidentally, he did not publish his works on the calculus of variations and mechanics in these *Notes*, being convinced that the versions he had sent earlier to Berlin must surely have already been published in the *Notes* of the Berlin Academy. Furthermore, counting on the support of the Berlin Academy—where he hoped soon to obtain a position, the material conditions of work in Turin being less than satisfactory—, he began around the same time to prepare a two-volume treatise on the calculus of variations and its applications to mechanics which towards the summer of 1759 was almost finished; this we can infer from his letters to Euler of July 28 and August 4, 1759, in the second of which he informs Euler that he is sending him the first issue of the *Notes* of the Turin Scientific Society, and also begs him to pass on the enclosed letter addressed to Maupertuis.

However between the autumn of 1756 and the summer of 1759 events had taken place with unpleasant consequences for Lagrange—events of which Euler informs him in a letter written on October 2, 1759. First, he learns from this letter of Maupertuis' death on July 27 of that same year in Basel on his way back to Berlin, so that the work of Lagrange that he had taken with him on his journey had not been returned to Berlin. Second, Euler advises Lagrange in the letter to find a publisher for his work somewhere else, for instance in Geneva or Lausanne, since financial straits resulting from the seven years' war had rendered its publication in Berlin infeasible. The financial difficulties of the Berlin Academy were indeed severe, as is indicated by the fact that the Berlin *Notes* did not appear for five years from 1759. An obvious further inference was that Lagrange's anticipated move to Berlin was not to be entertained. Thirdly—and this was simply insulting to Lagrange—he learns from the letter that Euler, taking into account the importance of the discoveries in the variational calculus communicated to him by Lagrange and tormented by impatience, had re-derived those results analytically "in the spirit of Lagrange"—although he promised not to publish anything on this topic until Lagrange has done so. He was silent, however, about the fact that he had written two memoirs on that theme: "An analytical explanation of the method of maxima and minima" (E297) and "Elements of the calculus of variations" (E296; it is here that the term "variation" makes its appearance) and that he had given himself leave to present them to the Berlin Academy on September 9 and 16, 1756 respectively—though it is true that they were published only in 1766 in the Petersburg *Novi Commentarii*. What is more, Euler's son Johann-Albrecht learned about Lagrange's reformulation of the calculus of variations from his father and applied it to the solution of a certain problem in an essay presented once again to the Berlin Academy on February 17, 1757 and published in the *Proceedings* of the Bavarian Academy (1764, EA10); here again one finds the term "variational" used, and the solution involves the double integral. Thus Euler publicly broadcast Lagrange's unpublished discoveries, albeit always acknowledging his priority.

Lagrange's bitter disappointment may easily be imagined; however he long maintained silence before expressing it delicately much later, in a letter of October 28(?), 1762 informing Euler of the dispatch of the second issue of the Turin *Notes* containing his two memoirs "Attempt at a new method for determining maxima and minima of indeterminate integral formulae" and "Application of the foregoing method to various problems of mechanics" (L7,8). He did not let slip the opportunity of stating that he had published these memoirs only because Euler had decided not to publish his own memoirs on the topic before him, and further that he had contented himself with a brief exposition, having destroyed the above-mentioned large treatise which had been in the final stages of preparation.

It has to be said that Lagrange's exposition was so difficult to read while Euler's was, as always, so accessible, that it was the latter that became the more widely known, until superseded by the completely new treatment given in the "Appendix on the variational calculus" to the third volume of Euler's *Integral calculus* (SPb., 1770; E385). Although in both of his earlier memoirs Euler had been careful to note Lagrange's priority, nevertheless many readers assumed that Euler was the inventor of the calculus of variations.

It is clear that from 1759 there was a natural cooling of relations between Euler and Lagrange, though they continued to correspond. More and more frequently the former pupil comports himself as an equal of his erstwhile teacher. All the same, Lagrange continued to develop themes set by Euler—like most mathematicians and mechanicians of the second half of the 18th century. Euler was indeed "our teacher in everything", in the words of P.-S. Laplace, one of the few mathematicians of the time not to enter into direct contact with Euler[12]. This chill in the formerly so warm relations between Euler and Lagrange was eased for Lagrange by the establishment of a close connection with d'Alembert, whom Lagrange met during his six-month-long visit to Paris in the summer and spring of 1763. Their acquaintanceship quickly evolved into a close friendship, soon to play a significant role in Lagrange's future career.

The correspondence renewed in 1759 between Euler and Lagrange was often interrupted during the next few years because of events relating to the war. Thus there were absolutely no letters written between the summer of 1760 and that of 1762, nor between autumn 1762 and winter 1765. And the basic theme of the correspondence is new: it is now devoted to a lengthy series of articles by Lagrange and by Euler, dealing directly with the difficulties surrounding the problem of a vibrating string. In the first two issues of the Turin *Notes* Lagrange published a memoir entitled "Investigations into the nature of the propagation of sound" and an "Appendix ..." to it (L4,6,9), and in his letter of July 28, 1759 informing Euler that he is sending him the first issue of the Turin *Notes*, he begs Euler to give his opinion of the work as a whole and in particular of his solution of the problem of the vibrating string contained in the first part.

Without dwelling on Lagrange's new method, we note only that, as he writes in his letter of August 4, 1759, he has concluded that the treatment of the problem by Daniel Bernoulli has crucial deficiencies while Euler's construction, disputed by d'Alembert, is closer to actuality. On October 23, 1759, having read Lagrange's memoir and so satisfied that Lagrange shares his opinion as to the type of functions admissible as solutions of the equations of mathematical physics, Euler expresses himself delighted with Lagrange's solution of the extremely difficult equations associated with the problem of the propagation of sound, adding that Lagrange's work has inspired him to pursue further his investigations into this circle of problems. (Euler had had occasion previously to work on the problem of sound propagation, publishing the results in an article in 1750 (E151).) In the same letter

---

[12]There is extant only one letter from Laplace to Euler, dated May 30, 1772, in which Laplace begs his help in publishing his first articles in the Petersburg *Novi Commentarii* in view of the lengthy delay in publication in Paris. Euler's reply remains unknown; it is possible that he did not reply, since Laplace's letter contains lavish praise of d'Alembert, of whom he was formerly a protégé.

Euler tells Lagrange of certain new results on the mechanics of rigid bodies, to which he will later devote his treatise of 1765 (E289).

Several of the letters following this one are devoted to discussions at a high mathematical level of various aspects of the problem, relating in particular to the propagation of spherical and cylindrical sound waves, and in November and December 1759 Euler produced in quick succession three memoirs (E305–E307) on that topic, all three of which were published in the Berlin *Notes* for that year, appearing, however, with considerable delay only in 1766. In the meantime, at the invitation of Lagrange he published a memoir on the propagation of vibrations in an elastic medium in the second issue of the Turin *Notes* (see Lagrange's reply of March 1, 1760), then, in the third issue, for 1766, articles on vibrations of strings (E317, E318) and the propagation of sound (E319)—where once again the question of the significance of non-smooth solutions is considered—, and two further papers, one on optical technology (E319)[13] and one on the integral calculus (E320). In all of the correspondence of this period these topics are all discussed animatedly, but in a respectful and once again friendly tone. However another circumstance arose to cause a further chill in their relations[14]: unbeknownst to one another they both entered essays in the competition of the Paris Academy of 1764 on the topic of lunar libration, and the prize was awarded to Lagrange (partly through d'Alembert's influence) for a work which would appear only in 1777 (L51). Euler was doubtless hurt by this decision but in any case the subject was not mentioned in his correspondence with Lagrange.

By then the Berlin period of Euler's life was coming to an end. D'Alembert, thoroughly *au courant* with the friction between Euler and Friedrich II, while suggesting to Euler that he might act as mediator of their discord, was behind the scenes simultaneously preparing to invite Lagrange to Berlin. On May 19, 1766 he wrote to the Prussian king as follows: "I dare to assure your Highness that Mr. de La Grange will be a good replacement for Mr. Euler in respect of his talents and work, in addition to which his character and behavior are such as will never give rise to any disagreements or the least trouble in the Academy". The subtext here obviously concerns Euler's quarrel with the king, which is discussed in the essays by K. Grau and K.-R. Biermann of the present collection. The king was fully persuaded by d'Alembert's recommendation. On the other hand Euler failed to react to d'Alembert's proposal; he had already reached an agreement with the Russian government to return to St. Petersburg under very advantageous conditions. He informed Lagrange of his imminent departure for St. Petersburg in a letter of May 3, 1766. He arrived in St. Petersburg with his household on June 28, 1766 and Lagrange came to Berlin on October 27, 1766 to take up the post Euler had vacated.

We shall only cursorily summarize the third stage in the correspondence between Euler and Lagrange. Both correspondents were burdened with a multitude of responsibilities, in addition to which, soon after returning to St. Petersburg, Euler almost completely lost his eyesight, thus requiring the aid of secretaries to read aloud to him and to write from his dictation or according to his instructions. Over the last 17 years of his life—or more precisely

---

[13] There would seem to be an error in quoting the Eneström indices here.—*Trans.*

[14] This is indicated by a letter from Lagrange to d'Alembert dated November 17, 1764 in which among other matters Lagrange declares that Berlin does not suit him "because Mr. Euler is there". However a little later, in connection with Euler's suggestion that they make the move to St. Petersburg together, Lagrange wrote to d'Alembert as follows: "You well understand that I thanked him for this".

the first ten of these since with his letter of April 3, 1775 he breaks off the correspondence as a result of increasing difficulty in comprehending orally many of the letters from Lagrange containing complicated calculations—the correspondence comprises only 15 letters, or one more if one includes the letter to Lagrange of March 16, 1772 written by the academician A. J. Lexell on Euler's instructions[15]. The correspondence deals with a variety of subjects relating to their chief preoccupations and most important recent publications. A significant portion of it is devoted to questions of number theory—a favorite subject of both correspondents—, mainly the solution of second-degree Diophantine equations, an area of number theory in which Lagrange extended and in a few cases completed work of Euler, for example on Pell's equation. (It is appropriate to mention here that Lagrange made important additions to the French translation (1773) of Euler's two-volume treatise on algebra (E387, E388), which had been published in German in 1770—though a Russian translation had appeared slightly earlier, in 1768–1769.) This sort of question is discussed in four letters of which the first is dated January 27, 1770 and the last December 30, 1770; however we shall refrain from describing their contents, which are of a highly specialized character, referring the interested reader instead to [15, pp. 105–106, 114–117]; see also Euler's last letter, so-called. We mention also a letter written by Euler in January 1775 and another written by Lagrange on February 10 of the same year, concerning paradoxical properties of certain improper integrals, which at that time were attracting the attention of mathematicians and which in the first quarter of the 19th century led to A. Cauchy's classical theory of the integral of a continuous function—a theory that embraces certain kinds of improper integrals.

Even such a short survey as this bears witness, we suggest, to the wealth of ideas contained in the three sets of Leonhard Euler's correspondence examined above and their interest for the history of the physico-mathematical sciences in the 18th century, for the biographies of the participants as well as the other scientists figuring in the correspondence, and finally for the history of the three Academies of St. Petersburg, Paris, and Berlin.

References

[1] Euler, L. *Opera omnia. Ser. I: Opera mathematica; Ser. II: Opera mechanica et astronomica; Ser. III: Opera physica. Miscellanea; Ser. IVA: Commercium epistolicum; Ser. IVB: Manuscripta.* Var. l., 1911 ff.

[2] *Correspondance de Leonhard Euler avec A.-C. Clairaut, J. d'Alembert et J.-L. Lagrange.* Eds. A. P. Juškevič, R. Taton. Basel: Birkhäuser. 1980; *Opera* IVA-5.

[3] *Die Berliner und die Petersburger Akademie der Wissenschaften im Briefwechsel Leonhard Eulers.* Berlin, 1959–1976. Vols. 1–3.

[4] Euler, L. *Correspondence. With annotated index.* L.: Nauka, 1967.

[5] Lagrange, J.-L. *Oeuvres.* Paris, 1867–1892. Vols. 1–14.

[6] Kopelevich, Yu. Kh. "Materials for the biography of Leonhard Euler". Ist.-mat. issled. (Research in hist. of math.) 1957. Issue 10. pp. 9–65.

---

[15]The letter deals with the so-called Lagrange inversion formula, i.e., with the power series used for the local inversion of homeomorphic functions which Lagrange obtained in 1770 (L18).

[7] *Manuscript materials of Leonhard Euler in the Archive of the Academy of Sciences of the USSR.* M.; L.: Izd-vo AN SSSR (Publ. Acad. Sci. USSR), 1962. Vol. 1: *Scientific description.*

[8] *Franco-Russian scientific connections.* L.: Nauka, 1968.

[9] "A chronological description of the works of A. Clairaut". Ist.-mat. issled. (Research in hist. of math.) 1976. Issue 21. pp. 240–260.

[10] Euler, L. *Institutiones calculi differentialis. Opera* I-10.

[11] ——. *Institutionum calculi integralis. Opera* I-11, 12, 13.

[12] Clairaut, A.-C. *Théorie de la figure de la terre, tirée de l'hydrostatique.* Chez David Fils, Paris, 1743.

[13] Euler, L. *A method for finding curves enjoying maximum and minimum properties.... Opera* I-24.

[14] ——. *Theoria motuum Lunae, nova methodo pertractata* .... Petropoli, 1772; *Opera* II-22.

[15] *A history of mathematics from ancient times to the beginning of the 19th century.* M.: Nauka, 1972. Vol. 3.

[16] Euler, L. *Introductio in analysin infinitorum. Opera* I-8.

[17] Truesdell, C. *The rational mechanics of flexible or elastic bodies. 1638–1788.* In: *L. Euleri opera omnia* II-11.

[18] Yushkevich, A. P. "History of the argument over a vibrating string". Ist.-mat. issled. (Research in hist. math.) 1975. Issue 20. pp. 221–231.

[19] ——. *History of mathematics in Russia to 1917.* M.: Nauka, 1968. p. 166.

[20] Demidov, S. S. "On the concept of a solution of a partial differential equation in the argument over a vibrating string in the 18th century". Ist.-mat. issled. (Research in hist. math.) 1976. Issue 21. pp. 158–182.

[21] Kolmogorov, A. N. *Mathematics.* BSE (Large Soviet Encyclopedia). 2nd ed., 1954. Vol. 26. p. 294.

[22] Taton, R. "Inventaire chronologique de l'oeuvre de Lagrange". Rev. hist. sci. 1974. Vol. 26. pp. 3–36.

[23] Lagrange, J.-L. *Méchanique analitique.* Paris: Veuve Desaint. 1788.

Letters to a German Princess and Euler's Physics

A. T. Grigor'ian and V. S. Kirsanov

Euler wrote his famous *Letters to a German princess* between 1760 and 1762, at the end of his 25-year stay in Berlin. They were addressed to the Brandenburg margravine Sophie-Charlotte, a relative of the Prussian king Friedrich II. Written in French, the letters first appeared in print only in 1768 [1][1], after Euler's return to St. Petersburg from Berlin following upon his quarrel with Friedrich II. Their publication was supported by the empress Catherine II. In January 1766 Catherine had written to Count Vorontsov: "I am certain that the academy will be resurrected from its ashes by such an important acquisition, and congratulate myself in advance in having restored this great man to Russia" [2]. Almost simultaneously a translation of the *Letters* appeared in St. Petersburg, carried out by academician S. Ya. Rumovskiĭ, a former student of Euler. (All quotations are from this translation.)

The publication of this three-volume work under the title *Letters on various physical and philosophical matters, written to a certain German princess* [3][2] was an event of great importance in the history of science and the enlightenment. In essence this was a unique encyclopedia of physical and philosophical knowledge, expounded in a popular style and hence accessible to the widest possible circle of readers. The success of these volumes was extraordinary, and interest in them continued not only during Euler's lifetime but also after his death. In this regard it suffices to note that the *Letters* were translated into many languages, including English, German, Italian, Spanish, Dutch, and Swedish, at that time went through 30 editions, and that by now the number of editions stands at 111 [4].

The *Letters* address a wide variety of problems: Apart from issues of physics—which constitute the main topic of the present article—one can find discussions of philosophical, theological, geographical, and other questions. Euler limits himself essentially to questions concerning optics and electricity and magnetism, although of course it is difficult to separate these topics from mechanics or astronomy since the interconnection of all natural

---

[1]The intensity with which Euler worked on the *Letters* is astounding: he wrote 234 letters (i.e., chapters) in all, so that on average it took him just three days to write a chapter!

[2]In what follows all quotations from the *Letters* are taken from this edition.

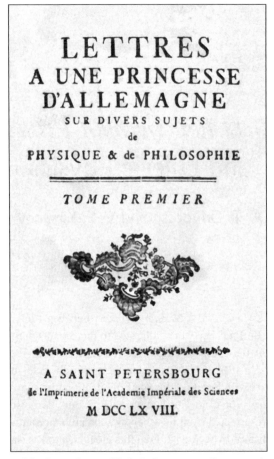

LETTRES
A UNE PRINCESSE
D'ALLEMAGNE
SUR DIVERS SUJETS
de
PHYSIQUE & de PHILOSOPHIE

TOME PREMIER

A SAINT PETERSBOURG
de l'Imprimerie de l'Academie Impériale des Sciences
M DCC LX VIII.

The title page of the first volume of Euler's *Letters to a German princess* of 1768.

phenomena is fundamental to Euler's conception, and the *Letters to a German princess* represents an attempt to present a picture of the world as completely unified as possible.

This view is confirmed by Euler's unfinished essay "A guide to the study of nature" [5], which attempts to reduce the variety of natural phenomena to the interaction of two of the simplest forms of matter. According to David Speiser [6], the "Guide" contains in systematized form the program worked out in detail in the *Letters* and other investigations.

Here we examine the fundamental physical ideas of Euler in the form in which he presented them in the *Letters*, bearing in mind that they represent the outcome of earlier investigations and reflections, which we shall likewise examine in the course of our exposition.

The key concept of Euler's physics was the ether, a rarefied form of matter filling all of the empty space in nature, and responsible for most physical phenomena—optical, electrical, magnetic—and also, it would seem, fundamental to gravitation. It was precisely the latter aspect of the ether that led certain investigators of Euler's physics to the conclusion that "the hypotheses which Euler used to explain physical phenomena were invariably Cartesian and were always based on the chief principle of Cartesian physics" (see, for example, [7]). However, as will be shown below, Euler's ether had little in common with that of Descartes; in fact it was in many ways the same as Newton's, since it was introduced to resolve the very same difficulties as had been encountered by Newton. However subsequently Euler extended the domain of application of this concept so that it also embraced optics.

The ether is first mentioned in Letter 19, where in the course of discussing the nature of optical phenomena Euler arrives at the following conclusion: "Thus one is forced to agree on two things: first that the spaces between the celestial bodies are filled with fluid matter; and second that rays do not issue from the sun and other luminous bodies, as Newton asserted. The rarefied matter filling the sky between [celestial] bodies is called the ether, of which one cannot doubt the unlimited rarefaction and fluidity" [3, Vol. 1, p. 73]. Recall that Descartes' ether was fundamental to his cosmology; in his world-picture the mechanical action of the ether moving throughout the universe caused the motion of celestial bodies along closed orbits. Euler's ether has nothing to do with cosmology, and while for Descartes light is represented as a kind of pressure propagated instantaneously by the medium of a so-called second element—the pressure of the ethereal wind—, in Euler's view "light is nothing more than the oscillation produced in the particles of the ether" [3, Vol. 1, p. 77, Letter 20]. The phenomenon of light is explained by Euler as analogous to sound: in precisely the same way as a sound is a vibration of the air, light consists of a vibration of the ether, an extraordinarily fine and elastic medium. He says: "... If suddenly the air were to become less dense and more elastic, then through these two changes the speed of sound would increase... If the air became as rarified and elastic as the ether, then sound would propagate as fast as light" [3, Vol. 1, p. 78]. Thus Euler was one of the few physicists of that time (including Johann II Bernoulli) who rejected Newton's corpuscular theory of light. Euler was decisively in favor of the wave theory of light. In the *Letters* he gave an explanation of that theory only in qualitative terms, but earlier, in 1744 (1746), he had carried out a detailed mathematical analysis of it in the work "A new theory of light and colors" [8]. S. I. Vavilov notes that in this work "Euler writes down, for the first time in the history of the study of light, the equation of a plane harmonic wave, i.e., creates the apparatus of elementary wave optics, in a form entirely sufficient for solving the simplest problems involving interference" [9]. Euler did not himself provide an explanation of the phenomenon of interference, although in the work "An attempt at a physical explanation of the colors of very thin surfaces" (1752) he gave an (incorrect!) explanation for the colors of thin films and Newtonian rings [10]. Although this explanation does not appear in the *Letters*, the original theory of light on which it is based is expounded there quite fully.

Euler thought that to each color there corresponds a definite frequency of vibration of the ether, and devised a special process intended to explain the origin of these vibrations. He proposes that any red body, for instance, contains particles which are able to vibrate only with the frequency of red light, so that when the body is exposed to white light, which contains the full spectrum of color vibrations, they begin to vibrate with the "red" frequency, and then transmit this vibration to the ether, causing us to see the body as red— and similarly for the other colors. Euler wrote: "The question of the properties of color has always exercised philosophers. Some have stated that they arise from some alteration of the rays, unknown to us. Descartes asserted that all colors are just a mixture of light and shade, and Newton looks to the sun's rays, which he claims emanate in fact from the sun, as causing them, considering that they are perhaps constituted of material of variable density producing rays of the different colors: red, yellow, green, blue, and violet. However since such a system is impossible, all that has been conjectured so far about colors serves only to reveal ignorance of their properties. Thus now, Your Highness, you must surely understand clearly that the character of each color depends on the number of vibrations which the particles presenting that color to us undergo in a given time interval" [3, Vol. 1, p. 109, Letter 27].

Euler returns to this question at the end of the second volume where once again using the analogy with sound, he states that: "The smallest particles constituting a basis for the surface of a body may be thought of as stretched strings since they are elastic and material; and if they are struck in the appropriate manner, vibrations occur at the rate of a definite number per second; the color we associate with the body then depends on this number [3, Vol. 2, p. 256, Letter 135].

Euler introduced the ether into the physical picture of the world because he refused in principle to accept the possibility of action at a distance. This is most clearly expressed in that part of the *Letters* devoted to the problem of gravitation. He says that there are here two points of view; according to one of these gravitation is to be considered as an inherent property of matter, whereas according to the other gravitation results from the action of external forces and depends on the ether. "The latter opinion", writes Euler, "has greater appeal to those preferring clear foundations in philosophy to obscure ones, since they cannot understand how two bodies at a distance from one another can act on each other when there is nothing between them [3, Vol. 1, p. 272, Letter 68]. This statement is almost word for word the same as that in Newton's celebrated letter to Bentley concerning his inability to imagine action at a distance through a vacuum [11]. We add that for Newton also the concept of the ether was merely a hypothesis invoked to explain many phenomena. As is clear from the "Questions" in his *Optics*, and also from many unpublished manuscripts, the concept of the ether allowed explanations of such widely differing phenomena as the transfer of heat in an airless chamber, the damping of a pendulum in a vacuum, various properties of light, the transmission of sensations from the sense organs to the brain, and many others [12]. In one version of his *Optics* Newton uses the concept of the ether to explain the force of gravitation, advancing the hypothesis that the ether is less dense in the interior of material bodies than in the surrounding space, and that this difference causes the bodies to move towards one another—as though from a denser to a less dense medium [13]. Contemporary research shows that for Newton, notwithstanding his many declarations about not making hypotheses, the ether remained an important feature of his picture of the physical world. Note also that Newton and Euler were in agreement that if the ether had density of the order of one millionth that of air, then its presence throughout the universe would have no observable effect on the motion of the planets. Newton also shows, in the second book of his *Principia Mathematica* (Proposition III, Theorem XII), that if celestial bodies were propelled by Cartesian winds, they could not describe closed orbits. The agreement in Newton's and Euler's approach to the problem of the motion of celestial bodies was first noted by Ya. G. Dorfman [14].

Like Newton, Euler resorted to the concept of the ether in order to avoid the introduction of "occult properties" into physics, such as gravitation, considered as an intrinsic property of matter, seemed to be. Thus the use of this concept fitted well with the materialistic tendencies of Euler's approach to nature.

We now turn to the problem of the nature of matter, where again we find a clear convergence of the views of Newton and Euler.

While Descartes identified matter with extension, Euler, following Newton, stressed their difference. He considered the chief property of matter to be its impenetrability. In Letter 69, after expounding Descartes' point of view on the issue, Euler concludes: "In summary, we see clearly the most general property of matter, and therefore of all bodies: this is impenetrability, that is, the impossibility of one body's passing through another, or of two bodies occupying the same place simultaneously" [3, Vol. 1, pp. 277–278, Letter 69]. And a body differs from empty space in that "while empty space has extension, it lacks impenetrability" [op. cit., p. 279, Letter 70].

In Letter 74 Euler elucidates the concept of inertia. He stresses the importance of the equivalence of a state of rest and a state of uniform rectilinear motion. The assigning of the same ontological status to these two states is equally the merit of Descartes and of Newton[3]. A. Koyré, the eminent historian of science, considered this insight one of the most revolutionary achievements of the science of the modern era [15]. With his characteristic perspicacity, Euler immediately recognized this, and emphasized the importance of this step in his *Letters*.

Euler understood inertia (called "grubost'" (coarseness) in Rumovskiï's translation) as the property of a body whereby it maintains a given state: "This property is essential to all bodies and, just like extension and impenetrability, is present in the same way in all bodies" [3, Vol. 1, p. 297, Letter 74].

Thus matter is defined by three fundamental properties: extension, impenetrability, and inertia. According to Euler, the quantity of inertia is determined by the mass of the body in question, and to overcome its inertia and change its state, force is necessary: "Force is the external cause of a change of state" [op. cit., p. 298]. Of course all three of these notions accord completely with those of Newton. Euler also assumed, again like Newton, the existence of absolute space and time, and that the structure of all matter is porous, with empty space between its particles. His viewpoint differs from Newton's in supposing that these empty spaces or interstices are filled with a rarefied material, or "ether", while Newton did not commit himself so categorically on this issue, making no mention of the ether in the *Principia*—although there is some evidence that he had intended to refer to it.

We now turn to electricity and magnetism. Of the problems of physics addressed in the *Letters*, the most space and attention are devoted to questions concerning these phenomena. Thus at least 17 letters are devoted to electricity (Letters 138–154 of the second volume) and 19 to magnetism (Letters 169–187 of the third volume).

After describing various electrical phenomena in the extensive Letter 138—including electrification by friction, sparks, thunder and lightning—Euler continues: "There is no doubt whatever that the source of all electrical phenomena is to be sought in some thin, fluid material; and we have no need to re-invent it. It is the same rarefied material we know as the ether" [3, Vol. 2, pp. 271-272]. The key to resolving the puzzle of electricity is the elasticity of the ether. Euler continues: "Elastic equilibrium is nothing more than a state of rest, when the forces annihilate each other which otherwise might have disturbed it.... From a lack of equilibrium in the air, wind is generated, which then carries the air

---

[3] And Galileo?—*Trans.*

from one place to another; in the same way some sort of wind occurs in the ether when its equilibrium is disturbed; however this wind, which carries the ether from a place where it is more compressed and stressed to another where it is more relaxed, is incomparably more rarified.

"On the basis of these assumptions, I make bold to claim that all electrical phenomena result from some disequilibrium in the ether... An electric force is nothing more than a disturbance of the ether's equilibrium" [*op. cit.*, p. 276, Letter 140]. Euler then goes on to describe in considerable detail the mechanism whereby a disturbance of the ether's equilibrium leads to visible manifestations of electricity, such as sparks associated with a discharge, thunder, lightning, and so on. To this end he postulates a hierarchy of structures of material bodies. In the first place the interstices or "holes" in a body may differ in relation to the ether contained in them: some are such that ether is released and taken in with difficulty, in other cases with ease, while yet others form an intermediate category in this respect. Thus there are two kinds of bodies: those with "closed" holes (where the ether is held in a more stressed state), and those with "open" holes (where the ether is less stressed). Bodies of the first kind are, in modern terminology, insulators, and of the second kind conductors. Next Euler introduces the concepts of positive and negative electricity, corresponding to whether the ether is compressed or rarefied in the interstices of a body.

Electrification by means of friction is explained as resulting from the compression of the interstices of those parts of the two bodies in contact, as a consequence of which the ether flows from one body to the other causing a disturbance in its equilibrium once the friction ceases.

Since, according to Euler, the interstices in air are more "closed", this leads to effects of light and sound when an electrical discharge takes place: "When ether moves from a body where it is more compressed to one where it is less so, then the air, which has closed interstices, interferes with this transference; as a result this ether is caused to vibrate violently, and, as we have seen, light consists in just such vibrations; the more violent the vibration, the brighter the light, to such an extent that the body may even be set alight" [op. cit., p. 150, Letter 150]. In a similar way a sound may appear in connection with an electrical discharge, since the vibrations of the ether may cause the surrounding air to vibrate also.

By means of his notion of the ether's elasticity Euler is able to explain in this part of the *Letters* practically all the phenomena of electrostatics. It should be mentioned that his application of the theory of the ether to electrostatics is entirely original; at least we are unaware of similar views being expressed prior to Euler. Also noteworthy is the remark in Letter 144 concerning "the highly significant situation arising in connection with both positively and negatively charged bodies, revealing a great deal concerning electrical phenomena" [*op. cit.*, p. 324]. Here he is speaking about an "electrical atmosphere" surrounding every charged body. Ya. G. Dorfman viewed this remark as tantamount to the introduction of the notion of the electric field into physics.

While Euler's explanation of electrostatic phenomena may be considered as originating with him, in his explanation of magnetic phenomena the influence of Descartes is clearly evident. Descartes imagined the earth pierced along its axes by fine screw-like canals, along which particles of a special material circulate continuously, forming a wind in the vicinity of the earth. He held the view that iron differed from all other matter in being pierced by screw-like canals similar to earth's, so that the flow of the material continuously circulating in the vicinity of the earth causes every particle of iron to take up an orientation allowing the material to flow through it unimpeded.

Euler changed Descartes' picture only by taking the canals to be like veins furnished with valves instead of being threaded like screws, in order to ensure that the magnetic fluid can flow only in one direction. The theory of magnetism presented in the *Letters* is a somewhat condensed version of his work "A new theory of the magnet" [16], awarded a prize by the Paris Academy of Science in 1777. In particular, the *Letters* contain no explanation of the nature of gravitation, whereas that given in "A new theory of the magnet" is very close to Newton's, as we have noted earlier.

In the *Letters* Euler begins by considering the properties of a magnetic needle and experiments involving a magnet and iron filings. He draws the following conclusion: "the configuration of the iron filings permits no doubt that a fine[4] invisible material caused them to assume this pattern. Moreover it is evident that this material must pass right through the magnet, entering at one pole and exiting through the other, so that by means of this continuous motion a wind arises in the vicinity of the magnet carrying the material from one pole to the other; and it is impossible to doubt that this motion is immeasurably fast" [3, Vol. 3, p. 107, Letter 176]. Thus Euler believes that the essence of magnetism in magnetized bodies consists in a continuous current of thin material passing through them, this material being an even more rarefied form of ether than that used by him to explain optical and electrical phenomena. The assumption of a specifically magnetic ether is necessary in order to explain the occurrence of magnetic phenomena in a vacuum—since the magnetic material must pass freely through the ether—and also the difference between magnetic bodies and non-magnetic ones—since the magnetic material passes freely through the latter in all directions, but through magnetic bodies only in one direction, namely from one magnetic pole towards the other.

Euler states that ordinary ether is less rarefied than the magnetic material; but then how is it that the finer[5] material encounters the greater resistance (since the coarser ether passes freely through the interstices in every direction)? Euler answers this as follows: "The magnetic material moves through the interstices at a much greater speed than the ether", and therefore meets with greater resistance.

In answer to the question as to where the magnetic material enters and where leaves a magnet, Euler first defines the different poles, and then, on the basis of notions from hydrodynamics, explains the attraction of opposite poles and the repulsion of like poles. His explanation of magnetization is interesting: the magnet imposes order on the interstices or "holes" in the iron being magnetized, and also the valves contained in them, until then in a state of complete disorder.

This completes our overview of the picture of physical phenomena given in Euler's *Letters to a German princess*.

What influence did Euler's views have on the development of physics? First we must acknowledge his merit as a critic of Newton's corpuscular theory of light. Most of his contemporaries and the next generation of scientists did not dare break with the corpuscular theory of efflux. Even Lambert, who is known to have preferred Euler's theory, was unable to do this. The decision in favor of the wave theory of light was taken only in the early

---

[4]The original has "non-fine".—*Trans.*

[5]The original has "coarser".—*Trans.*

19th century as a result of the discoveries of Young and Fresnel, both of whom were well acquainted with Euler's works.

As far as Euler's approach to action at a distance and his concept of the ether are concerned, the fruitfulness of these views came to be appreciated only long after his time. One of the first to acknowledge their importance was Faraday, who wrote in his *Experimental researches in electricity*[6]:

"There are at present two, or rather three general hypotheses of the physical nature of magnetic action. First, that of ethers, carrying with it the idea of fluxes or currents, and this Euler has set forth in a simple manner to the unmathematical philosopher in his Letters; in that hypothesis the magnetic fluid or ether is supposed to move in streams through magnets, and also the space and substances around them. Then there is the hypothesis of two magnetic fluids, which being present in all magnetic bodies, and accumulated at the poles of a magnet, exert attractions and repulsions upon portions of both fluids at a distance, and so cause the attractions and repulsions of the distant bodies containing them. Lastly, there is the hypothesis of Ampère, which assumes the existence of electrical currents round the particles of magnets, which currents, acting at a distance upon other particles having like currents, arranges them in the masses to which they belong, and so renders such masses subject to the magnetic action. Each of these ideas is varied more or less by different philosophers, but the three distinct expressions of them which I have just given will suffice for my present purpose. My physico-hypothetical notion does not go so far in assumption as the second and third of these ideas, for it does not profess to say how the magnetic force is originated or sustained in a magnet; it falls in rather with the first view, yet does not assume so much[7]".

How exactly did Euler imagine the ether, the concept at the center of his physics? For him it was not at all simply a variant of ordinary matter. The Eulerian ether was the medium for the propagation of force fields, and in this respect he was following Newton's ideas, which, as L. Rosenfeld phrases it, "were later on eliminated from his system by epigones with less feel for the subject" [18]. These ideas contained within themselves the possibility of explaining natural phenomena "by means other than mechanical laws[8]". One can follow the line of acceptance in the 19th century for example among English mathematical physicists—from Green and MacCullagh to William Thomson (later Lord Kelvin), for whom the elastic ether introduced into optics by Euler remained an essential feature of the physical picture of the universe, and the main focus of creative effort. To this branch of the evolution of physics belongs the work of Faraday and Maxwell, leading to the creation of classical field theory.

We conclude with a few words about the influence of Euler's physical ideas on science in Russia. It cannot be doubted that these ideas greatly influenced the formation of views on natural science in Russia around the turn of the 19th century. This influence can be traced along three branches. First, there was the well known similarity of views about physical phenomena between Euler and Lomonosov, which could hardly fail to be reflected in the subsequent development of Russian science and philosophy. Second, Euler had several

---

[6] Vol. 3, 1855, pp. 528–529.—*Trans.*

[7] It is interesting that Faraday appends to this passage the footnote: *Euler's Letters, translated*, 1802, Vol. I, p. 214, Vol. II, pp. 240, 242, 244".

[8] Draft of a letter from Newton to Leibniz. Quoted from [19].

students at the Petersburg Academy of Sciences (S. K. Kotel'nikov, S. Ya. Rumovskiĭ, M. E. Golovin, N. I. Fuss), who exerted a substantial amount of propagandistic effort in spreading the views of their teacher. Third and last, outside academia there appeared enthusiasts and spreaders of enlightenment (Ya. P. Kozel'skiĭ, A. N. Radishchev, P. I. Gilyarovskiĭ, M. M. Speranskiĭ), for whom the name of Euler signified the highest scientific achievements of that time, and this was reflected in their books, lectures, and activities generally [20].

In particular Euler's theory of light and color greatly influenced Lomonosov's "Discussion of the origin of light, furnishing a new theory of colors" (1756) [21]. We mention also the works of Fuss on dioptrics, which was based on Euler's work, and extended it. Especially noteworthy is that part of Fuss' work concerned with the construction of achromatic optical instruments. Traces of Euler's influence can also be seen in the philosophical works of Kozel'skiĭ and Radishchev [22,23]. Such examples might without difficulty be multiplied.

Finally, Euler's ideas furnished the basis for the reform of the teaching of physics in Russia in the era of Catherine II. The first physics textbook satisfying the new requirements, written by Gilyarovskiĭ, follows Euler's *Letters* very closely [24]. The great Speranskiĭ, future advisor to Alexander I and author of *Introduction to the codex of state laws*, containing ideas on constitutional monarchy, began his career modestly as a physics teacher at the Nevskiĭ Spiritual Seminary, and his lectures were also based wholly on Euler's *Letters*; this came to light only in 1872, when his lecture notes were discovered, whereupon they were published under the auspices of Moscow University [25].

References

[1] Euler, L. *Lettres à une princesse d'Allemagne sur divers sujets de physique et de philosophie.* SPb., 1768–1772, Vols. 1–3; *Opera* III-11, 12.

[2] Pekarskiĭ, P. "Catherine II and Euler". In: *Notes of the Imperial Academy of Sciences*, 1964, Vol. 6, Book 1, pp. 59–92.

[3] Euler, L. *Letters on various physical and philosophical matters, addressed to a certain German princess.* Translation from French to Russian by Stepan Rumovskiĭ. SPb., 1768–1774, Vols. 1–3.

[4] Fellmann, E. A. "Leonhard Euler: Ein Essay über Leben und Werk". In: *Leonhard Euler. 1707–1783*, Basel, 1983, p. 71.

[5] Euler, L. *Anleitung zur Natur-Lehre, worin die Gründe zu Erklärung aller in der Natur sich ereignenden Begebenheiten und Veränderungen festgestzt werden. Opera*, III-1, pp. 16–178.

[6] Speiser, D. "Eulers Schriften zur Elektrizität und zum Magnetismus". In: *Leonhard Euler. 1707–1783.* Basel, 1983. p. 226.

[7] Minchenko, L. S. "Euler's physics". Proceedings of the Institute for the History of the Natural Sciences and Technology, Academy of Sciences USSR, 1957, Vol. 19, pp. 221–270.

[8] Euler, L. "Nova theoria lucis et colorum" (1746). *Opera* III-5, pp. 1–45.

[9] Vavilov, S. I. "The physical optics of Leonhard Euler". In: *Leonhard Euler*, M.: Academy of Sciences USSR, 1935, pp. 32–33.

[10] Euler, L. "Essai d'une explication physique des couleurs engendrées sur des surfaces extrèmement minces" (1752). *Opera* III-5, pp. 264–268.

[11] *Four letters from Sir Isaac Newton to Dr. Bentley.* London, 1756, p. 25.

[12] Newton, I. *Optika.* M.: Gostekhteorizdat, 1954, pp. 264–268.

[13] Newton, I. *Optics.* Dover ed., based on the 4th ed. of 1730, New York, 1952, p. 348.

[14] Dorfman, Ya. G. "The physical views of Leonhard Euler". In: *Leonhard Euler.* M.: Academy of Sciences USSR, 1958, p. 383.

[15] Koyré, A. *Newtonian studies.* London, 1965, pp. 66–69.

[16] Euler, L. *Nova theoria magnetis. Opuscula varii argumenti.* Berolini, 1751, Vol. 3: *Opera* III-10.

[17] Speiser, D. "L'oeuvre d'Euler en optique physique". L'Histoire des sciences: Textes et études, Paris, 1978, pp. 218.

[18] Rosenfeld, L. "The velocity of light and the evolution of electrodynamics". Nuovo cim. suppl. Ser. 10, 1957, Vol. 4, p. 1638.

[19] McGuire, J. E. "Body and void". Arch. Hist. Exact Sci., 1966, Vol. 3, p. 203.

[20] Grigor'ian, A. T., Kirsanov, V. S. "Euler's physics in Russia". In: *Leonhard Euler.* Basel, 1983, pp. 385–394.

[21] Lomonosov, M. V. "Discussion of the origin of light, furnishing a new theory of colors" (1756). *Complete Collected Works.* L.: Academy of Sciences USSR, 1953, Vol. 3, p. 315.

[22] Kozel'skiĭ, Ya. P. *Philosophical propositions.* SPb., 1764.

[23] Radishchev, A. N. "On the mortality and immortality of man". *Selected Philosophical Works.* M.; L.: Academy of Sciences USSR, 1949.

[24] Gilyarovskiĭ, P. I. *A guide to physics written by Peter Gilyarovskiĭ, teacher of mathematics and physics at the teacher's college, physics at the Society for Young Noblewomen, and Russian stylistics and Latin at the noble Pages Corps.* SPb., 1793.

[25] Speranskiĭ, M. M. *Physics, taken from the best authors, organized and supplemented by Mikhail Speranskiĭ, teacher of physics and philosophy at the Nevskiĭ seminary;* St. Petersburg, 1799. M.: Society for Russian History and Antiquity at Moscow University, 1872.

Euler and I. P. Kulibin[1]

N. M. Raskin

A significant portion of Euler's *oeuvre* is devoted to the solution of technological problems. The range of technological questions that interested him was exceptionally wide. There is scarcely any area of the technological practice of his time that he did not submit to theoretical scrutiny in the light of the latest results of mathematics and mechanics. This allowed him to create a new theory of machines, which may be considered a branch of the theory of motion of a rigid body that he developed.

This situation enabled Euler to assess very quickly and accurately the various new technological proposals, contrivances, and inventions that appeared. In this connection he would suggest methods for testing proposed constructions, which were then also of subsequent use, and would continue to work on developing and improving these same methods.

The authority of the great scientist in the field of technological science—then in its formative stage—was enormous. It was not by chance that the Petersburg and Berlin Academies of Sciences were constantly requesting him to act as expert in connection with the appraisal of the various inventions and technological projects submitted for their consideration. In addition, from the very beginning of his scientific career Euler participated in the competitions on technological themes periodically organized by the Paris Academy of Sciences, winning prizes and approbation.

Sometimes Euler's work as consultant on projects where his expertise was sought served as the stimulus for the creation not only of new theories, but also of fundamentally new technological designs. In this connection it suffices to recall that Euler's reactive hydraulic turbine arose from attempts to perfect J. A. Segner's "water wheel".

The present essay is devoted to Euler's collaboration with the remarkable Russian inventor and mechanic Ivan Petrovich Kulibin, who from 1769 was employed at the Petersburg Academy of Sciences as chief technician and foreman of the technical workshops.

It is likely that Euler first encountered Kulibin's technical work in connection with his examination of the first draft of plans for a single-span bridge over the Greater Neva that the technician presented to the Academy in 1771.

The construction of permanent bridges over wide and fast-flowing rivers constituted one of the main problems occupying bridge-builders in the second half of the 18th century.

---

[1] Kulibin was cultivated in Russia as a peasant folk hero—especially in the Soviet era. Nowadays Russians of a certain age have been heard to lament the ignorance of the younger generation concerning him.—*Trans.*

The need for such bridges was especially keenly felt in St. Petersburg: for that large and fast-growing city, situated on the islands of the Neva's delta, the provision of new and more permanent bridges than the pontoon bridges then in existence had become a matter of urgency. However such was 18th-century engineering expertise that the competent design of a structure of the complexity of a permanent bridge over the wide and rapid Greater Neva[2] was not feasible. This was the situation with which Kulibin was faced when, soon after arriving in St. Petersburg, he embarked on the task of planning a single-span bridge over the Neva. Once he had his first version ready, Kulibin constructed a model of his bridge in order to infer from tests on the model conclusions about the strength and load-bearing capacity of its actual projected counterpart. However a scientific experimental methodology for validating this inference was lacking, so that reliable judgements as to the strength and other qualities of the planned bridge could not be obtained. Thus Kulibin was compelled to work on the development of a methodology for testing his model that would justify the required inference to the real bridge.

The idea of working on such a methodology occurred to the inventor as a result of the examination by the Academy of Sciences of the first draft of his design of a bridge over the Greater Neva. Based on this design, in 1771 Kulibin was able, after a few failed attempts, to construct a model on the scale 1:40 of the full size. This model was then tested in the Academy of Sciences. To this end they placed at the center of the arch of the bridge a load 15 times the total weight of the model. Although the model passed the test, nevertheless, as Kulibin himself wrote "the gentlemen academicians examined [the model] and based on their reasonings judged it doubtful". He goes on to give as "the chief ground for their doubt the fact that I could not proceed from this to the weight of the real bridge, but now with the help of the almighty Creator I have through certain experiments found that by means of a small model one can realize the weight of the real bridge..." [1, f. 236, op. 1, No. 38, l. 2, 5].

Undoubtedly Euler was one of the academicians who examined Kulibin's plans. This, then, would likely have been the first time they communicated. For after all Euler, although he was then almost completely blind, continued his intensive scientific activities with the aid of assistants and his students, in particular his work as consultant on technical proposals submitted to the Petersburg Academy of Sciences.

In his note on the preparation of a new model of a single-span bridge appended to a request addressed to the vice-director of the Academy of Sciences A. A. Rzhevskiĭ, dated December 9, 1772[3], Kulibin writes that by means of reasoning in conjunction with a series of experiments, he has deduced the "rule" for inferring the load-bearing capacity of the real bridge from the model [2, pp. 154–155].

Euler, who had earlier been requested to test and state his conclusions regarding models of other projected bridges, had of course also needed to develop methods for inferring from results of tests of a model to the actual construction. He wrote up his conclusions in a paper, which he presented to the Conference of the Academy (i.e., a regular academic meeting) on September 25, 1775. This article was published in Latin in 1776 under the title "A simple rule for determining the strength of a bridge or other similar body from the strength of a model" [4, pp. 271–285]. There he writes:

---

[2]Where the Neva passes through the center of St. Petersburg it is very broad and is for this reason called the Greater Neva. Some way upstream where it narrows, it is called the Lesser Neva.—*Trans.*
[3]All dates are "old style".

"§1. This problem arose recently in connection with the [proposed] construction of a bridge over the river Neva. Many had earlier tried their hand at this, constructing to that end scaled-down models of the actual projected bridge. Most had assumed that the bridge would be sufficiently strong if the model were strong to a prescribed degree. They believed that if the model was able to support a load proportional to that which the bridge would have to bear, then there could be no doubt that the actual bridge, built in proportion to this model, would be sufficiently strong. However, this conclusion is false, as becomes absolutely clear if one considers that of course such a bridge cannot be extended over an arbitrarily long distance—for example one or several miles—without being subject to bending under its own weight, independently of how strong the model might have seemed...For this reason I attempt here to carry out a precise investigation of the question of inferring from the strength of a model to that of an actual bridge of whatever size..." [4, pp. 271–272].

Euler's article then continues with a clear exposition, using mathematics, of the theory of modelling bridges, issuing in a confirmation of the "rule" that Kulibin had found by experimental means back in 1772. A brief and generally accessible exposition of certain of the results emerging from Euler's calculations later appeared as an article in one of the "Monthlies" for 1776, published in Russian by the Academy of Sciences [5]. Articles appearing in the "Monthlies" were aimed at a relatively wide readership, and were therefore written in non-technical language and without mathematical argumentation.

In the latter article it was noted that the model being tested should be "in every part exactly similar to the projected bridge [i.e., to scale]". Next the author proposes determining the weight of the model by actually weighing it. The weight of the actual bridge can then be obtained by "multiplying the weight of the model by the cube of the content [i.e., of the ratio n between the linear dimensions of bridge and model[4]]", yielding "the desired weight of the bridge". In order to determine the load-bearing capacity of the planned bridge, it is proposed that "the model be loaded with as much weight as it can bear; i.e., place on it lengthwise and transversely so much weight that it is on the verge of collapsing; next weigh all these weights and add them up. Then if the the sum of all of the weights is not greater than the weight of the model multiplied by the content less 1 [i.e., multiplied by $(n-1)$], it may truly be asserted that the projected bridge will likewise just barely hold up by itself and will collapse if to its own weight the smallest additional load is added". On the other hand, the article continues, if "the model can really support more, then from this [number] one can easily calculate how much the bridge itself will be able to support in addition to its own weight; and to do this one must multiply the excess of the weight that the model can actually withstand over the weight that it would at most need to bear, by the square of the content"[5].

The rule expounded by Euler for determining the load-bearing capacity of a bridge is easy to understand if one bears in mind that the weights of the model and the actual bridge, and hence the respective stresses arising in the cross-sectional elements of the model and the bridge, are in the same ratio as the cubes of their respective linear dimensions (assuming the density of the materials of model and bridge to be the same), while the cross-sectional areas of corresponding elements are in the same ratio as the squares of the linear dimensions

---

[4]That is, the scale.—*Trans*

[5]Thus the bridge could withstand a load $Q = n^2 \Delta q$, where Δq is the difference between the maximal tolerable load for the model and that corresponding to the maximal expected load of the bridge.—*Ed.*

of model and bridge. Hence the stresses set up in cross-sectional elements of model and bridge are in the same ratio as the respective linear dimensions.

Thus the Euler-Kulibin rule allows one to determine the load-bearing capacity of the planned bridge from the results of tests applied to a model constructed to a definite scale. Using their conclusions it was now possible to quickly and reliably assess the results of experiments with models of bridges and give an objective evaluation of their designs.

Very soon Euler, together with his colleagues and students, was confronted with the necessity of putting into practice the rule for inferring the load-bearing capacity of a bridge from the behavior of a model. This was occasioned by the proposal of a plan for a wooden single-span bridge across the Greater Neva made by Joseph de Ribas[6], a captain in the Shlyakhetnyĭ[7] Kadet[8] Land Corps. By edict of Catherine II the project had to be vetted by the Academy of Sciences. For the testing of de Ribas' model of his bridge—which till then had been demonstrated only to the empress—on February 22, 1776 a committee was appointed, headed by Euler and including J.-A. Euler, S. Ya. Rumovskiĭ, and the adjuncts N. I. Fuss and M. E. Golovin [3, p. 229].

Since the examination of the model had been decreed by Catherine II herself, the committee acted very quickly. Already by March 4, 1776 a report was presented to the Conference of the Academy [3, pp. 230–232] in which it was concluded that "this model cannot serve as a basis for the construction of a bridge over such a wide river as the Neva, since ...it [i.e., the model] could not support a load four times smaller than that which it would have to support according to our estimates". The committee noted further that the modest load-bearing capacity of de Ribas' model relative to its weight resulted from the use of oak rather than fir, which was recommended for bridge construction in particular because of its specific gravity. It may be surmised that it was this circumstance that prompted the author of the Russian version of the "rules" of model-building to emphasize that "...it [i.e., the bridge itself] should be in all respects similar to the model and made from the same kind of wood" [5, p. 138].

The failure of his first project did not stop de Ribas, and towards November 1776 he had produced a model of a newly designed, wooden, single-span bridge. He again succeeded in obtaining an edict from Catherine II for the testing of his model by the Academy of Sciences, and on November 11, 1776 the Academic Conference again appointed a committee led by Euler with the testing of the model as its mandate. The members of this committee included the academicians S. K. Kotel'nikov, J.-A. Euler, W. L. Kraft, A. J. Lexell, and the adjuncts N. I. Fuss and M. E. Golovin [3, pp. 265–266].

The committee participated in tests on de Ribas' second model twice: first on November 15, when, according to the minutes of the Conference, "...the method of carrying out the experiments used by Mr. de Ribas in connection with this trial, seemed to them [i.e., to the participating academicians] imperfect and even dubious..." [3, p. 267], and again on November 26 when certain of the academicians voiced decidedly negative opinions regarding the model. The results of these second trials, conducted this time using methods

---

[6] J. M. de Ribas (1749–1800) later distinguished himself in the Turko-Russian war of 1787–1791, and eventually attained the rank of admiral. The town and harbor of Odessa were built according to plans drawn up by him, in recognition of which one of the main thoroughfares of Odessa was named Deribasovskaya Street.

[7] From the Polish "szlachta", meaning Polish landed gentry.—*Trans.*

[8] A "kadet" was a student in a military school for officer-training.—*Trans.*

suggested by the scientists, again turned out unsatisfactory, and the committee rejected de Ribas' second project [6, pp. 543–546].

Now the committee could proceed with the examination of Kulibin's project. Incidentally, it is known that the Academy of Sciences and Euler himself had been studying plans prepared by Kulibin for a single-span, wooden bridge since 1771; however the consultations with him and the testing of the models built according to his specifications, were presumably carried out as part of the normal work of the Academy. This would seem to be the only possible explanation of the fact that there is no mention of any such tests, or for that matter of the testing of a known third such project of Kulibin's, in the Conference minutes; and information about these matters is likewise lacking in other official documents of the Academy.

Only in 1776, when the construction of a model to the specifications of Kulibin's third set of plans was nearing completion, does one find (two) references to his work in the Conference minutes. The first of these appears in the minutes of February 22, 1776 [3, p. 229] mentioned above. Following an enumeration of the members of the committee formed to examine and test de Ribas' first model, it is stated in these minutes that S. G. Domashnev, the then director of the Academy of Sciences, had delegated to this same committee the task of "examining the draft plans, description, and estimates regarding another bridge designed by the Academy's technician-artist Kulibin, a model of which should be completed in the near future". The second reference is contained in the minutes of the Conference meeting of April 18, 1776 and reads as follows: "Mr. Golovin presented a German translation of a memoir by the technician Kulibin, containing a description of a wooden, single-span bridge that he has proposed for construction over the Neva, a model of which will be completed imminently. The secretary has taken this translation away in order to read it and then communicate his opinion to the other members of the committee appointed to test the aforesaid project" [3, p. 237].

The members of the committee entrusted with vetting Kulibin's third project were acting both as appointees of the director of the Academy of Sciences and, obviously, as scientific experts participating at Euler's behest. The evidence for this is furnished by part of a document recording the opinions of the academicians Lexell, Kraft, and a third scientist (Fuss, judging by the handwriting) who left his report unsigned. This document was assembled in February 1776. Since the document is of such importance for establishing Euler's attitude towards Kulibin's project for a single-span bridge, we reproduce it here in full: "Although I very much doubt that Kulibin's model could support a load 12 times greater than its own weight, nevertheless I do not know if it is appropriate to express such an opinion without verifying it; it would suffice for the model to withstand a load of 485 poods[9] more than it is required to in order that the full-sized bridge might be loaded with almost 50,000 poods without the slightest risk; this seems to me to be the greatest load that it could bear in general. In any case I concur with Mr. Rumovskiĭ's opinion. *Lexell*".

"In order to be in exact accord with the truth it would seem to be appropriate to replace the words 'Having weighed...' with 'As a result of the calculations made by Mr. Kulibin, which are not entirely accurate but do not diverge too far from the required precision, the general weight should be estimated at...' *Kraft*".

---

[9] One pood ≡ 36 lbs.

"Your highly respected father [evidently the document was addressed to J.-A. Euler] considers that concerning this draft plan nothing needs to be added or subtracted; he finds that it answers fully to his own ideas. Was anything decided about Mr. Rumovskiĭ's proposal to charge certain members of the committee with investigating the weight indicated in this presentation? This matter is too crucial to be neglected" (*N. I. Fuss*) [1, f. 1, op. 2-1776, No. 2].

This document bears almost irrefutable witness to the high regard in which Euler held Kulibin's project at the beginning of 1776. There is nothing very surprising in this, since we know that the inventor had as early as 1771 built a model of a bridge to the specifications of his first project, which had been rejected by the academicians—among whom certainly Euler must have figured—, and towards the end of 1772 had prepared a second variant of his plan for a wooden, single-span bridge. In his description of the latter project Kulibin had adduced experimental grounds for the design of the bridge and the dimensions of its separate parts, and had also given an empirical derivation of the rule for inferring from the model's load-bearing capacity to that of the bridge. And this variant of his project, together with the corresponding model, had undoubtedly also been examined and tested with Euler's participation.

Towards the end of 1776 the construction of a model of a bridge to the specifications of Kulibin's third project was complete, and the testing of it was performed forthwith.

On December 27, 1776 Euler's committee—in which were included, perhaps at his initiative, the academicians A. J. Lexell, W. L. Kraft, and the adjunct P. B. Inokhodtsev—subjected Kulibin's model to a static load test. By decision of the committee the model, which was one-tenth actual size and weighed 330 poods, was weighted with the maximum estimated load: 2970 poods of "strip iron" (the combined weight of model and load being then 3300 poods). This load was distributed proportionately over the whole model. Next an additional 570 poods over and above the maximum estimated load was placed on the model. The committee then further arranged for 15 people to place themselves at intervals along the whole length of the model and remain there for a certain specified time. "... Standing under this iron load for 28 days", wrote Kulibin, "the model showed not the slightest signs of damage..." [2, p. 166]. One might infer from the fact that the structure was stressed over a substantial interval of time that Euler and Kulibin had in mind the tendency of wooden matter to creep under stress.

The issue of the *The St. Petersburg news* for February 10, 1777, reported on these trials and on the inventor himself as follows: "This excellent artist, whom nature endowed with a powerful imagination united with fairness of mind and an extremely logical reasoning ability, was both the inventor and executor of the model of a wooden bridge spanning 140 fathoms, i.e., the width of the Neva river... This model ... was vetted by the St. Petersburg Academy of Sciences on December 27, 1776 and to the Academy's unexpected pleasure was found on the strongest evidence to be completely right for the construction of a full-sized bridge... In 1773 the aforesaid Kulibin arrived on his own at the rules for determining from a model whether an actual bridge can support its own weight and how large an additional load it can withstand. These rules turned out to be similar to those that were *derived later*[10] on mechanical grounds by the great Mr. Euler, who is a member of our Academy,

---

[10] Author's italics.

which were printed in the Monthly of 1776 together with instructions concerning them, and entered into the Academic *Commentarii*" [2, p. 164].

This text furnishes convincing evidence that Kulibin had worked out completely independently the method for inferring the load-bearing capacity of a bridge from that of a model. Note that the *St. Petersburg news* was published by the Academy of Sciences, and the news items relating to that establishment were thus completely reliable.

Soon the news of the successful testing of Kulibin's model had spread far and wide through Russia; it also generated considerable interest among foreign scientists. It seems that the first person to draw the attention of the latter to Kulibin was Fuss, then an adjunct of the Academy of Sciences.

Fuss was working under Euler's guidance, and therefore naturally tended to be involved in the expert consultations on technological and other problems that Euler himself was occupied with, with the result that he was more than anyone else privy to his supervisor's opinions and evaluations.

Recently discovered copies of letters fron Fuss to Daniel Bernoulli [8, pp. 113–114] shed light on many hitherto unknown circumstances surrounding the successful testing of Kulibin's third model[11].

On January 5, 1777, ten days after the testing of Kulibin's model, Fuss wrote to D. Bernoulli as follows: "Over the last little while so many projects for building bridges over the Neva have been submitted here, that this undertaking has almost become a joke. However the academic technician Kulibin merits your knowing of him by the astonishing fact that from a simple peasant he has developed into a genuinely remarkable person through a happy predisposition for the art of mechanics bestowed on him by nature, so that without any outside help he has already created masterpieces and caused the public to be delighted with him and his model, on which he works incessantly. This is a model of a single-span bridge 1057 English feet long, to be built over the Neva. Kulibin, who is completely innocent of mathematics, found—I do not know by what means—that the outline of the arch should have the form of a segmented chain [i.e., of line segments], that his model weighs 333 poods, that the elements making up the bridge should successively and uniformly decrease in size [towards the center], and, finally, that the model of his bridge should be able to support a load weighing around 3300 poods for the bridge itself to be able to support its own weight. Mr. Euler arrived at the same conclusions *a priori* via arguments to be published in Volume XX of our *Commentarii*. He [i.e., Euler] worked on this for more than a year, completing it not long ago.

"*About ten days ago*[12] the model was loaded with 3500 poods; it did not bend, and left the committee, which is charged with studying it one of these coming days, in no doubt. He [i.e., Kulibin] is preparing the model to withstand a further 500 poods. The design of the model is not only clever—making clear the practicability of using such a cumbersome structure—but it is also in proportion, so pleasing to the eye, that from a distance one might mistake it for the arch of a stone bridge" [1. f. 40, op. 1, No. 189, 1. 4-4 ob.].

Daniel Bernoulli responded on June 7, 1777 with a letter containing the following remarks: "…What you tell me of your native Russian technician Mr. Kulibin and his pro-

---

[11] The following excerpts from the copies of Fuss' letters to D. Bernoulli are taken with slight changes from E. P. Ozhigova's article [8, pp. 113–114].

[12] Author's italics.

jected wooden bridge over the Greater Neva where its width is 1057 English feet, engenders in me a high opinion of this talented and artistic carpenter, raised among simple peasants and owing his higher knowledge only to his own intuition... It seems to me that the main artistry here consists in choosing wood of extreme precision in all dimensions... All the main elements must as far as possible mutually press against each other with the aid of large iron bolts, wedges, pintles, good dowel pins, and correctly carved excisions. You must have seen the work of Mr. Andrée, published in Zürich in 1776 in epistolary form. There you will find a detailed description of a wooden bridge 364 English feet long built in Schaffhausen; but there they took advantage of a natural support near the middle of the river, so that the longest span is only about 200 feet long, which is much less than 1057. The Neva's 1057-foot width seems to me extreme, and I confess that I myself would never feel I could vouch for the construction of such a bridge, unless between one shore of the Neva and the other two or three piles were erected to divide the bridge into three or four roughly equal parts. I arrived at this opinion only after carefully reading the whole of Mr. Andrée's description; in this connection I am not guided alone by the theory that is usually applied in such kinds of enterprises, because it is impossible to enumerate [beforehand] all circumstances that may have to be taken into account... The chief engineer must more often than not depend on his intuition. Thus I see the advantage in having such a person as Mr. Kulibin, whom I so thoroughly respect, even though I am unable to quell my doubts where so enormous a bridge as this is concerned. Can one be certain that the harsh frosts typical of that country will not disrupt the bridge's structure? After all, the slightest compression applied to every part might prove fatal. Please let me know the model's height at its center relative to the ends, and exactly how the great master distributed those 3500 poods with which he loaded the model. If the model was able to withstand the further 500 poods that he was preparing to place on it, then that would be needlessly weighty evidence of the possibility of success. In the past I have carried out many investigations into the strength and resistance of wood used in a variety of ways, and experiment has always confirmed my conclusions; however I remain in doubt as to the extent to which beams of rectangular cross-section and of given length can withstand powerful lengthwise compression before bending, or what load a perfectly vertical pillar can support before breaking under its weight. I would like your inspired technician to give his opinion about one or two examples; estimates are all I need from him" [9, pp. 671–673].

In his reply to D. Bernoulli, Fuss provides further details concerning Kulibin's model, and answers Bernoulli's questions as follows:

"I had foreseen that the brief mention of Kulibin's model made in my earlier letter would pique your interest. I also anticipated that you would perceive the difficulties arising in connection with such a complicated project.

"I knew there were various bridges of that type in Switzerland, for instance in Schaffhausen and Wettingeil[13], which they call 'gestreifte Brücken'[14]. I also recall seeing a model of a bridge designed by an alpine resident called Schwarzenbach, that had been approved in London for construction in Ireland. My impression then was that the model was not sufficiently strong, and indeed I heard later that the bridge that was—with great difficulty—built to the model's specifications, subsequently collapsed.

[13]The only town in present-day Switzerland with a name approaching "Wettingeil" would seem to be Wettingen.—*Trans.*

[14]Literally "striped bridges".—*Trans.*

"Kulibin's model is like nothing else of its kind that I have seen. It is too complex to be described in a few words, and whatever I might be able to communicate to you of his conception, imperfect as that would be, might very well destroy that good opinion of it that I would like to foster in you. As I have already told you, its shape is that of a low arch[15], reminiscent of a suspended chain turned over, whose height at its center relative to its ends is 8.4 English feet. It is made up of four vertical flat sections six feet high, each comprised of beams fastened together in the form of a lattice and buttressed against each other in such a way that in no situation could they ever bend. To ensure this each crossing is reinforced with iron bolts. All four of these flat sections are connected and interlinked in such a way that the effect is that of a single system. This is achieved through the planar sections..., which are oriented so that the ends of the bridge are two and a half times as wide as the middle, serving to provide resistance to the destructive shocks that might occur from the excessive pressure of the wind on these large flat surfaces. Such is the model that the Academy of Sciences considered worthy of approval, once it had seen that this subtle structure immediately withstood a load of 3600 poods of iron, more than half of which was placed on the middle third of the model.

However, notwithstanding the persuasiveness of this test, we were so cautious that we made no decisions concerning the realization of the project—though in fact this was not part of our mandate. In our presentation on Mr. Kulibin [and his project] we praised him lavishly for his persistence and artistry, even though these alone could not yield means for dealing with the host of eventualities that might arise in the course of realizing the project, and which might perhaps render it infeasible. After all, supposing that the construction had to be finished in 4–5 months, would there be enough time for the purveyance of a large amount of choice wood and the erection of scaffolding—which would have to be of considerable height since the bridge itself must be high enough to allow those vessels to pass under it that, as you know, enter the Lesser Neva between the fort and the Academy of Sciences in order to unload their freight at the market? And even if all these problems were solved, what advantage would follow? To be able to cross the river for two more months each year? Surely in winter people would prefer crossing the river on the ice to climbing up 84–90 feet to the level of the bridge, while in summer a pontoon bridge serves the same purpose, moreover costing, together with two other bridges over the Lesser and Greater Nevas, only eight thousand rubles. Hence a wooden bridge would be of use only when the Neva begins to freeze and ice on Lake Ladoga is drifting downstream, while the cost of maintaining it would more likely than not be just as great and its lifetime not very long.

"A stone bridge would grace the city and last for centuries. However ice constitutes an insuperable obstacle to the construction of such a bridge. One might protect it by means of frozen ice-cutters[16] if the ice were not so powerful—especially when drifting—and if the river flowed more swiftly so that it might with greater strength overcome obstacles.

"I have discussed the load-bearing capacity of the bridge with Mr. Kulibin, without obtaining a satisfactory answer from him. He has never conducted such experiments and does not fully trust his own estimates in that regard, but he promised me that he would carry out such experiments in the near future. I left it to him to choose what method to use

---

[15] The original letter has here "un arc surbaissé", meaning that the height of the arch is less than half the distance between its ends.—*Eds.*

[16] The original letter has "par des bois églacés".—*Ed.*

in performing them, and will be honored to communicate the results to you as soon as I myself learn of them..." [1. f. 40, op. 1, No. 189, l. 5–6 ob.].

Later, when he had heard in greater detail about the results of Kulibin's experiments, on March 18, 1778 D. Bernoulli wrote a letter to Fuss containing the following request: "...Would you be able to request Kulibin to confirm Euler's theory by means of appropriate experiments, in the absence of which his theory remains only hypothetically valid?" [9, p. 677].

The correspondence between D. Bernoulli and Fuss allows a significant reassessment and fresh understanding of the events unfolding around the testing of the model and the affirmation of the project for a single-span bridge over the Neva that Kulibin laid before the Academy of Sciences. In the letters to his teacher and patron cited above, the youthful Fuss—then all of 22 years old—clearly expressed not only his personal feelings of affection for the talented inventor, but also the atmosphere of benevolent attention and support emanating from the academic scientists surrounding Kulibin. It is clear also that although this was only the second time he had participated in such a consultative procedure in the Academy of Sciences, he nonetheless had a good understanding of the fundamentals of Kulibin's project. On the other hand it would appear that Fuss knew nothing—in fact had no inkling—of the extensive experimental work that the inventor had carried out on bridge-building, forming the basis of his projects. Fuss also knew little of the work-plan for the construction of the bridge elaborated by Kulibin.

In his letters Fuss confirms that the rule for inferring the load-bearing capacity of the bridge from that of the model was arrived at independently by Euler and Kulibin. This piece of evidence is especially important in view of the fact that as Euler's assistant, Fuss was continually involved in the work of the great scholar.

Fuss' descriptions of the conditions under which de Ribas' and Kulibin's models were tested are also of interest. As already mentioned, so many projects for a bridge over the Greater Neva were submitted that, as Fuss writes, "this undertaking almost became a joke". However even given these circumstances Fuss singles out Kulibin's project from the rest, emphasizing his liking for both the project and its author.

For a proper evaluation of the events that now followed, Fuss' statements that in the Academy of Sciences no one is even thinking of actually realizing Kulibin's project and that there is a marked preference for a stone bridge, are very important. (This preference was also expressed in government orders concerning projects for stone bridges in St. Petersburg prepared by the greatest architects and bridge-builders of the day Perrone and Flamani-Minozzi.)

The economic considerations mentioned in Fuss' letters to D. Bernoulli are also significant. Based on false notions of the height of the bridge—shared by certain of his contemporaries—Fuss believed that as a consequence of its great height the bridge would be used only during those two months of the year when ice is forming and drifting on the river. However it has since been established that the bridge planned by Kulibin would have had a rise normal for bridges with a suspended roadway ("the lower part for travelling on") [10, p. 216].

Fuss' evaluation of the architectural merit of Kulibin's bridge differed from that of several of his contemporaries. In this respect he considered the bridge to be completely satisfactory, similar in form to a stone bridge.

Daniel Bernoulli's assessment of I. P. Kulibin's labors are also interesting: he at once understood that the inferences and conclusions of the inventor were the fruit of extensive experimental work, and accordingly requested Fuss to question the technician on this score. Once he had the desired information, the Swiss scientist proposed to Kulibin that he verify Euler's theory experimentally. This proposal of Bernoulli's is proof of his high opinion of Kulibin's worth and talent.

Although as a rule Euler did not engage in experimental work, he had occasional meetings with Kulibin concerning the manufacture of new scientific equipment and instruments.

Kulibin, who had been appointed overseer of the academic technical workshops, occupies a prominent place in the history of Russian optical instrumentation in the 18th century thanks to his experience—in Lower Novgorod he had over the period 1764–1766 constructed on his own a reflecting mirror telescope, a microscope, and an electrical machine—and to the presence in the Academy of highly competent assistant technicians (I. I. Belyaev, I. G. Shersnevskiĭ, V. Vasilyev, and Z. Voronin).

Soon after his arrival in the academic workshops Kulibin organized the construction of new polishing moulds to be used in the manufacture of lenses for various optical instruments. He wrote to the advisory committee of the director of the Academy that he was starting on the construction of "several pairs of moulds of various sizes for the grinding and polishing of lenses and metal mirrors, from a tenth of an inch to an inch, from an inch to a foot, and from a foot to several feet, with in addition in some cases series of moulds of gradually increasing size, which might be used to make magnifying glasses and compound microscopes of various proportions, telescopes of different sizes, and other optical lenses of various focal lengths" [1, f. 3, op. 7, No. 36, l. 2–4].

Working in cooperation with I. I. Belyaev, the senior optician of the Academy of Sciences, Kulibin was able to raise the technical expertise of the optical workshop to a high level.

In addition to providing equipment to academic scientists, the workshop was permitted to fill orders for optical instruments from outside the Academy. The multitude of drawings and notes preserved in his personal archive bears witness to Kulibin's extraordinary efforts in the direction of equipping the optical workshop with new lathes and other instrumentation [2, pp. 378–428; drawings 108–140].

Among Kulibin's drawings there are three of microscopes. One of these consists of an outline typical of the tube of a Cuff microscope. Another is a diagram of a five-lens microscope with a doubly concave lens between the collective lens system and the two-lens ocular system. In the opinion of the well known historian of the microscope S. L. Sobol', "such a lens must to some extent magnify the image without the eyepiece needing to be moved further from the objective system, i.e., without unnecessary lengthening of the tube…It is obvious that Kulibin had a different goal: to compensate for the reduction of the image caused by the collective system. If this is indeed so, then this idea is original with him…It is highly likely that Kulibin arrived independently at this idea, which subsequently, beginning in the decade 1820–1830, was widely applied in achromatic microscopes" [12, p. 325].

Thus it is clear that one of Kulibin's spheres of invention had to do with the construction of new lens-systems for microscopes. He also worked hard at developing the process of technological production of optical glass [13, p. 93–95].

Although he paid most attention to the theoretical problems of optics, Euler also investigated practical questions of optical technology. One of the most important of his works in this area resulted from an attempt to build an achromatic microscope. Starting from Newton's research on the law of dispersion, Euler sought to discover the rule for combining two lenses so that the image is free of chromatic aberration—and ultimately he did produce a formula to this end. These investigations in turn provided the stimulus for the work of the English optician G. Dollond, who, ignoring Euler's formula, set himself the goal of seeing if some combination of crownglass and flintglass would produce an achromatic image. Dollond's work was crowned with success, and on that basis Euler worked out several optical systems for achromatic telescopes and microscopes.

It was in connection with the work of constructing an achromatic microscope that Euler once more came into contact with Kulibin.

Amongst the materials preserved in the Leningrad Section of the Archive of the Academy of Sciences of the USSR[17] illumining Kulibin's activities as overseer of the workshops of the Petersburg Academy of Sciences, one's attention is particularly drawn to a set of documents consisting of reports on workshop activity for 1771 and some years thereafter [2, pp. 134, 481–492]. These reports contain in particular a complete chronological record of the progress of work on Euler's achromatic microscope.

The first mention of this project occurs in Kulibin's report of January 8, 1773 on work carried out in December 1772: "…By authority of the forwarded copy of the committee's resolution and Professor Leonhard Euler's instructions, eight pairs of copper templates for producing copper moulds were made in connection with the construction of a new kind of telescope …". In January 1773 the manufacture of the copper moulds themselves was begun for the "grinding of lenses". It was not possible to use the usual moulds already present in the workshop for the production of the microscopes' lenses: special moulds had to be shaped having the appropriate radii of curvature. This work had been completed by March 1773: the eight pairs of moulds necessitated by the optics of Euler's telescope were ready [12, p. 309–320].

At the end of March 1773 "glass-grinding was begun", evidently first the crownglass, and in April and May flintglass. From April to June inclusive a "copper microscope tube with fittings … for the new microscope" was made—i.e., the tube with its parts and accessories. In June the student I. G. Shersnevskiĭ made a concave mirror "for the reflection of the rays", to which in August "mercury was applied" by master Belaev. The preparation of the flintglass lenses was begun in July 1773: "According to Professor Leonhard Euler's instructions concerning the new manner of telescope, the student Shersnevskiĭ is polishing ten lenses made from two kinds of flintglass". In his report of September 3, 1773 Kulibin writes that "the student Shersnevskiĭ has completed polishing the flintglass lenses for one microscope…".

There is now a hiatus in the manufacturing process of Euler's telescope lasting till May 1774; in the reports for May and September of that year, one reads that "on instructions of Professor Leonhard Euler copper fittings are being attached to the new kind of complicated

[17]Now the Petersburg Section of the Russian Academy of Sciences.—*Trans.*

microscope". It would appear that by "copper fittings" Kulibin means here the microscope's support, including table and other appurtenances. From October 1774 to January 1775 inclusive, there is no mention of work on Euler's microscope; only in the report for February 1775 is it mentioned that the "new kind of microscope" is in the polishing room. By March 1775 the work was complete, and in April Kulibin informed the Academic committee that the complicated microscope had been polished and varnished.

Kulibin's reports show that the basic work of preparation of Euler's microscope was carried out by Shersnevskiĭ, who had been the inventor's assistant back in Lower Novgorod. Of course he merely followed Kulibin's directions while Kulibin supervised the whole process. According to S. L. Sobol′ [12, p. 328] Kulibin designed the microscopes's tube, with its many new and original features.

Thus Euler's achromatic microscope—the most innovative of scientific instruments of the age—was constructed with the participation and under the supervision of Kulibin.

It is clear that the Greater Neva and its delta is an unfailing attraction to travelers, artists, writers, and poets. What is less obvious is that scientists and technician-inventors are also fascinated by the river. In fact almost from the moment of the founding of the Petersburg Academy of Sciences, its scientists began to study that uniquely capricious and mighty river flowing through the very center of Russia's young capital. The academic scientists analyzed its waters and systematically recorded their level and rate of flow over many years in an attempt to grasp the river's essential nature and the root causes behind its watery regime, which included terrible floods threatening the very existence of the city. The Academy was inundated by projects submitted to it by putative technician-inventors for permanent crossings over the Greater Neva.

Euler, who had in Berlin continually been involved in the examination of technological projects—including those involving hydrotechnology, such as the the project to level the Finow canal between the rivers Havel and Oder—, continued with this kind of work when he returned to St. Petersburg, in particular in connection with investigations into the behavior of the Greater Neva. In these investigations he was continually assisted by his colleagues and students, especially Lexell, Fuss, and Golovin [3, pp. 477, 480].[18]

On August 21, 1780 the director of the Academy of Sciences S. G. Domashnev ordered the appointment of a committee to look into the level and rate of flow of the Neva. The membership of the committee comprised W. L. Kraft, N. I. Fuss, and M. E. Golovin, with J.-A. Euler as secretary. It was noted in the minutes[19] that the committee was to work in consultation with, and under the guidance of, L. Euler [3, p. 484]. It would appear on the face of it that the appointment of this committee was ultimately at the behest of some government department, or perhaps even of the imperial court.

A week later the committee had a report ready for Domashnev [3, pp. 486–487], in which it was noted that they had met on August 26, 1780, and that "it had been decided unanimously to communicate the following reflections and proposals:

---

[18]Observations of the Neva's pattern of behavior were carried out also by other academicians (P. S. Pallas, W. L. Kraft).

[19]Of the Academic Conference?—*Trans.*

"1. The level and rate of flow of the Neva's current should be put under observation and ascertained separately ... [with the warning that] an understanding of the [rhythm of] the river's levels would require extensive and continuing labor, difficult and intricate...". It was also noted that winter was the best time of the year for carrying out such work, since the observations may then be made from the ice covering the river, and that good astronomical instruments would be essential. They particularly stressed that great accuracy would be necessary in positioning the observation posts and correcting for the effects of refraction.

Section 2 of the committee's report contains the observations that "The flow rate of the river changes from day to day. It depends on the direction and strength of the wind...", and continues by drawing attention to the great significance that the topography of the shores and the internal strength of the current would have for the determination of the rate of flow. Other factors affecting the accuracy of the observations are also mentioned, such as the accuracy of the measurements of the distance from the shore of the point observed and of the depth of the river.

Thus the report submitted by Euler's committee gave detailed instructions concerning the activities of those persons assigned the tasks of making the observations and investigating on that basis the changing level and rate of flow of the Neva. It might be supposed that the members of the committee themselves wished to decline carrying out the observations.

Who among the intimates of the government of Catherine II or its institutions needed these observations so urgently? And for what purpose? One may conjecture two reasons for such interest: 1) preparations for the competition for the greatest improvements in the building of river vessels, announced by the government in 1781; or 2) news of Kulibin's work on the construction of his "mechanical boat" going on at that time.

For the construction and testing of his "mechanical boats" Kulibin had rented a plot of land on the bank of the Neva situated "along the Schlüsselburg road, beyond the river Slavyanka". Naturally Kulibin did not advertise this project, and information concerning it may easily not have come down to us. A half-century later, in 1832, his son S. I. Kulibin wrote: "First he carried out small-scale experiments, then performed a trial with a small rowboat, but in order to really verify the successful action of it, he bought a 'tikhvin' boat [i.e., a barge] that would bear a load of up to 4000 poods, and on this boat he carried out the real trial. All this cost him more than 7000 rubles. Having tested his machine without witnesses and become convinced of its undoubted usefulness, he informed Prince Potemkin, who conveyed this news to Madame Empress, as a result of which there were issued commands from on high to the collegium of the Admiralty that they observe it in its genuine version" [1, f. 296, op. 1, d. 163, 1. 1–2].

The outcome of the inventor's intensive labors of many years was a "mechanical boat", in which the motive power supplied by barge-haulers was replaced by that of a machine drawing its power from the river current.

It is noteworthy that in realizing his mechanical boat Kulibin applied the experience acquired during the construction of a pontoon bridge across the Neva [2, pp. 179–180, 224–225]. Among other activities, he had then made measurements of the "strength of the river's surge" using simple equipment of his own contriving.

On November 8, 1782 Kulibin's mechanical boat was tested on the Neva by a special government committee whose members included, in addition to members of the collegium of the Admiralty, prominent imperial dignitaries. The committee concluded that the trials were successful, and on November 10, 1782 an award to Kulibin of 5000 rubles in paper money was announced, as a prize for his construction of the mechanical vessel.

It is important to point out that the Academy of Sciences took no part in these trials. However Euler, who clearly always paid close attention to any proposal associated with the name of Kulibin or relating to ship-building—a topic of perennial interest to him—, reacted quickly to news of the results of the trials of Kulibin's mechanical boat.

At the meeting of the Conference of the Academy of Sciences of April 28, 1783 a decision was taken to include in the physico-mathematical section of the first part of the issue of the scientific journal of the Academy for 1780[20] three selected essays of Euler's, one of which dealt with the exploitation of the force of river currents to power ships moving against the current [3, p. 668]. This work of Euler's [14] was undoubtedly inspired by Kulibin's "water-powered boat".

In that paper Euler gave an analysis of the functioning of the mechanism installed on the vessel, activated by the river's current. The results of his analysis were negative. Reviewing the content of this article in 1805, the academician S. E. Guryev wrote (see [15]): "§17... And then the vessel will move against the current with a speed of 0.0460, which is approximately 1/22 of the speed of the current... §18. As far as the practical use of such machines is concerned, we truly doubt that such a contraption could be put to use. For since its construction requires no little expense, it is best to employ people instead, [in any case] indispensable to the vessel, and most of all when such motion can be achieved using only a moderate number of people. However, in spite of that, the problem itself is of sufficient interest for its solution to be worth deducing from the elements of mechanics" [15, pp. 104–105]. Kulibin, who probably never learned of Euler's negative conclusions, continued to work on improving the construction of his water-powered boat and the mechanism for utilizing the current almost to the end of his life. Although in the process he invented much that was new and original, of course he was unable to eliminate the basic flaws of his first water-powered engine. Although in many ways he had managed to perfect his water-wheels, they were still not sufficiently efficient or reliable for regular use on water-powered vessels. And in addition to their having intrinsically extremely small efficiency, they depended entirely on the strength of the river current, which, as Euler had observed, was inconstant. For these reasons "water-powered vessels" were not useful in terms of increased load-bearing capacity or speed—they were in fact not even competitive on this score with vessels owing their motive power to barge-haulers. During the last part of his life the inventor had to endure the tragedy of having his hopes dashed that his "water-powered vessels" would ever be used as river transport.

However it should be remembered that this was the era when water-paddles were being replaced by steam as a source of motive power, and Kulibin also worked hard on problems associated with the use of steam engines in river vessels. That he was a pioneer of steam-powered transport in Russia cannot be doubted.

In his creative explorations I. P. Kulibin, like many another inventor of the time, was constantly confronted by the necessity of using machines. However very often the old-fashioned machines—water-wheels and windmills—were unsuited for the particular work

---

[20]This journal, as with the Academy's other publications, appeared with long delays.—*Author*

required of them, and the thoughts of inventors frustrated by their vain attempts to solve the problem to hand, might wander off in the direction of the *perpetuum mobile*. Kulibin also worked on the problem of perpetual motion. In a several notes written during the last years of his life, the inventor mentions that from the moment he entered into the service of the Petersburg Academy of Sciences he had begun working on the design and construction of various "self-moving machines" [1, f. 296, op. 1, d. 5, l. 2; d. 9, l. 1–2; d. 35, l. 2–3; d. 36, l. 3–4; d. 78, l. 2–4]. There he also refers to a consultation with Euler on that question in 1776, apparently obtaining from him the response that "he does not at all dismiss the idea that such a machine [i.e., a perpetual-motion machine] might be made to work", and the plea "May some fortunate person make such a machine and reveal it [to the world] in our time" [1, f. 296, op. 1, No. 5, l. 2].

It seems that Kulibin's first biographer P. P. Svin'in was thinking of this plea when he wrote that "It is curious that Kulibin was encouraged in this pursuit by the great mathematician L. Euler, who in answer to the question as to his opinion of perpetual motion, stated that he favors its existence in nature and thinks that it will be invented in some happy fashion, like other discoveries that had been thought impossible before being made" [16, p. 37]. Following Svin'in this was repeated in many biographies of Kulibin. However it is known that during the last quarter of the 18th century, in the Petersburg Academy of Sciences— where Euler and his school would have had the last word in the matter—they not only ceased discussing perpetual-motion projects but no longer accepted such submissions for examination.

In addition to the above-described encounters between Euler and Kulibin, the latter often communicated directly with the later students and colleagues of the great scientist. For we know that the last years of Euler's creativity were passed in the close-knit community of young scientists clustered about him, comprising his school in the Academy of Sciences. The intercourse between Kulibin and both the founder of this school and its members has had a profound influence on the evolution of scientific and technological thought in our country.

References

[1] LO Arkhiva AN SSSR (Leningrad Section of Archive Acad. Sci. USSR).

[2] *Manuscript materials of I. P. Kulibin in the Archive of the Acad. Sci. USSR: Scientific description with appended texts and drawings.* M.; L.: Izd-vo AN SSSR (Publ. Acad. Sci. USSR), 1953.

[3] *Minutes of the meetings of the Conference of the Academy of Sciences from 1725 to 1803.* SPb., 1900. Vol. 3.

[4] Euler, L. "Regula facilis pro dijudicanda firmitate pontis aliusve corporis similis ex cognita firmitate moduli". Novi comment. Acad. Sci. Petrop. (1775). 1776. Vol. 20. pp. 271–285; *Opera* II-17.

[5] [Euler, L.] "An easy rule for inferring from a model of a wooden bridge or other load-bearing machine whether the same can be done with the full version". In: *Collection of essays selected from the Monthly for various years.* SPb., 1792. Part 8. pp. 138–140.[21]

---

[21] This article appeared first in the *Monthly with recommendations, from 1776* (SPb., b.g., b.p.).—*Eds.*

[6] Raskin, N. M. "Technological questions in Euler's works". In: *Leonhard Euler.* M.: Izv-do AN SSSR (Publ. Acad. Sci. USSR), 1958. pp. 499–556.

[7] Lysenko, V. I. *Nikolaĭ Ivanovich Fuss.* M.: Nauka, 1975.

[8] Ozhigova, E. P. "On the correspondence of Daniel Bernoulli with Nikolaĭ Fuss". Vopr. istorii estestvozn. i tekhn. (Questions in the hist. of nat. sci. and technology). 1981. Issue 1. pp. 108–115.

[9] *Correspondance mathématique et physique de quelques célèbres géomètres du XVIIIe siècle.* SPb., 1843. Vol. 2.

[10] Yakubovskiĭ, B. V. "I. P. Kulibin's bridge projects. 1. A wooden arched bridge over the Neva". Arkhiv istorii nauki i tekhniki (Archive for hist. sci. and technology). 1936. Issue 8. pp. 191–252.

[11] Chenakal, V. L. "Optics in prerevolutionary Russia". Tr. In-ta istorii estestvozn. AN SSSR (Works Inst. for hist. nat. sci. Acad. Sci. USSR). 1947. Vol. 1. pp. 121–167.

[12] Sobol', S. L. *History of the microscope and investigations concerning microscopy in 18th century Russia.* M.; L.: Izd-vo AN SSSR (Publ. Acad. Sci. USSR), 1949.

[13] Raskin, N. M. *I. P. Kulibin.* M.; L.: Izd-vo AN SSSR (Publ. Acad. Sci. USSR), 1962.

[14] Euler, L. "De vi fluminis ad naves sursum trahendas applicanda". Acta Acad. sci. Petrop. (1780:1). 1783. pp. 119–131; *Opera* II-21.

[15] "On the force of a river's current applied to vessels moving upstream on that river. From the works of the famous Euler, communicated by Academician Guryev". Tekhnol. zhurn. (Technology journal). 1805. Vol. 2, Part 2. pp. 89–113.

[16] Svin'in, P. P. *The life of the Russian technician I. P. Kulibin and his inventions.* SPb., 1819.

Euler and the History of a Certain Musical-Mathematical Idea

E. V. Gertsman

In the edifice of the total scientific legacy of Leonhard Euler his music-theoretical works occupy a rather modest place. They include his treatise *Attempt at a new theory of music, clearly expounded on the most reliable principles of harmony*, published in 1739 [7], three articles of moderate length: "A proposal as to the reason for certain dissonances..." (1766) [8]; "On the true character of modern music" (1766) [9]; and "On the authentic principles of harmony..." (1774) [10], and also a series of letters (Nos. II–VIII) from his celebrated *Letters to a German princess* (1768) [11, Vol. 1]. The dates of publication of these works indicate that throughout his scientific life Euler often returned to the study of music. However these music-theoretical works were destined to make a much more muted impact than those in the natural sciences. Indeed, some of them have been completely forgotten, while others are known only to a few specialists, and then only by name or at best by their basic ideas.

There are many reasons for the neglectful attitude of later generations to Euler's music-theoretical works. The main reasons have to do with the active inculcation of temperament into European musicology, and the wide dissemination of the theoretical ideas of Jean-Philippe Rameau.

It seems clear that the penetration of the idea of uniform temperament[1] into European music was a consequence of the peculiarities of that historical stage in the development of modes of thought. Euler understood that because of the equivalence of all tones and semitones in the case of uniform temperament "it is easy (non incommode) to sing any melody higher or lower by a semitone or tone or any other interval" [7, p. 150]. However, he considered that such a temperament is unacceptable "because of a lack of a rational relation (rationem rationalem) between sounds, excepting the difference of an octave" [7, p. 149].

---

[1] "Uniform (even) temperament" signifies the subdivision of each octave into 12 intervals, equal in the sense that the ratio of the frequencies of the notes corresponding to the end-points of each of these intervals should be fixed (the "diatonic scale"). It follows readily from the fact that the ratio of the pitches corresponding to the end-points of the octave is 2, that the ratio for each interval should be $2^{1/12}$. The end-points correspond to the black and white keys on a modern piano. It is the ratio of the pitches of two notes that determines their musical relation.—*Trans.*

Title page of Euler's *Attempt at a new theory of music* of 1739

Judging by Euler's views, he was unable to go along with uniform temperament because it involved uniformization of the intervals which, although it offered certain conveniences, resulted—so he felt—in an impoverishment of the intonational variety of music-making. Hence he created his own system, abandoning that of uniform temperament which by the close of the 18th century had been adopted by an increasing majority of music-theorists. Thus Euler's work went "against the tide" and of course this did not help to make it popular. Moreover his views often differed from those of Rameau, whose theory was at that time beginning its triumphant progress through Europe. To be more precise, the musician Rameau and the mathematician Euler used "different languages", which did not tend to promote mutual understanding between them. (Two letters have survived, one from each to the other, written in 1752 [19, pp. 151–152; 25, p. 481]; for more on these letters see [12, 14].) Gradually the consolidating authority of Rameau stifled Euler's ideas.

In this connection one should also recall the negative attitude of many of the leading musical theorists of the last two centuries to the application of mathematics to musicology. This was after all a period when serious and successful attempts were made to establish fundamental theoretical methods of analysis of musical sound. These methods were radically different from the earlier, only partially applied, very simple mathematical means (fractions and proportion) used to express intervals and rhythmical relations. In this regard the more sophisticated mathematical apparatus applied by Euler also went against the general trend.

All of these circumstances militated in no small way against the acceptance of Euler's music-theoretical work. Only now is it becoming clear that this neglect was unjust. It is of course true that Euler was occasionally under the sway of notions that were dubious from the point of view of musical logic. However at the same time the facts show that often it is precisely in his works that a transition in the progress of European musicology can be seen, and in a few cases one finds the very first expositions of ideas that would later prove fruitful for its development—not to mention the numerous questions that he settles in new and original ways. For confirmation of this it suffices to turn to one of his smaller music-theoretical works, for example to the article "On the true character of modern music" [9]. One of the theses laid down there—an undoubtedly new and progressive idea for its time—is that musical dissonance is not a static phenomenon, but one undergoing an historical development via the evolution of the musical art.

Euler argues for this idea by contrasting "ancient" and "new" music [9, pp. 516–518][2]. One must remember that in musicology this idea was gradually accepted only much later, in the late 19th century and early 20th [3, pp. 1500–1516]. Historians of musical pedagogy have yet to give an appropriately appreciative evaluation of Euler's practical recommendations for the transposition into a solfeggio of the untempered note-sequence[3] described in this essay [9, pp. 530–533]. In the same article he produces evidence for the trend towards the use in contemporary musical practice of intervals expressed via the number 7, providing grounds for the appropriateness of their introduction [9, pp. 527–529].

It is well known that M. Mersenne was the first to come out in favor of "septimal" intervals. One then hears of the idea again from Descartes and Leibniz, and later Ch. Huygens invents a 31-note temperament made up of such intervals[4] [37]. These were scientists who appreciated the trend towards a broadening of the means of musical expression and wished to promote the activization of theory by their investigations. Euler was the first 18th-century scholar to join them in this endeavor [7, p. 118], to be followed later by G. Tartini, J. Serre, J.-J. Rousseau, and others. Thus among 18th-century intellectuals Euler was in the avant-garde with regard to this question.

Among the very promising ideas contained in Euler's article, there is one that may be provisionally called "the theory of acoustical substitution", the essence of which is as follows: When one is exposed to intervals and chords whose acoustical relations are expressed by complicated numerical ratios impeding their appreciation, the hearing "substitutes for one or two of the notes others only slightly different from them and expressible by means of numbers amongst which the relations are simpler" [9, p. 525]. For example, on exposure to the chord $D - d - f - h$, where the relation amongst the notes is expressed by the sequence of numbers 27, 54, 64, 90, the hearing, Euler maintains, might replace the note $f \equiv 64$ by the lower note $f \equiv 63$ "in order to arrive at the numbers 27, 54, 63, 90, all divisible by 9". In this case "the notes will be in the same relation as the numbers 3, 6, 7, 10, which are of course relatively small and will have a pleasant effect on the hearing" [9, p. 526].

Euler made another pitch for this theory in the article "A suggestion as to the reason behind certain dissonances...", claiming that "it is essential to distinguish between the relations perceived by the hearing at a given instant and the relations amongst the notes represented by numbers" [8, p. 511].

---

[2] The page numbers quoted in this and succeeding references are those of the *Opera omnia*.

[3] That is, analogous to the usual "do-ré-mi-fa-so-la-ti-do" of the white keys on a piano, starting from C.—*Trans.*

[4] That is, intervals expressed via the number 7?—*Trans.*

Euler considered that in the case of uniform temperament, a fifth, say, can be represented by various irrational ratios (for instance $\sqrt[12]{2^7} = 2^{7/12}$), which differ only very slightly from the simplest ratio $3/2$ expressing it. And although the instrument be tuned to uniform temperament, the hearing will nonetheless perceive that interval as given by $3/2$. Euler was convinced that "the perceived proportion is [always] simpler than the actual one, because the difference is so small as to be imperceptible. The organ of hearing is used to interpreting as a simple proportion any proportion differing only slightly [from such a simple proportion]" [8, p. 512]. Does "the theory of acoustical substitution" not anticipate the idea of the "zonal nature of hearing" [14]?

In fact both of these ideas are based on the concept of specific intervals of pitches—of course having definite limits—, each of which the hearing ultimately takes to represent a single "shape" for that interval. This comparison is rather suggestive for the clarification of Euler's contribution to musicology.

It is essential to add at this juncture that Euler was among the first to arrive independently at the idea of the indispensability of the logarithm to musical theory [7, p. 117]. (The very first to apply "musical logarithms" was Juan Caramuel y Lobkowicz, in 1670 in his treatise *The new science* (*Mathesis nova*) [1, p. 3].) The future development of musical acoustics was to demonstrate the long-range prospects of this method, when at the end of the 19th century such units of measure of intervals as the "cent" [5] were introduced by A. Ellis in the appendix to his translation of H. Helmholtz' monograph [16, pp. 446–451].

Among other original musical ideas of Euler is his proposed definition in precise mathematical terms of the "degree of agreeableness" ("gradus suavitatis") of intervals and chords, affording in essence a measure of consonance and dissonance. It is well known that music theorists had long been exercised by the problem of consonance and dissonance. What exactly distinguishes them? What should be considered a dissonance and what a consonance? What criteria should be applied? How can one measure the respective degrees to which they are present in a sound?

The evolution of musical thought and the—historically-speaking—continually changing language of music have constantly confronted music theorists with the problem of appropriately evaluating new means of artistic expression, new and unusual sound-combinations often being perceived as dissonant. The whole of the history of musical criticism bears witness to the perennial accusations brought against innovative composers for their "abuse of dissonance", "incorrect use of dissonance", or "neglect of consonance", etc.—which explains the abiding interest of musical science in the "consonance-dissonance" problem. Any conception of a theory of music with pretensions to seriousness had always to include a clear and precise definition of consonance and dissonance—as a matter of prestige. It follows that for Euler in particular his methodology for defining the "degree of agreeableness" was of crucial importance. However, in order to fully appreciate his achievements in relation to this question, it is first necessary to turn to the distant era when the first attempts were made to distinguish consonance and dissonance by mathematical means.

The fact is that in an absolute majority of cases of classification of sounds into consonantal and dissonantal ones in musical science, the criteria used have always been the

---

[5]A "cent" is a logarithmic unit of measure of musical intervals. The equally tempered semitone ($1/12$ of an octave) is taken equal to 100 cents, so that in terms of ratios of pitches of the ends of a semitone interval, a cent is equivalent to $2^{1/1200}$.—*Trans.*

overly subjective ones of individual acoustical perception. Observing the inappropriateness of such an approach, certain theorists sought to solve the problem using mathematics. In Chapter 6 of the first book of his *Harmonics*, Claudio Ptolemy (1st–2nd centuries) expounded the ideas of the Pythagoreans of calculating the degree of consonance of intervals in terms of integer proportions (2:1 for an octave, 3:2 for a fifth, 4:3 for a fourth[6]), corresponding to positions of the bridge on a monochord[7].

According to Ptolemy, the Pythagoreans, "in order to preserve the proportionality ('ὁμοιότητος) of relations, subtract unity from both numbers composing the proportion, and use the resulting numbers to define the coefficients (τῶν 'ανομίων). They call the proportions yielding smaller [coefficients] more consonant (συμφωνοτέραζ)" [4, p. 14]. Porphyry[8], a commentator on Ptolemy, explains the Pythagorean terminology used here as follows: "They called the subtracted unities identical ('ὁμοια), and the quantities remaining after the subtraction different ('ανόμια)" [5, p. 108][9].

Thus the Pythagoreans subtracted unity from both integers of each proportion, and then added the resulting numbers. The smaller this sum, the more consonant they considered the interval. Porphyry writes: "They say that the intervals for which the least coefficients (τα 'ανόμοια 'ελάσονα) are obtained are more consonant than the others" [5, p. 108]. Thus according to this criterion, an octave has the greatest consonance (since the result of this calculation yields 1), a duodecimo (3:1) is next (the calculation yielding 2), a double octave (4:1) and a fifth both have consonance indicated by 3, and a fourth by 5.

However according to Ptolemy himself this method is "ridiculous". He considers that the Pythagoreans' criterion encounters a "difficulty" first of all because they "apply [the concept of] consonance only to epimoric[10] and multiple proportions" [4, p. 13]. Hence the "completely obvious consonance of an undecimo evades the rule they invented for consonance", since the undecimo (8:3) is "neither an 'epimoria' nor a multiple proportion". Ptolemy supports this opinion with the following argument: "The sound of the undecimo is made up of an octave and a fourth[11]. But if the consonance of the octave, consisting of sounds indistinguishable in significance (κατα την δύναμιν), is united with that of any other [interval], then the form of that [i.e., the latter] interval is inevitably preserved...It follows immediately that: if a fifth is consonant, then an octave together with a fifth must [also] be consonant; if a fourth is consonant, then an octave and a fourth [i.e., an undecimo] will also be consonant" [4, p. 13].

Ptolemy also objects to the Pythagorean criterion on the ground that according to them a duodecimo (3:1) is assigned the coefficient 2, while a fifth (3:2) and a double octave (4:1) both have coefficient 3, yet, writes Ptolemy, "it is totally obvious that the latter two are much more naturally consonant (συμφωνοτέρου) than the duodecimo. For a fifth is

---

[6]A "fourth" is approximately the note of the fourth white key (or sixth if the black keys are included) starting from C, and is thus ideally $2^{5/12} \approx 4/3$ times the pitch of that C. Similarly, a "fifth" has pitch $2^{7/12} \approx 3/2$ times the pitch of the C.—*Trans.*

[7]A "monochord" is an instrument of ancient origin for measuring and demonstrating the relations of musical tones, consisting of a single string stretched over a sound box and a movable bridge set on a graduated scale. *Webster's ninth new collegiate dictionary*, 1990.

[8]Porphyry (232/4–305) studied in Rome under Plotinus.—*Trans.*

[9]For this reason I permitted myself the translation of τα 'ανόμοια as "the coefficients".—*Author*

[10]According to a certain Greek-English lexicon, this means "containing an integer and one part more".—*Trans.*

[11]Starting from any C on a piano, the eleventh white key is the fourth of the next octave up, and so ideally has pitch $2^{17/12} \approx 8/3$ of the pitch of that C.—*Trans.*

simpler, and has a sound of purer consonance than the duodecimo. The double octave is in such a relation to the duodecimo, i.e., 4:1 [is in such a relation] to 3:1, as a single octave to its fifth, i.e., as 2:1 is to 3:2...Hence an octave is more consonant than its fifth to the same degree to which a double octave is more consonant than a duodecimo" [4, pp. 14–15].

Ptolemy finds yet another inadequacy in the Pythagoreans' criterion, namely that it does not take into account that one and the same interval may be expressed via ratios of different numbers, and not just in lowest terms: "For the the relation remains the same ('ίδιοζ) not only for the smallest numbers comprising it, but in general for all numbers similarly related to one another, since for such [relations] the similarity is preserved (το παραπλήσιον)" [4, p. 14]. Indeed, for different pairs of numbers representing the same proportion, the Pythagorean coefficients will be different.

Thus Ptolemy's criticism of the Pythagoreans is based on two arguments, one music-theoretical, the other mathematical. The first relates to acoustical perception or, as he himself writes, "derives from manifest experience" ('από τῆζ εναργοῦζ πείραζ). The musical thought of antiquity took as given the melodic identity of the notes in an octave in the case when those notes served the same function in different tetrachords [17, p. 219]. The extension of any interval by a full octave was not considered as affecting the perception of it. Thus in the quoted text, Ptolemy is not concerned with the identity of the acoustical perception of such intervals as a fourth and an undecimo or a fifth and a duodecimo—indeed any musician can without difficulty distinguish between them. He is talking rather about the melodic essence of a fourth and a fifth after they have been extended by an octave, claiming that this is why they remain consonances.

Ptolemy's mathematically based criticism is also clear. In the first place he is advocating that the concept of consonance not be restricted just to epimoric and multiple proportions, and in the second place he objects to the Pythagorean definition of the coefficient of consonance. Generally speaking the Pythagoreans had but a single aim in view: to demonstrate that the octave was the most consonant of all intervals. Ptolemy could scarcely have disagreed with this. He was exercised only by their proposed methodology, which failed to take into account the various equivalent ways of expressing an interval. For example the proportion representing an octave can be expressed not just by 2:1, but in an infinity of other equivalent ways, and each of these will give rise to a different Pythagorean coefficient for the consonance of an octave.

It seems that I. Düring was mistaken in his opinion that the Ptolemaic criticism of the Pythagoreans derives from Didim of Alexandria [6, p. 180]. From the text of Porphyry to which Düring refers in this connection, it is clear that Didim either shared the Pythagoreans' views, or was simply neutral in his exposition of it: "Certain Pythagoreans, such as Archytas and Didim say that in order to establish relations of consonance, they [i.e., the Pythagoreans] compare them with each other..." [5, p. 107]. There is therefore no reason to consider the views expressed in Ptolemy's critical essay as not original with him.

Ptolemy expounds his own method in the following words: "For the definition [of the coefficient of consonance] it seems more expedient to replace the smallest number [of each proportion] by a fixed number[12], for example 6, and, subtracting it (and not a number proportional to it) from all larger [members], compare the remainders as coefficients. In

---

[12]Presumably then all larger numbers in the proportion would be increased by the same factor.—*Trans.*

the case of a double proportion [1:2 = 6:12] we obtain 6, for a three-halves proportion 3, and for an epitrite[13] 2, and then the coefficients [will turn out] larger for more consonant [intervals]" [4, p. 14]. Thus Ptolemy proposes replacing the smallest member of every proportion by a number fixed once and for all, and then taking the differences between that number and the larger numbers of a proportion as the coefficients of consonance. In this text he clearly alludes to the "harmonic" sequence 6:8:9:12 so famous in antiquity, where the octave is represented by the proportion 12:6, its fifth by 9:6, and its fourth by 8:6. According to Ptolemy's method the coefficients of consonance are here 6 (= 12 − 6) for the octave, 3 (= 9 − 6) for the fifth and 2 (= 8 − 6) for the fourth.

Notwithstanding the difference between the Pythagoreans' and Ptolemy's methods, their evaluation of these intervals coincides: the octave turns out to represent the most consonant chord, next the fifth, and then the fourth. Recall that for the Pythagoreans the greater the degree of consonance the smaller the coefficient, while for Ptolemy it is the other way round, so that the order of the indices of consonance is inverted as between the two methods:

	Pythagoreans	Ptolemy
octave	1	6
fifth	3	3
fourth	5	2

This indicates that in effect both methods really only aimed at providing a "mathematical foundation" for the preordained natural reactions of the hearing. It is in addition completely obvious that in purely mathematical terms neither method is valid. This more likely than not explains the fact that historians of antiquity have almost always skirted in silence the Ptolemaic-Pythagorean dispute. There is no mention of it, for example, in C. Stumpf's painstaking investigations of ancient conceptions of consonance [34, pp. 6–8, 55–64]. How are we to explain the fact that such a serious mathematical school as that of the Pythagoreans and such a solid mathematician as Ptolemy committed an obvious *absurdum in adjecto*?

An answer is provided by the circumstance that in antiquity it was not a question of measuring the consonance of absolutely *all* intervals. In ancient musicology the term $\sigma \upsilon \mu \varphi \omega \nu \iota \alpha$ was applied only to certain particular intervals, namely the fifth, fourth, octave, and their derivatives (the undecimo, duodecimo, double octave, etc.) [22, pp. 307–308]. Consequently, for the Pythagoreans it was a matter of indifference that the dissonant second (9:8) should by their method have the same coefficient of consonance as the interval representing the ideal of consonance, namely the octave, since the second was not one of the two-note chords covered by the term $\sigma \upsilon \mu \varphi \omega \nu \iota \alpha$.

Similarly, for Ptolemy it was immaterial that the "harmonic" proportion 6:8:9:12, used by him to motivate his concept of the degree of consonance, was not adapted to providing such mathematical expressions for other intervals, except for "symphonic" ones, i.e., those contained within a single octave. Ancient musicology was concerned with classifying by degrees of consonance only those intervals considered to be "symphonic" and no others. Ptolemy criticized the Pythagoreans precisely because the application of their method to

---

[13] From the Greek for the proportion 4:3, that of a perfect fourth.—*Trans.*

the undecimo resulted in a partial contradiction of accepted music-theoretical tenets. In this case theory dictated the norms of musical practice.

Thus the actual music-theoretical problem before the Pythagoreans and Ptolemy, as they understood it, was first partially solved by the former, and then completely by the latter—although both methods are mathematically dubious. However any one-sided criticism of these ancient means of defining degrees of consonance, i.e., from the mathematical point of view alone (see for instance [24, p. 74]), fails to give an historically accurate evaluation of these first scientific attempts to quantify the aesthetic-artistic tendencies of the musical art of antiquity.

The middle ages and the Renaissance have left no evidence of any such similar attempts. These arise again only 16 centuries later in Euler's main music-theoretical work *Attempt at a new theory of music...*, where he does not limit himself to intervals and the simplest chords evoking a clear "degree of agreeableness", but strives to define this notion precisely for complex sequences of chords and whole musical themes. He even expresses the belief that one can define the "index of agreeableness" for a complete musical opus, i.e., express in mathematical terms an aesthetic-artistic evaluation of an opus. Closer acquaintance with Euler's method reveals that it in fact constitutes a new approach to the definition of consonance. Before expounding this method, I shall describe briefly Euler's attitude to the versions of the Pythagoreans and Ptolemy and to the latter's criticisms of the Pythagorean definition[14].

In Book IV of his *Attempt* (§§16–19) Euler describes this divergence, recalling Ptolemy's defense of his assignment of a high degree of consonance to the undecimo: "I do not find anything doubtful (nihil reprehendum reperio) in this refutation by Ptolemy, since one should take into account not just the form of the relations but also their simplicity or complexity". However Euler's opinion of Ptolemy's own definition is that it is "no more reliable (neque...magis est firmum)" than the Pythagoreans'. He closes his discussion of the position taken by the Alexandrian scholar with the following words: "I do not consider that consonantal agreeableness is at all defined by such an unreliable relation (rationem...precarium)". It is perhaps interesting that in his discussion of the Pythagorean-Ptolemaic differences Euler gives no details of the ancient definitions of consonance, contenting himself merely with expressing his opinion as to the ability of any proportion to express consonance or dissonance.

Euler's own method is based on notions in terms of which he explains the perception of a mutual affinity between sounds: "We perceive an affinity between two given sounds if we understand their relationship in terms of numbers containing their vibrations per unit of time (si intelligamus rationem, quam pulsum eodem tempore editorum numeri inter se habent)" [7, p. 34]. From this he immediately draws a far-reaching conclusion: "All pleasure in music arises from the grasping of the relations that connect these numbers to one another". In this regard Euler was following in the footsteps of Leibniz, who considered that music is "a hidden exercise in arithmetic of the arithmetically innocent soul (exercitium arithmeticae occultum nescientis se numerare animi)" [20, p. 240]. This opinion had been widely circulated in the scientific world. For instance Ch. Gol′dbach, Euler's elder colleague in the Petersburg Academy, also considered that "music is the manifestation of

---

[14]Note that at the beginning of the 18th century there was available not only an excellent edition—for the time—of Ptolemy's "Harmonics", but also a good Latin translation of that work [27].

concealed mathematics" [18, p. 180]. And Euler also was convinced that with increasing difference in the frequencies[15] comprising any chord the consonance decreases, so that the simplest and most consonant chord should be a unison, where the two notes of the chord have the same pitch (1:1). Euler denotes the unison schematically by two identical rows of dots arranged one under the other:

$$\begin{matrix} \bullet & \bullet & \bullet & \bullet & \bullet & \bullet & \bullet \\ \bullet & \bullet & \bullet & \bullet & \bullet & \bullet & \bullet \end{matrix}$$

He concludes that the unison has "for us the first and simplest degree of order (primum et simplicissimum nobis... gradum ordinis)", and consequently the first "degree of agreeableness". On the same grounds the octave (1:2) is of the second "degree":

$$\begin{matrix} \bullet & \bullet & \bullet & \bullet & \bullet & \bullet \\ \bullet & & \bullet & & \bullet & & \bullet \end{matrix}$$

Next follow the duodecimo (1:3) and the double octave (1:4), to both of which he assigns the general third "degree of agreeableness":

$$\begin{matrix} \bullet\bullet\bullet\bullet\bullet\bullet\bullet\bullet\bullet & & \bullet\bullet\bullet\bullet\bullet\bullet\bullet\bullet\bullet\bullet\bullet \\ \bullet\quad\bullet\quad\bullet\quad\bullet\quad\bullet & & \bullet\quad\bullet\quad\bullet\quad\bullet \end{matrix}$$

Euler justifies these assertions as follows: the proportion 1:3 is very simple since "it is expressed by smaller numbers". The proportion 1:4 "is perceived more easily since it consists of a doubling of the double proportion and is therefore not much more difficult to recognize than the double relation (ideo facilius percipi videtur, quod sit rationis duplae dupla, hincque non multo difficilius discernatur quam dupla ipsa)" [7, p. 37]. Judging by the qualification "one may speak for and against this (in utramque partem potest disputari)", he understood that his reasoning was questionable, but nevertheless did not renounce it. Next after the duodecimo and the double octave, decreasing "degrees of agreeableness" are assigned in order to the fifth (2:3), the fourth (3:4), the third (4:5), etc. Euler's argument then proceeds as follows.

Note first that he considers only proportions between whole numbers since those not so expressible will involve irrationalities, and intervals expressed by irrational proportions are indistinguishable acoustically from appropriately close rational approximations of them. He first defines the consonance degree (or "degree of agreeableness") of intervals expressed by ratios of the form $1/2^n$ to be $n + 1$, so that the consonance degree increases—whence consonance decreases—along the sequence $1/2, 1/4, 1/8, 1/16, \ldots$. Now on the basis of his assignment of consonance degree 1 to the unison 1:1, degree 2 to the second 1:2, and degree 3 to the duodecimo 1:3, Euler assigns consonance degree p to the interval expressed by the ratio $1/p$ for every prime number p, and then degree $p + 1$ to $1/2p$, and in general degree $p + n$ to intervals expressed by the ratio $1/2^n p$. Since the intervals expressed by $1/p$ and $1/2p$ are thus assigned degrees p and $p + 1$, yet one is roughly twice as difficult to apprehend as the other interval, the difference of 1 in their degrees is taken as indicating this. More generally then, the appropriate assignment of degree to intervals expressed by a

---

[15]The original has "vibrational motions" here.—*Trans.*

ratio of the form $1/pq$, p, q prime, is $p + q - 1$, and to those expressed by ratios of the form $1/pqr$, p, q, r distinct primes, $p + q + r - 2$, and so on. Ultimately Euler arrives at the following general formula for the consonance degree $C(n)$ of an interval expressed by a ratio $1/n$, where the decomposition of n as a product of distinct prime powers is given by

$$n = p_1^{\alpha_1} p_2^{\alpha_2} \cdots p_m^{\alpha_m}, \qquad p_1, p_2, \ldots, p_m \text{ distinct primes,}$$

namely

$$C(n) = \alpha_1 p_1 + \alpha_2 p_2 + \cdots + \alpha_m p_m - (\alpha_1 + \alpha_2 + \cdots + \alpha_m - 1).$$

For example

$$C(72) = C(2^3 \cdot 3^2) = 2 \cdot 3 + 3 \cdot 2 - (3 + 2 - 1) = 6 + 6 - 4 = 8.$$

This formula is easily seen to satisfy $C(kl) = C(k) + C(l) - 1$ for all positive integers k, l.

The following example indicates how the definition of consonance degree is extended to intervals expressed by proportions of more general form: The consonance degree of the major three-chord represented by 4:5:6 (comprised of a major third 4:5 and a minor third 5:6) is defined to be[16]

$$C(60) = C(2^2 \cdot 3 \cdot 5) = 2 \cdot 2 + 1 \cdot 3 + 1 \cdot 5 - (2 + 1 + 1 - 1) = 9.$$

As already mentioned, Euler extends his notion of the index or exponent (exponens) of the "degree of agreeableness" to various higher conceptual levels of musical sound and to all the musical structures of a composition: "In order that a musical composition give pleasure, it is necessary, first that the indices of the individual chords (singularum con-sonantiarum) be known, second, the indices of the sequences of chords (consonantiarum successionum), third, the indices of the separate passages (singularum periodorum), fourth, the indices of pairs of successive passages or changes in tune (modorum mutationes), and fifth, the index of all phrases taken together, i.e., of the whole work" [7, p. 94].

Thus for Euler any musical composition can be transformed by the appropriate analysis into a system of coefficients expressing the "degree of agreeableness" of its musical elements. It is interesting that for him even harmony (or tunefulness) amounts "to nothing more than the index of a sequence of chords (nil aliud sit nisi exponens seriei consonantiarum)" [7, p. 175].

This shows that Euler's method heralds a new historical stage in the enterprise of achieving "mathematical mastery" of the musical art, on an incomparably higher level than all previous attempts. However ten years after the appearance of Euler's work, the Cambridge professor Robert Smith was still persisting with the Pythagorean method of defining the degree of consonance [33, p. 22].

Most European music theorists gave Euler's treatise a highly critical reception. In criticizing the *Attempt...*, the famous J. Mattheson expressed himself as quite generally opposed to the use of any kind of mathematical apparatus to investigate musical questions

---

[16]In the original there is no explanation as to why precisely this number is associated with this continued proportion, and not, say, $120 = 2^3 \cdot 3 \cdot 5$.—*Trans.*

[21, p. 539]. L. Miller, editor of the *Musical library*, also had hard words [23, p. 328]. J. Scheibe, likewise deploring any incursion of mathematics into musical territory, remarked sarcastically that "the chords of the great Euler ... could never touch and excite the hearts of listeners" [32, p. XVIII]. Alone the famous German composer and theoretician J. Kirnberger was favorable to Euler's theory (see [36, p. LVI]).

Debate on Euler's mathematical conception of music continued in the 19th century, though somewhat abated. For instance, H. Riemann wrote that Euler's fundamental work demonstrated the "inadequacy of mathematics as a foundation for music" [29, p. 1473], described his method as an "extremely fanciful construction" [30, p. 10] and his *Attempt...* as a "cautionary example for all times" [31, p. 60]. This view was shared by C. Stumpf [35, p. 22]. On the other hand, F. Fétis expressed a high estimation of Euler's theoretical investigations [13, p. 90] and H. Helmholtz also regarded them favorably, considering that in Euler's theory "a multitude of particularities emerge with astounding verity" [15, pp. 326, 327].

Again in the 20th century no unanimous estimation of Euler's contribution to musical theory emerged. Some rejected it entirely [38, pp. 1616–1617], while others saw in his investigations a rational basis for a genuinely scientific theory of music [36, pp. LIII–LIV]. Doubtless the latter view was based more on Euler's music-theoretical work as a whole; however it is nonetheless true that his definition of consonance degree was decisive in determining it. The overall rather tempestuous reaction to his theory is perfectly natural, given that the notion of expressing the "degree of agreeableness" of musical chords and phrases goes to the heart of the most complex problems of artistic thought, many of which remain unresolved to this day.

A modern explication of the mathematical side of Euler's method has been given by Richard Busch in [2, pp. 43–44]. As far as its music-theoretical aspects are concerned, it should first be noted that in comparison with the Pythagoreans and Ptolemy, Euler very significantly enlarged the sphere of application of the definition of consonance degree: the ancients restricted their definitions to just "symphonic" intervals, while Euler took his index to apply at any conceptual level of a musical composition—not to mention its application to arbitrary intervals and chord formations. Furthermore his method is distinctive in that it is founded on physico-acoustical parameters of the musical material under investigation, i.e., the frequencies of the notes, rather than merely musical ones, i.e., harmonies. Consequently, for Euler, notes, say, an octave apart, are not to be regarded as harmonically identical, but are rather to be expressed by different mathematical symbols indicating their pitch[17]. (At the time this aspect of Euler's theory prompted Rameau to write the critical article [28].) Although Euler knew very well that "in music, notes differing from one another by one or several octaves are considered similar (pro similibus)" (see the *Opera* III-1), he did not incorporate this similarity of perception into his theory, considering instead all intervals, including the octave, from a purely mathematical standpoint, i.e., as "acoustic distances" of various sizes, to be denoted, therefore, by various numerical proportions. As a result, Euler's method does not comprehend the idea of the "musical identity" of notes, so important from the viewpoint of musical artistry, in which the harmonic range predominates—and, more generally, Euler's theory does not register the perceptual continuum in which the

---

[17] Rather than "pitch" the original has "acoustical-height characteristics".—*Trans.*

basic musical "events" occur. The same applies of course to other "identical" intervals, which, like the octave, are represented by various mathematical relations not reflecting the harmonic essence of the notes comprising them. However the very attempt to introduce mathematics into the domain of musical art on such a grand scale constitutes a memorable event in the history of musicology—even if we agree that it was unsuccessful. Nevertheless, in the future, when at last mathematical methods of analysis that do not contravene music's intrinsic logic have become part and parcel of musical science, the Eulerian attempt will have left its trace.

References

[1] Barbour, M. J. *Tuning and temperament. A historical survey*. New York, 1972.

[2] Busch, H. R. *Leonhard Eulers Beitrag zur Musiktheorie*. Diss. Regensburg. 1970. (Köln. Beiträge zum Musikforsch. Vol. 58.)

[3] Dahlhaus, C. "Konsonanz-Dissonanz". In: *Die Musik in Geschichte und Gegenwart: Allgemeine Enzyklopädie der Musik*. Basel, 1958. Vol. 7. pp. 1500–1516.

[4] Düring, I. (Ed.). *Die Harmonielehre des Klaudios Ptolemaios*. Göteborgs Högsk. Ärsskr. 1930. 36/1.

[5] ——. (Ed.). *Porphyrios Kommentar zur Harmonielehre des Ptolemaios*. Göteborgs Högsk. Ärsskr. 1932. 38/2.

[6] ——. "Ptolemaios und Porphyrios über die Musik". Göteborgs Högsk. Ärsskr. 1934. 40/1.

[7] Euler, L. *Tentamen novae theoriae musicae ex certissimis harmoniae principiis dilucide expositae*. Petropoli, 1739; *Opera* III-1.

[8] ——. "Conjecture sur la raison de quelques dissonances généralment reçues dans la musique". Mém. Acad. Sci. Berlin (1764). 1766. Vol. 20. pp. 165–173; *Opera* III-1.

[9] ——. "Du véritable caractère de la musique moderne". Mém. Acad. Sci. Berlin (1764). 1766. Vol. 20. pp. 174–199; *Opera* III-1.

[10] ——. "De harmoniae veris principiis per speculum musicum repraesentatis". Novi comment. Acad. Sci. Petrop. (1773). 1774. Vol. 18. pp. 330–353; *Opera* III-1.

[11] ——. *Lettres à une princesse d'Allemagne...*; *Opera* III-11, 12.

[12] ——. *Correspondence. Annotated index*. L.: Nauka, 1967.

[13] Fétis, F. *Esquisse de l'histoire de l'harmonie, considérée comme art et comme science systématique*. Paris, 1840.

[14] Garbuzov, N. A. *The zonal nature of the perception of pitch*. M.; L.: Izd-vo AN SSSR (Publ. Acad. sci. USSR), 1948.

[15] Helmholtz, H. *Die Lehre von den Tonempfindungen als physiologische Grundlage für die Theorie der Musik*. Braunschweig, 1863.

[16] ——. *Sensations of tone*. 2nd ed. London, 1885.

[17] Gertsman, E. "The functional theory of harmony of antiquity". In: *Problems of musical science*. M.: Musyka, 1983. Issue 5.

[18] Yushkevich, A. P., Kopelevich, Yu. Kh. *Christian Goldbach*. M.: Nauka, 1983.

[19] Jacobi, E. R. "Nouvelles lettres inédites de Jean-Philippe Rameau". Rech. musiq. franç. 1963. Vol. 3. pp. 151–152.

[20] Leibniz, G. W. *Epistolae ad diversos... auctores.* Ed. Ch. G. W. Kortholt. Lipsiae, 1934. Vol. 1.

[21] Mattheson, J. "Die neue Zahl-Theorie. 1739." *Plus ultra.* Hamburg, 1755. Vol. 3.

[22] Michaelides, S. *The music of ancient Greece: An encyclopaedia.* London, 1978.

[23] Mizler, L. *Musikalische Bibliothek.* Leipzig, 1746. Vol. 3. p. 2.

[24] Münxelhaus, B. *Pythagoras musicus. Zur Rezeption der pythagoreischen Musik-theorie als quadrivialer Wissenschaft im Lateinschen Mittelalter.* Bonn–Bad-Godesberg, 1976.

[25] Pelseneer, J. "Une lettre inédite d'Euler à Rameau". Bull. Cl. sci. Acad. Belg. Sér. 5. 1951. Vol. 37. p. 481.

[26] Πτολεμαίου Κλαυδίου 'Αρμονικων βίβλια Γ. Ex Codd. MSS Undecim, nunc primum Graece editus Johannes Wallis. Oxoniae, 1682.

[27] Πτολεμαίου Κλαυδίου 'Αρμονικων βίβλια Γ. Ex Codd. MSS editi; nova versione Latine, et notis, illustrati. In: Wallis, J. *Opera mathematica.* Oxoniae, 1699. Vol. 3.

[28] Rameau, J.-Ph. "Extrait d'une réponse à M. Euler sur l'identité des octaves". *Mercure de France.* Paris, 1753.

[29] Riemann, H. *Musiklexicon.* Leipzig, 1882.

[30] ——. *Musicalische Syntaxis.* Leipzig, 1877.

[31] ——. *Grundriss der Musikwissenschaft.* Leipzig, 1908.

[32] Scheibe, J. *Über die musikalische Composition.* Leipzig, 1773. Vol. 1.

[33] Smith, R. *Harmonics or the philosophy of musical sounds.* Cambridge, 1849.

[34] Stumpf, C. *Geschichte des Konsonanzbegriffes. 1. Die Definition der Konsonanz im Altertum.* Abhandl. Philos. Philol. Cl. Kgl. Bayer. Akad. Wiss. 1897. Vol. 21. Issue 1.

[35] ——. "Konsonanz und Dissonanz". *Beiträge zur Akustik und Musikwissenschaft.* Berlin, 1898. Vol. 1.

[36] Vogel, M. "Die Musikschriften Leonhard Eulers". *Opera* III-11, 12.

[37] ——. "Die Zahl Sieben in der spekulativen Musiktheorie". Philos. Diss. Bonn, 1954.

[38] Winckel, F. "Euler". In: *Die Musik in Geschichte und Gegenwart: Allgemeine Enzyklopädie der Musik.* Basel: Kassel, 1954. Vol. 3.

Euler's Music-Theoretical Manuscripts and the Formation of his Conception of the Theory of Music

S. S. Tserlyuk-Askadskaya

Leonhard Euler's scientific activity embraced many areas of enquiry. Although it is certainly true that the problems of mathematics and its possible applications in other, primarily scientific, disciplines, were at the center of his interests, his works in music do not at all represent incidental episodes in the scientific career of the great scientist. He was involved in the musical art throughout his life. This is mentioned by Euler's first biographer, his student and son-in-law N. Fuss, who during the last years of the renowned scholar served as his secretary. According to Fuss [1, 16] Euler devoted all his spare time to music-making, and this love of music has subsequently often been linked to his investigations in the theory of music.

No information has come down to us concerning any musical training that Euler may have had. In the absence of any surviving recollections or other documentary evidence, we can only make guesses as to the extent that interest in music and its theoretical conceptualization might have been stimulated in Euler in his parents' home or when he was attending his Latin school. However firmer conjectures can be made regarding his student years at Basel University. It is known that Basel was then an important center of European musical thought. The work of the great 16th-century music-theorist Glarean had been associated over many years with that city [2]. Musical studies formed part of the curriculum at Basel university [3, 273], and it is highly probable that within its walls Leonhard Euler was educated in the basics of musical theory.

One thing can be asserted with confidence: even before he began working on his "Dissertation on sound"—the first significant work of the youthful scientist—Euler's knowledge of musical theory was already quite extensive and varied. The evidence for this comes from the notebooks—unique scientific diaries—of the young Euler, preserved in the Leningrad Section of the Archive of the Academy of Sciences of the USSR[1] [4] and described in de-

---

[1] Now the St. Petersburg Section of the Archive of the Russian Academy of Sciences.

tail by G. K. Mikhaïlov [5]. Notes on musical theory appear in these notebooks earlier than any referring to his "Dissertation on sound".

Some of the material on the theory of music from the early notebooks was published by G. Eneström [6]. Eneström selected for this purpose Euler's rough plan for an ambitious work on the theory of music, as well as certain notes dealing with a numbered base scale (cadent sequences in various keys). The questions basic to Euler's planned large project concerned the composition of passages of chords and phrases for single voice in major, minor, and "middle" keys, as well as the regularities observed by the various musical genres and forms. The contrast between the subject matter of these early notes and that of the later published work on musical theory indicates a marked change in the direction of his researches in that domain.

Apart from the material published by Eneström, the notebooks contain many other interesting notes on music-theoretic topics. Some of these describe experiments with rhythmical combinations, attempts to calculate the relation between the length of the string of a monochord and the twelve notes of the chromatic scale, and there are notes with titles such as "The rules of composition".

As Euler understood them, the rules of composition were those concerned with the preparation and resolution[2] of dissonant intervals, and he strove to discover regularities in such resolutions based on the voice range of the notes of the intervals. This approach reveals a definite connection with the musical practice of polyphony in strict style, although the possibilities of resolution of dissonances found by Euler actually conform to the norms of a later time.

Thus for example the notes relating to the minor seventh fix on the following means for resolving it: "1. A minor seventh prolonged in the descant should be resolved into the fourth. 2. A minor seventh prolonged in the bass should be resolved into a major sixth, or very rarely into an octave. 3. A prolonged descant immediately following an octave or a minor sixth or a fifth or a minor third, goes over into a minor seventh. 4. A prolonged bass immediately after an octave goes over into a seventh" [4, l. 35 ob.]. For major seconds the "rules of composition" are as follows: "1. A second or ninth prolonged in the bass is to be replaced by an octave or decimo. 2. A prolonged descant immediately after a minor third goes over into a major second or octave. 3. A prolonged bass immediately following a third goes over into a second. 4. A major second prolonged in the descant is to be replaced by a minor third" [4, l. 36]. These observations are fragmentary and obviously not intended to have meaning independently. However notes of this type show very clearly that Euler was not opposed to methods of empirical musicology, and that in fact he strove to make theoretical generalizations on the basis of his observations of the regularities of musical composition.

In the notes from the same period, we find Euler's first searchings after objective sources of musical harmony, and the preconditions for these to be perceptible by the hearing —the theme that would subsequently become central to his lasting music-theoretical work. Already in Basel the novice scientist had arrived at the idea of relating the character of two- and three-note chords and sequences to relations among the frequencies of the sounds comprising them. In the relevant notebook this idea is illustrated by appropriate sketches

---

[2]The passing of a voice part from a dissonant to a consonant tone or the progression of a chord from dissonance to consonance. *Webster's ninth new collegiate dictionary.* 1990.

[4, l. 40–40 ob.]. It was precisely from this seed that the branching tree of the Eulerian music-theoretical conception was to grow.

It has often been claimed that Euler's music-theoretical investigations owed not a little to the influence of the music-theoretic views of G. W. Leibniz [8]. Euler himself more than once refers to Leibniz in his works on musicology [9, p. 163; 10, pp. 173, 177, 184] since he was very impressed by the latter's description of music as a "hidden arithmetical exercise of the arithmetically innocent soul" [11, p. 179]. Nevertheless the contents of the aforementioned notebooks allow us to exclude the possibility that Leibniz' idea influenced the Eulerian conception of a theory of music in the beginning stage of its formation. Euler can only have become acquainted with the correspondence between Leibniz and Ch. Gol'dbach—where the bulk of Leibniz' musical musings were concentrated—when he was in St. Petersburg. This correspondence was published only in 1734 [12], though the possibility cannot be ruled out that through his friendship with Gol'dbach, Euler may have been privy to it four or five years earlier [11, p. 118]. In any case the notebooks tell us that Euler's music-theoretical ideas were conceived independently of Leibniz' influence, so that his learning of Leibniz' musical ideas later, when he was a Petersburg academician, can at most have had the effect of lending support to his own.

It is highly likely that Euler's absorption with the theoretical underpinnings of the musical art also led him to his researches in acoustics. His "Dissertation on acoustics" represents the first of a series of such investigations, and its connection with his music-theoretical quest is demonstrated by his inclusion of a summary of the dissertation's conclusions in the first chapter of his treatise *Attempt at a new theory of music*. Those conclusions furnished the essential scientific presuppositions on which he based his music-theoretical conception, without which it would have taken a different form.

Euler's music-theoretical investigations took on a qualitatively different character after he moved to Russia and began to work in the Petersburg Academy of Sciences. The exceptionally favorable conditions provided to the academic scientists by the Russian government allowed for the possibility of working on a variety of problems. Although occupied with his multitudinous investigations in mathematics, physics, mechanics, and astronomy, Euler did not abandon his youthful plans to produce a substantial music-theoretical work.

He himself considered his first attempts at writing an extensive music-theoretical work as less than satisfactory. In the Euler archive there are two incomplete manuscripts [13, 14] whose pages reveal just how far from simple he found the process of crystallizing and expounding his music-theoretical idea. Although Euler was able to write out fluently, with a minimum of preparation and materials to hand, the lengthiest of mathematical investigations, requiring the most complicated calculations and intricate chains of argument, in the case of his theory of music he made two unsuccessful attempts to expound his idea, and was satisfied with the third variant only to the extent of deciding to leave the final judgment to his readers.

Although the first two music-theoretical manuscripts were described in the 17th issue of the *Proceedings of the Archive of the Academy of Sciences of the USSR* [15, p. 103], no researchers on Euler's legacy had hitherto been attracted by the task of examining them in detail. Also relevant here is H. R. Busch's opinion that the *Attempt at a new theory of music* was constituted from the congeries of unrelated fragments, concerning whose contents he, Busch, made several rather arbitrary conjectures [7, pp. 13–16]. This opinion is not supported by any documentary or otherwise substantial evidence; even the early two manuscripts show Euler's point of view on music-theoretical questions to have been thoroughly thought through, and the different sections to be in mutual agreement.

It is highly likely that of these two music-theoretical manuscripts the one entitled *Treatise on music* was written first. There is a great deal of evidence supporting this conjecture. In the first place, the exposition of his music-theoretical idea contained in this manuscript differs more substantially from that which Euler ultimately deemed publishable. The textual structure of the *Treatise on music*, with its numbering of the paragraphs of its introductory section and its four chapters, is quite different from that of the *Attempt at a new theory of music* and indeed from that of the other manuscript, which is much closer to the published version. The four chapters of the *Treatise on music* are as follows:

Chapter 1. On the sounds employed in music.
Chapter 2. On the agreeableness that may inhere in sounds.
Chapter 3. On the euphony of sequences of notes and chords.
Chapter 4. On musical composition.

As Euler's chronologically earlier—most likely first—attempt to expound his music-theoretical conception in its entirety, the *Treatise on music* uses provisional terminology, intended to be superseded. By tracing the path of development of the terminology in that manuscript, one is able to obtain a more precise understanding of certain key concepts in Euler's published works on music. This possibility is all the more valuable in the case of those concepts which Euler himself failed to define unambiguously and precisely, as a result of which they were later incorrectly interpreted by critics of his works on music.

Thus one of the central categories in Euler's music-theoretical investigations, namely the degree of euphony (gradus suavitatis), was identified by critics with the degree of consonance [16–20, 7]. As a consequence of the inappropriate substitution of one concept for another, many of Euler's arguments were deemed erroneous since their conclusions seemed to be in complete disagreement with actual perceptual experience of consonant and dissonant relations. The critics were not discountenanced even in the face of Euler's unambiguous contrasting of the concepts of euphony (suavitas) and consonance (consonantia) in his *Attempt at a new theory of music*.

According to Euler, elucidating the concept of the degree of euphony allows one to determine the boundary separating consonances from dissonances only "to a certain extent" [9, p. 62]. As he understands it, the degree of euphony represents an acoustical-mathematical characteristic of pitch, determining the degree of agreeableness and simplicity of evaluation of the perception. "The difference between consonances and dissonances consists not only in the greater or lesser ease of their apprehension, but also in the system of composition. Chords whose employment is less convenient [i.e., in relation to the stylistically prescribed requirements of preparation and resolution] are called dissonances even if they are easier to apprehend than those usually relegated to the consonances" [9, p. 62].

In the *Treatise on music* the concept of the degree of euphony is not immediately introduced. In the pages of the manuscript that concept appears together with the evident synonyms "degree of agreeableness (gradus gratitudinis)", "degree of ease or difficulty of apprehension of order (gradus facilitatis vel difficultatis, quo ordo percipitur)", "degree of beauty (gradus venustatis)", and "degree of concordance (gradus consonantiae)".

In his later investigations in music theory, these terms no longer appear, since the shades of meaning that they may have expressed have been swallowed up in the term "degree of consonance".

But then the concept "gradus consonantiae" turned out to be unsatisfactory for the following reason. Along with the general meanings "concordance, harmony" that the word "consonantia" has in Latin, another meaning had been added in the specialized music-theoretical literature, namely "consonance", with meaning solely determined by its opposition to "dissonance". It is not unreasonable to suppose that Euler, fearing a terminological tangle, resolved to avoid using the term "gradus consonantiae". The fact that in the critical literature on Euler one term has been replaced by another should weigh heavily on the conscience of the critics of his music-theoretical works.

As may be concluded from just a perusal of the names of the main sections of that manuscript, the *Treatise on music* addresses many important questions of the Eulerian conception of a theory of music. At the same time, certain crucial questions of aesthetics, musical acoustics, and the realization of Euler's views in practice, are scarcely touched upon, and indeed the set of music-theoretical problems considered is far from complete in comparison with his published work. On the other hand in the *Treatise on music* the calculations are given in much more detail, the acoustical-mathematical foundations of his investigations are fully worked out, and considerable attention is paid to the aesthetic basis of his conception. Thus one may say that the underlying kernel of Euler's music-theoretical investigations was formed and began its development at this initial stage.

Euler left the second of his music-theoretical manuscripts untitled. It is possible that it represents merely a rough draft of an intended music-theoretical work—whence the lack of a title. On the first page, after the words "First chapter of a theory of music", the title of that chapter is given: "On music and sounds in general". There are six more chapters, with the following titles:

Chapter 2. On the fundamentals of euphony.
Chapter 3. On concordances.
Chapter 4. On successions of chords.
Chapter 5. On series of chords.
Chapter 6. On various intervals subdividing an octave.
Chapter 7. On intervals [found] in the represented genres.

In the description of this manuscript appended by the Euler Archive [15], it has been given the name *Seven chapters of a theory of music*, and in what follows we shall refer to it thus.

Much of the *Seven chapters of a theory of music* repeats the conclusions of the *Treatise on music*, but while in the earlier manuscript the reader becomes witness to the birth and initial development of Euler's conception, here the exposition gives the impression of a retailing of established facts. All the same, this manuscript contains much that is new. *Seven chapters of a theory of music* is interesting in that it contains an explicit plan laying out the theses that according to Euler's main idea would have to be examined in due course. From this plan it is evident that the manuscript was projected to contain a further five chapters beyond the seven that have survived, to be devoted to the investigation of problems of changes in harmony and systems in the context of simple composition, rules for the creation of melody (polyphonic composition), questions of temporal—i.e., rhythmical—relations abstracted from those of pitch, the peculiarities of the use of various musical times and measures, and, finally, the technique for producing a connected composition, uniting within itself arrangements of pitch and time. With the exception of the topic of changes in harmony and system, which is dealt with in Euler's *Attempt at a new theory of music*, none of these questions was ever to be considered by Euler again. Their resolution would have represented the logical completion of the Eulerian quest for a theory of music, and would have made more explicit the practical implications of his conception, which Euler's contemporaries found so hard to discern in the questions he actually did discuss.

Compared with the *Treatise on music*, the *Seven chapters of a theory of music* shows the greater maturity both in content and the fleshing out of Euler's basic idea. Here almost all of the questions considered in the earlier manuscript are considered, but in a more compact and elegant exposition. Only certain rather special, but admittedly interesting, topics escape mention here—for instance the question of the euphonic equivalence of a three-note chord played in minor and major keys, and that of the identity of different octaves. Also absent from the manuscript are discussions of time relations, since in all likelihood these were to form the subject-matter of a special chapter on that topic, in the end remaining unwritten. However the manuscript shows clearly how Euler's music-theoretical investigations had deepened, become enriched with new ideas, and his terminology become more precise. Here he has a large section dealing with the acoustics of musical sounds, elaborates on his theory concerning the passage from one harmony to another via chords common to both, and considers various ways of tempering instruments. In this manuscript he concentrates especially on the acoustical-mathematical foundations of three kinds of music, namely the diatonic, chromatic, and anharmonic; the last of these is, he maintains, formed by employing an interval of the natural minor seventh, i.e., the seventh overtone of the natural scale. It is interesting that the arguments in the manuscript concerning the prospects for using the natural minor seventh have been crossed out by Euler [14, 1. 33], although he actively investigated questions relating to this chord in later music-theoretical articles [10].

Euler's treatise *Attempt at a new theory of music* is central to his music-theoretical heritage. As a 32-year-old Petersburg academician, it is likely that Euler set great store by this work since he clearly prepared it with extreme care. Here he gave the most complete exposition of his views on the subject of music—its theoretical foundations and problems in practice.

The treatise, whose full title is *Attempt at a new theory of music, expounded on the most reliable principles of harmony*, consists of a preface by the author, an appendix on musical notes, and 14 chapters:

Chapter 1. On sound and the hearing.
Chapter 2. On the beautiful and the foundations of harmony.
Chapter 3. On the origins of music.
Chapter 4. On chords.
Chapter 5. On successions of two chords.
Chapter 6. On the kinds of chords.
Chapter 7. On the names given to various intervals.
Chapter 8. On the kinds of music.
Chapter 9. On the diatonic-chromatic kind.

Chapter 10. On kinds of music more complex than the diatonic-chromatic.

Chapter 11. On chords in diatonic-chromatic music.

Chapter 12. On harmonies and systems in diatonic-chromatic music.

Chapter 13. On rational composition in a given harmony and system.

Chapter 14. On changing the harmonies and systems.

Only on comparing Euler's *Attempt at a new theory of music* with the two earlier manuscripts, does it become clear just how much effort he invested in the final preparation of his music-theoretical researches for publication. While not altering the inviolable ideas basic to his original conception, Euler worked assiduously on perfecting it through new insights and consequent conclusions. His laborious searchings were directed at a single goal: the creation of a musical theory that, being based on objectively existing physical-acoustic and aesthetic regularities, might prove useful to practicing musicians.

As already noted, the problems addressed in the *Attempt at a new theory of music* do not exhaust the full set of questions intended for resolution according to the incomplete manuscript *Seven chapters of a theory of music*. The questions left out of consideration in the published treatise are precisely those that might have influenced the results of Euler's ruminations in the direction of actual contemporary compositional practice—though this in no way indicates that he underestimated the role of musical practice. Frequently, and with alacrity, he notes the coincidence of his theoretical conclusions with the accepted creative norms of contemporary musical composition, and generally appeals to actual practice as evidence of the correctness of his reasoning—preferring, however, to seek a basis for his arguments outside the musical art.

In Euler's treatise one sees the views of the baroque age peculiarly refracted. Comparison with the manuscripts shows that he must have been consciously searching for such points of contact, since the relevant ideas appear only in the final published version. In the pages of the *Attempt at a new theory of music* one encounters echoes of the theory of effects, the study of musical rhetoric. Euler links his conclusions with the use of the thorough-bass and questions of the tempering and tuning of keyed instruments. He also subjected the work of his predecessors to a searching examination, paying particular attention to the Pythagorean ideas, and those of Aristoksen[3] and Ptolemy, and stressing the radical differences between their views and his own.

Among the influences on Euler's music-theoretical research, one should also mention the work of A. Boetius [21], which he refers to in the manuscripts [14, 1. 29]. There is also reason to believe that Euler had studied A. Kircher's *Musurgia universalis* [11, pp. 54–55]. (Euler's friend and long-time correspondent Ch. Gol'dbach was greatly interested in Kircher's work.) It was precisely in this work that the term "suavitas" was first used in relation to music[4], a term that was to take on the significance of a music-theoretical category in Euler's hands. There is a further piece of evidence pointing to the connection with Kircher, namely the perpetuation of an inaccuracy in the manuscript *Treatise on music* [13, 1. 2 ob.] first perpetrated in the *Musurgia universalis* [22, p. 217]: In their reasonings concerning the duration of sounds both scholars employ the terminology of mensural[5]

---

[3] Aristoksen of Tarenta (4th century B.C.) was a philosopher, a student of Aristotle, and a representative of the peripatetic school. His musical synthesis had lasting influence on European musical theory.

[4] This observation is due to H. R. Busch [7, p. 3].

[5] Relating to polyphonic music originating in the 13th century with each note having a definite and exact time value. *Webster's ninth new collegiate dictionary.* 1990.

rhythmics, but mistakenly use the terms "chroma" and "semichroma" for the two smallest units of duration—whereas they are usually called "fusa" and "semifusa". Euler also refers to J.-Ph. Rameau, J. Sauveur, and G. W. Leibniz.

The ideas that are furthest developed in the *Attempt at a new theory of music* are precisely those that he had been in the process of working out in the manuscript versions. The section devoted to musical acoustics has become considerably more subtle with the addition of treatments of the principles of note-formation and production for various musical instruments, and the laws governing the vibrations of strings. In connection with the aesthetic basis of his conception he formulates the very important principle of equivalence of the objective and subjective in the evaluation of music, reducing musical perception to the perception of order in relations between vibrations. The categories of agreeableness and euphony are sustained in this work not only in their narrow music-theoretical meanings, but also in their broader aesthetic ones. In the *Attempt* the investigation into harmony is much further developed, and provides grounds for the possible means for passing from one harmony or phrase to another via a common three-note chord. Distinguishing various levels of arrangement of pitch, Euler notes that their perception moves from the lower to the higher levels. In this connection he takes the ability to evaluate the whole pitch structure to depend on the level of development of the hearing and acoustical experience of the listener.

In the *Attempt* many theoretical positions taken up in the earlier manuscripts are elaborated on and supplemented by new material: there are treatments of complete and incomplete chords, intervals, kinds of music, the peculiarities of musical perception, and perceptive remarks concerning the stylistic determinacy of consonances and dissonances. For measuring the size of intervals Euler here introduces[6] logarithms to the base 2. In the *Attempt* he examines diatonic-chromatic music in great detail for the natural reason that that kind of music was dominant in contemporary musical art. One finds many penetrating observations made in the course of his description of the regularities of the changes in complexity of his index in passing from one formal part of a musical composition to another.

Comparison of Euler's *Attempt at a new theory of music* with the two manuscript versions that preceded it also reveals the falsity of a formerly widely accepted interpretation of the content of a letter from Euler to his teacher Johann I Bernoulli of May 25/June 5, 1731 [24]. It was usual for this letter to be taken by researchers as definitely pertaining to Euler's work on the *Attempt* [25, p. 37], and in the historical literature on Euler an excerpt from the letter is adduced that putatively provides an exhaustive characterization of his project, namely: "... I have striven to represent music as a part of mathematics" [26, p. 48]. Two inferences usually made following an explication of this statement, are in need of correction.

In the first place, on the basis of this letter, the researchers in question draw the conclusion that towards the middle of the year 1731 work on the *Attempt* was to a significant degree complete. However here they have somehow lost sight of the circumstance that the content of the music-theoretical project that Euler described in detail in his letter to Bernoulli does not coincide with that of the *Attempt*. Of course we do not mean here the

---

[6]It had long been accepted that Euler was the first to introduce base-2 logarithms for measuring the size of musical intervals, until an anonymous manuscript dated 1705 was discovered in the New York Public Library by J. Yasser, in which base-2 logarithms had already been used for that purpose [23, p. 21].

general plan that Euler had formulated in outline from the very first, but the character of exposition and degree of elaboration of various of the themes and their anticipated application. The description of the Eulerian music-theoretical conception contained in his letter to Bernoulli corresponds to the content of the manuscript *Treatise on music*, and for this reason in all probability 1731 was the year when he began to work actively on his set of music-theoretic problems. There is a very simple reason behind the erroneous linking of Euler's letter to the *Attempt* in this way: since the earlier manuscripts on music theory had never been carefully examined, it was natural to connect the letter to the published Eulerian research in music following it chronologically.

Furthermore, any interpretation of Euler's conception reducing it merely to an attempt to make music theory a part of mathematics cannot be considered satisfactory. Euler does indeed express himself as to his designs in that vein: "My ultimate aim in this work has been to try to represent music as a part of mathematics and derive in the appropriate order from correct premises everything that makes the union and mixing of notes pleasant" [24, p. 383]. However on reading his actual exposition one sees that he takes his "correct premises" to be aesthetic ("metaphysical") and acoustic regularities, and that the role of mathematics is reduced to that of a language of description of the phenomena under study as well as a way of thinking whereby ultimate conclusions are derived by means of logically rigorous reasoning from a few indubitable postulates. Shorn of context, the above bare quotation from Euler's letter does not adequately convey Euler's actual intentions as eventually realized, giving only a distorted version of them.

After the appearance of his *Attempt at a new theory of music*, for a considerable time Euler ceased working on music-theoretical problems[7]. The treatise stimulated much discussion. In particular, music theorists of an empirical bent did not find among the Eulerian investigations instructions for applying them in practice to the creative process, and were sharply critical of the work. The few favorable responses generally pertained to particular aspects of Euler's conception, far from providing the full picture. One thing is certain: the crux of Euler's music-theoretical conception was left out of account by all critics, a fact noted with sorrow by the author himself [28, p. 273].

The appearance and reception of the *Attempt* coincided with Euler's move to Berlin, where the empirical school of music-theoretical thought was very strong. The camp of music-theorists of empirical orientation was led by the prominent musicologist and brilliant publicist F. W. Marpurg, whose musical opinion no one at the court of Friedrich II dared dispute. Clearly such a situation did not auger well for the reception of Euler's new music-theoretical work.

Only at the end of the 1750s do notes on music theory reappear in Euler's notebooks [29, l. 62 ob.–63]. The content of these notes presaged a new direction in Euler's music-theoretical investigations, related to the examination of questions only partially touched on in his earlier works. Euler shed new light on these questions, although their resolutions were rooted in the groundwork provided in the *Attempt*. It is true that Euler now sought to adapt the style of his exposition to suit the tastes of practicing musicians by reducing the mathematical framework used to express the ideas to a minimum, indicating a strong desire to be understood by his contemporaries.

---

[7]O. Spiess [27, p. 175] mentions Euler's attempts at composition, also roundly criticized. The manuscript notes of these compositions have to date not been found.

Euler's music-theoretical articles of the 1760s and 70s [10, p. 30] are devoted to two central problems. The first of these concerns the role of the natural minor seventh, and the second—closely related to the first—the peculiarities of the perception of pitch. Here he formulates an important thesis concerning the accuracy of pitch perception and the tendency of the hearing to alter the perception of what is heard in the direction of greater simplification of the relation between the notes (this relation being expressed by a set of frequencies). Euler conjectures that human hearing must have evolved to the point where natural minor sevenths are perceived as consonant intervals, since this requires neither training nor resolution. In this connection he considers the relation in general between the concepts of consonance and dissonance.

Euler certainly did not look upon his music-theoretical investigations as purely intellectual constructions. He strove insistently to have his ideas realized in practice, believing that the "true foundations of music" he had revealed must serve to promote the perfection of musical creation and performance. To this end he proposed a system of training of the hearing, based on the natural intervals of the natural scale enjoying the simplest acoustical relations. He also carried out calculations relating to a 24-interval temperament, including that of the natural minor seventh. On instruments so tempered it would be possible to produce an enormous number of new chords, thereby enriching the resources of expression while at the same time satisfying the criterion of genuine musical harmoniousness.

Having arrived at his important thesis of a qualitative jump in the development of human hearing, relating to its interpretation of the natural minor seventh as a consonant interval, Euler saw in this a motive for a fundamental revision of the whole system of pitches. However on this score the scholar was to a significant extent mistaking the desire for the reality, and the basic features of the principles of pitch organization that had become entrenched in musical art by the beginning of the 18th century were to remain viable for a long time thereafter. The time for such radical changes had not arrived, and during the next century and a half the language of music evolved gradually, realizing possibilities inherent in the European pitch organization of the first half of the 18th century. The situation changed abruptly around the turn of the 20th century, when the principles of pitch, which had for so long seemed unassailable, began to be radically challenged. Examination of current musical practice reveals that much in the organization of pitches has evolved in the direction foreshadowed by Euler, so that his music-theoretical legacy now becomes especially topical.

Euler's works in music theory were created in an age of radical shifts in musical thought—in some ways similar to the present. In such an historical period it is especially difficult to discover any regularity subsisting in the changes taking place or to determine the main tendencies of the developmental process. Under such conditions only a very few would dare attempt, like Euler, to realize a music-theoretical conception based on broad aesthetic and theoretical assumptions. Most music-theorists limit themselves to a mere reporting on the situation as they find it or generalizing from observations of contemporary musical composition, assuming in the best case the role of interpreter of the latest musical regularities.

Euler's experience demonstrates that the creation of a music-theoretic conception of significant scope is possible only by going outside the framework of empirical observations and generalizations based on them. His approach lent his investigations a prognosticatory character and led him to conclusions not only valid for the music-theoretical phenomena he was analyzing, but also relevant to the future development of music and its theoretical basis.

References

[1] Fuss, N. "Éloge de Monsieur Léonard Euler". SPb., 1783.

[2] Fritzsche, O. F. *H. Glarean, sein Leben und seine Schriften.* Frauenfeld, 1890.

[3] "Basel". In: *Musical encyclopedia.* M., Vol. 1. 1973. Col. 273.

[4] LO Arkhiva AN SSSR (Leningrad Section Archive Acad. Sci. USSR), f. 136. op. 1, No. 129, Notebook No. 1.

[5] Mikhaïlov, G. K. "The notebooks of Leonhard Euler in the Archive of the Acad. Sci. USSR". Ist.-mat. issled. (Research in history of math.) 1957. Issue 10. pp. 67–94.

[6] Eneström, G. "Bericht an die Eulerkommission der Schweizerischen naturforschenden Gesellschaft über die Eulerschen Manuskripte der Petersburger Akademie". Jahresber. Dtsch. Math. Ver. 1913. Vol. 22. pp. 191–205.

[7] Busch, H. R. *Leonhard Eulers Beitrag zur Musiktheorie.* Regensburg, 1970.

[8] Haase, R. *Leibnitz und die Musik: Ein Beitrag zur Geschichte der harmonikalen Symbolik.* Hommerich, 1963.

[9] Euler, L. *Tentamen novae theoriae musicae ex certissimis harmoniae principiis dilucide expositae.* Petropoli, 1739; *Opera* III-1.

[10] ——. "Conjecture sur la raison de quelques dissonances généralement reçues dans la musique. Du véritable caractère de la musique moderne". Mém. Acad. sci. Berlin (1764). 1766. Vol. 20. pp. 165–199; *Opera* III-1.

[11] Yushkevich, A. P., Kopelevich, Yu. Kh. *Christian Goldbach.* M.: Nauka, 1983.

[12] Kortholt, Ch. *G. W. Leibnitii epistolae ad diversos...auctores.* Lipsiae, 1734. Vol. 1.

[13] Euler, L. *Tractatus de musica.* LO Arkhiva AN SSSR (Leningrad Section Archive Acad. Sci. USSR), f. 136. op. 1, No. 241.

[14] ——. *Tractatus de musica.* LO Arkhiva AN SSSR (Leningrad Section Archive Acad. Sci. USSR), f. 136. op. 1, No. 242.

[15] *Manuscript materials of Leonhard Euler in the Archive of Acad. Sci. USSR.* M.; L.: Izd-vo AN SSSR (Publ. Acad. Sci. USSR), 1962. Vol. 1.

[16] Mizler, L. *Musikalische Bibliothek.* Leipzig, 1746–1752. Vol. 3.

[17] Riemann, H. "Euler Leonhard". In: *Musiklexicon.* 4th ed. Leipzig, 1894. p. 285.

[18] Chevalier, L. *L'Histoire de l'étude de l'harmonie.*

[19] Fétis, F. J. *Esquisse de l'histoire de l'harmonie, considérée comme art at comme science systématique.* Paris, 1840.

[20] Vogel, M. "Die Musikschriften Leonhard Eulers". *Opera* III-11, pp. XLIV–LX.

[21] Paul, O. *Boetius und die Griechische Harmonik. Des Anicius Manlius Severinis Boetius fünf Bücher über die Musik aus der lateinischen in die deutsche Sprache übertragen und mit besonderen Berücksichtigung der griecher Harmonik sachlich erklärt von O. Paul.* Leipzig, 1872.

[22] Kircher, A. *Musurgia universalis.* Roma, 1650. Vol. 1.

[23] Yasser, J. *A theory of evolving tonality.* New York, 1932.

[24] Eneström, G. "Der Briefwechsel zwischen L. Euler und J. I Bernoulli". Bibl. Math. 1903. Vol. 4. pp. 344–388.

[25] Euler, L. *Correspondence. Annotated index.* L.: Nauka, 1967. 391 pages.

[26] Thiele, R. *Leonhard Euler.* Leipzig, 1982.

[27] Spiess, O. *Leonhard Euler.* Frauenfeld; Leipzig, 1929.

[28] *Leonhard Euler und Christian Goldbach. Briefwechsel, 1729–1764.* Hrsg. A. P. Juškevič, E. Winter. Berlin, 1965.

[29] LO Arkhiva AN SSSR (Leningrad Section Archive Acad. Sci. USSR), f. 136, op. 1, No. 136, Notebook No. 8.

[30] Euler, L. "De harmoniae veris principiis per speculum musicum repraesentatis". Novi comment. Acad. sci. Petrop. (1773). 1774. Vol. 18. pp. 330–353; *Opera* III-1.

An Unknown Portrait of Euler
by J. F. A. Darbès

G. B. Andreeva and M. P. Vikturina

1. In the art collection of the State Tret′akovskiĭ Gallery, amongst the paintings of the German artist Josef Friedrich August Darbès there is a portrait of "an unknown old man", dated 1778[1]. In the course of preparing materials for a new catalogue, it was established that this is a portrait from life of Leonhard Euler.

The work of discovering this fact began with the following natural question: Might it be that an actual historical person is concealed behind the strong-willed, intelligent, highly individual face in the representation? In order to answer this question it was necessary to consult the biography of the artist to see whose portraits he had painted towards the end of the 1770s.

The name J. F. A. Darbès is today known only to specialists, and appears only rarely in overviews of the history of German art[2]. From rather meager information gleaned chiefly from various reference works and encyclopedias of the representational arts, the following very brief outline of his life was assembled.

The artist was born in Hamburg in 1747 and died in Berlin in 1810. From 1748 he lived with his family in Copenhagen, studying in the Academy of Arts[3]. On completing his studies, he visited Holland, Germany, and France[4]. In 1785 he settled in Berlin, and in 1796 attained professorial rank at the Berlin Academy of Arts. Among his contemporaries he was well known for his portraits of the Prussian king Friedrich Wilhelm II, the queen, and especially for his portrait of Goethe of 1788. Darbès won fame as a master of portraiture sketched in pencil and miniature representations using silver crayon and sanguine[5][6].

During the period of interest to us, Darbès was living in Russia: from 1773 to 1785

---

[1] Oil on canvas, 61.3 × 47.3cm. An oval inscribed in a rectangle. Inv. 13144.

[2] In Russia and Latvia paintings by Darbès are preserved in art museums in Moscow, Leningrad, and Riga.

[3] Occasionally it is erroneously asserted that he was Danish-born (see for example [1, p. 167]).

[4] According to some sources he also visited Poland [2, p. 273].

[5] The self-portrait of the artist preserved in the Engravings Room in Dresden, was produced by such means (paper and crayon, 11.5 × 8.5; after 1797 inv. 355)

[6] Sanguine is a deep red-brown crayon or pencil containing iron oxide. *Oxford Dictionary of English. 2nd ed., 2005.* For more detailed information about the the artist's life, see [3, pp. 31–32; 4, pp. 391–392; 5, p. 359].

he worked in Kurlandia[7] and St. Petersburg. Over those years Darbès produced portraits on canvas of, among others, August Hupel, doctor of philosophy and theology at Dorpat University, Ignatius Fesler, a professor of oriental languages and philosophy, and Carl Otto and Carl Dietrich von Loewenstern. From literary sources it appears that the most celebrated of the male portraits from Darbès' brush was a likeness of Leonhard Euler. In D. A. Rovinskiĭ's *Detailed dictionary of engraved Russian portraiture* it is stated that in Mitava[8] in 1780 an engraving was made by one Samuel Kütner from a portrait by Darbès of Euler in his declining years, already blind [6, p. 2180]. We infer from this that the portrait itself had been painted in or prior to 1780. From this portrait, considered as representing its original most faithfully, several other engravings were made subsequently[9]. Since the portrait in the Tret'akovskiĭ Gallery reproduced here depicts an old man completely blind in his right eye, and the date 1778 is painted on the canvas, we conjectured that the picture before us was indeed of Leonhard Euler. Comparison of the canvas with the engraving by Kütner then confirmed this conjecture[10].

It is interesting that indirect evidence for Euler's being the subject of the portrait by Darbès is also provided by surviving documents of Johann III Bernoulli, nephew of his famous namesake. In 1778 Johann III visited Euler in St. Petersburg, and watched him posing for his portrait [9, p. 245].

Thus we may take it as established that the portrait by Darbès in the Tret'akovskiĭ Gallery is indeed of Euler. The portrait is of exceptional value in that it is from life and was painted during Euler's second Petersburg period.

The investigative work did not end with the identification of the face in the portrait. Until recently[11] it had been thought that the portrait of Euler by Darbès was in the Museum of Art and History in Geneva[12]. The origin and later fortunes of the Genevan portrait are fairly clearly established by surviving documents [10, p. 59, No. 179][13]. The portrait was painted in St. Petersburg where it belonged to a certain Pierre Joseph Amey, who presented it to François Duval[14], court jeweler to Catherine II. After Duval's return to Geneva, his collection was absorbed into that of Étienne Dumont, who in 1829 donated the portrait to the Arts Society. It has been in the Museum of Art and History since 1908. Thus there is absolutely no doubt that the portrait of Euler in the Genevan museum was indeed painted in St. Petersburg.

How are the two portraits related? Here it is important to note the following: The canvas

---

[7] Or Courland, now part of Latvia.–*Trans.*

[8] In Latvia.

[9] By Bartolozzi, Riedel, and other masters [7].

[10] The engraving in question is preserved in the Pushkin State Museum of the Representational Arts. The engraving is incised, 24.6×17.8, oval inscribed in rectangle. Inv. R. 8608. There is also an engraving from F. F. Vigel''s collection, now housed in the section of the Scientific Library of Moscow University devoted to rare books and manuscripts [8, p. 87, No. 151]. (Note that in [8] the artist's surname is transcribed incorrectly.)

[11] That is, a little before 1986.

[12] Oil on canvas, 62×45.5, oval. Inv. 1829–8.

[13] The author would like to thank René Loche, the caretaker of old manuscripts at the Museum of Art and History in Geneva, for providing information concerning this portrait.

[14] François Duval (1776–1854) was well known in St. Petersburg artistic circles as a passionate collector and friend of many artists. Thus in 1816 the well reputed Russian portrait painter O. A. Kiprenskiĭ took "the good Mr. Duval" as his traveling companion on a journey to Italy via Germany and Switzerland. On arrival in Geneva, Kiprenskiĭ stayed in the home of Duval's elder brother. There are several known portraits of Duval and his family painted by Kiprenskiĭ [11, p. 418; 12, pp. 114, 228].

in the Tret'akovskiǐ Gallery is both dated and signed by the artist, while that in the Genevan museum lacks both date and signature, although most of the known paintings by Darbès are signed. The existence of two almost identical paintings, only one of which is dated and signed, prompted a more painstaking investigation of the portrait in the Tret'akovskiǐ Gallery. Thus the painting was subjected to scientific analysis: the painter's technique was appraised, and the signature and condition of the paint on the canvas were closely examined.

2. The investigation into Darbès' manner of painting was carried out at the Tret'akovskiǐ Gallery using a stereoscopic microscope, textural photographs, and X-ray technology. Such technological aids allowed the formation of a substantially complete idea of the artist's method. X-ray photographs established that the canvas on which Euler's portrait was painted is of a linen weave of medium density, stretched and fastened by hand to a subframe, as was indicated by the discovery on the left edge of an X-ray photograph of some minor random deformation of the threads of the canvas. The white primer was almost opaque to X-rays, indicating the presence of chemical elements of high atomic weight in its composition. The presence in the primer of ceruse[15], barium, and other compounds, produced additional shadows on the X-ray photographs, reducing the quality of the skiagrams[16] of the thin colored layer of the actual portrait. As a result the radiographs picked out satisfactorily only the more complete right side of the face. Not having been painted over, the left side did not produce a satisfactory image. Most striking in the X-ray images of the face in the portrait is the absence of a sense of depth resulting from the lack of modeling in light and shade, notwithstanding the apparent solidity of the execution and the quantity of ceruse applied to the better-lit right half of the face. The facial features are scarcely discernible. The colored layers, which are picked out on the X-ray photographs as whitish spots of varying intensity, are finishing touches. There are no painterly brush-strokes on the face, so that the X-rays do not reveal peculiarities of the painter's style. It should be mentioned that usually such X-ray images reveal a colored layer of no great thickness with a low textural relief.

Examination of the system of artistic construction of the face under the microscope and by means of textural photographs, allowed a more detailed analysis of the multi-layered work, and the determination of the character of the painter's touch. The presence of the dark under-shading tells us that the picture was painted in the order "from dark to light". The composition of pigment varies with the intensity of light falling on the face. The more the illumination, the greater the quantity of a thick paste of white lead that has been applied, and, conversely, in the less illuminated areas of the face the paint contains proportionately less ceruse. The highlights, for example the tip of the nose, have almost pure ceruse applied to them. From the textural photograph—which was taken using oblique lighting and which confirms observations of the colored layer made with the microscope—one sees that the final layers of ceruse have been extended by means of short, somewhat narrow brush-strokes. The illuminated arc of brow towards the temple, the temple itself, and the more prominent part of the jaw are more thickly daubed. The shell of the ear is also partially highlighted, as is the right side of the nose and the raised areas of flesh just above and below the lips. The short brush-strokes visible do not significantly contribute to the definition of the shape of the face. As a rule finishing touches applied with a more textured brush are perceived as

---

[15] White lead used as a pigment. *Webster's Ninth New Collegiate Dictionary.* 1990.
[16] That is, pictures formed from outlines of shadows. *op. cit.*

Portrait of Euler painted by J. F. A. Darbès in 1778
State Tret′akovskiĭ Gallery, Moscow

giving off the most light. In the present portrait the gradual increase of ceruse in the colored
layers creates an impression of smoothness of transition from light to dark, which in turn
suggests a plasticity in the moulding of the facial topography. The layer of paint applied
to the inner corner of the eye and along the nearer edge of the nose is likewise lacking in
any structural touches that might give more definition to the nose. The brush-strokes are
very short and difficult to distinguish. It may be the case that in some places—for example above the upper lip—the paint was applied with the flat of a soft brush. The touches

Textural photograph of Euler's portrait of 1778

used to convey the illuminated parts of the cheek and the light-colored neck-scarf are more expressive. It should be mentioned that the final layer of paint has worn thin. The lacquer finish is of variable thickness, and when exposed to polarized ultraviolet rays the older and newer films of lacquer show different degrees of illumination. The artist's signature lies under an old and grimy layer of lacquer, and has been executed in black pigment on the already dry colored layer. Under luminescence the signature was illegible, merging with the craquelure[17] of the colored layer. All of this testifies to the signature's considerable age, and therefore that in all probability it is in fact the artist's[18].

---

[17] Network of fine lines in the paint or varnish of a painting. *Oxford Dictionary of English.* 2nd ed. 2005.

[18] The above-described investigations were carried out by the X-ray and art expert M. P. Vikturina. The prints of the radiographs and textural photographs were made by the X-ray technician I. N. Fatyukhin.

3. The above-described examination of Euler's portrait using X-ray and other technology affords no reason for believing that Darbès was not the artist. Of the artistic peculiarities and manner of execution of the portrait in the Genevan museum we can only form an approximate idea from photographs—and that does not suffice to determine whether that work is definitely by Darbès or a copy from the original contemporary with him. We are obliged to leave this question to the appropriate specialists in the Genevan museum. We feel that it is completely reasonable to suppose that the portrait in the Tret′akovskiǐ Gallery is the original one painted by Darbès, and that the one in the Genevan collection is a copy made by Darbès himself from his successful original. It is indeed possible that the copy was made at the behest of Kütner for his engraving: there is mention in the literature of a request by Kütner for a portrait of Euler by Darbès.

Unfortunately, at present we know precious little of the history of our portrait. The back of the canvas of the painting helped to establish a few facts. The stamp and number appearing there show that the portrait was at one time held by the State Museum Trust. In Book No. 2 of that trust's inventory, which is kept in the Central State Archive of Literature and Art, Euler's portrait is labeled "Portrait of an unknown old man in a hat". It is mentioned that the portrait came to the Trust from the collection of the Naryshkin family in Moscow, and that in 1926 it was transferred to the Pushkin State Museum of Fine Arts [13]. The portrait was moved to the Tret′akovskiǐ Gallery in 1928.

One would like to believe that in due course it will be discovered just who the Naryshkins were, and how the portrait came to be in their possession.

We can however already assess the importance of the discovery in the Tret′akovskiǐ Gallery of a hitherto unknown portrait of Leonhard Euler. For one thing, Euler's contemporaries considered precisely this portrait as one of the best and most true to life. For instance, the academician N. Fuss, who knew Euler intimately, asserted that the image in the painting was a very accurate likeness of the great scientist [14, p. 301]. We have before us the face of an elderly person, always of the utmost modesty, conscientiousness, and unswerving benevolence towards others, evoking respect and affection in all who knew him personally. Abel Burja, a mathematics teacher and friend to Euler, recalled: "I saw him on the second last day of his life ... happy and affectionate as always. He complained only of dizziness. When one of his grandsons came into the room at around five o'clock in the evening, he began to joke with him, sitting on the couch and smoking tobacco" [15, p. 606]. As portrayed by Darbès a mere four years before his death we see the great scholar in a light-brown dressing-gown with a fur collar, and a velvet beret of muted green tones, with a pouch suspended from his neck; we note the high forehead, the strong facial features, the deep folds between his eyes and on the lower part of his face around the firmly closed—and perhaps slightly smiling—lips, and the steadfast gaze, fixed on an invisible interlocutor, of the left eye—the ailing one being hidden in shadow by the artist, perhaps out of delicacy. One senses in this portrait of one of the most gifted of men—notwithstanding his advanced age—his penetrating mind, tenacious memory—Euler could recite the *Aeneid* by heart—, energy, and his astonishing capacity for work.

It is difficult to overestimate the importance of this find. After all, the Academy of Sciences of the USSR[19] has no original portrait of Euler from life. The sculptured bust standing

---

[19] Now of Russia.

Leonhard Euler
Engraving by S. G. Kütner from the portrait by J. F. A. Darbès

in the Presidium of the Academy of Sciences in Moscow, the work of J. D. Rachette, was created posthumously.

It was a pleasant coincidence that this discovery was made at a time when the worldwide community of scientists was marking the 200th anniversary of the great mathematician's death. This represents a modest contribution by Soviet art experts to the preservation of the memory of one of the world's most outstanding scientists.

References

[1] Brangel', N. N. *A wreath for the dead.* SPb., 1913.

[2] Nagler, G. K. *Neues allgemeines Künstler-Lexicon.* München, 1836.

[3] Neumann, W. *Lexicon Baltischen Künstler.* Riga, 1908.

[4] Thieme, U., Becker, F. *Allgemeines Lexicon der bildenden Künstler.* Leipzig, 1911.

[5] Bénézit, E. *Dictionnaire des peintres, sculpteurs, dessinateurs et graveurs.* Paris, 1976.

[6] Rovinskiĭ D. A. *Detailed dictionary of engraved Russian portraits.* SPb., 1888. Vol. 3.

[7] Glinka, M. E. "Leonhard Euler (attempt at an iconography)". In: *Leonhard Euler.* M.: Izd-vo AN SSSR (Publ. Acad. Sci. USSR), 1958. pp. 569–589.

[8] Saprykina, N. G. *Portraits from the collection of F. F. Vigel': Annotated catalogue.* M.: Izd-vo MGU (Moscow State Univ. Press), 1980.

[9] Bernoulli, J. *Reisen durch Brandenburg, Pommern, Preussen, Curland, Russland und Pohlen in den Jahren 1777 und 1778.* Leipzig, 1779. Vol. 3.

[10] *Le catalogue manuscrit de la collection Duval de 1808.* Arch. Mus. art et hist. Genève.

[11] *Olden times.* 1908. July-September.

[12] Atsarkina, E. N. *O. A. Kiprenskiĭ.* M.: *Mol. gvardia (Young guards),* 1948.

[13] Central State Archive of Literature and art, f. 686, op. 1, No. 77, l. 17, No. 3529.

[14] Pekarskiĭ, P. P. *History of the imperial Academy of Sciences in St. Petersburg.* SPb., 1870. Vol. 1.

[15] Pavlova, G. E. "Forgotten testimony of contemporaries concerning the death of Leonhard Euler". In: *Leonhard Euler.* M.: Izd-vo AN SSSR (Publ. Acad. Sci. USSR), 1958. pp. 605–608.

Eulogy in Memory of Leonhard Euler

composed in French and delivered by

Nikolaï Fuss[1]

at the meeting of the Academy of October 23, 1783

To portray the life of a great man who has brought renown on his age through the enlightenment of humanity is the same as praising human reason. Whoever takes on this task will labor in vain to fulfill it if he does not combine a perfect knowledge of science with an agreeable style appropriate to a eulogy—which, so they say, has no place in the abstract sciences. Although on the one hand, by virtue of the importance of my subject, there is no need of embellishments, on the other hand it is essential that the writer who would describe certain events present them appealingly, clearly, and forcefully. He must show how nature produced the great individual, examine the circumstances through which his gifts were brought to fruition, and, in the course of enumerating his scientific achievements, not neglect to consider the prior state of the relevant domains of enquiry so as to define his points of departure.

Having assumed the duty of portraying to this assembly the life of Leonhard Euler, a man immortalized by his discoveries, I felt the weight of all these responsibilities, and saw that, apart from a self-confession of weakness multiplied by a grief at Euler's death that at this very moment wells up anew, the strict limitations on speeches made before such assemblies prevent me from playing the role of author. Thus I shall describe only certain

---

[1] N. Fuss, Euler's closest student and assistant, delivered the eulogy in memory of his teacher at the meeting of the Academic Conference of November 3/October 23, 1783. The eulogy was read in French, at that time the language used in the Academy of Sciences, and published in St. Petersburg as a separate book with a list of Euler's published works, compiled by Fuss, appended. Later Fuss prepared a somewhat revised German version of his "Eulogy", which was published in Euler's motherland, in Basel, in 1786. In the late 1790s the Academy of Sciences decided to publish a collection of selected works by its academicians and to include in the Russian translation of this collection a translation of Fuss' eulogy to Euler. According to the minutes of the meeting of the Academic Conference of June 28/17, 1799, Academician S. Ya. Rumovskiĭ announced that "In gratitude to his former honored teacher he takes on the task of translating the 'Eulogy' into Russian". Rumovskiĭ's translation from the French original of 1783 was published in 1801 in the collection *Academic papers, selected from the first volume of the Proceedings of the Imp. Academy of Sciences* (SPb., 1801, Part 1. pp. 97–167). It should be mentioned that in certain places of his translation Rumovskiĭ used Fuss' German version of 1786.—*Eds.* (The present English translation has been made from Rumovskiĭ's version of 1801, which in the original of the present book was reproduced with modernized spelling, but without changes to his late 18th-century Russian.—*Trans.*)

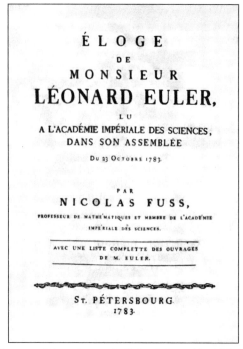

Title page of the eulogy to L. Euler delivered by N. I. Fuss

aspects of the life of the great man, and to any who may feel strong enough to praise him according to his merit I will supply the necessary facts, for my part placing but a few flowers on the grave of my dearest and most illustrious mentor.

Leonhard Euler, professor of mathematics, member of the St. Petersburg Imperial Academy of Sciences, former director of the Royal Berlin Academy, member of the Royal Paris Academy, fellow of the London Society, etc., was born in Basel on April 4/15, 1707. His father, Paul Euler, was a pastor in Riehen, and his mother, Margareta Brucker, came from a family well reputed in educated society.

He passed his earliest years in Riehen. For the simplicity of his morals, which distinguished him throughout his life and which in all probability facilitated the formation of conditions favorable to the career that would render his name immortal, he was obliged to the example of his parents and to pastoral life in a country where the worm of moral change had always proceeded by slow stages.

He received his first instruction from his father who, as a lover of mathematics and former student of Jakob Bernoulli, began to teach his son mathematics as soon as he was old enough. Intending that his son pursue a clerical profession, he did not expect that what had initially served merely as a useful pastime would later become his constant and most important occupation. The seed sown in the young geometer quickly took firm root. With his happy mathematical bent and outstanding gift, he was content only when his mental powers were focused on mathematics.

Happily, his father did not persist in trying to distract him from working on mathematics

—which he himself also liked—and knowing that mathematics greatly stimulates the development of the ability to reason and that it is useful in all branches of knowledge, became reconciled to the situation. Hence there was ample time for the young Euler's mind to develop, and this growth proceeded at a pace that often signals exceptional gifts and heralds future greatness.

Sent to Basel to study philosophy, Euler was assiduous in following the professors' instruction. The astonishing memory that he was endowed with allowed him to speedily absorb all the non-mathematical subject-matter and devote the rest of his time to the scientific subjects that he favored. His strong mathematical penchant and his ardor, inflamed by continuing successes, soon caught the attention of Johann Bernoulli, the greatest mathematician of the time, who, noting how Euler stood out from the rest of his students but not having time to give special tutelage to the young mathematician, acceded to his request to come on Saturdays for help in resolving any difficulties encountered in reading the most abstruse mathematical works. This is a fine way to learn, although one where success might be anticipated only in the case of a passionate mind of untiring concentration—conditions met by Leonhard Euler, destined even then to surpass his teacher, acclaimed for several of his works.

In 1723 Euler was awarded a master's degree. On that occasion he gave a speech in Latin in which he compared the Newtonian and Cartesian philosophies[2]. Having achieved this distinction, in submission to his father's will he began to study theology and oriental languages. Although these studies—essential to the calling his father had prescribed for him—did not correspond with Euler's inclinations, they were not without benefit. Not long afterwards, having been allowed to change to mathematics—from which in any case nothing could ever have wholly distracted him—he became passionately engrossed in that subject. He continued to attend to Johann Bernoulli's advice, and became friends with his sons Nicholas and Daniel. This connection, founded on a coincidence of interests, was to be the cause of our Academy's satisfaction at numbering Euler among its members.

In founding the Academy of Sciences in St. Petersburg, Catherine I was realizing the intentions of Peter the Great. In 1725 the two young Bernoullis were summoned to St. Petersburg. Before leaving they promised Euler, who had expressed a strong desire to follow them, that they would do all in their power to obtain a decent position for him. The next year they informed him that they had found such a position, and in this connection advised him to start applying his mathematical knowledge to physiology. A person of exceptional gifts undertakes nothing in vain. All that was needed for Euler to become a physiologist was

---

[2]This date is incorrect. L. Euler was enrolled in the Philosophy Faculty of the University of Basel on October 20, 1720. This faculty considered itself the focus of university education. In the university library there are preserved to this day three little essays by the young Euler: "A discussion of arithmetic and geometry" (1721), "On moderation" (1722), and "The resolution of an apparent contradiction in the analytic concept of negative quantities" (1722?). Also preserved are two dissertations defended in January 1722 by candidates for the vacant Chair in logic, in connection with which the student Leonhard Euler of the Philosophy Department acted as official "respondent". And in another dissertation—on the history of Roman shipbuilding—it is recorded that in November 1722 L. Euler again acted as "respondent". On June 9, 1722 Leonhard Euler received his "first laurels", the lowest scholarly degree given by the Philosophy Department (corresponding to the unofficial title of laureat or baccalaureat). In 1723 he completed the philosophy program and sat the examinations, and at an official ceremony held on June 8, 1724 delivered the lecture in Latin comparing the Newtonian and Cartesian philosophies, the text of which has been lost. The award of the master's degree represented the culmination of his philosophical education at the University of Basel. In obedience to his father's wishes, on October 29, 1723 L. Euler had already enrolled in the senior theological faculty, where among other things he studied ancient Greek and Hebrew.—*Eds.*

Nicholas Fuss
Silhouette by J. F. Anthing 1784

the desire to do so. Euler registered as a medical student and worked on the lessons taught by the most skillful doctors in Basel with the intensity appropriate to a cheerful spirit[3].

Far from absorbing all of his mental powers, as vigorous as they were copious, this work did not prevent Euler from publishing a discourse on the diapason of the voice and solving a problem proposed by the Paris Academy concerning the most efficient placement of the masts on a ship. His solution was the closest among all others to that which was awarded the prize in 1727. This essay and the proposal that he defended in competing unsuccessfully for the position of professor of physics in Basel show that Euler had long before turned his thoughts to the theory of seafaring, which he was to enrich later with so many discoveries.

Fortunately for the St. Petersburg Academy fate favored Euler neither with a posting in the civil service nor a position in the university[4], and for this reason he left his fatherland to

---

[3] Euler officially enrolled in the medical faculty as a student of Professor J. R. Zwinger only on April 2, 1727, the day following the announcement that B. Staehelin had been chosen to fill the vacant Chair in physics, and just three days before his departure from Basel for St. Petersburg on April 5.—*Eds.*

[4] In fact Euler was not subject to a whim of fate in 1727. A protocol for filling vacant Chairs at the University of Basel had been established on February 2, 1718. Once a vacant Chair had been announced, applicants had to register their candidacy and present a dissertation (Specimen disputatorium pro vacante cathedra, Dissertatio or Theses). They were then required to give one or several test lectures and a defense. L. Euler lectured on his dissertation "On sound" on February 18, 1727, being opposed mainly, perhaps, by the official "respondent" E. L. Burckhardt, whose name appears on the title page of the dissertation. However the discussion was open to all. A month later Euler gave a second lecture "On the cause of gravity", the text of which has been lost.

There were eight candidates for the Chair of physics in 1727: Daniel Bernoulli, Jakob Hermann, B. Staehelin (a doctor of medicine), A. Birr (a medical candidate), A. J. Buxtorf and P. Ryhiner (theological candidates), L. Wenz (a lawyer), and Leonhard Euler. Three groups of six representatives of the professoriat and the higher administration were each to choose one name for the drawing of lots. In one group there were three votes for Hermann. In another there were three votes for Staehelin, and one each for D. Bernoulli, Ryhiner, and Euler. Thanks to an examination of the ballots by handwriting experts, we now know that the vote for Euler in this group was cast by Johann I Bernoulli (see: Burckhardt, A. "Über die Wahlart der Basler Professoren im 18. Jahrhundert". Basler Ztschr. Geschichte und Altertumskunde. 1916. Vol. 15. pp. 28–46). The third group chose Birr. The neutral casting of lots then resulted in Staehelin being chosen. In addition to these eight applicants there had been initially a further four: E. Hess, J. Thellusson, L. Wolleb, and E. König. However these withdrew their candidatures when it became clear that none of those participating in the selection process wished to vote for them.—*Eds.*

settle in St. Petersburg, which he found a congenial place for demonstrating his surpassing gifts to the scientific world. His first articles fulfilled the expectations of the Academy and his compatriots Hermann and the Bernoullis.

On being made adjunct in mathematics, Euler could abandon physiology and devote himself to his true vocation from which neither parental designs nor the modest remuneration usually associated with the pursuit of mathematics could deter him. Soon he had enriched the early volumes of the *Commentarii* with many articles, leading to a rivalry with Daniel Bernoulli which was to persist throughout their lives, not, however, spoiling their friendship or causing envy—an emotion unworthy of a noble soul, obscuring the other virtues.

At the time when Euler began to immerse himself in mathematics, a mathematical career was not considered attractive. A person of average gifts could not hope to distinguish himself in mathematics. The names of the great men who had become famous at the end of the previous century or in the current one were fresh in everyone's memory: Newton and Leibniz, who had transformed mathematics, had died not long before; the important services rendered all branches of mathematics through the inventions of Huygens, Bernoulli, de Moivre, Tschirnhausen, Taylor, Fermat, and many others, were still present to the collective memory. In that celebrated era what was left for Euler to achieve? Could he hope that nature, so sparing of her gifts, having just produced several great mathematical minds might perform another miracle for his benefit? Sensing what in fact she had done in creating him, he set out on his chosen path with a courage drawn from a sense of his own excellence, and demonstrated that his mathematical forbears had not exhausted the treasures contained within geometry and analysis.

And in fact the calculus of infinitesimal quantities was very well suited to his youthful abilities, not having been adequately perfected by its inventors. In mechanics, dynamics, and most of all in hydrodynamics and the science of the motion of celestial bodies the state of the relevant calculus was notably imperfect. The method of applying the differential calculus in these areas was well known; however when it became necessary to pass from the elements to the quantities themselves[5] difficulties arose on all sides. As far as the properties of the integers are concerned, the works of Fermat, whose labor was crowned with considerable success, had been lost, and with them the profound investigations they contained. Artillery and seafaring were based on dubious rules derived from sets of observations often contradicting one another, rather than on a well-founded, genuine theory. The irregularities in the motion of celestial bodies, and especially the complex combination of forces affecting the moon's motion, drove mathematicians to despair. Practical astronomy struggled with the imperfections of telescopes, and reliable rules for building them were lacking. At various times Euler turned his thoughts to all of these concerns. He greatly improved the integral calculus and introduced a new form of algebraic computation applying to both sines and achromatic telescopes; he simplified the countless analytic operations and with their potent aid, combined with his astounding ability to analyze the most abstruse of algebraic expressions, shed new light on all of the mathematical sciences.

Not long after entering the Academy Euler narrowly escaped having to change to a field of endeavor different from that of his inclinations. The demise of Catherine I threatened the disbanding of the Academy, which by virtue of its newness had not become firmly

---

[5] That is, to solve the appropriate differential equations.—*Trans.*

entrenched. The Academy was regarded as an institution without palpable use but incurring nontrivial expenses. At that time the standpoint from which such associations should be viewed—namely that they have been established with the single aim of gathering, disseminating, and improving upon all useful scientific discoveries—was unfamiliar. The foreign academicians saw that they should take the intended measures under advisement, and Euler in particular decided to enter into naval service. Admiral Sivers, considering such a person as Euler a find, offered him a posting in the navy at the rank of lieutenant with the promise of rapid promotion[6]. Fortunately circumstances changed in favor of the Academy, and when in 1730 the academicians Bilfinger and Hermann left to return to their respective fatherlands, Euler became professor of physics, and then when in 1733 his friend Daniel Bernoulli also left, Euler was appointed his successor.

The great number of articles that Euler presented to the Academy at that time shows the astonishing fertility of his mind, his prodigious ability to solve the most difficult problems, and his unbounded diligence, particularly demonstrated in the unusual situation that arose when in 1735 the Academy had urgent need of certain calculations that other mathematicians were insisting would require several months to complete. Euler agreed to take on the task, and to the amazement of the Academy completed it in three days. But at what personal cost? He fell into a fever and was for some time at death's door. Although he eventually recovered, he lost the sight of his right eye as a result of an abscess that had appeared during his illness. At the loss of so valuable an organ any other person would have been sparing of himself so as to preserve the remaining eye; Euler, however, would not hear of resting, sooner renouncing food than work, which through an unrelenting mental ferment had become absolutely essential to him[7].

The great change occasioned by the introduction of the integral and differential calculus into almost all branches of mathematics, affected mechanics in particular. At various times Newton, Bernoulli, Hermann, and Euler himself had enriched this sophisticated and indispensable branch of applied mathematics with a multitude of new discoveries. All the same a complete treatment of motion was lacking, if one excludes two or three works that Euler considered imperfect. He saw with regret that in Newton's *Philosophical principles* and Hermann's *Phoronomia*, at that time the finest works on dynamics in existence, those great men had used the synthetic method to cover the tracks by which they had arrived at results of great importance to mechanics; to lay bare their approach Euler used all the analytic means at his disposal, thereby solving many problems that none before him had dared to tackle. He organized his discoveries, together with those of other mathematicians, into a book which he published in 1736.

Clarity of concepts, precision of statement, and orderliness of exposition are all qualities that every author must strive for in his writings if he wishes them to be instructive to his readers. These qualities comprise but the smallest part of the merit of Euler's *Mechanics*. Obscurity and disorder are qualities for which one with a gift for presenting the most profound of his investigations clearly and in fullest accord with reason could never be reproached. This work confirmed Euler's reputation and placed him among the foremost

---

[6]This story of L. Euler's negotiations in connection with possible service in the navy is confirmed by no other surviving documents.—*Eds.*

[7]In 1735 L. Euler did in fact suffer a serious illness. However he lost the sight of his right eye as a result of an infection in 1738. (See: Bernoulli, R. "Leonhard Eulers Augenkrankheiten". In: *Leonhard Euler.* Basel: Birkhäuser, 1983. pp. 471–487.)—*Eds.*

mathematicians at a time when Johann Bernoulli was still alive. Only to superior minds is it vouchsafed to proceed by such rapid strides as to be comparable at the very beginning of their careers with a man with so many triumphs to his credit over all the English and French mathematicians brave enough to try their strength against him.

I have already mentioned that on entering the Academy Euler immediately began to enrich the *Commentarii* with a multitude of articles demonstrating his insight. Among these are works on the theory of important curves such as tautochrones, brachistochrones, and other trajectories, deep investigations into the integral calculus, the properties of numbers, the motion of celestial bodies, the attraction of spheroidal bodies, and many other topics, a hundredth part of which would suffice to render anyone famous. And the solution of the problem of isoperimetries—become to some extent famous by virtue of the disagreement between the brothers Jakob and Johann Bernoulli, each of whom thought he had solved it while in fact neither had obtained a completely general solution—made Euler famous and established his superiority in analysis. The quantity and quality of all these works are to be marveled at, and it is beyond comprehension that one person could publish so much.

It is true that this most industrious of persons did not take part in the entertainments that such a highly respected and renowned scholar would inevitably have been tempted with—indulgences which he might have been forgiven because of his youth and happy disposition. Euler's chief relaxation was music, but even here his mathematical spirit was active. Yielding to the pleasant sensation of consonance, he immersed himself in the search for its cause and during musical performances would calculate the proportions of tones. One may say that his *Attempt at a new theory of music*, published in 1739, was the fruit of his relaxation.

This deeply pondered work, full of new or newly presented ideas, did not meet with the success anticipated, perhaps because it contains too much mathematics for a practitioner of music or too much music for a mathematician. In any case, apart from the theoretical portion, based partly on the first principles enunciated by Pythagoras, it contains a great many rules for the guidance of composers of music and manufacturers of musical instruments. The science of voices is also expounded with the clarity and precision that distinguish all of Euler's works.

In this theory—whose physical basis is undisputed—Euler derives [his definition of] the agreeableness of music from the principle that the appreciation of any kind of perfection causes us to feel pleasure, and since order is a form of perfection stimulating a pleasant feeling in us, the whole of the pleasure we obtain from pleasant music consists in the perception of the measure that voices conform to by mutual accord both as regards the organization of the lengths of notes and the number of vibrations[8] propagated through the air. Euler applies this psychological rule, on which his theory is based, to all levels of music.

It is true that this is not completely adequate: since it is not in a mathematician's power to submit the soul's intuitions to calculation, it is difficult to establish the truth of Euler's basic principle. However if that principle's truth is conceded, then it must be admitted that a better application of it is not possible. Moreover all objections that might be brought against this principle, even if irresolvable, would scarcely diminish the value of the work. It represents a kind of knowledge that, although perfect in every part, rests on unsure foun-

---

[8]That is, frequencies.—*Trans.*

dations, and, while marvelling at the architect's art one only regrets that he was unable to raise his edifice on a firmer foundation.

Prior to the publication of his opus on music, Euler wrote an *Introduction to arithmetic*. At the behest of the director of the Academy, many academicians undertook to compose elementary textbooks on science for the training of the young, and the great mathematician did not at all consider it beneath him to work on such a textbook which, though an inferior occupation for one with his powers, was important in relation to its purpose. The zeal and care with which he executed any task not properly one of his duties, explain why he was assigned so many additional responsibilities—such as that assigned to him by the Governing Senate in 1740 of supervising the Geography Department.

When in 1740 the Paris Academy—which had earlier, in 1738, awarded him a prize for his essay on a property of fire—proposed the important topic of the ebb and flow of tides, Euler was given a fresh opportunity to demonstrate his reasoning powers. His solution of the problem—a very difficult one requiring complicated calculations—is a model of analytical and geometrical reasoning. It is true that he had to share the prize with Daniel Bernoulli and Maclaurin, rivals worthy of him. Among the competitions of the Paris Academy this was rare for the brilliance of the entries; and it might be added that never before had that Academy received three such satisfying and sound solutions of a single problem. Euler's essay was distinguished especially by its comprehensible explanation of the combined action of the gravitational forces of sun and moon on the sea, by the elegance with which he introduced the water's inertia[9] into the argument—having initially been forced to neglect it—, by the great number of clever integrations, and, finally, the subtle elucidation of the main features of ebbing and flowing tides.

Nothing could be more convincing of the validity of Euler's results on this problem than the coincidence of his ideas with Bernoulli's. Notwithstanding the differences in the assumptions on which these two great men based their arguments, their conclusions were in agreement in many respects, for example in their determination of the tides in cold latitudes. Thus does truth sometimes present herself in different guises to her lovers, whatever may be their means of striving to gain possession of her.

I mentioned earlier that Euler and Bernoulli's applied mathematical interests often coincided. In some such cases Bernoulli's solution was superior to Euler's because his arguments were based on more accurate physical assumptions. Bernoulli had the patience to obtain the postulates basic to the deductions from experiments performed with great care and cogency, while Euler, whose passionate spirit spurred him to finish his calculations, seldom resorted to experiments. With his powerful intuitive ability to distinguish between truth and falsity by means of analogy and imagination, Euler sometimes made bold assumptions, balanced by his facility in performing algebraic computations and his artfulness in simplifying complicated analytical formulae, comparing the results with experiment, and inferring reliable conclusions, thus surpassing not only Bernoulli but all of his mathematical contemporaries.

Some scholars seek to establish their reputations through an extensive correspondence, while for others their reputation is the source of that advantage—if indeed letter-writing may be so termed. Euler figured among the latter sort, since all of the greatest mathemati-

---

[9]The original has the Russian for "coarseness" or "roughness".—*Trans.*

cians of the time considered it an honor to correspond with him. However it should be noted that Euler's correspondence with Johann Bernoulli began as early as 1728 and continued till the latter's death in 1748, and this Nestor among mathematicians did not consider it beneath him to often ask advice of his former student and submit his works to his scrutiny[10].

We now come to a notable juncture in Euler's life. The many outstanding successes with which his works were crowned had made him well known throughout Europe, and in 1741 the Prussian minister Count von Mardefeld made him an offer on behalf of his ruler. The old Royal Society founded by Leibniz was being revived under the patronage of Friedrich II. On ascending the throne Friedrich expressed the intention, worthy of his name, of dissolving the former Society and establishing an Academy of Sciences, for this purpose inviting Euler to his country. The unstable situation of the St. Petersburg Academy of Sciences at that time and the advantageous conditions of the offer from Berlin weighed heavily, and in June 1741 he left St. Petersburg for Berlin with his family to enhance the reputation of an Academy founded under the patronage of the king of philosophers.

Shortly after arriving in Berlin Euler received from the king, busy with affairs of war at an encampment near Reichenbach, a gracious letter demonstrative of the esteem in which he held Euler. The conduct of war, always fatal for science, did not permit Friedrich II to fulfill immediately his good intentions with respect to the Academy. In the meantime the new Society was to be made up of former members of the disbanded one together with certain new members. Euler was one of the latter, and the most recent volume of the *Notes* of that Society was adorned with five of his articles, perhaps the best of those appearing. Following these there appeared with extraordinary rapidity a multitude of works published in the issues of the Academy's annual *Notes*.

This flood of works on every mathematical topic of importance, difficulty, and depth, is the more astonishing in that at the same time Euler was sending articles to be published by the St. Petersburg Academy, which in 1742 recognized his contribution by granting him a pension. His articles published in the *Commentarii* accounted for around half of the total. Such was the immediacy of his reasoning that—one may surmise—the most profound deductions and complex calculations cost him no more effort than to write them out, and posterity will find it hard to believe that a single lifetime sufficed for the production of so many works.

Of particular note among Euler's writings of that period are those concerned with a method of finding curves for which some quantity or other is a maximum or minimum. While working earlier on problems of this type, related to isoperimetries, he had even then seen the tremendous applicability of his investigations not only in pure mathematics but also to questions of physics. He observed that all curves thrown up by the answers to physical questions maximize or minimize some quantity or other, and that many others can be found by isoperimetric means. He even asserted that all natural phenomena could be

---

[10]In order to convey the faith that Bernoulli long had in Euler's insight, it is enough to quote an excerpt from one of his letters: "De caetero gratissimum mihi fuit intelligere, quod ad admirationem usque Tibi placuerint, quae scripsi de oscillationibus verticalibus, propter simplicitatem expressionis et insignem usum, quem praestare possunt in explicandis navium ponderibus; maluissem autem, ut ipse quoque calculum fecisses ex Tuo ingenio, quo mihi potuisset, annon in ratiocinando erraverim. Nam ingenue fateor, me Tuis luminibus plus fidere quam meis. Quae uberius affers, Vir excell. de Isoperimetricis, credo equidem, Te omnia probe ruminasse atque ad veritatis trutinam expendisse, ita ut vix quicquam restet, quod acerimam Tuam sagacitatem subterfugere potuerit etc."

explained in terms of maximization and minimization, starting from initial or acting causes, if only one were able to tell what quantity nature is maximizing or minimizing. Daniel Bernoulli attempted to determine the shape of a compressed elastic strip by such means but, arriving at a differential equation of degree four, was unable to deduce the general equation giving all curved shapes that an elastic strip might assume. He communicated this to Euler together with his guess that all trajectories described about one or several points of attraction can be determined in the same way. This conjecture stimulated Euler to apply himself to this question, and in 1744 he published a long essay on isoperimetries of which one might say that it used all the treasures of higher analysis and that it contains the preliminary foundations of the calculus of variations, which he and the illustrious Lagrange later brought to a greater degree of perfection.

In that same year the Berlin Academy was reconstituted, Euler was appointed director of the mathematics section, and his theories of the motion of planets and comets and of magnetic force, awarded a prize by the Paris Academy[11], were published.

The investigation of the magnetic force contained in this essay is so well known that there is no need to dwell on it very long. Taking as basic Descartes' opinion that all phenomena associated with magnetism are produced by the circular motion of an infinitely rarified substance in microscopically small interstices of magnetized bodies, Euler imagines pores in the magnet in the form of adjacent, parallel, vein-like tubes, and so narrow that only the thinnest quantity of ether can pass along them, proposing that the ether's elasticity propels it through these pores and that on being expelled from them returns to the place where it entered, thus forming a kind of vortex. On the basis of these hypotheses, so subtly elaborated, Euler explains all magnetic phenomena, and the agreement with experiment has won him many followers.

In that year also, the king asked Euler for his opinion as to the best work on artillery. A book on that subject had been published in England by that same Robins who had criticized Euler's *Mechanics* without understanding it. Praising Robins' book, Euler took on the task of translating it into German, with necessary comments and explanations appended. This appendix contains a complete theory of projectiles, and over the intervening 38 years nothing has been published on this theme that might be considered an advance on what Euler then wrote on that difficult branch of mechanics. The merit of this appendix was soon recognized. Turgot, an enlightened minister of the king of France, ordered it translated into French and its use in schools of artillery as a textbook. While giving Robins his due, Euler in modest fashion corrected his errors, thus avenging Robins' disparagement of the *Mechanics* by making his book famous. I will say nothing further about this magnanimous gesture, which though consonant with his greatness still provokes astonishment.

Following the appendix to Robins' book Euler wrote various works on physics, among which his theory of light and colors is particularly noteworthy. Euler sought the causes of fire, the weight of bodies, and electric and magnetic forces in the ether, and calculated the small resistance that that rarefied substance must oppose to the motion of celestial bodies. Clearly, he found the explanation of light in terms of the Newtonian efflux unsatisfactory. His examination of Newton's theory forms the introduction to his *A new theory of light*

---

[11] Euler's essay "On the magnet" was awarded a prize only in 1746. It is true, however, that this theme had been proposed for the competition of 1742, but because the entries were considered unworthy it was set again in 1744 and yet again in 1746.—*Eds.*

and colors, published in 1746. Here he demonstrates clearly that the emptiness of space is incompatible with the radiation of light from the sun and the fixed stars. The rays they emit must then fill the whole of the subcelestial region and celestial bodies would then encounter greater resistance than from the ether, whose existence Newton rejected precisely for this reason[12]. Euler shows that[13] it is impossible for celestial particles to move at such unimaginable speeds without mutual interference of their trajectories, and calculates the decrease in the sun's matter, concluding that the whole of that huge body must exhaust itself in a few seconds through radiation. Finally he adduces a different—but nonetheless solid—objection based on the structure of transparent bodies, which if they are to admit the passage of the substantial rays of light must [for the same reason] themselves lack substance, so that they cannot really be bodies at all.

Descartes thought that light reaches us in the same way as sound or the voice. And indeed, if we consider how sight and hearing act over greater distances than the other senses, that sound and light both propagate rectilinearly, and that both are subject to reflection, then the similarity between those two senses becomes striking. Taking this resemblance as fundamental, Euler inferred that light must be caused by a vibration of the ether since sound is caused by a similar motion of the air, that the differences in the colors, as in the case of voices, depends on the number of vibrations [per unit of time], and that on impinging on a resonant object a sound may change its direction and also in some fashion undergo refraction, much like the sun's rays. Having substantiated this view with all the rigor available to physical reasoning, Euler goes on to elucidate by the most persuasive and natural means all the phenomena of light and sound, including the various kinds of refraction of rays not susceptible to explanation using Newton's system, but following as a matter of course from the Eulerian theory, so that one might bring them to light by reason alone were they not already known through experience.

At the same time as Euler was refuting the theory of the efflux of light, Wolffian philosophy was all the rage in Berlin. The only topics of conversation were monads and the principle of sufficient reason[14]. Wolff and his adherents had recourse to this principle in all situations, and its extremely broad range of application prompted Euler to make jokes at its expense; however he respected monadology as an important though false theory, of sufficient worth for him to make his opinion of it generally known. This opinion is contained in his essay on the elements of bodies, where he shows that the simple parts[15] of a body cannot be unimaginably small unless they are either infinitesimal of nothing at all, and that the mass or inertia[16] of bodies is as general a property as extension and solidity, incompatible with the power ascribed to the simple parts of bodies of being able to continually change their states. Hence the simple parts of bodies, like the Epicurean atoms, cannot exist, and all conclusions as to the different forces originating initially from indistinguishable ones become null and void. Repudiating this system, subject to the same fate as many another

---

[12] That is, because of the resistance of the ether to the motion of the celestial bodies?—*Trans.*

[13] Presumably under the assumption that space is empty.—*Trans.*

[14] The principle that there must be a sufficient reason—causal or otherwise—behind the how, when, where, and why of whatever exists or occurs. It was used by Parmenides, but is most famously associated with Leibniz who used it to exclude all arbitrariness and account for "truths of fact" as opposed to "truths of reason", which derive from the law of contradiction.—*Anon.*

[15] Monads, perhaps.—*Trans.*

[16] The original has "coarseness or laziness".—*Trans.*

impressive though false theory, Euler proposes the force of inertia[17], introduced earlier by Leibniz, in place of the properties that Leibniz and Wolff attributed to the monads, considering it the source of all change. In terms of this concept he explained pressure and the mutual interaction of colliding bodies, and proved that matter alone cannot have the power of thought. The defenders of monadology wrote many essays rejecting this opinion; however together with monadology itself they have now all been consigned to oblivion[18], and their tracts are now recalled as examples of the delusions to which human reason is sometimes subject.

The property of inertia that Euler proposes as underlying all forces represents a substantial idea according with the simplicity that nature observes in all her laws. Although it is true that it is based on a metaphysical notion, it is subject to calculation, and if a concept suffice to explain phenomena then nothing further need be required of it.

This would be the appropriate point at which to discuss the many other philosophical investigations of Euler, where with pleasure and surprise one finds robust physics combined with lofty mathematics; however the limitations on my speech compel me to pass over in silence both these and his reasonings on the tails of comets, the northern lights, the zodiacal light, space and time, the origins of force, etc. So prodigal was he of ideas, so apposite in his discoveries of the most important mathematical truths, and so subtle in his explanations of physical phenomena! He adhered to bold assumptions if calculations were seen to confirm them, but was cautious otherwise. He also created profound and beautiful theories; some of these have already received their due while the others will be judged by posterity. It is enough for the chronicler to indicate in what respect they are new.

From the philosopher we turn directly to the mathematician. Of all the applicable knowledge that analysis combined with geometry had raised to a certain level of perfection, seafaring alone had till then failed to benefit from the successes of the physico-mathematical sciences. Apart from hydrographics and knowledge relating to the heading of vessels, no investigations of a mathematical nature had been made—the unsatisfactory experiments of Huygens and the chevalier Renau on navigation and the determination of a ship's speed being hardly worthy of note[19]. Euler was the first to venture to transform

---

[17]The original has "force of coarseness".—*Trans.*

[18]Leibniz' monadology has in fact retained its interest—at least among philosophers.—*Trans.*

[19]F. Rudio has noted that here Fuss has neglected to mention Johann Bernoulli's celebrated work "Essai d'une nouvelle théorie de la manoeuvre des vaisseaux" published in Basel in 1714 (see: Bernoulli, J. *Opera omnia.* 1742. Vol. 2. pp. 1–96), which followed the publication in 1689 of Renau's *Théorie de la manoeuvre des vaisseaux*, and his disagreement with Huygens.

Three years prior to the appearance of Euler's *Scientia navalis* (SPb., 1749) P. Bouguer published his well known work *Traité du navire, de sa construction, et de ses mouvements* (Paris, 1746). However Euler had completed his treatise well before the appearance of Bouguer's work. The delay in the appearance of the *Scientia navalis* was due exclusively to difficulties in the publication process. Although this circumstance has been documented, Euler nevertheless forfeited unconditional priority in the discovery of several important results. He was obliged to note this fact in 1749 in the preface to his treatise (which was published simultaneously also in M. V. Lomonosov's Russian translation). In connection with a certain question concerning oared vessels, Euler wrote: "However, the illustrious Mr. Bouguer, fellow of the Paris Academy, disagrees completely on this point in the excellent French essay on the construction of ships that he published not long ago... Yet this mistake is not a great one, and is compensated for by the exceptional quality of the other parts of that essay". And further: "In this connection it is essential to note that in this book of mine I do not claim the works of others as my own, since many of the articles that we have composed are in agreement, which circumstance might without foundation seem to show that I copied them from a work that appeared almost four years before my book. However the whole of the Imperial Academy of Sciences, on whose orders I began to compose this book as long ago as 1737, is perfectly aware that the first part had been completed prior to my departure from St. Petersburg in 1740 and the other

seafaring into a complete and exact science. The impetus to undertake these investigations was supplied by an essay on the vibrations of floating vessels by de la Croix, published in 1735. Euler's investigations into the equilibrium conditions of ships allowed him to calculate their stability, and the success achieved in this respect encouraged him to embrace the full compass of seafaring. This is how that great two-volume work, published by the Academy in 1749, came to be. The first volume contains a mathematical exposition of all that equilibrium theory, the motion of floating bodies, and the resistance of liquids contain, that is both difficult and profound.

However such general rules of seafaring are insufficient; that discipline deals with floating bodies of a particular shape, and requires calculations not only of the resistance to, and forces acting on, a ship, but also means for reducing the former and increasing the latter as much as possible, so that a ship can be given a shape combining all possible advantages while averting oscillations and fulfilling the original intentions concerning its construction. Thus apart from general rules as to the construction and navigation of a ship, the theory must show how to adjust against each other all the properties that a good ship should possess. Some of these properties are such that without some sacrifice of others they cannot be secured. For example, maximum stability and maximum speed cannot be achieved in one and the same vessel. In this respect it is necessary to know to what extent each property can be sacrificed for the others. This is the concern of the second volume, where Euler gathered together all that the art of a navigator and shipbuilder might demand of a theory. Somewhat later he enriched this important area of applied mathematics with new and subtle reasonings, distributed among the *Notes* of the St. Petersburg and Berlin Academies. Two of these stand out: one on the means of compensating for insufficient wind, and the other on the action of pitching and yawing of a vessel, which in 1759 was awarded a prize by the Paris Academy.

In the absence of accurate and reliable rules, shipbuilding was for a long time carried out perforce on the basis of custom alone, and notwithstanding accumulated experience, the construction and fitting out of ships suffered from many inadequacies. However with the appearance of Euler's treatise, shipbuilding was suddenly furnished with a complete theory, achievable by other arts only gradually through manifold experimentation.

However this theory is expounded in a language not used by professional shipbuilders, assuming a knowledge of mathematics that it would be unreasonable to demand of a shipbuilder or navigator. Hence Euler's important discoveries might find practical use only when they had been separated out from the the profound argumentation and difficult analytic derivations. This want was felt by Euler himself, and the frequent conversations that he had with Admiral Knowles after returning to St. Petersburg, prompted him to exclude from the theory everything not of direct relevance to seafarers and insufficiently comprehensible to them, and in 1773 he published his *Complete theory of the building and navigation of ships, accessible to all who practice seafaring*.

part half finished, which I completed in Berlin soon after my arrival, so that all parts could have been published the following year. To bear witness to this I might bring many friends to whom I then communicated the most important chapters of my work; however the Academy of Sciences suffices for my justification. Moreover during all the time I was laboring over the composition of this book, the renowned Mr. Bouguer was living in America, where, so they tell me, he wrote his essay, so that there could not have been any correspondence between us on scientific topics that one or the other of us might have used". (Taken from: Lomonosov, M. V. *Complete collected works*. L.: Nauka, 1983. Vol. 11. pp. 177–178.)—*Eds.*

Never has a work by a mathematician enjoyed such brilliant success. Soon a new edition was published in Paris and was used as a textbook in the Royal Naval Colleges[20], and in recognition of the advantages that his many discoveries bestowed on France and all its enlightened peoples (to quote the exact words of the Parisian publishers of the Eulerian theory), the king sent its creator 600 livres. At almost the same time this elegant work was published in Italian, English, and Russian, for which Her Highness the Empress Catherine II granted Euler 2000 rubles.

Although the original work and its adaptation were published at different times, the latter only after Euler's return to St. Petersburg, I have brought together in one place the main results of our great mathematician touching on one and the same theme in order that one might form a clearer estimate of just how great his services were to navigation and shipbuilding, two of the most useful and profound branches of human knowledge.

In 1749 Friedrich II instructed Euler to examine the Finow canal connecting the rivers Havel and Oder. From the fifty-four royal letters addressed to Euler between 1741 and 1777, including many in the king's own hand, I see that he requested Euler's advice on many occasions, and Euler, examining the accounts of the Schönebeck salt works, of the water works at Sans Souci, and of many other projects involving revenues and expenses, was of no little service to the king in preserving him from large and wasteful expense, and was often entrusted with business relating to the Berlin Academy and the University of Halle[21].

The time has come to gather together and impose systematic order on the great mass of important discoveries in the differential and integral calculus made by Euler over a thirty-year period, and scattered among the *Notes* of various academies. He himself long ago intended to do so; but before realizing this intention, he felt he must prepare for those able to understand this lofty subject-matter a special treatise where all the preliminary con-

---

[20]The letter from Turgot to Euler on this occasion does honor to both men, and for that reason I was unable to resist the temptation to include it here:

à Fontainebleau, le 15 Octob. 1775

Pendant le tems, Monsieur, que j'ai été chargé du département de la Marine, j'ai pensé que je ne pouvois rien faire de mieux pour l'instruction des jeunes gens élevés dans les écoles de la Marine et de l'Artillerie, que de les mettre à portée d'étudier les ouvrages que Vous avez donnés sur ces deux parties des Mathématiques: j'ai en conséquence proposé au Roi, de faire imprimer par Ses ordres Votre traité de la construction et de la manoeuvre des vaisseaux et une traduction françoise de Votre Commentaire sur les principes d'Artillerie de Robins.

Si j'avois été à portée de Vous, j'aurois demandé Votre consentement, avant de disposer d'ouvrages qui Vous appartiennent; mais j'ai cru que Vous seriez bien dédommagé de cette espèce de propriété par une marque de la bienveuillance du Roi. Sa Majesté m'a authorisé à Vous faire toucher une gratification de mille Roubles, qu'Elle Vous prie de recevoir comme un témoignage de l'estime, qu'Elle fait de Vos travaux et que Vous méritez à tant de titres.

Je m'applaudis, Monsieur, d'en être dans ce moment l'interprête, et je saisis avec un véritable plaisir cette occasion de Vous exprimer ce que je pense depuis longtems pour un grand homme qui honore l'Humanité par son génie et les sciences par ses moeurs. Je suis etc.

Turgot

[21]When the position at Halle left vacant at Wolff's death needed to be filled, the king turned to Euler for advice. Euler at first recommended Daniel Bernoulli, and when Bernoulli declined suggested inviting Segner, who took up the position on very advantageous terms. It was Euler and no other who persuaded the king to purchase for that university the laboratory equipment remaining after Wolff's death. He was also commanded by the king to enter into negotiations with von Haller to entice him into Prussian service. Von Haller's demands were not to the king's liking, and the question was dropped.—*Note added by N. Fuss in 1786.*

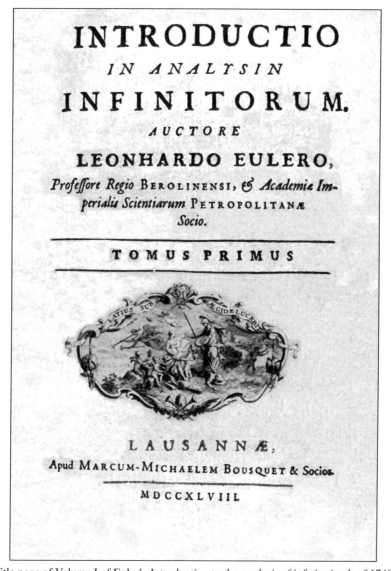

INTRODUCTIO

IN ANALYSIN

INFINITORUM.

AUCTORE

LEONHARDO EULERO,

Profeſſore Regio BEROLINENSI, *& Academiæ Imperialis Scientiarum* PETROPOLITANÆ *Socio.*

TOMUS PRIMUS

LAUSANNÆ,

Apud MARCUM-MICHAELEM BOUSQUET & Socios.

MDCCXLVIIL

Title page of Volume I of Euler's *Introduction to the analysis of infinitesimals* of 1748

cepts they might require could be found. To this end he composed the *Introduction to the analysis of infinitesimals*, published in Lausanne in 1748; in this work he expounded the whole subject of algebraic and transcendental functions, their transformation, resolution, and analysis; here he assembled all that is useful or that one should know concerning infinite series and their summation; he proposed here a new and striking treatment of exponential quantities, deriving from it the clearest exposition of the riches of logarithms and their applications; and here also he explains his new algorithm of quantities related to circles or angles. In the second part he gives a general presentation of the theory of curves and their subdivision, together with the theory of solids and their boundary surfaces, explaining how their dimension leads to equations in three variables. This important work concludes with a treatment of curves of double curvature obtained as intersections of two curved surfaces.

Following the *Introduction* Euler began to compose his texts on the differential and integral calculus, published by the Academy in 1755, 1768, 1769, and 1770. The chief merit of the first of these, devoted to the differential calculus as perfected by its inventors Newton and Leibniz and their successors the Bernoullis, consists in the means by which Euler expounds the true elements of the calculus, the systematic organization of the material, and the clarity with which he demonstrates its applications to the study of infinite series and the theory of maxima and minima. Although his own discoveries are blended in with those of the original inventors, the imprint of his intellect is ineffaceable. Where his powerful mind finds nothing further to add, he improves the discoveries of others by bringing greater clarity to the known elements and the associated rules, and derives new corollaries. Who could fail to see the improvements Euler has made? Everywhere in these works one finds new discoveries—however there is not room enough here for a detailed enumeration of these.

The integral calculus, whose origins are hidden in the genesis of the differential calculus, remains far from the state of perfection attained by the latter. There are rules for proceeding from numbers to the elementary theory, but there are as yet no general rules for going from the elements to numerical evaluations[22], but if such rules should be discovered in the future then posterity must acknowledge its debt to Euler for having prepared the way, for having invented methods for evaluating many integrals that defeated all others. Great honor is due him for having extended the limits of this higher-level computation further than its inventors had dreamed possible, and Newton himself, should he be able to return to us, would be astounded at the extraordinary difficulties that this great man had the strength to overcome.

The third volume of the *Integral calculus* contains an exposition of a new discipline with which he enlarged analysis, namely the calculus of variations. I mentioned earlier that isoperimetric problems led him to the basic idea of curves differing infinitesimally from a given curve. This idea was grasped immediately by Lagrange, Euler's worthy successor, who by eliminating all geometrical reasoning transformed the problem into an analytical one that could be solved using a new kind of calculus. Subsequently this was greatly improved by Euler, who called it the calculus of variations because the differences between the curves are taken to be variable[23].

We have already seen that Euler's capacious intellect was not always occupied with mathematics pure and simple, so wide was its compass. Anything having even a miniscule connection with mathematics might serve as an object for his contemplation; he subjected to calculation everything that could possibly be calculated. We shall now see how much physics, optics, and astronomy owe to his theory of light and colors alone.

His examination of Newton's theory provided him with the opportunity of considering the differences in refraction of light rays[24] and the harmful effects for dioptric telescopes caused by the dispersion of colors. Their use was all but abandoned in favor of telescopes using mirrors, where these effects were mitigated. Examination of the marvelous structure of the eye gave him the idea that by means of a combination of various transparent bod-

---

[22]That is, for evaluating integrals?—*Trans.*

[23]The original has: "The content between the varying quantities is taken as varying". It would perhaps more appropriate to say that the curves themselves are subject to variation.—*Trans.*

[24]Of different colors, presumably.—*Trans.*

INSTITVTIONVM
CALCVLI INTEGRALIS

VOLVMEN PRIMVM

IN QVO METHODVS INTEGRANDI A PRIMIS PRIN-
CIPIIS VSQVE AD INTEGRATIONEM AEQVATIONVM DIFFE-
RENTIALIVM PRIMI GRADVS PERTRACTATVR.

AVCTORE

LEONHARDO EVLERO

ACAD. SCIENT. BORVSSIAE DIRECTORE VICENNALI ET SOCIO
ACAD. PETROP. PARISIN. ET LONDIN.

PETROPOLI
Impenfis Academiae Imperialis Scientiarum
1768.

Title page of Volume I of Euler's *Integral calculus* of 1768

ies this defect might be eliminated. To this end in 1747 he proposed building an objective piece from two lenses with the space between them filled with water. The well known English artist Dollond disputed this idea, maintaining that such a compound lens was at variance with Newton's ideas. Euler wasted no time in demonstrating the truth of his own opinion; a few experiments carried out using meniscuses[25] separated with various liquids confirmed his opinion, and Dollond, having found two types of glass meeting Euler's specifications most precisely, in 1757 crowned his happy intuition with the construction of so-called achromatic telescopes, thereby beginning a new epoch in astronomy.

Dollond's success in constructing a telescope using Euler's idea, against which he had armed himself as if to ward off an attack by an adversary, prompted Euler to intensify his investigations into dioptric instruments. He tried extremely hard to eliminate from telescopes all defects arising from the shapes of the lenses and the dispersions of different rays, and proposed general rules for the construction of telescopes and microscopes, the accuracy of which was attested to by experiments and by telescopes built according to his specifications[26].

---

[25] Concavo-convex lenses.—*Trans.*

[26] King Friedrich II, to whom Euler sent some telescopes made to his specifications, expressed his pleasure in these instruments by means of the following letter in his own hand:

Thus we are obliged to the exchange between Euler and Dollond for one of the most important inventions of the century, of such consequence and usefulness for astronomers, easing their observational labor and leading to the discovery of new celestial phenomena.

The disagreement among Euler, d'Alembert, and Bernoulli over the motion of a vibrating string is of significance only to mathematicians. Daniel Bernoulli was the first to investigate the physics of the sounds produced by such a string, and believed that Taylor's solution sufficed for its elucidation. Euler and d'Alembert, bringing to bear all available resources of the deepest and highest analysis on the question, showed that Bernoulli's solution, which was derived from Taylor's trochoids[27], was inadequate. This debate continued for some time, ending only with Daniel Bernoulli's death[28].

Another debate, not lasting so long but quite heated on both sides, was carried on with Professor König, who claimed that Maupertuis was not the originator of the principle of least action. However since this disagreement did not concern Euler's research, it suffices to mention here only that he took part in it out of sincere friendship with Maupertuis, and that it provided an opportunity for publishing some supremely elegant discursions on the topic[29].

The first solution, by d'Alembert, of the important problem of the precession of the equinoxes and the wobble in the earth's axis, prompted Euler to present his own arguments on this topic in Volume V of the *Notes* of the Berlin Academy—the same volume containing the happy resolution of the conflict that originally arose between Leibniz and Johann Bernoulli over logarithms of negative and imaginary numbers. The problem of the precession of the equinoxes led Euler to the study of unsteady rotational motion of bodies, where there is no single axis of rotation. Approaches to the investigation of such motion were lacking at that time, so that he was forced to go back to the fundamental laws of motion and derive from them general rules for determining the motion of bodies rotating about an unstable axis. These investigations led the way to the discovery of a new law enabling him to solve the problem in its broadest formulation.

Je vous remercie des petites Lunettes d'approche qui me sont arrivées à la suite de votre lettre du 14 de ce mois; et je loue le soin que vous prenez de rendre utile aux hommes la Théorie que vous fournit votre étude et votre application aux Sciences. Comme mes occupations présentes ne me permettent pas de les examiner avec l'attention, que mérite tout ce qui vient de votre part, je me réserve à le faire, quand j'en aurai plus de loisir. Sur ce je prie Dieu, qu'il vous ait en Sa sainte et digne garde.

Waldou ce 15 Septembre 1759   Féderic

[27] An early name for cycloids.—*Trans.*

[28] In 1766 I apprised Bernoulli of a new Eulerian method of determining the vibrations of a string. The following excerpt from Bernoulli's letter shows how these two great men responded to one another: "L'esquisse que Vous me faites de la méthode de Mr. Euler m'a fait plaisir, mais elle n'a changé en rien mes idées sur cette matière; je suis toujours persuadé, que ma méthode donne *in abstracto* tous les cas possibles; j'avoue cependant que dans certains points de vue, celle de Mr. Euler est fort préférable à la mienne; mais il y a aussi d'autres points de vue pour le contraire, puisque ma méthode peut être appliquée à tel nombre de corps fini qu'on propose, lors même que dans le Système il n'y a point de retour parfait ou période à attendre. Quoiqu'il en soit de mes prétensions, je suis toujours prêt de baisser Pavillon devant mon Admiral".

[29] Long before Maupertuis announced the principle of least action Euler had discovered many minimizations in nature, for example in the motion of celestial bodies, the motions of bodies acted on by several attractive central forces, in many curves, etc. I indicated earlier in connection with isoperimetric problems just how close he came to this universal law. Moreover in view of his prior applications of the principle to a great many problems of mechanics—as its discoverer acknowledged publicly in one of his works—Euler had a definite claim to the principle as his own which, however, he always dismissed out of magnanimous modesty.—*Note added by N. Fuss in 1786.*

These arguments, shedding new light on all of mechanics, were worth presenting in full. In his *Mechanics*, mentioned earlier, Euler had expounded the motion of infinitesimal bodies[30], and now proposed publishing a special treatise on the motion of rigid bodies, which appeared in 1765. This work may be considered a complete exposition of mechanics since it contains all the laws of motion of infinitesimal bodies presented in a fresh and clear manner, and, following these, all of his important discoveries, hitherto scattered, concerning the motion of bodies, formulated in a way that was to greatly advance celestial mechanics and be of such significant use in astronomy and seafaring.

While in Berlin Euler continued to perform important services for the St. Petersburg Academy, communicating to that Academy the larger and more important portion of his output, jealously attending to the uses it was put to, and undertaking the supervision of students sent to him from the Academy[31]. He remained a member of the St. Petersburg Academy, and it would seem that the Russian regiment stationed in Berlin must have had this in view in providing him with a bodyguard, and likewise the Imperial Court in compensating him generously for his expenses in its service.

Given Euler's strong attachment to Russia, where he had spent the best years of his youth, and to the society in which he had sown the seeds of his renown, he must surely have felt a desire to return forthwith to Russia. The accession of Catherine II to the All-Russian throne, her temperate rule and her patronage of the sciences and its practitioners, raised the Academy to greatness and confirmed Euler in his intention to end his days in Russia.

In May 1766 he was nearing the fulfillment of his desire. Prince Vladimir Sergee-vich Dolgorukiĭ, a Russian minister at the Prussian court, announced, in the name of his Monarch, agreement to all the conditions that Euler had stipulated in respect both of himself and of his family[32]. He secured permission for his eldest two sons to leave with some difficulty, but the youngest, a lieutenant in the artillery, was refused permission to follow his father to St. Petersburg.

In June of that year Euler left Berlin, where he had spent twenty-five years greatly honored according to his due. The royal princes, especially the sovereign Margrave of Magdeburg and Sweden[33] [34] viewed his departure with regret expressed in the most tender fashion.

Just prior to his departure Euler received an invitation from the Polish King through prince Chartoriskiĭ to visit him in Warsaw on the way to St. Petersburg. There he spent ten days as agreeably as the kindness and generosity of that sovereign might allow the philosopher[35]. Thus after so long an absence he once again saw St. Petersburg on the 16th

---

[30]Point-masses, presumably.—*Trans.*

[31]For several years the academicians Kotel′nikov and Rumovskiĭ stayed at Euler's home and benefited from the instruction of this incomparable teacher.

[32]It is known that these conditions were very favorable. Over and above a yearly salary of 3000 rubles and a guaranteed pension of 1000 rubles for his widow, three of his sons were to be advantageously provided for (and were so).—*Note added by N. Fuss in 1786.*

[33]The original has "Markgraf Magdeburg-Schwedskiĭ".—*Trans.*

[34]The Margrave's regret at the termination of his friendly and frank interaction with Euler was made keener by his gratitude to the great man for tutoring two of his daughters. When the royal family was staying in Magdeburg Euler wrote letters to the elder of these, now Abbess at Herforden, on various physical and philosophical subjects, which he published on returning to St. Petersburg.

[35]For the rest of his life he recalled with gratitude the king's kindness and devotion—the outward manifestation

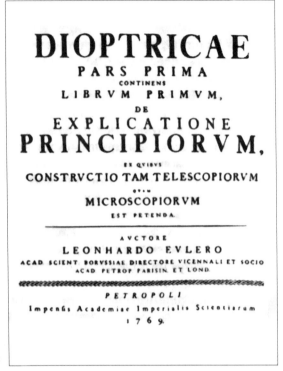

Title page of Volume I of Euler's *Dioptrics*, published in 1769.

day of 1766, where he was forthwith presented to Her Imperial Highness. The first favor he received from her was her intercession on his behalf for the release of his youngest son to reunite with his family in St. Petersburg. No sooner had Euler concluded the business of organizing their new home, towards which Her Highness kindly contributed 8000 rubles, than he fell seriously ill, recovering only at the cost of losing his vision completely. The cataract that had grown over his left eye, damaged earlier by incessant strain, deprived him of the last vestiges of sight.

What a harsh contingency to befall one for whom the habit of hard labor had become essential and whose untiring mind, continually occupied with some new idea, must now contemplate the possibility of not being able to continue as before! A person not endowed with such sterling qualities might have succumbed to a fate of complete inaction; however Euler's extraordinary memory and powers of imagination, free of all distractions, very soon

---

of his intellectual and spiritual qualities—which was strengthened by the ensuing correspondence between them. I cannot resist embellishing my speech with one of the letters the King wrote to him in 1772:

Monsieur le Professeur Euler. En répondant à votre lettre du 4 Août dernier, J'aurois bien souhaité de pouvoir confirmer l'opinion que vous avez des circonstances plus heureuses, sur lesquelles votre amitié pour Moi vous a dicté l'expression d'un coeur vertueux et sensible. Mais ... Je vous remercie cependant de votre bonne volonté à cet égard, et Je passe à la reconnoissance que Je dois à vos soins, pour me communiquer les observations que les habiles Astronomes de votre Académie ont faites à Bender et vers les embouchures du Dniestr et du Danube, avec les positions de quelques endroits également importans pour la Géographie. Je tâche de les mettre à profit pour perfectionner celles qui se font dans ce pays-ci avec assez d'application et de succès, malgré les troubles qui mettent un grand obstacle au progrès des Sciences. Je vous en demande la continuation, autant pour l'utilité publique que pour Ma propre satisfaction particulière, et désirant d'avoir des occasions pour vous en donner des marques effective, Je prie Dieu, qu'Il vous ait, Monsieur le Professeur Euler, en Sa sainte et digne garde.

Fait à Varsovie, le 7 Juin 1772.   Stanislas Auguste Roy

showed how the loss could be retrieved to a large extent—a loss that might have seemed to place limitations on the work of the great man.

The first work produced by the blind Euler was a textbook on algebra, written out to his dictation by a young man from Berlin employed for this purpose who, beyond an ability to do simple arithmetic, had no real understanding of mathematics. The upshot was the well-known *Introduction to algebra* which amazes as much by the circumstances in which it was composed as by its order and clarity. The inventive spirit of its creator shines out of this textbook, and as far as I know this is the only work in which Diophantine equations are formulated and solved in appropriately systematic fashion. The book was very soon brought out in Russian and French translations.

The arrival of Krafft in St. Petersburg provided the stimulus for Euler to set to work on an undertaking he had long had in mind, namely the assemblage in a single treatise of everything that had been discovered during the past thirty years concerning the theory of telescopes and microscopes and means for improving them. With his habitual indefatiga-bility he set about fulfilling this goal, and over the years 1769, 1770, and 1771 published his three-volume *Dioptrics*.

The first volume is devoted to the general theory of this branch of science—which name it scarcely merited prior to being systematized by Euler. Because of the length of the tubes that were formerly used in the construction of compound objectives together with the lack of definition of the images, astronomers had essentially been obliged to abandon them in favor of mirror telescopes. In both cases the rules for calculating the best arrangement of lenses were utterly chaotic; and although this problem is actually one of simple elementary geometry, requiring very little knowledge of the differential calculus, the available solutions were very inadequate, so that success in this science must be reckoned from the time that Euler began to work on it.

The second and third volumes contain rules for the optimal construction of catoptric[36] telescopes and microscopes. The calculation of the blurring caused by the spherical sur-faces of the lenses provides a splendid example of the subtlest algebra, and one is bound to marvel at the artistry of the means by which he was able to combine in every kind of optical instrument both definition, maximal field of vision, and moderate length at all mag-nifications, independently of the number of lenses it might be appropriate to use.

While this important opus was in the process of being published, the following works also appeared: *Letters to a certain German princess*; *Lessons in the integral calculus*; *Computations concerning the comet seen in 1769*; *The appearance of Venus and the so-lar eclipse of that year*; *A new lunar theory*; and *Theory of the construction and navigation of ships*—not to mention a large number of articles published in the *Proceedings* of the Academy[37].

Scarcely had the first of these works appeared, than academician Rumovskiĭ translated it into Russian, a new edition came out in Paris, and a German translation was published

---

[36]That is, involving mirrors.—*Trans.*

[37]The three volumes of Euler's *Dioptrics* appeared over the period 1769–1771. Of the works mentioned here, the *Letters to a German princess* (E343, E344, E417) was published in St. Petersburg in 1768–1772, the *Lessons in the integral calculus* (E342, E366, E385) in 1768–1770, the *Investigations and computations of the orbit of the 1769 comet* (E389) in 1770, the *New theory of the moon* (E418) in 1772, and the *Theory of the construction and navigation of ships* (E426) in 1773. The work on the determination of solar parallax using data from observations of the passage of Venus across the face of the sun (E397) was published in the *Novi commentarii* of the Academy of Sciences for 1769 (SPb., 1770).—*Eds.*

in Leipzig[38]. As regards its content it is enough to note that in view of its accessibility to a wide readership—in particular the fair sex—it succeeded in making the name of Euler known to those unable to judge him by his more important works.

The year 1769 will always be remembered in the history of science—especially of astronomy—as the year when all great nation-states cooperated in aiding astronomers in their observations of the transit of Venus across the face of the sun. The Russian empress and the French, English, and Spanish kings sent their astronomers to the four corners of the earth to observe this extremely rare and important event, serving to determine the size of the solar system. Seven astronomers[39], encouraged by the patronage of the All-Russian sovereign and burning with desire to take part in the undertaking, were dispersed over the expanses of the Russian empire. In the meantime Euler was thinking of the means for using their observations to determine the solar parallax. To this end he invented a new method utilizing not only the observations of the transit of Venus but also those of the solar eclipse occurring a day prior to the transit and furnishing the means for determining the longitude of the observation posts of the transit. Thus one may say that astronomy is indebted to Euler in particular for the advantage it obtained from an extremely accurate determination of solar parallax.

Euler devoted a considerable portion of his time to investigations relating to the moon. As early as 1746 he published his *Lunar tables*[40] and then in 1753 the *Lunar theory*. Based on this work the late Tobias Mayer compiled tables still in use by astronomers, for which he received an award instituted by the English government for the demonstration of a method for determining longitude at sea. The English parliament simultaneously allotted 300 pounds sterling to Euler for the research that led Mayer to his successful solution of this important problem[41].

The Paris Academy, having included Euler among its foreign members[42] and awarded

---

[38] Here N. Fuss had in mind, apparently, the new French edition of 1770 issued "in Mitava and Leipzig", since the first Paris edition appeared only in 1787.—*F. Rudio*

[39] In both published texts of N. Fuss' speech of 1783 and 1786, he talks of ten rather than seven astronomers.—*Eds.*

[40] Euler's *New tables for calculating the position of the moon* (E76) was published in Berlin in 1745.—*Eds.*

[41] The acknowledgement of genuine merit by an enlightened people is flattering to him on whom it is bestowed and encouraging to those following in his footsteps, and for this reason I cannot let the opportunity pass of quoting an excerpt from the letter sent by the secretary of the commission formed in England to examine the longitude question:

Admiralty Office. London, 13 June 1765.

Sir, The Parliament of Great Britain having, by an act passed in their late sessions (a printed Copy of which I herewith transmit to you) been pleased to direct, that a sum of money not exceeding Three hundred pounds in the whole, shall be paid to you, as a reward for having furnished Theorems, by the help of which the late Mr. Professor Mayer of Göttingen constructed his Lunar Tables, by which Tables great progress has been made towards discovering the longitude at Sea. I am directed by the Commissioners of the Longitude to acquaint you therewith and to congratulate you, uppon [sic] this honorary and pecuniary Acknowledgement, directed to be made you by the highest Assembly of this Nation, for your usefull and ingenious labours towards the said discovery, etc.

[42] It is well known that the number of honored members of the Royal Academy of Paris was limited to eight and rarely did anyone not of very exceptional merit dare aspire to such membership. Euler was accepted at a time when there was no vacancy in the Academy. The circumstances surrounding his acceptance did him great honor, and therefore without hesitation I include here by way of explaining them the following letter from the royal state minister the marquise d'Argenson:

à Versailles, le 15 Juin 1755

Le Roi vient de Vous choisir, Monsieur, d'après les voeux de Son Académie, pour remplir une place d'Associé externe dans cette Académie; et comme Elle a nommé en même tems Mylord Maclesfield, Président de la Société

prizes to three of his essays on planetary motion, in 1770 and 1772 proposed as the theme of a new competition that of improving lunar theory as it then stood, and Euler, with the help of his eldest son, entered a solution that was awarded a double prize[43].

In the last of these works Euler expounds a method for calculating the many irregularities in the moon's motion, not obtainable using the earlier method in view of the complexity of the calculations arising from it. This led to a new attack on lunar theory, and with the help of the academicians Albert Euler, Krafft, and Lexell, he developed it further to the extent of being able to compile new lunar tables, which were published together with the new theory in 1772. Rather than attempting, as in the earlier theory, to find sterile solutions of the three second-order differential equations arising from the laws of mechanics, he referred the irregular motions of the moon to a system of three coordinates of its position, and since all of the lunar irregularities depend on the mean distance of the sun from the moon, or the lunar eccentricity or parallax, or the inclination of the lunar orbit to the ecliptic, he subdivided them into corresponding classes. By this method, applied with great care, and by means of subtle arguments peculiar only to a first-rank mathematician, he achieved success beyond all expectations, and if one considers the plethora of calculations, the means used to shorten the arguments, and the application of the results to lunar motion, one cannot fail to be astounded.

But the patience and spiritual tranquility that such immeasurable labor required provoke even greater astonishment, when we recall the prevailing circumstances and the time it took to complete the work. Deprived of sight, without any means of carrying out the prodigious calculations other than memory and imagination, devastated by the fire that bereft him of most of his family[44] and property, forced to leave the home where, since he knew every nook and cranny, habit might compensate for blindness, burdened by the care that this sad and unexpected event and the reestablishment of his household demanded, Euler was nonetheless able to complete a project that alone would have sufficed to immortalize the name of anyone who could have achieved it even under favorable conditions and in a state of spiritual calm. Who would not marvel at the strength of his soul, at his heroic spirit, at his imperturbability and unfailing cheerfulness in the midst of such misfortune[45]?

Royale de Londres, pour remplir une pareille place, qui vaque par la mort de M. Moivre, Sa Majesté a décidé que la première place de cette espèce qui vaquera, ne sera pas remplie. L'extrême rareté de ces sortes d'arrangements est une distinction trop marquée pour ne pas Vous en faire l'observation et Vous assurer de toute la part que j'y prens. L'Académie desiroit vivement de Vous voir associé à Ses travaux, et Sa Majesté n'a pu qu'adopter un témoignage d'estime que Vous méritez à si juste titre. Soyez persuadé, Monsieur, qu'on ne peut Vous être plus devoué que je le suis.   M. D'Argenson

Perhaps no justification is needed for my having included here this and certain other letters from the copious Euler correspondence with remarkable people. Although, it is true, these documents add nothing to the reputation of the great man, they may nevertheless serve to show that he was appreciated by his own generation. Incidentally, in connection with the above, one may mention also the following fact, of some relevance to Euler's standing: the king appointed Euler's eldest son as honored member of the Paris Academy to succeed Euler both in honor of his father's memory and to recognize his own merit in equal measure.—*Note added by N. Fuss in 1786.*

[43] These are not the exact facts. In the preface to the ninth volume of the *Recueil des pièces, qui ont remporté les prix de l'Académie des sciences*, which contains both prize-winning essays, the authors of the first of these ("Théorie de la Lune", 1770) are given as L. and J.-A. Euler, and the author of the second ("Nouvelles recherches sur le vrai mouvement de la Lune") as just L. Euler.—*F. Rudio*

[44] Does this mean that some members of his family perished in the fire?—*Trans.*

[45] Through this accident, in addition to the many books and works in manuscript lost, the essay containing a solution of the problem proposed by the Paris Academy also perished. Albert Euler was compelled to rederive the whole of the lunar theory, including all calculations, from memory.

Following this unhappy event, whose impact was mitigated somewhat by a gift from the Monarch in the amount of 6000 rubles, Euler requested of the well known eye doctor, baron Wenzel, that he remove the cataract from his eye. To his ineffable joy and that of his family the operation restored his eyesight. However the rejoicing was short-lived: whether the operation itself was at fault or Euler's failure to nurse his eye out of impatience to use his restored eyesight, he soon became blind for a second time, amid intense suffering.

Hence the unfortunate old ascetic was compelled to resort to others for help with his work. Two of his children—the academician and the lieutenant colonel—and the academicians Krafft and Lexell took turns in helping him both in completing the great projects he had begun, and in the composition of many articles eventually published in the *Proceedings* of the Academy, which for the sake of brevity I shall not dwell on here—except to note those concerned with the equilibrium and motion of fluids and with the perfecting of achromatic telescopes and microscopes.

Euler's improvements on the dynamics of Daniel Bernoulli, allowing superior calculations to be carried out, prompted him to revisit that part of mechanics, as he had hinted he might in his *Mechanics* and also indicated in the *Notes* of the Berlin Academy. Euler realized this intention by producing four works containing everything of the theories of hydrostatics and hydrodynamics that is most profound.

These works are abundant in indescribably felicitous applications of general rules and satisfying explanations of many natural phenomena. For example, in discussing the loss of equilibrium of the air arising from differences in temperature and density, he explains the underlying cause of winds, especially those prevailing over the Indian Ocean; and in discussing the equilibrium of fluid bodies acted on by one or several attractive forces, he provides an explanation of the shape of the earth and the state of the fluid bodies surrounding it, thus elucidating the phenomenon of the ebbing and flowing of tides. Passing from the study of the equilibrium of fluid bodies to their dynamics, he shows how the whole theory of motion of such bodies is encapsulated by two second-order differential equations, and gives applications of his general rules to fluids contained in vessels, pumps, and tubes. His investigations into the motion of air led him finally to a theory of the propagation of sound and the production of notes in flutes.

Such are the various important things explained by Euler in his hydrodynamics. Not much had been written on this difficult part of applied mathematics, and since Euler's writings on the subject surpass to such a great extent anything written before, it was desirable that his treatment of hydrodynamics be published separately from the *Proceedings* of the Academy for the sake of those wishing to penetrate this important branch of science.

In composing his *Dioptrics* Euler, in his theory of compound objective lenses, neglected the distance between the lenses and their thickness, considering these of little importance; nevertheless there are cases where the space they take up is not so small that it may be dismissed as not contributing to the blurring of an image. His treatment of compound objective lenses and their use in every type of telescope or microscope, which was published as a series of articles in Volume XVIII[46], serves to remove this imperfection. There he describes various means for making such instruments shorter and with wider fields of vision. I myself wrote a textbook for optical artisans based on the principles set out in these

---

[46] Of the *Proceedings* of the Petersburg Academy, presumably.—*Trans.*

articles, which was published under the auspices of the Academy in 1774; the translation of this guide into German by Professor Klügel was appended to his own *Dioptrics*, which appeared in Göttingen in 1778[47].

In many places in Germany widows' funds and other mortality funds have been set up. The censure they attract for being excessively profitable either to those participating or to the founders, prompted Euler to reflect on the means of providing institutions of this sort with a firm financial basis, at least as far as the completeness of lists of deceased permits. Such were the origins of his *Explanations of widows' funds, etc.*, published in 1776. This work contains everything of any importance in the theory of probability.

More than once Euler promised count Vladimir Orlov that he would present enough papers to the Academy to suffice as the contribution of the mathematics class to the *Proceedings* for twenty years following his demise. Not blindness, nor the attacks[48] due to age, nor the great number of his inventions[49] could diminish his industry or exhaust his teeming intellect. Over seven years he presented more than seventy works to the Academy written out by the adjunct Golovin, and around two hundred and fifty with calculations done by me[50].

There is not a single one among these numerous works that does not contain some new discovery or a subtle take on a question opening fresh vistas. Here we find solutions of the most difficult of differential equations, a great many excellent new methods using higher analysis, profound investigations elucidating properties of numbers, elegant proofs of many of Fermat's theorems, answers to many questions relating to the equilibrium and motion of rigid, flexible, and elastic bodies, and the demystification of many seemingly paradoxical phenomena. Everything that the theory of celestial bodies embraces that is difficult and diverting—such as their mutual interactions and irregular motions—is here brought to the highest degree of perfection that analysis as wielded by the mind of the greatest of mathematicians permits. There is no branch of mathematics that does not owe its present state of perfection to him.

Such are the works of Euler, such the feats worthy of perpetual remembrance. Posterity will join his name to those of the great Galileo, Leibniz, Newton, and all who have honored humankind through their intellect; his name will be remembered when those of so many others who owe their fleeting moment of renown to the vanity of our age are gone to everlasting oblivion.

There have been few scientists who have written so much as Euler, but there are none to compare with him in number and variety of mathematical discoveries.

In pondering the good that those born to extend the horizons of knowledge are capa-

---

[47] According to G. Eneström's inventory the German translation of Euler's treatment of telescopes and microscopes (E446) as reworked by N. Fuss, was published in Leipzig in 1778 by Professor G. S. Klügel of the University of Helmstedt as a separate publication with his comments included.—*Eds.*

[48] Spells of feeling unwell?—*Trans.*

[49] It may be thought that because of the great number of Euler's discoveries the exhilaration that the soul experiences on learning of new truths and which the mathematician has occasion to taste in a purer form and more often, perhaps, than any other scientist, would be dulled in him. However Euler always remained extremely receptive to this pleasurable feeling and expected his passion to be shared by everyone. How often was he offended by the expression of indifference with which I, out of modesty, habitually communicated my little discoveries to him!—*Note added by N. Fuss in 1786.*

[50] The earlier articles among these were subsequently published as a collection for the use of mathematicians interested in Euler's works; two volumes of articles were issued with the title *Opuscula analytica.*—*Note added by N. Fuss in 1786.*

ble of doing for humankind, and in considering nature's frugality in providing the means for exceptional gifts to be realized for the enlightenment of humanity, one is scarcely able to refrain from wishing that such gifted people be exempt from the general law to which everyone is inescapably subject, or that they might live longer than average. However on the evidence of Euler's activities and works he might indeed be considered long-lived, and except for his blindness, his life was free of the usual ordeals that cause great psychological stress[51]; till his last hour he preserved that mental resilience that distinguished him throughout his life, and which shines forth even from the very last of his works.

Early last September he began to experience a certain dizziness, which nonetheless did not prevent him from calculating the motion of air balloons and searching for the solution of the extremely difficult equation that these calculations threw up. This dizziness presaged his death, which occurred on September 7. On that day he conversed at the table with Lexell about a new planet discovered around that time, and other matters, all with his usual penetration. After lunch he lay down to rest, and afterwards, while drinking tea and joking with his grandchildren, he was suddenly struck as by a blow, and said: *I am dying*. These were his last words, and in a few hours the glorious flow of life, lasting 76 years 5 months and 3 days, came to an end.

Thus died our senior academician, for fifty-six years the glory and adornment of the Academy, assisting at its birth and fostering its rise to prominence. The influence of the great man on the work of the Academy is especially visible in the contents of the Academy's *Proceedings*, where, notwithstanding his participation in the affairs of the Academy *in absentia* from Berlin, one can see when he departed and when returned to St. Petersburg— as if his presence alone were enough to encourage and enliven the whole organization.

Euler had a strong and robust build, and indeed if he had not been so endowed would have been incapable of withstanding the severe illnesses that he endured in the course of a life replete with so much mental and physical exertion.

His last days were passed peacefully and without illness, except for certain age-related spells of unwellness. His general health was such that the time that old age is usually forced to spend resting, could be devoted in his case to his work, and, while so dedicating his remaining days to science, he delighted in his renown as the fruit of his intellect, in the general honor shown him as the fruit of his virtue, and in the sweetest relaxation vouchsafed him in the bosom of his family.

Knowledge that we call erudition was not inimical to him. He had read all the best Roman writers, knew perfectly the ancient history of mathematics, held in his memory the historical events of all times and peoples, and could without hesitation adduce by way of examples the most trifling of historical events. He knew more about medicine, botany, and chemistry than might be expected of someone who had not worked especially in those sciences.

The great fame, and even greater respect deriving from good deeds—not always consonant with a scholar's sense of his own worth—, often brought travelers to his door. By the time they left many of these were overwhelmed by wonder and amazement, incapable of comprehending how one who had for more than a half-century carried out research in mathematics and physics had been able to absorb so much general knowledge outside his

---

[51] This seems at odds with the fact that he was predeceased by ten of his thirteen children—not to mention the loss of his home by fire.—*Trans.*

specialty. The source of this ability lay in his retentive memory, which registered everything it was exposed to; he who could recite by heart the whole of the *Aeneid* and knew what the first and last lines were of every page of his edition of that classic, might with ease preserve in his memory everything that he read at that time of life when impressions are most strongly felt[52].

Perhaps this also explains his inability to imitate the pronunciation of the milieu in which he lived. To the end of his life Euler kept the accents of a citizen of Basel. Often in the course of conversation he would entertain his listeners with expressions used in Basel whose meaning had long vanished from my memory.

There was no one like him for his unfathomable readiness, without the least sign of displeasure, to leave aside his calculations in order to engage in conversations of no significance, and then return directly to his profound ruminations. The ability to shrug off the scholar's mien, dissimulate his superiority, and adjust himself to the intellectual level of the person he was dealing with, was so exceptional in him that it cannot but be considered a merit. His constant gentleness of spirit, his quiet and unforced cheerfulness, pleasure in certain harmless jokes, and the gift of being an entertaining raconteur, all made dealings with him agreeable and amiable.

Euler's extraordinary vivacity—without which his astonishing spiritual activity would be incongruous—sometimes caused him to overstep the limits of mildness; but his anger was quickly tamed, soon wore itself out, and never did he nurse a grudge against anyone.

He was fair and good-hearted to the highest degree. An implacable enemy of all oppression, he had the spiritual firmness to judge it and arm himself against it irrespective of persons or circumstances. And many still remember how happily he sometimes contrived to console the downtrodden by deflecting abuse[53].

He was a devout religious believer; his piety was sincere and his prayers were reverential; he was zealous in the fulfillment of his Christian duties. He loved his nearest and dearest, was of the utmost patience, and if he occasionally showed his indignation, then only towards the enemies of the faith, and most of all to the preachers of atheism—as expressed in his *Defense of revelation against the objections of the godless*, published in 1747.

He was a good husband, a good father, a good friend, a good citizen, and fulfilled all of his obligations to society faithfully. Everything about him conspires to justify our regret and righteous sorrow at his loss[54].

Euler married twice. In 1733 he married the maiden Catherine Gsell, daughter of a

---

[52] A further example showing the tenacity of his memory and imagination is worth mention here. To pass the time he taught his grandchildren algebra and geometry. In connection with the extraction of roots he needed to give them numbers that were perfect powers; to this end, one night when afflicted with insomnia he calculated mentally all powers up to the sixth of the numbers from 1 to 19, and several days later to our great amazement was still able to reel them off by heart.

[53] Euler was fair in his recognition of others' merits, even those of his adversaries—a quality far from characteristic of all great personages. How often have I heard him, in a demonstration of undoubted satisfaction, sincerely praise Daniel Bernoulli, d'Alembert, Lagrange, and many others. Every new discovery gave him as much joy as if he had made it himself—which shows that he was more concerned with the extension of the empire of knowledge than with worldly approval.

[54] It is a pleasure for me to inform the reader of this speech that the Prussian and Swedish Kings, the crown prince of Prussia, the Swedish Margrave, and the Duke of Courland have all affirmed their participation in the loss to the Academy occasioned by Euler's death, and have conveyed expressions of regret to his eldest son in the form of letters containing the greatest praise of his intellect and virtue.

Swiss-born artist whom Peter I took into his service when in Holland, and sister of the famous president von Loen. On the death of his first wife domestic circumstances compelled him in 1776 to enter into a second marriage, with the maiden Salome Abigail Gsell, a cousin of his first wife, daughter of Maria Graff and granddaughter of Sybil Merian, well known for their drawings of the insects of Surinam.

Of the thirteen children of the first marriage eight died in infancy, and of the remaining three sons and two daughters only the sons survive. The eldest of these, long following in his father's footsteps, is well known for his own articles and the collaboration in his father's last works, and has been recognized by awards from the St. Petersburg, Paris, Minsk, and Göttingen Academies. The second son, a court doctor and collegial councillor, has won merited praise for his skillful and thorough treatments. The third son, lieutenant-colonel in the artillery and director of the Sesterbetsk[55] factories, is well known for his astronomical observations. In 1769 he was sent to Orsk to observe the transit of Venus. The elder daughter, wife to major von Bell, died in 1781; and the younger daughter, who was married to baron van Delen, to his great grief died in 1780 in the duchy of Yulsk. These five children produced thirty-eight grandchildren for Euler, of which twenty-six survive[56].

The image of the honored and venerable ascetic will long come before my gaze, surrounded by his multitudinous family all trying to make his old age easier to bear and gladden his last days in various ways. Never again shall I see a sight so striking in its agreeableness as that I was till now favored with every day.

My dear sirs, any attempt of mine to portray to you that delightful scene of domestic bliss would be in vain. Many of you were, like me, eyewitnesses! Above all those of you assured of fame through having had such a teacher[57]! There are five such former students here; what scientist can boast that he united in a single collective so many of his students? Let us express before all our eternal and most fervent gratitude and so demonstrate that our incomparable teacher evokes astonishment as much for his rare goodness as for his extraordinary intellectual power. Friends! Academicians! Mourn him with the sciences, which have never suffered such a loss, with his family of which he was adornment and support! My tears and yours flow together; his benefactions, especially to me, will to the end of my life remain ineradicably with me.

---

[55] Elsewhere called Sestroretsk.—*Trans.*

[56] Such was the state of Euler's family at his death. However by the time a Russian translation of this speech was proposed, the eldest son Albert Euler had been promoted to councillor of state, the middle son was deceased, and the youngest had retired from the army at the rank of lieutenant general of artillery.

[57] In the Academy there were eight mathematicians who at various times had the good fortune to profit from his tutelage, namely Albert Euler, Kotel'nikov, Rumovskiĭ, Krafft, Lexell, Inokhodtsev, Golovin, and I. Three of these were absent at the time of my speech.

O friends and confreres, whom I saw before me weeping piteously! When grief stopped my voice I could only shake your hands; but your sincere expression of regret will never be extinguished from my memory, and here before everyone I pay due respect to the sensitivity and love that on that occasion you demonstrated towards our most dear and incomparable teacher.

Leonhard Euler's Family and Descendants

I. R. Gekker and A. A. Euler

Introduction

The aim of this article is to describe briefly what is currently known about Euler's family and descendants.

A fundamental work by K. Euler on the genealogy of the Eulers entitled *The Euler-Schölpi genealogy: History of an old family* was published in 1955 [1]. This book provides information about representatives of the various branches of the Euler family tree who lived or are still living in Switzerland, the USSR, the Federal Republic of Germany, France, England, and other countries. The author traces the family's forbears as far back as 1287. A single chapter (45 pages out of 320) is devoted to the Russian branch of the Euler family, and includes a depiction of a portion of the family tree labeled with the names and dates of birth of up to a hundred direct descendants of the great scientist. Short biographical sketches of some of these are also supplied. However the author restricted himself to those descendants with the surname Euler, thereby excluding those on lines branching at a female descendant and therefore having different surnames.

There were in fact earlier attempts at a history of the Euler family. Thus in 1908 F. Burckhardt published his *Towards a genealogy of the Euler family in Basel* [2]. Information about Euler's family and its background may also be found in his autobiography, written to his dictation in 1767, and in articles about him and his three sons included in the biographical directory published in Basel in 1780. (See [3] for a Russian translation.) The Russian lieutenant-general N. P. Euler, one of Euler's great-grandsons, became interested in his family's history and in 1864, while on a tour of duty in Germany, examined the documents extant there. Death put an end to his planned sketch of the family. In 1880 extensive notes that had been made earlier by one of Euler's grandsons, A. Kh. Euler, a general in the artillery, were published [4]. Using the available documentation [5, p. 238] the Soviet historian L. B. Modzalevskiĭ prepared in the 1930s a sketch of the social history of the Euler family in Russia, but the project remained uncompleted.

The authors of the present article, direct descendants of two of Leonhard Euler's sons—Johann-Albrecht and Christofor—, have used both published and archival materials in preparing it, resulting in a significantly expanded and more accurate version of the history of the Russian branch of the Euler family. We have also made a point of providing information about descendants on the distaff side as well as the male lines. The list of Euler's descendants [6] with its extensive bibliography, included[1] in the present jubilee collection, has also been a valuable aid. However, the present paper was written before that list was complete, so that unfortunately not all of the material it contains was available to us.

We note that two lines of direct male descendants of L. Euler's eldest son Johann-Albrecht preserving the surname Euler[2], intersect at the beginning of the 19th century at the level of grandchildren, and, similarly, two male lines through Euler's son Carl intersect in Russia in the mid-20th century. (There are a few representatives of these lines in Switzerland.) There are male descendants of the youngest son Christofor living in the USSR and in Switzerland.

Obviously to dwell equally on every one of Euler's descendants who lived in Russia at some time during the two-century period of interest to us (or still lives there) is in practice impossible. There are extant portraits and detailed biographies of some of them, brief sketches of the lives of those of a second group, and for a third group we have only their names. And in many cases where the descent is through a female it has proved impossible to establish even the names of descendants, since after multiple changes of surname through marriage they quickly become untraceable. At present the number of Euler's descendants known by name is about a thousand. Here we shall relate the histories of only a few, for the most part those about whom we have more substantial information—often by virtue of their greater creative activity in one form or another. For further details we refer the reader to the bibliography included with the list [6].

The history of the Euler family in Russia reflects in part the history of Russia itself. Among Euler's descendants there were several scholars, including three mathematical members of the Petersburg Academy of Sciences (J.-A. Euler, P. N. Fuss, and E. D. Collins), professors and doctors of science (the literary historian K. K. Foigt, the zoologist A. V. Chernaĭ, the chemist N. A. Chernaĭ, the jurist E. N. Berendts, the paleontologist R. F. Gekker, the astronomers E. N. Fuss and V. E. Fuss, the historian K. V. Elpat′evskiĭ, the architectural academician V. F. Gekker, and the diplomats E. I. Euler and K. F. Gekker); there were also a significant number of doctors, teachers, engineers, geologists, alpine engineers, etc. Several of Euler's descendants worked, or are at present working, in the Academy of Sciences. Among the many who served in the army or navy, there were nine generals and an admiral with the surname Euler. In prerevolutionary times the women of the Euler family were usually occupied with domestic duties and the raising of children. Two were Fräuleins at the tsarist court, and a few worked as teachers or servants. Following the Great October Revolution, female doctors, geologists, engineers, philologists, interpreters, etc. begin to appear in the record.

---

[1] Omitted from the English translation.—*Trans.*
[2] Or in the Russianized form Eiler.—*Trans.*

I. Leonhard Euler's ancestry

The most complete source of information about L. Euler's forebears is the aforementioned monograph by K. Euler [1]. The roots of the Euler family tree stretch into the distant past—as far as the 13th century and the German town of Lindau on the shores of the Bodensee, situated in a northern spur of the Swiss Alps where the modern borders of Switzerland, Austria, and the German Federal Republic meet. As early as the 13th century, in Lindau, with its traditional town hall and gothic cathedrals, surrounded by meadows, fields, and vineyards, and in neighboring villages, there were to be found bearers of the surname Euler-Schölpi. It was only with the move of a member of this family to the Swiss city Basel in the 16th century that the written form of the surname stabilized as "Euler". Among those working the land on large or small holdings near Lindau were the first possessors of the name Euler, then spelt in a variety of ways: in the 14th and 15th centuries, Öwler and Äweler; in the 16th century, Öwbler, Ewbler, Ouwler, and, finally, Euler. The Russianized spelling "Eiler" stems from the stable last of these. One plausible version of the origin of the name has it deriving from the word "Au", meaning meadow or watered field. In this connection it is appropriate to note that the name of the town Lindau and other neighboring towns end in "au". Thus it is possible that "Äuler" was originally the name of a proprietor of a meadow or field of modest size (Äule), pointing to Euler's ancestors as tillers of the soil. Another, less likely, version proposes that the name Euler comes from the Latin word "ulla", meaning pot, transformed into the German "Aul". The Latin word for little pot[3], "ullarius", may then have evolved into the German Aulner, Auler, Uller, Üllner, Iller, and, finally, Euler. The second half of the surname, Schölpi, has an even more remote etymology. The German word "Schelpin" signifies a curve or scythe[4], and has the figurative meaning of "little cheat", i.e., has the character of a nickname. At various times this part of the surname was also spelt in different ways: in the 13th century, Schelbelin; in the 15th, Schölpi, Schulpi, Schilpi, Schölpin, Schulpin, Schelpi; and in the 16th, Schelppe, Schelpi, Schölbi, Schulpe. In the oldest surviving document, from 1287, there is a reference to a certain bailiff by the name of Schelbelin. In 1553(?) one Hans Euler-Schelpin (1510–1568), a practical-minded person descended from a family of landowners, was made a Bürger of Lindau. He owned a house with a courtyard, a field, an orchard, and a vineyard. Following his purchase in 1535 of the seigneurial holding Edelin in Schachen, he was called Schelpin von Schachen. His wife, Barbara Kaes, was from an old family known since 1400. In 1555 their son Jörg (or Georg) Euler (1532–1598), also called Schelpin, married Catherina Schnell (1534–1584) who came from a family possessing a coat of arms depicting a unicorn on a blue field. This image was later adopted as part of the Euler coat of arms. Jörg was also elected Bürger of Lindau. He belonged to the evangelical church.

The scion of the Basel branch of the Euler family tree was Hans-Georg (or Jörg) Euler (1573–1663), a member of the next, i.e., third, generation. In 1594 this great-great-grandfather of L. Euler moved at the age of 20 to Basel, a town with many-sided political and economic ties to Lindau, situated 200 kilometers lower on the Rhine. (Basel grew from the Roman fortified settlement of Basilia, first appearing in the records in 372 B.C.[5])

---

[3] Or potty.—*Trans.*

[4] Or perhaps "braid" or "spit of land" since the Russian for all of these is "kosa" (stressed on the "a").—*Trans.*

[5] When there was a Celtic settlement there.—*Trans.*

On arriving in Basel, Hans-Georg acquired a handloom and became an artisan-carder. In that same year 1594 he married Ursula Ringsgewandt (1573–1611), the daughter of a Basel brush maker, who bore him nine children. Following her death he married for a second time, with Eva Reck, who produced six more children. All of the sons from the first marriage— including L. Euler's great-grandfather Paul I Euler (1600–1673)—took up their father's profession. In 1623 Paul I Euler married Anna Hoch (1606–1673), who presented him with eight children. One of these, Euler's grandfather Paul II Euler (b. 1635), continued in his father's and grandfather's profession. In 1669 he married Anna-Maria Gassner (1642–1712), who bore him a son.

In those times it was not only the nobility that displayed coats of arms as a mark of distinction but also people of the less exalted classes[6]. The first Euler to acquire a coat of arms was a nephew of Paul II Euler, also Paul Euler (1654–1731), for many years a pastor and councillor in the consistory at Zweibrücken (now in Rheinland-Pfalz) [1]. His coat of arms consisted of a shield with a deer rampant on an azure field, surmounted by an owl volant with gold and azure dexter chief and gold and maroon sinister chief. It is important to note that the coats of arms of the various branches of Eulers, scattered over several countries, were often very different from one another. Thus while one crest might depict a deer rampant on an azure field surmounted by a deer in the chief, another might show two deer and an owl, symbols of family continuity and wisdom. It is no accident that the words for deer (Eillum) and owl (Eilchen) in middle-high-German resemble the name Euler. The coats of arms were engraved on household items and on seal rings used for sealing letters.

The letter seal used by Leonhard Euler showed his initials in Latin together with their mirror image on a shield surmounted by a crown (1727–1729) or shell (1755–1759) or a fabulous beast reminiscent of a rampant steed (1755–1765) [7]. This "beast" had a beard, an erect mane, a lean body, and a docked tail. Over the shield there was a helmet and a similar mythical animal. In the escutcheons of his descendants (and possibly his own during his last years) there appear rampant white or black unicorns on a blue field. For example the escutcheon displayed at the funeral of Euler's daughter Charlotte van Delen in 1780 shows a unicorn in one half of the field supporting a shield [1], and on the coat of arms acquired in 1846 by Euler's grandson general A. Kh. Euler, a black unicorn, again supporting a shield, occupies a quarter of the field, the remaining three quadrants being taken up with artilleristic images.

The son of Paul II, Paul III Euler (February 16, 1670–March 11, 1745) [8], L. Euler's father, did not follow in the footsteps of his artisan forebears, instead entering in 1688 the theological faculty of the University of Basel (founded in 1460), from which he graduated in 1700. There he also attended the lectures of the famous mathematician Jakob Bernoulli. In 1701 he obtained a position as pastor in an orphanage and in St. Jakob's church "on the Birs". Finally, in 1708 he secured a modest living in the village of Riehen, five kilometers outside Basel, where he remained till his death. In 1706 Paul Euler married Margareta Brucker (1677–1761), daughter of a hospital pastor, and from a background of clergy and scholars. They had four children: boys Leonhard and Johann-Heinrich, and girls Anna-Maria and Maria-Magdalena. The family's material and domestic conditions were difficult; their dwelling consisted of only two rooms. In order to support his family Pastor Euler also gave lessons at a children's religious school.

---

[6]Compare Shakespeare's efforts to acquire the official title of "gentleman".—*Trans.*

L. Euler's personal seals: 1727–1729, 1755–1759, 1755–1765
Photographs by W. Obolensky, Basel

In 1731 the elder of Euler's sisters Anna-Maria (August 19, 1708–May 29, 1778) was married to Christian Gengenbach (1706–1770), an organist from Münster, and the younger, Maria-Magdalena (November 11, 1711–July 30, 1799), to the pastor Johann-Jakob Nörbel (d. 1758).

Leonhard's younger brother Johann-Heinrich (December 7, 1719–September 8, 1750) became an artist. From 1735 to 1740 he studied in St. Petersburg with L. Euler's father-in-law Georg Gsell, and then in 1741 went to Berlin with his brother and lived there for some time. Unable to find appropriate work in Berlin, he returned to Basel, where in 1746 he married Catherina Imhof, who died in giving birth to their first child. In 1750 he remarried. His second wife, Anna-Maria Hugelschopfer, died three months after the death of her husband in that same year 1750.

2. Leonhard Euler and his family

The great mathematician Leonhard Euler was born in Basel on April 15, 1707. His first name was given him in honor of his godfather the privy councillor Leonhard Reschpinger. In fact the name Leonhard was common in Basel since St. Leonhard's was one of the most highly reputed churches of that city. The childhood of the future scientist was spent in the village of Riehen, not far from the city. He then went to study at the Latin gymnasium in Basel, staying with his maternal grandmother. In 1720 he entered the University of Basel, graduating as baccalaureat in philosophy in 1722, and as Master of Arts in 1724. His predominantly humanistic education notwithstanding, Leonhard was most assiduous in his mathematical studies. His mathematical training was supervised by Johann Bernoulli, who since 1705 had occupied the Chair left vacant by the early death of his brother Jakob. Unable to find suitable work in tiny Switzerland in spite of several attempts, at the age of 20 he left for Russia at the invitation of the newly founded St. Petersburg Academy of Sciences. His position there had been arranged through the good offices of two friends who had preceded him, the brothers Nicholas (1695–1726) and Daniel (1700–1782) Bernoulli. In St. Petersburg he shared an apartment with Daniel for six years. In Russia in 1727 the brilliant scientific career of Leonhard Euler was thus launched. He also became the founding father of the Petersburg branch of the Euler family, bringing to a close the Basel period of the family history. His salary increased in line with his scientific achievements: in 1727, as an adjunct of the Department of Higher Mathematics he received 300 rubles annually plus an apartment, firewood, and candles; in 1731 as professor of theoretical and experimental physics, 600 rubles annually plus 60 rubles towards an apartment, firewood, and candles; from 1733 he was professor of higher mathematics, and in 1735 was paid 860 rubles annually, rising to 1200 rubles by 1741. At that time these were very decent salaries.

On January 7, 1734 Euler married Catherina Gsell (1707–1773), a daughter of the academic artist Georg Gsell (d. 1740), also of Swiss origin, who had come to Russia from Amsterdam at the invitation of Peter I. Catherina Gsell's mother, née Maria-Gertrude de Loen, was Georg Gsell's second wife. In the 1730s Georg Gsell taught drawing at the art school attached to the Academy of Sciences together with his third wife the artist Maria Graf (1678–1744). Prior to the wedding Euler acquired a 1200-square-meter plot of land situated on the 10th Line of Vasil′evskiĭ Island, between Bol′shoĭ Avenue and the Neva (not far from the Academy of Sciences), where he had a spacious wooden house built on a brick foundation.

The first-born of the Euler family was a son Johann-Albrecht, followed by three girls, all of whom died in infancy, and somewhat later a second son Carl, subsequent children all being born in Berlin. L. Euler had altogether 13 children, eight of whom died in infancy. The third son Christofor and the two surviving daughters, who later married, were born in Berlin.

Euler was a good family man. Family relations were patriarchal, as had been the way when he was a child in the family of his father the pastor. Although Euler devoted most of his time to his scientific and other work, he found time to give instruction to his children and later grandchildren, teaching them mathematics in particular, and was constantly concerned about the progress of his nearest and dearest. Frequently he supported his or his wife's relatives. Euler was good-natured and a religious moderate. He was usually cheerful, and liked to joke in a benign manner. Conversation with him was pleasant and instructive. All the same he was easily irritated, although his anger would pass almost as soon as it appeared, and he never bore a grudge. As a descendant of many generations of artisans, Euler preserved the characteristic circumspection and thrift of a German-Swiss Bürger. Apart from mathematics and physics, he had a lively interest in the other natural sciences, and a thorough knowledge of the writers of the ancient world and the history of all times and peoples. He showed no interest in contemporary theater or literature, except for the puppet theater. He was a great lover of music and listened to it by way of relaxation. He was a passionate smoker. In addition to German, which he always spoke in his native dialect, he knew French, Latin, and other "ancient oriental and European languages", and both spoke and wrote in Russian quite freely.

We shall not describe here the active and highly successful scientific side of Euler's life, since this is better left to specialists. In this and other respects all went well till 1738, when at the age of 31 through overwork he contracted an infection resulting in the loss of his right eye.

During the "Bironovshchina"[7] and immediately following it conditions in Russia were difficult and the political situation unstable. The Academy of Sciences was forced to introduce measures of economy because of its wretched financial situation and even the payment of salaries became irregular. The situation became even more complicated when upon the death of the empress Anna Ioannovna the infant Johann VI was proclaimed emperor with his parents Anna Leopoldovna[8] and the German prince Carl-Heinrich as regents, and the minister Münnich[9] as the power behind the throne. The Russian Guard, consisting mostly of Russian nationals, began to vent its frustration over the influence wielded by foreigners at court; in the capital the populace was restive. These conditions determined Euler on accepting the invitation of the Prussian king Friedrich II, who had come to the throne on May 31, 1740, and was determined to carry out a reorganization of the operations of the Berlin Academy, which had been stagnating for some time. (The Scientific Society, as it was then called, had been founded by G. W. Leibniz in 1700.) Naturally Euler could not have foreseen the turn that events would soon take in Russia, and the coming to power of Elizabeth I, a daughter of Peter I, resulting in an improvement of the situation for the country as a whole and for the Academy of Sciences in particular. In 1741 Euler moved to Berlin with his family of ten, comprising, apart from himself, his wife, two sons, his sister-in-law

---

[7]That is, the period of Biron's rule. Biron, or Biren, was one of the empress Anna's Baltic German ministers— she herself being the daughter of the Lithuanian serving girl who became Peter the Great's second wife. Biron was also her lover, and "for ten years (1730–1740) Russia suffered so that Anna might be amused and Biren might become rich". Following a period of political instability between 1740 and 1741, in the early stages of which Biren was "kidnapped and sent off to Siberia", in December 1741 a *coup d'état* organized by Elizabeth, daughter of Peter the Great by his first wife (the empress Catherine I) was successful. Elizabeth ruled till 1762. (From: D. M. Sturley. *A short history of Russia*. Harper and Rowe, New York, etc. 1964.)—*Trans.*

[8]A niece of Anna Ioannovna.—*Trans.*

[9]Commander-in-Chief of the army.—*Trans.*

with her four sons (Anna Gsell, his wife's sister, had married Ludwig Wermelen in 1720), and his artist-brother Johann-Heinrich. The king paid the expenses associated with the removal from St. Petersburg to Berlin, including the servants. In Berlin he was appointed director of the mathematics class of the Academy and paid a salary of 1600 thalers, which was equivalent to the 1200 rubles he had been getting in St. Petersburg at the time of his departure.

In 1742 Euler purchased for 2000 thalers a three-storey house with cellars, and seven windows in its façade. This house has survived in somewhat reconstructed form to this day (Berenstrasse 21/22, not far from the present Comic Opera); in 1907 a memorial plaque was attached to the façade. In 1755 Euler also acquired in Charlottenburg, near Berlin, a holding of moderate size for 6000 thalers, including an attractive house, large garden, ploughland, and meadows, where he kept six horses and ten cows. There his children lived with a tutor, and the property was managed by his mother, who moved to Berlin on the death of her husband in 1745. The house was free of billeted soldiers. (A billet of eight soldiers had been posted permanently in Euler's home in St. Petersburg.) In 1760, during the seven years' war (1756–1763), Euler's property was pillaged by Saxon soldiers in the command of Russian officers. However his losses were generously indemnified by Russia, and he was paid 4000 thalers compensation. In 1763, following the death of his mother in 1761, and in anticipation of leaving Berlin, Euler sold his estate. The live-in students who came from Russia, France, and Switzerland to study under him, lived in his Berlin house. During all the years of his sojourn in Berlin (1741–1766) L. Euler remained an honored member of the Petersburg Academy of Sciences, which paid him the substantial annual pension of 200 rubles (270 thalers). He regularly sent his works to St. Petersburg for publication and acted as editor of papers sent to him, supervised adjuncts of the Academy (S. K. Kotel'nikov, S. Ya. Rumovskiĭ—a future vice-president of the Academy—, and M. Sofronov), acquired technical literature and equipment on behalf of the Academy, and chose suitable candidates to fill vacancies in St. Petersburg. In the period 1743–1744 count Kirill Grigor'evich Razumovskiĭ—brother of Elizabeth Petrovna's favorite Alexeĭ Razumovskiĭ[10]—studied under him, and in 1745, at the age of 18, was appointed president of the Petersburg Academy of Sciences.

During his Berlin period Euler was more than once invited to return to St. Petersburg. For some time he vacillated—this would mean a third uprooting of his household. However in 1766 a severe worsening of his relations with Friedrich II tipped the balance, Euler accepted the Russian goverment's invitation, and with his household of 16, including servants, returned to St. Petersburg, the removal expenses being paid this time by Catherine II. The departure from Prussia, which Euler had been thinking about since 1762, was made easier by the circumstances that Euler had kept his and his family's Basel citizenship (his

---

[10]"Elizabeth I's illnesses may excuse her some of the charge (by Platonov) of laziness, but she was also ignorant, coarse, capricious, and superstitious. She was known for her untidiness and extravagance; at her death she left 15,000 dresses, 5,000 pairs of shoes, and piles of unpaid bills. She was also immoral, and her lovers ranged from the illiterate cossack shepherd and choirboy Alexeĭ Razumovskiĭ, whom she married and raised to great wealth, to Ivan Shuvalov, a patron of art and letters who did much for education and founded Moscow University in 1755. Elizabeth encouraged the arts, even if at extravagant cost: the Winter Palace was built by the architect Rastrelli for ten million rubles; the theatre was established and French fashions and the French language were adopted by the court and nobility. Elizabeth was more able than she is sometimes allowed to have been; she knew how to pick able men to advise her, and she did her best for Russia". D. M. Sturley. *A short history of Russia*. Harper and Rowe, New York, etc. 1964.—*Trans.*

wife and children having been made citizens of Basel in 1752), and that the king Friedrich II had not too long before been defeated by the coalition—to which Russia belonged—in the seven years' war, so that the provocation of any further conflict with Russia would have been highly inconvenient. The youngest son Christofor, an officer in the Prussian army, was at first refused permission to leave for St. Petersburg but this was soon granted.

In St. Petersburg Euler was received with great honor: soon after arriving he had a lengthy audience with Catherine II, during which questions concerning the activities of the Petersburg Academy were discussed. He was allocated an annual salary of 3000 rubles. All of his sons were provided for according to their specialities—more will be said on this score below. Euler was also granted 8000 rubles towards the purchase of a large two-storey stone house with 13 façade windows overlooking the Neva. (By way of special dispensation no soldiers were billeted on the home.) This house is still in existence at the address 15 Lieutenant Schmidt Quay, though in somewhat reconstituted form: a third floor has been added and the façade extended. To mark the 250th anniversary of L. Euler's birth, on April 15, 1957 a memorial plaque of white marble sculpted by Yu. G. Klugge was affixed to a wall of the house. In 1983 the house underwent radical repair for use as a highschool with a philological bias. In the school's mathematics office there is an exhibit dedicated to L. Euler's life and scientific activities.

In 1767 the sixty-year-old Euler almost completely lost sight in his remaining, left, eye. In 1771 an operation to remove a cataract by the well known oculist J. B. Wenzel restored sight to the eye; however owing to the lack of antiseptics, as well as Euler's overhasty resumption of active work, it was soon lost again. From this time on Euler was unable to recognize anyone by sight or read without help, and could write only large letters with chalk on a blackboard laid on a large table. His assistants, J.-A. Euler, N. I. Fuss, and M. E. Golovin would then copy from the blackboard into exercise books. In 1772 the region of St. Petersburg between the 7th and 21st Lines of Vasil′evskiĭ Island caught fire, and over 500 houses, including Euler's, burned down. All of the family's belongings were consumed by the flames, together with a large portion of the library, and some manuscripts. However most of Euler's manuscripts were saved. The Russian government came to Euler's aid with a grant of 6000 rubles towards the construction of a new home.

In 1773, after 40 years of conjugal life, Euler's wife died. After three years as a widower, in 1776 at the age of 69 he married again, considering that a house is not a home without a mistress, and only a wife can fill that role. His second wife, whom he had long known, was Salome-Abigail Gsell (1723–1794), stepsister of his first wife. Her mother was the above-mentioned artist Maria Graf. In the years 1780 and 1781 Euler's surviving daughters died; both had married and produced families. Around 1778 Euler's hearing became noticeably poorer. In September 1783 he began to experience attacks of dizziness. Euler died on September 18, 1783 in his 77th year, and was buried in the Smolensk Evangelical Lutheran Cemetery on Vasil′evskiĭ Island. To the gravestone there was affixed a stone plaque inscribed in German with the words: "Here abide the earthly remains of the wise, just, and famous Leonhard Euler, born in Basel on April 4, 1707 and dying on September 7, 1783". Several decades later, at the time of the funeral of his daughter-in-law Emily, wife of Carl Euler, this stone plaque was discovered overgrown and almost indecipherable. It was only in 1837 that there was erected over Euler's grave a massive sarcophagus of polished pink Finnish granite carrying the inscription: "To Leonhard Euler, Petersburg Academi-

cian. Born in Basel on April 4/15, 1707. Died in St. Petersburg on September 7/18, 1783".
In 1957, in connection with the celebration of the 250th anniversary of Euler's birth, his
remains—together with the monument—were removed to the Lazarian Cemetery of the
Aleksander-Nevskiĭ Monastery (now the Leningrad Necropolis), where they were reburied
not far from the grave of the first Russian academician M. V. Lomonosov.

Euler did not like to travel. After leaving Basel for St. Petersburg in 1727, he never
visited his motherland again, although he kept his Swiss citizenship to the end of his life.
He used to relax at his *dacha* at Duderhof on the outskirts of St. Petersburg and at his estate
near Berlin. In 1748 he went with his wife and eldest son to Frankfurt-am-Maine to meet
his mother on her way to him after the death of his father. In 1761 he went with his son
Carl to Halle to pay a visit to J. A. Segner.

L. Euler had 45 grandchildren, of which 26 were alive at his death. Following the return
to St. Petersburg, his sons Johann-Albrecht and Carl and their families lived with him in the
house on the bank of the Neva. Christofor, as a professional soldier, lived where he served—
initially in Sestroretsk, and then in Vyborg[11]. After Euler's death, his widow received an
annual pension of 1000 rubles and Johann-Albrecht 200 rubles. Soon the house and library
were sold. The academician J.-A. Euler, who by 1769 had become conference-secretary
of the Academy of Sciences, moved to a house on the bank of the Neva not far from the
Academy, at the corner of the 7th Line. In 1785 Carl, a doctor, bought a stone house of
moderate size on Sredniĭ Avenue (No. 25) which survived till sometime after 1957, but has
now been replaced by another building.

L. Euler was of average height, sturdily built, broadshouldered, with a heavy tread, and
light-colored eyes; he was rarely ill (except for his eye problems). Right up to his death
he had a phenomenal memory allowing him to carry out the most complex calculations
in his head, a superhuman ability to work, the facility of quick concentration, and an iron
will. It was said that he could work with a cat perched on his back and in the midst of his
grandchildren.

3. Leonhard Euler's children

The eldest of Leonhard Euler's sons, the mathematician and academician Johann-Albrecht
(1734–1800), was his father's closest helper, secretary, and student. Born in St. Petersburg,
at the age of seven he moved to Berlin with his parents. He was successful in his studies of
the humanities at his gymnasium and of mathematical subjects with his father: arithmetic,
geometry, trigonometry, algebra, mathematical analysis, physics, and astronomy. In 1749
he participated together with his father in the project to level the course of the Finow canal
linking the Oder with the Havel. In 1754, at the age of 20, Johann-Albrecht was elected to
membership of the Berlin Academy of Sciences (at an annual salary of 200 thalers, rising
by 1764 to 400 thalers). In 1754 he was also appointed inspector of the Berlin astronomical
observatory. In this connection he observed and described Halley's comet when it appeared
in 1758. On returning to Russia in 1766, he was assigned the rank of professor of physics
and academician in the Petersburg Academy of Sciences, in accordance with his position
as director of the Physics Office (1766–1769), which came with an annual salary of 1000

---

[11] A town about 200 kilometers from St. Petersbug in the direction of Finland.—*Trans.*

Johann-Albrecht Euler
Portrait by E. Handmann 1756

rubles. The majority of J.-A. Euler's works, numbering more than 70, are devoted to questions in mathematics, mechanics, physics, celestial mechanics, and astronomy. In essence a mathematician, he worked in physics as a theoretician; he carried out research in electricity, hydrostatics, the motion of the moon, planets, and comets and made calculations relating to optical systems. Seven of his works, which like many of the others were produced under the direct supervision of his father, were awarded prizes by the Petersburg, Paris, and other Academies of Sciences. In view of the difficulty of separating Johann-Albrecht's work from that of Leonhard Euler, they have all been included in the *L. Euleri Opera omnia*.

In 1769 J.-A. Euler became conference-secretary of the Petersburg Academy of Sciences, a position he held for over 30 years, till his death. During this time he essentially ceased doing scientific research, although he regularly made meteorological observations of importance for the systematic study of Russia's climate. He was put in charge of the Academy's publications, participated in the compilation of lunar tables, carried on an extensive academic correspondence (of about 1000 letters over a period of 15 years) with foreign scholars, and from 1766 to 1774 was a member—along with his father and four other academicians—of an economic commission formed to examine the administration of the Petersburg Academy of Sciences. He was also permanent secretary of the Free Economic (Scientific) Society, and from 1776 inspector (director of the scholastic section) of the Shlyakhetnyĭ[12] Kadet[13] Land Corps. He was also a member of the Munich (1762), Stockholm (1771), Fliessingen (1775), and other Academies of Sciences. He held the civil-service rank of councillor of state and was awarded the degree-four order of St. Vladimir,

---

[12]From the Polish "szlachta", meaning Polish landed gentry.—*Trans.*
[13]A "Kadet" formerly denoted a student of a military highschool.—*Trans.*

providing him an *entrée* into the hereditary Russian nobility. According to his contemporaries, J.-A. Euler fulfilled his duties with the greatest conscientiousness, ability, and fairness, was exceptionally hospitable, personable, and benevolent in his relations with people, and had many friends. In 1760 he married Anne-Charlotte-Sophie von Hagemeister (1734–1805), daughter of a Prussian royal counsellor and chief linen-keeper, who presented him with ten children.

Leonhard Euler's second son Carl (1740–1790) was a doctor. He was taken by his parents to Berlin as a one-year-old infant. There he attended school and was taught by private tutors and his father, from whom he learned the elements of philosophy and mathematics. Later he attended lectures on botany, anatomy, and physiology. In 1756 he took part in a botanical and mineralogical expedition through Germany, and in 1760 through Belgium. In 1762, having completed his studies at Halle, he qualified as a medical doctor, and in 1763 began to work as a general practitioner in the French community in Berlin. On returning to St. Petersburg with his father, Carl Euler was appointed court doctor to Catherine II ("imperial archiatric") and member of the State Medical Office, and from 1772 served as doctor to the Academy of Sciences ("ordinary[14] medic"). In 1779 he attained the civil-service rank of collegial councillor. His contemporaries commented on his erudition, conscientiousness, and experience. Incidentally, it should be noted that C. Euler's research on planetary motion, awarded a prize by the Paris Academy of Sciences in 1760, was unquestionably carried out with the direct involvement of his father.

In 1766, ten days before leaving Berlin, C. Euler married Anne-Emilie (Amalia) von Bell (1741–1830), daughter of a Prussian royal counsellor, who bore him 11 children. Carl Euler was buried with his wife in the Smolensk Lutheran Cemetery (in his father's tomb) [9].

L. Euler's elder daughter Catherine-Elena (1741–1781) was born in Berlin. In 1777 she was married to Carl-Josef von Bell (1744–1830), a first major in the Russian army, who had earlier been a Prussian army officer, but had moved to Russia in 1771, where he became chief quartermaster. He subsequently rose to the rank of colonel and was admitted into the Russian nobility.

The youngest of L. Euler's sons, Christofor (1743–1808), was also born in Berlin. In addition to attending school, he was taught by his father and eldest brother, after which he embarked on a career in the artillery. He served in the the seven years' war as a Prussian *Oberleutnant*, taking part in some of the campaigns. At first Friedrich II prevented his leaving for Russia with his father on the grounds that he was Prussian-born, and even had him imprisoned in the Küstrin fortress as punishment for the renewal of the petition for his release, but through the intercession of Catherine II in 1767 he was finally allowed to leave and reunite with his father in St. Petersburg. There he was enlisted in the Russian army at an appropriate rank. In 1769, at the behest of the Academy of Sciences, he was sent to Orsk fortress in the Urals to assist in the observation of the transit of Venus across the face of the sun; he also carried out astronomical observations in the town of Yaitsk, and in Zaporozhskaya Sech'[15], Kremenchug, Samara, Glukhov, Orenburg, Cherkassk, Taganrog, and other places. Christofor Euler fought in the Russo-Turkish war of 1770–1772, where he distinguished himself and was awarded a medal. In 1778, at the rank of artillery major,

---

[14]That is, regular, or State, doctor.—*Trans.*
[15]A former South Russian Cossack community.—*Trans.*

he was put in charge of the construction of the Sestroretsk arms factory. On the completion of this project in 1789, Catherine II had him promoted to major-general of artillery, and appointed him commander of artillery in the army then in action against Sweden. At the conclusion of peace with Sweden in 1790 he was kept on as commander of all artillery units disposed in Finland and fortifications along the Russian border. He was stationed in Vyborg. In 1783 a large fire broke out, but thanks to his presence of mind the powder magazine, containing around 11,000 poods[16] of gunpowder, was prevented from exploding, and the whole garrison and others were saved. In 1794 he was transferred to St. Petersburg, where in 1799 he retired at the rank of lieutenant-general. In retirement Ch. Euler lived at first on the estate Kozlov-Bereg that he had purchased in Minsk Province, and on selling it in 1803 acquired a house in Vyborg and, 10 kilometers away, the country estate Rakalaioka, where he died in 1808.

In 1776 Ch. Euler married Anne Winterstedt, who died in 1777 following her first childbirth. In 1778 he remarried. His second wife, Anna Sergeevna von Krabbe (1755–1813), was the daughter of an Estland[17] noble, and niece of the commander of Russian artillery General-in-Chief baron J. J. Meller-Zakomel'skiĭ. They had 11 children.

Euler's younger daughter Charlotte (1744–1780) was likewise born in Berlin. Following the family's return to Russia in 1766 she was married to baron Jakob van Delen (1743–1786), a former officer in the Prussian army who had sought her hand earlier in Berlin in 1763. However since his rank was that of cornet, it would then not have been possible to obtain permission to marry from Friedrich II, who had prohibited junior officers—ensigns and cornets—from marrying. Van Delen owned an estate near Aachen on the river Rur, a tributary of the Maas (or Meuse), and in 1770 he took Charlotte and their two sons to live there. She lived out the remainder of her life on her husband's property and was buried in the little church at Hückelhoven. Charlotte was the only one of Leonhard Euler's children to leave Russia.

It should be added that all three of Euler's sons took out Russian citizenship.

4. Leonhard Euler's grandchildren

The majority of L. Euler's 45 grandchildren were born while he was still living, many of them lived under his roof, and he taught some of them mathematics.

Four of J.-A. Euler's ten children were boys. The eldest, Johann-Leonhard (1762–1827) became a colonel of artillery in the Russian army; at one time he served as an economist in the Ekaterinskiĭ Institute in St. Petersburg. He in turn had two sons, the colonel Ivan Euler (1790–sometime after 1842), and the staff-captain Fëdor Euler (1796–sometime after 1832), neither of whom, it seems, married.

The second of J.-A. Euler's sons, Egor (or Georg) (1770–1831) served from 1784 to 1811 in the Ministry of Foreign Affairs as an archivist, translator, and ambassadorial secretary in Berlin and Hamburg, as well as Copenhagen, Altona, and Lübeck. He took part in M. I. Kutuzov's diplomatic mission to Constantinople. In 1812 Egor was appointed to the Arkhangel'sk customs office of the Department of Foreign Trade. He died childless from cholera in Arkhangel'sk.

---

[16]One pood = 36 pounds.—*Trans.*

[17]Formerly a province in northern Estonia.—*Trans.*

Johann-Albrecht's third son Christofor (1772–1847) never married and ended life as a retired sublieutenant. The youngest son of the four, Pavel (Paul) Euler (1778–sometime after 1815), was a retired ensign in the Guard, and worked in the Department of Forests (in 1800 he was *Forstmeister*[18] in Orël, and in 1814 collegial assessor[19] in the Karachaevsk District of Orël Province). The only thing known about his family is that his wife was Russian.

The eldest daughter of Johann-Albrecht Euler, Ekaterina Ivanovna (1761–1809), affectionately called "Trinetta" by her family, worked as class supervisor, or governess, in the Kindergarten of the First Kadet Corps in St. Petersburg, and one of her sisters, Anna Ivanovna ("Emilia") (1767–1831), held a similar position in the Smol′nyĭ Institute. Neither married.

Two other daughters of Johann-Albrecht, Albertina and Charlotte, married scholars. Albertina Euler (1766–1829) was born in Berlin and was taken in that same year to St. Petersburg. In 1784 she married Nikolaĭ Ivanovich Fuss (1755–1826) in St. Petersburg; Fuss had been invited in 1773 to come to St. Petersburg from Basel to work as Leonhard Euler's assistant. In 1783 he was elected academician in mathematics and from 1800 to 1826 was permanent secretary of the Petersburg Academy of Sciences. For ten years following his arrival in St. Petersburg he lived in Euler's home; he helped him prepare around 300 articles for publication. N. I. Fuss himself authored over 50 works in astronomy, geometry—in particular spherical geometry—, series summation, differential equations, mechanics, and fortification. He obtained Russian citizenship in 1799 and was granted the civil-service rank of councillor of state, which according to the "Table of ranks" was equivalent to the military rank of major-general. Both Fuss and his wife were buried in the Smolensk Lutheran Cemetery.

Charlotte Euler (1773–1831), named after her mother, was born in St. Petersburg in 1789, and married Daniel Bernoulli's nephew, the young academician and mathematician Jakob II Bernoulli (1759–1789), who had come to St. Petersburg in 1786. (Daniel Bernoulli had worked in St. Petersburg from 1725 to 1733.) Charlotte's marriage lasted only a very short time, as her husband drowned while swimming in the Neva in that very year 1789. After a decent interval Charlotte remarried, taking as her second husband the pastor of the Reformed Church in St. Petersburg John David Collins (1761–1833), whose background was Scottish, though he came to Russia from Prussia. He subsequently ran a private boarding school.

Of the 11 children of Carl Euler only one of the boys, Leontiĭ (1770–1849), survived to adulthood. In 1779 he enlisted in the Imperial Guard, attached to the Preobrazhenskiĭ Regiment, and retired from service in 1791 at the rank of captain. He then taught in the Kadet Land Corps, worked as an accountant in the State Bank of the Municipal Administration of Pavlovsk[20], and as an inspector in the Riga Customs. During the years 1811 and 1812 Leontiĭ Karlovich Euler worked as a supervisor in the economics department of the *lycée* at Tsarskoe Selo[21]. He is mentioned in the "national songs" of the first-year *lycéens* who were

---

[18] Forestry superintendant.—*Trans.*

[19] A civil-service rank at the eighth level in Tsarist Russia.—*Trans.*

[20] A town about 30 kilometers outside St. Petersburg.—*Trans.*

[21] The "Tsar's Village" outside St. Petersburg. Aleksander Sergeevich Pushkin, Russia's greatest poet, attended the *lycée* there, founded by Aleksander I with the object of educating youths from the best families.—*Trans.*

Aleksander Christoforovich Euler
From a portrait in the War Ministry's collection

classmates of A. S. Pushkin. Later he worked as a translator in the Petersburg Customs. He had a reputation as a most honest person.

Carl Euler's daughter Anna (1770–1822) was married in 1793 to the Russian army major baron Wilhelm van Delen (ca. 1768–1810). Her sister Elizabeth (1780–1852) married Karl Ambrosievich Foigt (1762–1811), a philosophy professor at the University of Kazan'. Following his death she took as second husband J. O. Braun (1774–1819), another professor at that university, and then a third, Professor F. I. Erdman (1793–1862).

Ekaterina Karlovna Euler (1772–1845) married Carl-Friedrich Boltenhagen (1768–1831), councillor of state and director of the Commercial Bank. They have direct descendants living today. Note that Ekaterina Euler-Boltenhagen is indeed a daughter of Carl, and not of Johann-Albrecht, as has sometimes been claimed; this correction eliminates certain contradictions that had arisen concerning various dates in the lives of J.-A. Euler's children.

Among the children of Ekaterina-Elena (or Catherine-Elena) Euler-Bell, there was only one son, Andreĭ Bell (1779–1865); he became a colonel in the artillery.

The best known of the children of Christofor Euler was his eldest son Aleksander Euler (1779–1849), a general in the artillery. He received his early mathematical training from his father. He then attended Wedemeier's boarding school in St. Petersburg. N. I. Fuss coached him in mathematics. In 1790 he enlisted in the army as a sergeant. He served in the war in Finland as his father's adjutant. In 1791 he began studying at the gymnasium in Vyborg, and from 1794 to 1796 served in the Second Gunship Regiment and the artillery galleys of the Baltic fleet, and later in an artillery battalion of the Imperial Guard. With that same battalion he took part in the campaign against the French, being awarded the third-degree Anna[22] for distinction in the battle near Austerlitz. In the campaign of 1807 he

---

[22] A medal.—*Trans.*

took part with his battalion in many battles in Prussia, including those of Gutenstadt, Heilsberg, and Friedland, earning the fourth-degree St. Vladimir with ribbon. From 1810 he was
commander of the artillery battalion of the Imperial Guard, as well as all artillery companies stationed in St. Petersburg, Moscow, Smolensk, and Arkhangel'sk, and the gunpowder
works at St. Petersburg, Bryansk, and Okhtensk. In the Patriotic War of 1812 he fought
with his battalion in the campaigns of Vilna (now Vilnius), Drissa, Dvinsk, Mozhaisk, and
Tarutina[23]. At the battle near Borodino[24] he commanded M. I. Kutuzov's artillery reserve
of 30 companies and 360 cannon. For valor, verve, and organizational ability he received
the third-degree St. Vladimir, and in December 1812 was promoted to major-general. He
took part in the battles near El'nya, Krasnyĭ, Vilna, and at the river Berezina[25]. He continued to command the Guard artillery during the ensuing foreign campaign. For his part
in the battle of Leipzig he was awarded the second-degree Anna with diamonds. In Basel
he was received triumphantly as a grandson of the great Leonhard Euler. In 1826 he was
promoted to lieutenant-general, in 1831 became a member of the council of Chief Head-
quarters concerned with military settlements, in 1833 vice-director and in 1834 director of
the Artillery Section of the War Ministry and general of artillery, and in 1840 a member of
the War Council of the Ministry. In addition to all that A. Ch. Euler served on several committees and commissions relating to the workshops, the lathes used in arms manufacture,
the vetting of metals, etc. According to his contemporaries, he was possessed of a quick
and clear intelligence and exceptional administrative ability, knew his business, was very
experienced, zealous in service, just, respected by his superiors, and was concerned that the
soldiers be properly fed and equipped.

In Kiev in 1804 Aleksander Christoforovich Euler married his cousin Elizaveta Niko-
laevna Gebener (1785–1844), daughter of the commander of the Okhtensk gunpowder
works major-general Nikolaĭ Andreevich Gebener and Elizaveta Sergeevna Krabbe, his
mother's elder sister.

Christofor Euler's second son Fëdor (1784–sometime after 1835) served as major-
general in the artillery. The next younger brother, Pavel (1786–1840), graduated from the
First Kadet Corps in 1803 and began military service as an ensign in the Ryazan' Infantry
Regiment. In 1805 he took part in a military expedition to Rügen Island in the Baltic and
in a march through Swedish Pomerania as far as the river Bezer. In 1806 he fought against
the French near the rivers Bug and Narev and in 1807 at a crossing of the Kaltflüss; he
fought in the battles of Preussisch-Eylau, Scharpick, Heilsberg, and Friedland, for which
he was awarded the fourth-degree Anna. In 1808 he was again in Swedish Pomerania at
the seige and capture of Sveaborg (the fortress of Svea), and in 1809 took part in the expul-
sion of the enemy from the Aland Islands. During the Patriotic War of 1812 he fought in
the battles of Gedianov and Borodino, earning promotion to captain, and then at Tarutina,
Maloyaroslavets, Vyaz'ma, and Krasnyĭ. In 1813 he fought in the Duchy of Warsaw, and
took part in battles near Kalisch, in Saxony, and near Lützen (for which he received the

---

[23] Places on Napoleon's route to Moscow.—*Trans.*

[24] A village about 70 miles west of Moscow, near which on September 7, 1812 Napoleon's Grande Armée
fought a bloody but largely indecisive battle with the Russian army under General M. I. Kutuzov, prior to entering
Moscow and waiting in vain for the Russians to formally capitulate. The estimated 45,000 Russian casualties
included prince P. I. Bagration, Commander of the Russian Second Army—and of course prince Andreĭ Bolkon-
skiĭ.—*Trans.*

[25] In crossing which (in November 1812) the French suffered around 36,000 casualties.—*Trans.*

fourth-degree St. Vladimir with ribbon), at Hesselbach in the land of the Czechs, at Fort Konstein, and Gross-Kotte, at Kulm (where for distinguished service he was promoted to major), and near Elsin, Estenbeim, and Leipzig (for which he received the second-degree Anna). In 1814 he was in France. In 1815 he participated in the second French campaign, passing through Galicia, Austria, Moravia, and Bavaria as far as the town of Vertus. In 1817 he was made battalion commander, and in 1823 commander of the Eletsk Infantry Regiment (and was awarded the fourth-degree order of St. George). In 1827, by reason of illness, he retired at the rank of colonel in uniform. However three years later, in 1830, he again enlisted, this time in the Gendarmerie. In 1839 he was transferred to the cavalry as a major-general attached to the Ministry of Internal Affairs.

Christofor Euler's fourth son, Constantine (1788–1863) served as an officer in the hunter regiments, took part in military campaigns, was awarded a gold sword carrying the inscription "for valor" and a medal for his part in the taking of Paris in 1814. In 1823 he was allowed to resign from military service "because of wounds", at the rank of colonel. He then held the position of governor of various towns in succession in the Vitebsk and Minsk provinces, and in 1840 finally retired permanently.

Of Christofor's four daughters, two remained unmarried, and the other two, Blandina (1780–1868) and Aleksandra (1789–1843) both married colonels in the artillery: Aleksander Petrovich Beckmann (1772–1851) and Fëdor Karlovich Reren (1788–1851).

The following is all that is known about the children of Charlotte Euler and Jakob van Delen. Their eldest son Leonhard-Albrecht-Carl (1767–1821) was initially his uncle Christofor's adjutant, and then in 1794 was sent to serve in Holland. Later he worked as chief engineer building roads and bridges in Westphalia, and in 1814 held the rank of major in the Prussian army. The year 1825 finds him again serving in Holland, this time as inspector of waterworks construction. In 1809 be became corresponding member, and in 1816 full member, of the Academy of Sciences of the Netherlands. In 1803 he married his cousin Klasine van Delen.

Charlotte's second son, Johan-Kristofel-Wilhelm van Delen (1768–1810) served as a military engineer in Russia at the rank of major.

The youngest son, Johan-Casper-Ferdinand Delen (1779–1872) was at first a company commander, and then retired from the army and became a Dutch Burgomeister.

One of Charlotte's daughters, Sophie (b. 1777), married a company commander in the Russian cavalry by the name of Friedrich-Carl Lobedank, and moved to Russia.

(In the original there follow six further sections, omitted from the present translation. Their titles are respectively: "Leonhard Euler's great-grandchildren"; "Leonhard Euler's great-great-grandchildren"; "L. Euler's descendants: the Eulers"; "The Fusses, Struves, Ottens, Chernaïs, and Elpat'evskiïs"; "The Collinses, Berendtses, and Gekkers"; and "Other branches of L. Euler's descendants".)

Conclusion

It may well be asked why two of L. Euler's distant descendants—members of the Soviet scientific and technological intelligentsia and far from having any expertise in history or

A group of L. Euler's descendants in the Conference Hall of the
Leningrad building of the Academy of Sciences, October 27, 1983

First row (left to right): A. A. Shestakov, A. A. Euler, N. N. Euler;
Second row: D. A. Euler, S. N. Euler, D. A. Shestakov;
Third row: M. N. Afanasov, I. R. Gekker, K. G. Manuilov

literature—would spend so much time searching in archives and libraries for relevant doc-uments, including the results of the searches of other equally distant descendants, and then ordering the materials thus unearthed. Of course this project was initially stimulated by a personal desire to sort out our confused family tree. Later the broader interest arose of attempting to understand the mutual historical connections between generations.

Once transferred to Russia, Euler's family—originally of peasant stock (in Lindau), then craftsmen, petit-bourgeois, and clergy (in Basel)—became one of scholars and teach-ers, civil-servants and military men—all Russian citizens from the second generation on. The occupation of important scientific, social, government, and military posts, and their patriotic service, helped to consolidate the Eulers' position in Russia and draw them closer to other members of our native intelligentsia. Since they did not, as they say, seek brides or grooms abroad, they married with representatives of a variety of Russian families. Almost all of Leonhard Euler's descendants lived in Russia, became thoroughly assimilated, and actively participated in Russian life.

The number of Leonhard Euler's descendants that are known to us is above a thousand. Over a span of nine generations, around 160 of these were given the name Euler at birth. We know the names of about 400 descendants now living, of whom over half are in the USSR.

In our work on this article many of Leonhard Euler's descendants helped us with advice and documents: M. N. Afanasov, S. N. Bruns, R. F. Gekker, S. F. Gekker, W. Obolensky,

N. N. Savko, E. S. Skalon, S. N. Skalon, M. V. Shestakova. A. B. Euler, N. N. Euler, T. V. Euler, G. G. Enkel, and others; to all of these we express our gratitude. We acknowledge also the help given us by E. N. Amburger, Yu. Kh. Kopelevich, E. P. Ozhigova, V. E. Pavlov, and the editors of the present collection of essays.

Postscript. On June 9, 1987, following the completion of the manuscript of this article, one of its authors, and the last living representative of the sixth generation of Leonhard Euler's descendants, Aleksander Aleksandrovich Euler, died. He was born on November 22/9, 1913 into the family of A. N. Euler, a postal and telegraph inspector. In 1938 he graduated from the Leningrad Institute of Railway Engineers, and in 1948, after working as a section engineer in signals and communications along the Yaroslavl line, began teaching in that institute, preparing specialists in automata and computing technology. He had a candidate of technological science degree[26] (1949), was promoted to *docent* (1954), and then to professor (1966), chaired the departments of Automata in Rail Transportation and Electronic Computing Machines, and was Dean of the Electrotechnological Faculty of his institute. A. A. Euler carried out scientific research on the construction of automatic systems in rail transport, wrote several textbooks and manuals, and taught courses at technical colleges in Peking and Prague. Aleksander Aleksandrovich took part in the jubilee celebrations held in honor of Leonhard Euler in 1933, 1957, and 1983.

References

[1] Euler, K. *Das Geschlecht Euler-Schölpi: Geschichte einer alten Familie.* Giessen: W. Schmitz, 1955.

[2] Burckhardt, F. "Zur Genealogie der Familie Euler". Verhandl. Naturforsch. Ges. Basel, 1908. Vol. 19. pp. 122–138.

[3] Kopelevich, Yu. Kh. "Materials for a biography of L. Euler". Ist.-mat. issled. (Research in Hist. Math.) 1957. Issue 10. pp. 9–65.

[4] "Notes of A. A. Euler". Rus. arkhiv. 1980. Book 2. pp. 333–399.

[5] Chernov, S. N. "Leonhard Euler and the Academy of Sciences". In: *Leonhard Euler.* M.; L.: Izd-vo AN SSSR (Publ. Acad. Sci. USSR), 1935. pp. 163–238.

[6] Amburger, E. N., Gekker, I. R., Mikhaĭlov, G. K. "Genealogy of L. Euler's descendants". In the present collection[27].

[7] Obolensky, W. "Reh oder Einhorn? Das Wappentier Leonhard Eulers". In: *Jahrb. Schweiz. Ges. Familienforsch.* 1986. pp. 141–154.

[8] Raith, M. "Der Vater Paulus Euler". In: *Leonhard Euler.* Basel: Birkhäuser, 1983. pp. 459–470.

[9] Shestakova, M. V. "Leonhard Euler's descendants through his son Carl". In the present collection[28].

[10] *A. S. Popov in the characterizations and recollections of his contemporaries.* M.; L.: Izd-vo AN SSSR (Publ. Acad. Sci. USSR), 1958.

---

[26] Approximately the same as a PhD.—*Trans.*

[27] Omitted from the English translation.

[28] Omitted from the English translation.

Index

Abel, N. H., 28, 144
Abu Kamil, 138
Academy of Sciences
  Berlin, 5, 22, 25, 55, 75–86, 88, 90n, 245,
    246, 290, 293, 294, 301, 302, 305, 317,
    370, 377, 378, 381, 403, 406
    Mémoires of, 177
    Notes of ,139, 284, 294, 296, 297, 302,
      377, 381, 386
    observatory of, 284, 285, 406
    and Euler's collected works, 56, 57
  of the USSR, 39, 86, 366, 398
    Archive of, 122, 128, 328, 349
  Göttingen, 396
  Minsk, 396
  Paris, 4, 10, 23, 30, 43, 44, 46, 69, 168n, 172,
    247, 254, 255, 256, 289, 291, 292, 293,
    294, 305, 313, 317, 342, 370, 372, 376,
    390, 392n, 396, 408
    Mémoires of, 273
  Petersburg, 3, 10, 12, 14, 15, 25, 39–51, 81,
    121, 125, 128, 168, 172, 176, 178, 245,
    246, 249, 263, 270, 285, 301, 305, 317,
    318, 321, 326, 332, 370, 373, 375, 377,
    386, 396, 398, 402, 403, 404, 406, 407,
    410
    Academic Conference of, 42
    and Euler's collected works, 53-75
    Archive of, 53, 64
    structure of, 41
    observatory of, 269, 270, 271, 272, 274,
      275, 284, 285
      Time Service provided by, 275, 276, 277
  Turin, 302
action at a distance, 310
Adami, J., 295
Adodurov, V., 47
Aepinus (or Epinus), F. U. T., 25, 48, 76
aerodynamics, 23
algebra, 34, 100–101 (in the notebooks), 137–
  152, 153, 154, 159, 161, 162, 389
  commutative, 150
  geometrical, 137
  numerical, 137

Alhazen's problem, 111
al-Karaji, 138
Ampère, A. M., 314
analysis, 24, 28, 29, 76, 168, 183, 192, 292, 375,
  383, 393
  complex, 15, 29, 34, 109
  Diophantine, 100, 124, 130, 132, 139, 147,
    153–165
Anna Ioannovna (empress of Russia), 17, 41, 47,
  403
Anna Leopol'dovna (mother of Ioann VI), 17,
  403
anthropic principle, 267–268
Apollonius' tangent problem, 111
Arbogast, L., 28
Archimedes
  law, 23
  snail, 113
  spiral, 113
argument
  over logarithms, 27, 294–295,
  over the Leibniz-Wolff philosophy, 82, 83,
    379–380
  over the principle of least action, 82, 83, 189,
    190, 386
  over the vibrating string, 27, 180, 295–297,
    303, 386
Aristoksen, 355
Aristotle, 184, 277, 278
arithmorrhoea, 208n
astronomical tables, 63, 256, 390, 391
  Cassini's, 256
  Halley's, 256
astronomy, 6, 17, 115–116 (in the notebooks),
  128, 245–262, 263–268, 307, 373, 384
  applied, 259–260
Augustine, St., 281
aurora borealis, 248, 271, 277, 278
Auwers, A., 57

Backlund, O. A., 58, 59, 60, 61, 66, 68, 127
ballistics (artillery), 22, 23, 114, 239, 241–244,
  378
  trials, 42

Basel University, 3, 8, 9, 349, 371, 402

Bauschinger, J., 61

Bayer, T. S., 283

Belyaev, N. M., 180, 231, 327, 328

Bernoulli, Daniel, 3, 10, 11, 12, 23, 26–28, 37, 40, 43, 45, 115, 176, 177, 179, 213, 214, 226–228, 229n, 247, 271, 274–276, 281, 284, 290, 291, 293, 294, 297, 303, 323, 324, 326, 371, 374, 376, 382n, 386, 392, 402, 410

Bernoulli, Jakob, 4, 8, 11, 29, 185, 213, 217, 370, 375, 400, 402

   second problem of (on isoperimetries), 113

Bernoulli, Jakob II, 410

Bernoulli, Johann I, 3, 4, 9, 10, 26, 27, 29, 45, 53, 174, 176, 177, 185, 193, 243, 290, 294, 301, 356, 357, 371, 375, 377, 386, 402

   bending law of, 179

Bernoulli, Johann II, 9, 309

Bernoulli, Johann III, 89n, 362

Bernoulli, Nicholas I, 9

Bernoulli, Nicholas II, 3, 9, 10, 11, 40, 54, 67, 291, 371

Bertrand's postulate, 124, 133

Bessel, F. W., 54

Bessel's equation, 231

beta function, 17

Bettstein, K., 77

Bézout, E., 143

Bilfinger, G. B., 43, 81, 374

Biron (or Biren) (Duke of Courland), 17, 403

Blumentrost, L. (president of the Petersburg Acad.), 12, 22, 40, 41

Bolotin, V. V., 225, 231

Bobylev, D. K., 127

Boetius, A., 355

Bombelli, R., 138, 145, 146

Bougainville, L.-A., de 75

Bouguer, P., 291, 380n

brachistochrone, 114, 185, 373

Bradley, J., 76

Brouncker, W., 104n

   interpolation method of, 104

Brown, E. W., 253

Brucker, Margareta (Euler's mother), 370, 400

Bruckner, I. (technician at Petersburg Academy), 45

Buffon, G. de, 75, 267

Bunyakovskiĭ, V. Ya., 54, 55, 56, 120, 127

Burja, A., 366

Busch, H. R., 345, 351, 355n

calculus, 373, 374, 384

differential, 7, 26, 29, 67, 76, 105 (in the notebooks), 130, 291, 384

integral, 7, 26, 29, 49, 105–109 (in the notebooks), 130, 291, 294, 303, 304, 373, 384, 385, 389

variational (isoperimeties), 7, 17, 29, 76, 98, 113, 114, 176, 183–212, 219, 292, 300, 302, 303, 375, 377–378, 384

calendars (or almanacs)

   of the Berlin Academy, 22, 80, 83, 90–94

   history of, 269, 283

canonical (or contact) transformations of phase space, 197, 198, 199

Capon, R. S., 203

Cardano's rule, 100

Carnot's thermodynamical principle, 207

cartography, 5, 15, 45, 116, 269, 282

Cassini, J., 273

Catherine I (empress of Russia), 371, 373

Catherine II (empress of Russia), 25, 31, 307, 315, 320, 330, 382, 387, 408

Cauchy, A. L., 179, 200, 305

Cauchy-Lagrange integrals, 178

Cauchy-Riemann equations, 7, 34, 295

Cauchy's theory of light, 201

Cavendish H., 76

Celsius, A., 76, 277

Chaplygin, S. A., 202

Chebyshev, P. L., 37, 38, 55, 56, 120, 127

Chelomeĭ, V. N., 180, 231

Chetaev, N. G., 200, 203, 207

chromatic aberration, 24, 247, 274, 384–385

Clairaut, A.-C., 8, 12, 24, 37, 98, 112, 113, 172, 174, 177, 252, 269, 289, 290, 296, 299

   correspondence with Euler, 291–294

Cotes-Euler formula, 7

Cotes' theorem, 106

combinatorics, 101 (in the notebooks)

comets, 6, 59, 247, 248, 250, 251, 264, 279, 389

   masses of, 267

   tails of, 278

Commentarii of the Petersburg Academy, 11, 13, 14, 17, 25, 34, 46, 168, 270, 276, 323, 373, 375

Condorcet, M.-J. A.-N. de, 63

continued fractions, 17, 101–105 (in the notebooks)

coordinates

   Cartesian, 172, 173, 174, 239

   generalized, 192n, 193

   heliocentric, 256

   natural, 169

Copernicus, N., 281

Courtivron, G. de, 189
Cramer, G., 8, 12
Crelle, A. L., 54
critical load (of a column), 220
Croze, M. V. de la, 283
curvature, 7, 112, 214
curves, 383
  algebraic, 153
  birational equivalence of, 162, 163
  cubic, 144–148, 153, 154
  elliptic, 147, 163
    arithmetic of, 147, 154, 159
  rectifiable, 106
  rectification of, 112

d'Alembert, J. le Rond, 8, 12, 23, 24, 26, 27, 30,
  35, 37, 54, 95, 110, 115, 139, 170, 172, 173,
  174, 183, 186, 189, 190, 192, 252, 269, 289,
  290, 293, 299, 303, 304, 386
  correspondence with Euler, 294–298
d'Alembert-Lagrange principle, 192, 206, 207
d'Alembert's
  paradox, 177n
  principle, 170, 171, 193, 207, 226, 229n
Darbès, J. F. A., and his portrait of Euler, 361–
  368
D'Arcy, P., 189
Dashkova, E. R., (princess and director of the
  Petersburg Academy) 50
Dedekind, R., 150
Delacroix, 45, 381
Delaunay, C.-E., 253
Delisle, J. N., 12, 15, 43, 116, 270, 271, 273,
  274, 275, 276, 277, 281, 282, 283, 284, 285,
  289
Descartes, R., 9, 35, 38, 80, 82, 106, 139, 184,
  185, 278, 279, 280, 309, 311, 312, 337, 378,
  379
Dickson, L. E., 130, 131
Diderot, D., 190
Dinnik, A. N., 221
Diophantus, 137, 138, 145, 146, 153, 154, 155,
  158, 160, 162, 163
dioptrics, 24, 388, 389, 392, 393
Dirichlet, P. Lejeune-, 150
Dolgorukiĭ, V. S. (prince), 87, 387
Domashnev, S. G. (director of the Petersburg
  Academy), 50, 321, 329
Dorfman, Ya. G., 310, 312
Duval, F., 362

earth, 280

artificial satellites of (ASEs), 258–259, 266
atmosphere of, 274
shape of, 259, 280, 281, 291
orbit of, 265, 284
Einstein, A. 82, 209
elastic
  curve, 216, 218, 222, 229
  (or bending) rigidity, 214, 217
  stability, 219–226
Elizabeth I (empress of Russia), 18, 403, 404
energy
  conservation of, 191, 193n
  kinetic (or "live force"), 188, 191, 194
  potential, 213
Eneström, G., 59, 60, 61, 62, 121, 123, 128, 130,
  289n, 350, 393n
enlightenment, 8, 80, 83, 298, 299, 307
ephemerides, 263
Epicurean atoms, 379
equations
  algebraic, 130, 100–101 (in the notebooks),
    137
    resolvent of, 144
    solution by radicals of, 143–144
  difference-differential, 109
  differential, 7, 17, 23, 26, 37, 109–111 (in the
    notebooks), 171, 239, 240, 247, 292, 298
    metrical theory of, 258
  Diophantine, 27, 35, 99, 130, 137, 139n, 146,
    151, 153–165
  hydrodynamical and continuity equations for
    an incompressible fluid, 177
  integro-differential, 230
  of a rotating rigid body, 174, 175, 386
  of mechanics, 174
  Riccati's, 109
Eratosthenes, 133
Erman, J. P., 90
ether, 308–315
Euclid, 149
Euclidean space, 258
Eulerian
  coats of arms, 400
  numbers, 7
Euler
  angles, 7
  characteristic, 7
Euler, Albertina (a granddaughter, and wife of
  N. I. Fuss), 410
Euler, Aleksander Christoforovich (or A. Kh.) (a
  grandson), 397, 400, 410–412
Euler, Aleksander Aleksandrovich, 415
Euler, Anna-Maria (elder sister), 400, 401

Euler, Carl (or Karl) (second son), 17, 398, 402, 406, 408, 410

Euler, Catherine-Elena (elder daughter), 408

Euler, Charlotte (younger daughter), 409

Euler, Christofor (third son), 30, 31, 398, 402, 405, 406, 408–409

Euler, Johann-Albrecht (eldest son), 17, 25, 31, 32, 35, 49, 50, 62, 84, 95, 302, 320, 320, 329, 391, 392n, 396n, 398, 402, 405, 406–408, 410

Euler, Johann-Heinrich (brother), 17, 400, 401, 402, 404

Euler, Maria-Magdalena (younger sister), 400–401

Euler, N. P. (a great-grandson), 397

Euler, Paul (father), 370, 400

Euler-Lagrange equation (of the calculus of variations), 190, 191, 204, 215, 300

Euler-Lambert formula, 264

Euler, Leonhard
  ancestry, 399–402
  and music, 14, 117, 335–347, 349–360, 375
    critical reception of his theory of music, 344–345, 357
  and science, 6, 87, 123, 245, 254, 260, 263, 278, 281, 317
  as astronomer, 269–288
  children, 406–409
  background, 92, 399–402, 414
  Berlin period, 75–86, 87–96, 245, 289, 307
  blindness, 17, 248, 304–305, 374, 388, 391, 392, 394, 405
  character, 22, 38, 83, 87, 93, 122, 123, 260, 263, 366, 374, 391, 393n, 394, 395, 402, 406
  contretemps
    with d'Alembert, 294–297
    with Friedrich II, 31, 93, 84, 298, 404
    with Lagrange, 302–303
    with Lambert, 87–96
  correspondence, 12, 35, 45, 53, 55, 83, 85, 122, 123, 127, 128, 131, 260, 282, 289–306, 376–377
  expository style, 29, 178, 246, 299, 303
  first St. Petersburg period, 3, 11, 21, 39–47, 129, 237–240, 245, 289
  grandchildren of, 409–413
  gratitude towards Petersburg Academy, 39, 93, 285
  hundredth anniversary of his death, 120
  library of, 124
  mathematical symbolism, 7
  notebooks, 14, 97–118, 119–126, 129
  classification of, 97, 123
  portrait in the Tret′akovskiĭ Gallery, 361–368
  prolific output of, 5, 13, 25–26, 49, 80, 120, 167, 298, 307n
  residences
    in Berlin, 77, 404
    in St. Petersburg, 402, 405
  second St. Petersburg period, 13, 33, 48–50, 129, 248, 289
  theorem on regular polyhedra, 111
  textbook on arithmetic, 4, 15, 376
  two-hundred-and-fiftieth birthday, 86, 245
  two-hundredth anniversary of his death, 84
  two-hundredth birthday, 56
  unpublished manuscripts, 35, 119–126, 127–135, 156
    on the theory of music, 349–360
  world-view, 5, 281
  youth in Switzerland, 3, 8, 370, 402

Euler-Schölpi family, 399

Euler's
  constant, 7
  function, 7, 124, 130
  identity, 132

Faber, G., 67

Fagnano, G. C., 55

Famintsyn, A. S., 56, 57

Faraday, M., 314

Fermat, Pierre de, 37, 145, 146, 153, 155, 158, 160, 162, 184, 185, 292, 373, 393

Fermat's
  last theorem, 130, 139, 148–151
  little theorem, 130, 131
  principle, 184, 195

Fétis, F., 345

Foncenet, D., de 142

Fontaine, A., 97, 291

force, 238

Formey, S., 84

Fourier, J. B. J., 7, 28, 34, 36, 264, 297, 298

Frenicle, F., 111

Fresnel, A. J., 200, 314

Friedrich I, 90n

Friedrich II (the great), 17, 22, 47, 76, 78, 79, 80, 83, 88, 95, 189, 245, 285, 290, 294, 298, 307, 357, 377, 382, 385n, 403, 404, 405, 408

Friedrich-Wilhelm I (father of Friedrich II), 22

Friedrich-Wilhelm II, 361

Frobenius, G., 56

Fueter, R., 67, 69

Fufaev, N. A., 203

functions

algebraic, 383
analytic, 300
abelian, 147
complex, 106
elliptic, 55
exponential, 103, 295
gamma, 17
hyperbolic, 105
logarithmic, 27, 104, 105, 294, 295, 296, 383
    application to musical theory, 338, 356
symmetric, 140
transcendental, 383
trigonometric, 105
Fundamental theorem of algebra, 26, 138, 139-143, 294
Fuss, N. I., 24, 32, 34, 38, 49, 53, 63, 119, 127, 315, 320, 321, 322, 323, 324, 326, 329, 349, 366, 405, 410, 411
    eulogy in memory of Euler 119, 369–396
Fuss, N. N., 49, 53, 56, 127
Fuss, P. N., 32, 34, 49, 53, 54, 55, 56, 61, 62, 64, 67, 84, 120, 127, 398

Galileo, 242, 311n, 393
Galileo's law of inertia, 194
Galois, E., 140, 144
Gauss, C. F., 54, 131, 134, 142, 143, 150, 206, 246, 250, 251, 258
    principle of, 207
geodesy, 115–116 (in the notebooks), 259, 269, 282
geography, 117 (in the notebooks), 307
    department of, Petersburg Academy of Sciences, 260, 282, 376
geometry, 111–112 (in the notebooks)
    algebraic, 137, 147
    differential, 98, 112–114 (in the notebooks)
Girard, A., 139
Glarean, 349
Goethe, J. W. von, 361
Goldbach, Ch., 11, 26, 27, 40, 53, 98, 99, 102, 103, 105, 131, 132, 148, 163, 290, 342, 351, 355
    identity of, 101
    theorems of, 102, 103
Golitsyn, B. B., 56, 58, 60
Golovin, M. E., 17, 49, 315, 320, 321, 329, 393, 396n, 405
Gottignies, G. F., de 98
Graewenitz, F., 243
Grube, F., 68
Gsell, Catharina (Euler's first wife), 17, 396, 402
Gsell, Georg (Euler's father-in-law), 401, 402

Gsell, Salome-Abigail (Euler's second wife), 32, 396, 405
Gutzmer, A., 66, 67

Hagen, J., 60, 61
Haller, A. von, 76
Halley, E., 273
Halley's comet, 267, 406
Hamiltonian (of a dynamical system), 196, 198
Hamilton-Jacobi partial differential equation, 199, 205
Hamilton-Ostrogradskiĭ principle, 201
Hamilton's
    principle, 195–201, 204, 205, 208
        second form of, 196, 198
    dynamical equations, 196, 197
Hamilton, W. R., 184, 191, 196, 197, 198
Harriot, T., 100
Heinsius, G., 15, 82, 260, 284
Helmholtz, H., 209, 338, 345
Helvetic (or Swiss) Society of Natural Scientists, 35, 57, 58, 69, 70, 84, 128
Hermann, J., 11, 14, 42, 43, 81, 168, 170, 193, 207, 238, 243, 271, 291, 373, 374
Hertz, H., 201, 207
Hilbert, D., 145, 150
Hill, G. W., 171, 252, 263, 265
Hipparchus, 116
histories of the Euler family, 397–398
Hölder, O., 201, 203, 204
Hooke's law, 179, 213
Hooke, R., 213
Hurwitz, A., 145
Huygens, Ch., 184, 240, 276, 337, 373
Huygens'
    experiments on navigation, 380
    principle, 184, 196
    wave theory of light, 200
Huygens-Fresnel principle, 184n
hydrostatics, 23, 115, 392

infinite products, 101 (in the notebooks)
Inokhodtsev, P. B., 49, 322, 396n,
integers
    algebraic 148–151
integral
    theory of, 305
    double, 301
    elliptic, 26, 147, 161, 223, 294
    improper, 305
integration (quadrature), 112, 113, 291
    numerical, 259–260, 266
Ivan V (brother of Peter I), 40

Jacobi, C. G. J., 54, 55, 64, 84, 120, 147, 161, 183, 191, 193, 198, 199, 258
  principle of, 194, 195
Jacobi, M. H., 54
Jeffries, H., 203
Jupiter, 5, 172, 255, 265, 296
  eclipse of moons of, 271

Kant, E., 35, 76, 80, 82
Kantemir, A. D., 271
Keplerian
  ellipse, 247, 253
  motion, 265
Kepler, J., 249, 272, 280, 281
Kepler's equation, 249
Kerner, M., 202
Kies, J., 80, 285
Kirch, Ch., 285
Kircher, A., 355
Kirchhoff, G. R., 218, 223
Kirnberger, J., 345
Kleist, G. von, 76
Klügel, G. S., 393
Knobelsdorf, G. von, 78
Kobold, H., 61
Köhler (or Koehler), D., 83, 84, 91, 95
Kolmogorov, A. N., 258, 299
König, S., 82, 189, 190, 386
Korff, J. A. (president of the Petersburg Academy), 41, 43, 44, 45, 282
Kotel′nikov, S. K., 25, 47, 48, 315, 320, 387n, 396n, 404
Kraf(f)t, G. W., 11, 35, 271, 272, 275, 276, 277, 278, 279, 283, 284
Kraf(f)t, W. L., 49, 254, 320, 321, 322, 329, 389, 391, 392, 396n
Krazer, A., 59, 60, 61, 67
Kronecker, L., 143, 150
Krylov, A. N., 36, 64, 68, 70, 71, 222, 293
Kulibin, I. P., 5, 24, 49
  and plans for a bridge over the Neva, 317–327
  and manufacture of Euler's achromatic telescope, 327–329
  and his water-powered boats, 330–332
Kummer, E., 150
Kütner, S., 362, 366
Kutuzov, M. I., 412

Lagrange, J.-L., 8, 12, 26, 27, 32, 36, 53, 98, 103, 105, 106, 114, 133, 140, 141, 142, 143, 148, 170, 175, 176, 183, 184, 185, 186, 188, 190, 191, 192, 193, 222, 223, 250, 251, 253, 255, 257, 258, 289, 290, 292, 297, 378, 384
  correspondence with Euler, 298–305
  principle of, 191, 201
Lagrange multiplier, 205, 208, 214
Lagrangian, 195, 203
  variables, 178
  symbolism, 192
Lambert, J.-H., 8, 12, 76, 87–96, 98, 105, 111, 250, 290
Laplace, P.-S., 30, 76, 142, 246, 251, 253, 255, 269, 303
Laplacian determinism, 264
Lavoisier, A. L., 75
Lavrent′ev, M. A., 224
Law, J., 280
Lebedev, P. N., 248
Leibniz, G. W., 4, 8, 14, 22, 26, 27, 37, 38, 82, 99, 113, 185, 188, 190, 228, 294, 300, 301, 337, 342, 373, 377, 380, 384, 386, 393, 403
  philosophy (monadology) of, 80, 82, 89, 379
  views on music, 342, 351, 356
Lejeune-Dirichlet, P., 150
Lenin, V. I., 70
Leonhardi Euleri opera omnia, 53–73, 84, 101, 121, 128, 134, 167, 246, 289n, 407
Leonardo da Vinci, 213
Le Verrier, U., 253
Lexell, A. J., 32, 35, 49, 95, 247, 248, 254, 255, 305, 320, 321, 322, 329, 391, 392, 396n
l'Hospital, 185
libration, 257, 265, 304
Lidov, M. L., 268
Liebert, J., Ch. 284
Lie, S., 259, 267
Lie groups, 184, 199, 209
Linnaeus, C., 75
Linus, F., 98
Lobachevskiĭ, N. I., 38
Lobkowicz, J. C., 338
Lomonosov, M. V., 8, 15, 17, 25, 47, 59, 75, 76, 117, 122, 314, 315, 406
Lorentz group, 200
Louis XV, 280
Louville, J. E., 273
lunar
  atmosphere (absence of), 272, 274
  eclipse, 271, 272, 284
  motion, 24, 35, 76, 116, 171, 172, 252, 253, 254, 259, 265, 293, 294, 373, 389, 390, 391
    Hill-Brown theory of, 253
  occultations, 271

Lyapunov, A. M., 35, 56, 58, 60, 61, 64, 65, 66, 67, 68, 127, 251, 258, 292
Lyav, A. E., 222

machines, theory of 6, 45, 115, 317
Maclaurin, C., 8, 174, 239, 247
    theorem of, 106
Maclaurin-Euler summation formula, 7
Magnitskiĭ, L. F., 4
Malygin, S., 44
Markov, A. A., 56, 58, 60, 61, 64, 67, 68, 70, 127
Marpurg, F. W., 357
Mars, 116
Mascheroni, L., 67
mass (inertia), 238, 311, 379, 380
Mattheson, J., 344
Maupertuis, P. L. M., 22, 29, 30, 79, 80, 81, 83, 88, 188, 189, 193, 290, 294, 298, 300, 301, 386
Maxwell, J.-C., 200, 314
Mayer, J. T., 24, 76, 253, 390
mechanics (dynamics), 6, 36, 46, 62, 64, 128, 167–181, 183–212, 237–240, 243, 373, 374, 393
    analytical, 183, 184, 185, 191, 239, 299, 301
    celestial, 5, 23, 35, 59, 115, 167, 171, 239, 245–262, 263–268, 269, 284, 293, 295, 296, 373, 393
        integration of equations of, 257–259
        Euler-Cole method of, 259
    classical, 183, 237
    differential principles of, 206–207
    of an "elastica", 214–219
    of a point-mass, 14, 167, 168, 169, 186, 238, 239
    of a rigid body, 6, 17, 26, 71, 76, 114, 167, 169, 172, 173, 175, 239, 299, 317, 387, 393
    of continuous media, 167, 174, 176, 206, 207, 209
    of dynamical systems, 167, 171, 181, 183–212
        holonomic, 194
        nonholonomic, 201–205
    of elastic or flexible bodies, 23, 115, 169, 179, 180, 181, 191, 213–236, 393
    of fluids (hydrodynamics), 6, 17, 23, 167–181, 176, 180, 189, 291, 295, 373, 392
    oscillatory (vibrational), 114, 171, 180, 186, 200, 226–232, 295-297, 303, 304
    variational principles of, 183–212, 213, 228, 302
Merian, J. B., 89, 91

Mersenne, M., 337
meteorology, 260, 271, 280
Meyer, F.-Ch., 11, 271, 275, 277
Mises, R., 222
Modzalevskiĭ, B. L., 60, 61, 397
Moivre, A. de, 8, 106, 373
momentum, 238, 241
    angular, 174, 175, 176
    conservation of, 171, 176, 177
    generalized, 195
Müller, G. F., 46, 79, 86, 275, 277, 278, 279, 283
Musschenbroeck, P. van, 76

Napoleon I, 412n
naval science (seafaring), 4, 16, 17, 23, 44, 76, 116, 172, 173, 248, 279, 289, 372, 373, 380–382, 389
n-body problem, 171
Neĭmark, Yu. I., 203
Newcombe, S., 253
Newton, I., 4, 9, 14, 35, 37, 80, 81, 116, 154n, 167, 171, 238, 242, 243, 250, 272, 273, 280, 281, 301, 309, 310, 311, 314, 373, 374, 378, 384, 393
    and Principia mathematica, 14, 81, 167, 168, 171, 185, 238, 301, 310, 311, 374
Newton's
    corpuscular theory of light, 309
    laws of motion, 172, 173, 174, 183, 238–239
    theory of efflux, 279, 378, 379
    universal law of gravitation, 24, 252, 291, 293
Nikolaï, B. L., 225
Nikolaï, E. L., 218, 221, 225
Notes on the news, 46, 270, 272, 274, 276, 277, 279, 280
Novi commentarii of the Petersburg Academy, 47, 178, 301, 303n
Novoselov, V. S., 203
numbers
    amicable, 99, 124
    Bernoulli, 99
    distribution of primes, 99
    perfect, 99, 124
    polygonal, 124, 130, 132
    representation as sums of squares, 99, 130, 132, 133
number theory, 6, 17, 27, 37, 54, 68, 98–100 (in the notebooks), 127–135 (manuscripts), 292, 305, 375
    algebraic, 139, 150
nutation of the earth's axis, 76, 173, 266

optical-mechanical analogy 185, 195–201

Orlov, V. G., 31, 50, 393
Ostrogradskiĭ, M. V., 53, 54, 120, 195
Otto, J. C. F., 243
Ozanam, J., 99

Pappus, 101
Pars, L. A., 203
Pascal, B., 23
Pell's equation, 130, 139n, 305
perpetual motion, 332
perturbations, 172
Peter I (the great), 10, 371, 396, 403
Peter II, 40
philology, 117–118 (in the notebooks)
philosophy, 35, 36, 80, 87, 117, 307
    Cartesian, 81
    Leibniz-Wolff, 81, 117, 379, 380
physics, 6, 26, 34, 35, 80, 116–117 (in the note-
    books), 128, 269, 289, 297, 307–316, 384
    Cartesian, 308
    of electricity 117, 307, 311–313
    of light 378, 379, 384
    of magnetism 117, 307, 311–313
Planck, M., 57
planetary motion, 115, 171, 247, 248, 250, 253,
    254, 255, 256, 279
Poincaré, H., 147, 162, 209, 251, 258, 292
Poisson, S.-D., 175
Pontryagin's maximum principle (in control the-
    ory), 207–209
Popov, E. P., 219
Porphyry, 339, 340
Prandtl, L., 223
precession of the equinoxes, 172, 173, 192, 266,
    295, 386
principle
    of least action, 29, 188–195, 300, 301
    of sufficient reason, 379
probability and statistics, 6, 24, 101 (in the note-
    books), 393
Prokopovich, T., 281
Ptolemy, C., 116, 184, 339, 345, 355
    musical theory of, 339–342
Pushkin, A. S. 411
Pythagoras, 375
Pythagoreans' theory of music, 339–342, 344,
    345, 355

quadratic reciprocity, 63, 153

Rachette, J.-D., 33, 368
Rameau, J.-Ph., 336, 345, 356
Razumovskiĭ, K. G., 32

Renaldini, C. (count), 111
Renau, B., 380
Ribas, J. de, 320, 326
Riemann, H., 345
Riemannian metric, 194
Robins, B., 22, 177, 241, 378
Rousseau, J.-J., 337
Royal Society of London, 30, 370
Rudio, F., 57, 59, 60, 62, 64, 65, 67, 121
Rudolff, Ch., 8
Rumovskiĭ, S. Ya., 25, 35, 47, 48, 307, 315, 320,
    321, 369, 387n, 389, 396n, 404
Rykachev, M. A., 58
Rzhevskiĭ, A. A., 318

Salemann, C. G. H., 56, 58, 58
Sarazin, F., 70
Saturn, 5, 6, 172, 249, 255, 265, 271, 296
Sauveur, J., 356
Schooten, F., van 99
Schottky, F., 56
Schumacher, J. D., 41, 46, 82, 92, 277, 279, 284
Schwartz, L., 27, 298
Schwarz, H. A., 55
"scientists' heaven", 40
Sedov, L. I., 206
    variational equation of, 206
Segner, J. A., 23, 76, 100, 101, 111, 175, 296,
    406
    water wheel of, 23, 115, 317
series, 7, 17, 28, 98, 101–105 (in the notebooks),
    292, 383
    convergent, 7, 28
    divergent, 7, 28, 37, 63, 68, 263, 298
    Fourier, 224, 264
    Lie, 267
    trigonometric, 28, 34, 296, 297, 298
seven years' war, 22, 83, 88, 91, 299, 301, 408
'sGravesande, W. J., 81
Shersnevskiĭ, I. G., 327, 328, 329
Siacci, F., 243
Siegel, C. L., 258
Smirnov, V. I., 122, 124, 128
Smith, R., 344
Snellius, W., 184
Snell's law, 184
Sobolev, L., 27, 298
Sofronov, M., 25, 47, 404
solar
    eclipse, 271, 274, 389, 390
    noon elevations, 276
    parallax, 249, 390
    service, 277

Sonin, N. Ya., 57, 58, 59, 60, 127
sound propagation, 303, 304, 379, 392
Speiser, A., 68
Speranskiĭ, M. M., 315
spherical trigonometry, 112
Stäckel, P., 58, 60, 61, 62, 64, 71
Stählin, J. von, 32, 48, 119
Stanislas Auguste (king of Poland), 387, 388n
Steklov, V. A., 66, 68, 70, 71
St. Petersburg news, 46, 270, 271, 272, 322, 323
Stevin, S., 23
Stifel, M., 8
Stirling, J., 8, 12, 104
Stumpf, C., 345
sunspots, 274, 275, 277, 278
surfaces, 240, 301, 383
    geodesics, on 240
    minimal, 114, 301
    theory of, 113
Sulzer, J. G., 89, 91
Suslov, G. K., 202, 204

Tartaglia, N., 241, 242
Tartini, G., 337
tautochrone, 113, 375
Taylor, B., 296, 373, 386
Teplov, G., N. 18
Thomson, W. (Lord Kelvin), 314
three-body problem, 64, 171, 254, 257–259, 265
    case of two fixed centers, 258, 265
    equations of, 264
tides, 5, 46, 247, 259, 279, 280, 376, 392
Timoshenko, S. P., 222, 224, 231
topology, 17
Torricelli, E., 23, 242
Trediakovskiĭ, V. K., 283, 284
Truesdell, C., 170, 176, 191n, 228
Tschirnhausen, W. K., 143, 373
Turgot, A. R. J., 378, 382
two-body problem, 264–265
    Newton's geometric solution of, 264

typology of mathematicians, 37

Vandermonde, C. A., 143
Varignon, P., 113, 168, 193
Vavilov, S. I., 263, 309
Venus, 265, 272, 389
    influence on earth's orbit, 256
    occultation by the moon, 273, 274
    transit of, 50, 248, 290, 396
Viète, F., 138, 145, 146
    theorem of, 156, 157
Vignoles, A. de, 79
Vignon, P., 271
Voltaire, F. M. A., 22, 76, 80, 82, 117, 189
Voronets, V. P., 202, 204
Vorontsev, M. I. (count), 87, 307

Wallis, J., 98, 107
    product of, 104
Walther, C. T., 283
Weierstrass, K., 146,
    conditions of, 208
Weierstrass-Frobenius theorem, 143
Weil, A., 147, 164
Whittaker, E. T., 61
winds, 392
Winkelmann, J., 76
Wolff, C. F., 25, 40, 47, 76, 79, 80, 103, 382n
Wordsworth, W., 191n

Yasinskiĭ, F. S., 222
Young's modulus, 180, 214n
Young, Th., 314
Yushkevich, A. P., 83, 122, 128, 180, 298

zeta-function, 17, 63, 101–105 (in the notebooks), 131, 132
zodiacal light, 248, 278
Zolotarev, E. I., 150